Fundamentals of Optomechanics

Optical Sciences and Applications of Light

Series Editor
James C. Wyant
University of Arizona

Fundamentals of Optomechanics, *Paul Yoder and Daniel Vukobratovich*

Optics Manufacturing: Components and Systems, *Christoph Gerhard*

Femtosecond Laser Shaping: From Laboratory to Industry, *Marcos Dantus*

Photonics Modelling and Design, *Slawomir Sujecki*

Handbook of Optomechanical Engineering, Second Edition, *Anees Ahmad*

Lens Design: A Practical Guide, *Haiyin Sun*

Nanofabrication: Principles to Laboratory Practice, *Andrew Sarangan*

Blackbody Radiation: A History of Thermal Radiation Computational Aids and Numerical Methods, *Sean M. Stewart and R. Barry Johnson*

High-Speed 3D Imaging with Digital Fringe Projection Techniques, *Song Zhang*

Introduction to Optical Metrology, *Rajpal S. Sirohi*

Charged Particle Optics Theory: An Introduction, *Timothy R. Groves*

Nonlinear Optics: Principles and Applications, *Karsten Rottwitt and Peter Tidemand-Lichtenberg*

Numerical Methods in Photonics, *Andrei V. Lavrinenko, Jesper Lægsgaard, Niels Gregersen, Frank Schmidt, and Thomas Søndergaard*

Fundamentals of Optomechanics

Daniel Vukobratovich
Paul Yoder

CRC Press is an imprint of the
Taylor & Francis Group, an informa business

CRC Press
Taylor & Francis Group
6000 Broken Sound Parkway NW, Suite 300
Boca Raton, FL 33487-2742

First issued in paperback 2023

Publisher's Note
The publisher has gone to great lengths to ensure the quality of this reprint but points out that some imperfections in the original copies may be apparent.

Visit the Taylor & Francis Web site at
http://www.taylorandfrancis.com

and the CRC Press Web site at
http://www.crcpress.com

ISBN 13: 978-1-032-65238-2 (pbk)
ISBN 13: 978-1-4987-7074-3 (hbk)
ISBN 13: 978-1-351-21086-7 (ebk)

DOI: 10.1201/9781351210867

I dedicate this book to my wife Suzanne;
without her support, this would not have been possible
and
to Paul Yoder, Jr., my late good friend and coauthor, who introduced me
to optomechanics and pioneered the discipline in the United States.

Contents

Authors

Paul Yoder (BS physics, Juniata College, Huntingdon, Pennsylvania, 1947, and MS physics, Pennsylvania State University, University Park, Pennsylvania, 1950) learned optical design and optomechanical engineering at the US Army's Frankford Arsenal (1951–1961). He then applied those skills at Perkin-Elmer Corporation (1961–1986) and served the optical community as a consultant in optical and optomechanical engineering (1986–2006). A fellow of the Optical Society of America (OSA) and International Society for Optics and Photonics (SPIE), Yoder authored numerous chapters on optomechanics, published more than 60 papers, was awarded 14 US and several foreign patents, and taught more than 75 short courses for SPIE, US government agencies, and the industry.

Daniel Vukobratovich is senior principal multidisciplinary engineer at Raytheon Systems, Tucson, Arizona, and an adjunct professor at the University of Arizona, Tucson, Arizona. He has authored more than 50 papers, taught short courses in optomechanics in 12 different countries, and consulted for more than 40 companies. An SPIE fellow, he is a founding member of the optomechanics working group. He holds international patents and received an IR-100 award for work on metal matrix composite optical materials. He led development on a series of ultralightweight telescopes using new materials and worked on space telescope systems for STS-95, *Mars Observer*, *Mars Global Surveyor*, and Far Ultraviolet Spectroscopic Explorer (FUSE).

1 Introduction

1.1 INTRODUCTION AND SUMMARY

Optomechanics is defined as the maintenance of the shape and position of the surfaces of an optical system. Optical engineering is the control of light by the interaction of light with surfaces. Optomechanics is associated with the design and support of the surfaces of optical systems, normally under a wide variety of conditions.

This brief introduction includes a discussion of the development of optomechanics as an engineering discipline. Although optomechanics is important in the design of any optical system, recognition of the field is relatively recent, with the first professional optomechanics conference being held in 1980. A significant lack until now is the absence of a textbook on optomechanics; this book is intended to meet this need and serve as a reference for self-study by working professionals.

1.2 DEVELOPMENT OF OPTOMECHANICS

Any optical system requires a mechanical structure to support its elements. Optomechanics is concerned about the design of the mechanical structure of the optical systems. Hence, it can be argued that optomechanics is as old as optics. When Galileo designed the tube of his first telescope, optomechanics was created.

However, recognition of optomechanics as a separate discipline is relatively recent. One of the first texts to discuss the mechanical engineering aspects of optical systems was *Fundamentals of Optical Engineering*, by Donald H. Jacobs,[1] published in 1943. A famous quote from this book is the following: "in the design of any optical instrument, optical and mechanical considerations are not separate entities to be dealt with by different individuals but are merely two phases of a single problem."

More formal recognition of optomechanics had to wait until 1980, when the first international conference on optomechanical system design was held at an SPIE meeting in San Diego, California.[2] SPIE was originally known as the Society of Photo-instrumentation Engineers (although continuing to use the acronym SPIE, the society is now called the International Society for Optics and Photonics). It is not too surprising that its members recognized the need for better communication between the optical and mechanical engineering communities. Professional conferences served to improve the interchange of information between those working in the field. SPIE typically published proceedings within a few months after these conferences. Rapid publication enhanced the value of the conference proceedings, and these proceedings became a valuable resource for those working in optomechanics.

In 1985 SPIE began presenting short courses in optomechanics. Initially these were held within the United States. In 1987 SPIE and the British Scientific Instrument Research Association held a very successful two-day course on optomechanics as part of an optical engineering conference in London. The success of this course led to others that same year in Europe and Taiwan. In the next five years, optomechanical courses were held all over the world. Short courses on optomechanics continue to be presented by the SPIE and other organizations and are now available online as well as in video format.

These short tutorial courses in optomechanics were useful in introducing experienced engineers to the discipline. However, these short courses did not provide the in-depth exposure associated with a university program. After the success of the SPIE short courses, universities began to provide dedicated courses in optomechanics. Some of the pioneers were the Optical Sciences Center at the University of Arizona, the University of Rochester, Georgia Tech, and the TNO in the Netherlands.

Up until now, one problem with university courses in optomechanics was the absence of a textbook. This book is intended to be a textbook for university courses in optomechanics. It can also be used for self-study by practicing optical engineers. As a textbook, the emphasis is on basic principles, backed up with tutorial examples and exercises to allow the student to gain familiarity with the material. Since textbooks often serve as references for engineers starting out in a new field, reference material is also included.

In 1986 Paul Yoder published one of first reference books in the field, *Opto-mechanical Systems Design*.[3] This book is now in its fourth edition and has grown to two volumes. Much of the material in this textbook is taken from the fourth edition of *Opto-mechanical Systems Design*. While the fourth edition of *Opto-mechanical Systems Design* is intended as a desk reference for practicing engineers, this textbook is intended as an introductory work. The contents of the two books do overlap; one difference between this textbook and *Opto-mechanical Systems Design* is that the latter work contains many more design examples.

1.3 GENERAL CONSIDERATIONS

A new worker in the field needs to understand that there are significant differences between classical mechanical engineering and optomechanics. As discussed in the following text, these differences require a different approach to design, with emphasis on performance factors that are usually not important in conventional engineering. However, the statement that optomechanics is similar to ordinary mechanical engineering, just with the "decimal point moved a few places to the left," is also true. Mechanical engineering principles apply but are used to evaluate parameters such as deflection rather than strength.

Optomechanics is concerned about the shape and position of the surfaces of an optical system. Optical engineering is defined as the control of light. Light is generated, directed, and detected through interactions with surfaces. Examples are refraction and reflection. Hence, based on the preceding definition, optomechanics is the discipline associated with maintaining the geometry and alignment of optical surfaces, thus ensuring the design performance of an optical system.

Since optomechanics is concerned about the shape and position of surfaces, strain or deflection is more important than stress. The emphasis on strain rather than stress is an important difference from ordinary mechanical engineering. While stress is still a factor in optomechanics, deflection, which is defined as departure from a stress-free shape, is of considerably greater interest. Very small strains, sometimes as small as one part per million or less, are important in optomechanics. In comparison, the smallest strain of interest in conventional mechanical engineering is about 1 part in 500, which is associated with the 0.2% plastic yield criterion of metallic materials.

One necessary mental transition when new to optomechanics is to think in terms of strain rather than stress. This transition is particularly important for those with mechanical engineering experience. Emphasizing strain rather than stress often leads to designs which appear wrong, or overbuilt, when compared to conventional mechanical engineering practice. For example, a support structure for an optic may seem to be much heavier or more massive than necessary. In reality, when considering stability at the level of fraction of a wavelength of light, even a very massive optical support may be marginal.

The emphasis on strain rather than stress complicates other aspects of the transition from mechanical to optomechanical engineering. Material selection is based on properties that influence the ability of a material to maintain its shape under a variety of loading conditions. Yield stress is almost never an issue in the design of optomechanical support structures. As an example, for metals used in optomechanics, microyield stress (abbreviated as MYS), which is the stress necessary to cause a permanent deformation of one part per million, is important.

An exception to the rule that stress is usually not important in optomechanics is the strength of brittle optical materials. Glasses and optical ceramics fail by brittle fracture. Typically the fracture strength of optical materials is a small fraction of the strength of even a low-strength structural metal. For example, the strength of ordinary optical glass is on the order of 10 MPa; in comparison, the strength of a low-grade structural steel is about 250 MPa. Evaluation of the strength of optical materials is important in optomechanics and is discussed in Chapter 3.

1.4 PREREQUISITES

It is assumed that the reader of this text is familiar with basic optical and mechanical engineering. In particular, some knowledge of the principles of geometric optics is necessary for full comprehension of the material. Knowledge of more advanced topics in optics, such as optical design or radiometry, is not required. Similarly, basic knowledge of mechanical engineering principles is desirable. This background should be on the level of a basic first-year university course in mechanical engineering.

As an introductory work, the emphasis is on "first-order analysis" and approximations useful for beginning the design process. With the advent of the programmable pocket calculator and personal computer, much greater mathematical sophistication is possible in first-order analysis. For example, numerical solution of equations is much faster and easier than in the slide rule era. Although computer-aided design is discussed, algorithms are not given. For advanced discussions of computer-aided optomechanical design, there are several good references, such as that of Doyle et al., *Integrated Optomechanical Analysis*, second edition.[4] Familiarity with mathematics through algebra is assumed. There is little use of advanced mathematics in the text.

With some exceptions, the units in this textbook are from Système International (SI) and follow practices from *ISO Standards Handbook 2: Units of Measurement*.[5] An additional reference is ASTM E380-91a, *Standard Practice for Use of the International Systems of Units (SI)*.[6] US customary units (in.–lb–s) are occasionally employed and, where necessary, important measurements are given in both US and SI units. Typically temperature is given in kelvins to simplify analysis over wide temperature ranges. Mass is given in kilograms, requiring multiplication by g to produce force units when calculating deflection or stress.

REFERENCES

1. Jacobs, D.H., *Fundamentals of Optical Engineering*, McGraw-Hill, New York, 1943.
2. Bayar, M., Optomechanical systems design, *Proc. SPIE*, 250, 1980.
3. Yoder, P.R. Jr., *Opto-mechanical Systems Design*, Marcel Dekker, New York, 1986.
4. Doyle, K.B., Genberg, V.L., and Michels, G.J., *Integrated Optomechanical Analysis*, 2nd ed., SPIE Press, Bellingham, WA, 2012.
5. ISO (International Organization for Standardization), *ISO Standards Handbook 2: Units of Measurement*, 2nd ed., ISO, Geneva, 1982.
6. ASTM International, *Standard Practice for Use of the International Systems of Units (SI)*, E380-91a, ASTM International, West Conshohocken, PA, 1991.

2 Optomechanical Design Process

Contributed in part by David M. Stubbs, Kevin A. Sawyer, and David Aikens

2.1 INTRODUCTION AND SUMMARY

The *optomechanical design* of optical instruments is a tightly integrated process involving many technical disciplines. It begins when the requirement for a particular hardware item is established by the potential user, such as the military, other governmental organizations, or commercial representatives who seek ways to expand sales with a new or improved product. Once approved, funded, and staffed, the design effort proceeds through a logical sequence of major steps and concludes only when the instrument is awarded a pedigree establishing its ability to meet all its technical specifications and capable of being produced, within cost limits, in the required quantity—whether that is as a one-off (such as the highly successful Hubble Space Telescope [HST]) or as a large number of a much simpler item (such as a new spotting scope with an integrated digital camera for nature study).

In this chapter, we treat each major design step in a separate section. Admittedly, our approach is idealized since few designs develop as smoothly as planned. We endeavor to show how the process *should* occur and trust that those planning, executing, reviewing, and approving the design will have the ingenuity and resourcefulness to cope with the inevitable problems and bring errant design activities into harmony with minimal effect on schedule and cost.

The driving forces behind the methodology applied in the design process include schedule constraints; availability of properly trained personnel; facilities, equipment, and other resources; perceived demands from the marketplace; and the inherent costs of accomplishing and proving the success of the design. These we consider to lie within the province of *project management*, a subject clearly beyond the scope of this book.

A great influence on the optomechanical design process is the degree of maturity of the technology to be applied. For example, not many years ago, the design of the 2.4 m (94.5 in.) aperture HST capable of being lifted into the orbit of Earth by a space shuttle would have been impossible for a variety of reasons. One mechanical reason was the then nonavailability of structural materials with the required blend of high stiffness, low density, and ultralow thermal expansion characteristics. To have used aluminum, titanium, or Invar in the telescope truss structure in lieu of the less familiar, but promising new types of graphite–epoxy (GrEp) composites that were actually employed would have severely limited the performance of the instrument in the varying operational thermal environment.* Further, the strict telescope weight limitations imposed by the National Aeronautics and Space Administration could not have been met.

Complex optomechanical systems generally consist of many subsystems, each having its own specifications and constraints, as well as a unique set of design problems. Subsystems usually consist of several major assemblies that, in turn, consist of subassemblies, components, and elements. By dividing the overall design problem into a series of related, but independently definable parts, even the most complex system will yield to the design process.

* According to Krim,[1] the temperature stabilization requirements for the HST would have been ±0.027°C, ±0.06°C, and ±0.35°C with Al, Ti, or Invar structures, respectively. The actual structure, with a GrEp truss, maintained optical performance over a more realistic range as large as ±13°C.

No one design can be cited in this chapter to illustrate all the various steps of the optomechanical design process. We therefore utilize a variety of unrelated examples involving military, aerospace, or consumer instruments for this purpose. In real life, the magnitude of the effort required in any given step would be tailored to that appropriate to the specific design problem. The general approach to each step and to the overall design process might well be expected to follow the guidelines established here.

2.2 ESTABLISHING THE REQUIREMENTS

As pointed out by Petroski[2] in one of his series of interesting books on engineering design, many requirements for new hardware arise "out of the failure of some existing thing, system, or process to function as well as might be hoped, and they arise also out of anticipated situations wherein failure is envisioned." Alternatively, the availability of new technology that makes feasible the design and development of an instrument with new capabilities can lead to a "requirement" to put that technology to use in entirely new hardware. These requirements typically define the goals for the configuration of the item and its physical characteristics, performance in a given application environment, life cycle cost, etc.

The achievement of a successful state-of-the-art instrument design utilizing new materials requires more theoretical synthesis and analysis, experimentation, and qualification testing than would a design involving the application of only tried and proven materials and technologies. Applying a higher level of technology or entirely new technology to make a system perform better, weigh less, or last longer may increase cost over less capable, but available technology. Paraphrasing Sarafin,[3] who wrote from the vantage point of much aerospace experience, we should not just ask, "Can we make the system do …?" because the answer probably is, "Yes, we can." The more appropriate questions are "At what cost can we make the system do …?" "What are the technical risks of failure?" and "What would it cost in time and dollars to recover if we fail?" Careful consideration of these deeper issues will help balance the advantages and disadvantages of such alternate pathways.

Key elements that minimize risk and facilitate completion of assignments throughout the optomechanical design process are expedited communication between all involved individuals and easy access to required technical information. The former is greatly facilitated today by electronic means such as e-mail, teleconferencing, the use of cellular phones, and rapid transmission of document images measuring gigabytes. Further, information gathering is facilitated by worldwide access to a vast number of excellent reference libraries and technical data files via the Internet. The detailed design itself can now be computer based rather than in the form of paper drawings and other hardcopy documents. Computer-aided design and engineering (CAD and CAE) technologies allow access throughout multiple networks for information exchange yet limit design change privileges to the proper authorities. Communication between design, engineering, manufacturing, and test groups now can be accomplished by electronic means, thereby reducing transit time and enhancing the accuracy of data transmittal. Direct data entry into the computer of a machine, i.e., computer-aided manufacture, then facilitates making parts by eliminating many manual setup chores and reducing the possibility of human errors during data entry. Testing also can often be facilitated by computer control of the test sequence and automatic data storage, retrieval, and analysis.

2.3 CONCEPTUALIZATION

The first step in the evolution of the design of an optomechanical system is recognition of the need for a device to accomplish a specific purpose. Usually, the suggestion of a need brings to the minds of inventive design engineers at least a vague concept for instrumentation that might meet that need. Knowledge of how similar needs were met to some degree by prior designs plays an important role at this point. Experience indicates not only how the new device might be configured, but also how it should *not* be configured.

Functional block diagrams relating the major portions of the system are valuable communication tools throughout the design process. Figure 2.1 shows one such diagram for a high-performance photographic system to be applied in a downward-looking, surveillance application from a spacecraft in orbit about the Earth. This system is envisioned as several major assemblies: imaging optics, a fold mirror, a focal plane assembly, a mechanical structure, and a protective housing. Ancillary subsystems accomplish data processing, storage, and downloading of images to receiving station(s) on Earth.

The optomechanical makeup of one concept for the imaging optics block of Figure 2.1 is shown in Figure 2.2. Here, a second-level block diagram shows that the optical system consists conceptually

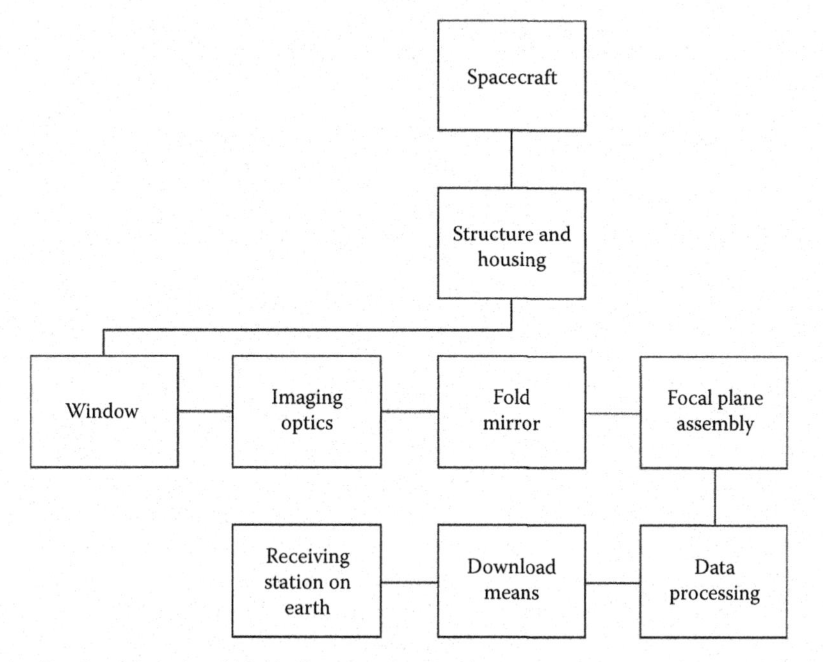

FIGURE 2.1 Top-level functional block diagram of a spaceborne camera.

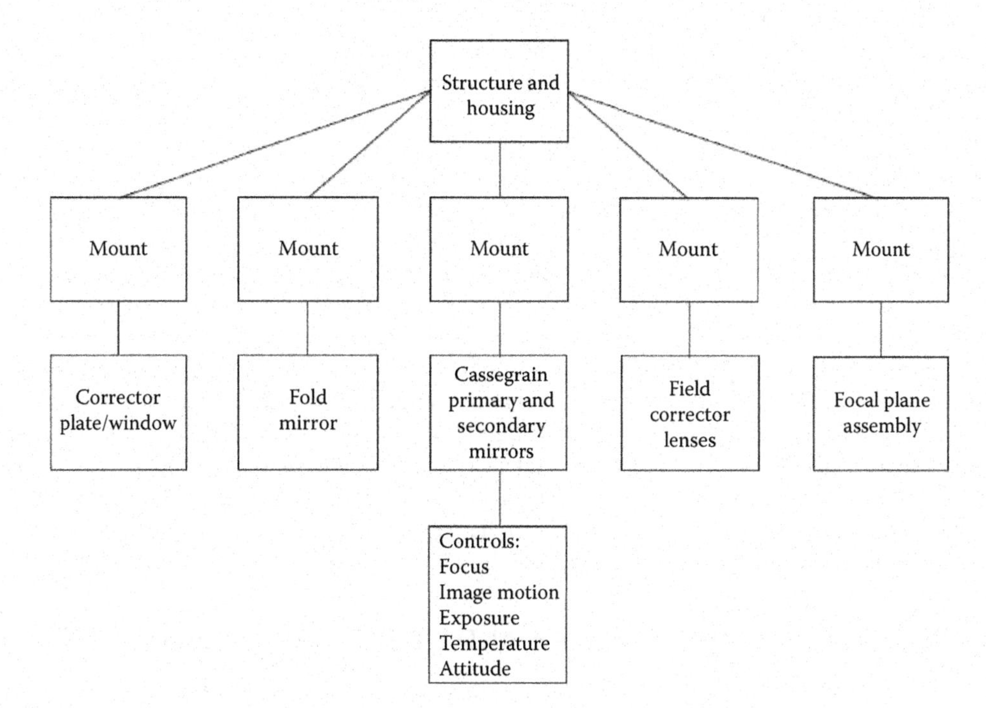

FIGURE 2.2 Second-level block diagram for the spaceborne camera in Figure 2.1.

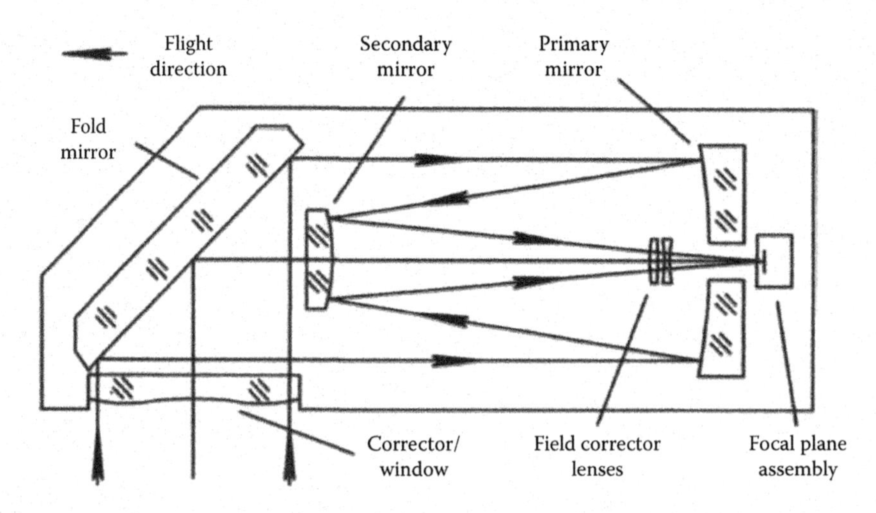

FIGURE 2.3 Conceptual optical schematic diagram for the spaceborne camera in Figure 2.1.

of (1) an aspheric corrector plate that reduces optical aberrations and also serves as a window, (2) a folding mirror capable of redirecting the vertical input beam by 90° to the horizontal axis of the optical system, (3) image-forming optics comprising primary and secondary mirrors in the Cassegrain configuration and aberration-compensating field lenses, and (4) a focal plane assembly (envisioned as a large multipixel detector array). Each of these major components is attached to the camera structure by its own mount. An electromechanical subsystem provides means for controlling exposure and focus, for compensating for image motion, and for stabilizing the temperature and a computer system that coordinates required operational functions.

At the top of Figure 2.2, a mechanical structure is provided to support and maintain alignment of all the camera components as well as to interface the camera with the spacecraft. A protective housing encloses the optics. This cover also must preserve the clean and dry internal environment established during assembly.

Figure 2.3 is a preliminary schematic for the optical system. At this stage in the conceptualization process, the detailed designs of the individual components making up this system would usually not be known.

As the function of the device to be designed is examined in more detail and the subsystem technical specifications begin to take form, the relative advantages and disadvantages of this and other potential concepts can be established and weighed. Parametric trade-off analyses are often performed at this time to develop approximate interrelations between design variables. This helps disclose incompatibilities between specific requirements. Rough estimates of the physical size and weight of the instrument if built along alternative lines may also prove helpful in pointing out the more favorable of alternative concepts. Preliminary material choices made at this time need be no more specific than to assume that optical glass would be used in lenses and windows; that mirrors probably would be lightweight glass-ceramic; that the refractive and reflective optical component thicknesses would be ~10% and ~17% of their diameters, respectively; and that the relative aperture and field of view in object space of the system would be some reasonable but specific values. Conceptual layouts of the most viable concept(s) can then be prepared for evaluation, comparison, and choice of the apparent best configuration. After appropriate review and approval, this then would serve as the starting point for a detailed preliminary design.

2.4 PERFORMANCE SPECIFICATIONS AND DESIGN CONSTRAINTS

Two of the most important inputs to the design process are the performance specification and the definition of externally imposed constraints. The former sets forth the prospective user's definition of what the end item must do and how well it must work to be judged acceptable, whereas the latter defines the physical limitations, such as size, weight, configuration, environment, and resource consumption, that affect optical, mechanical, and electrical interfaces with the surroundings. In the case of a scientific payload for a space probe, these generally would consist of many separate, complex, and lengthy documents. In the simplest cases, the specification could consist of one much shorter document giving a few general requirements. Parameters, in this case, would be left to the discretion of the optical and mechanical designers and engineers. In almost all cases, the preparation of at least one drawing to specify the optomechanical interfaces of the item would be appropriate.

A suggested list of items to be considered in the typical performance specification and constraint definition for an optomechanical system may be found in Table 2.1. These items are not necessarily in order of importance nor all-inclusive. Careful consideration of these features (and others that may be unique to the design in question) would help the design teams create a satisfactory end item or product. It is advisable also to indicate clearly the intended purpose of the instrument at the beginning of the specification.

Figure 2.4 illustrates an optomechanical interface drawing for a lens assembly. It defines the required external configuration for a 9 in. (22.9 cm) focal length, *f*/1.5 objective lens assembly with coaxial laser output and image-forming input channels. The interface drawing sets limits on the overall package size, defines critical dimensions, states the requirements for the perpendicularity of the optical axis of the imaging system (datum -A-) and of the image plane to the mounting flange

TABLE 2.1
Checklist of General Design Features Typically Included in Specifications and Constraint Definitions for Optical Instruments

- Performance requirements such as resolution, modulation transfer function at specified spatial frequencies, radial energy distribution, encircled or ensquared energy at specific wavelengths, or numerical aperture
- Focal length, magnification (if system is afocal), magnification, and object-to-image track length (if system has finite conjugates)
- Angular or linear field of view (in specified meridians if anamorphic)
- Entrance and exit pupil sizes and locations
- Spectral transmission requirements
- Image orientation
- Sensor characteristics such as dimensions, spectral response, element size and spacing, and/or frequency response
- Size, shape, and weight limitations
- Survival and operating environmental conditions
- Interfaces (optical, mechanical, electrical, thermal, etc.)
- Thermal stability requirements
- Duty cycle and useful life requirements
- Maintenance and servicing provisions (access, fits, clearances, torquing, etc.)
- Emergency or overload conditions
- Location of center of gravity and lifting provisions
- Human–instrument interface requirements and restrictions (including safety aspects)
- Electrical requirements and restrictions (power consumption, frequency, phase, grounding, etc.)
- Material selection recommendations and limitations
- Finish/color requirements
- Corrosion, fungus, rain, sand/dust, and salt spray erosion protection requirements
- Inspection and test provisions
- Electromagnetic interference restrictions and susceptibility
- Special markings or identifications
- Related consumables

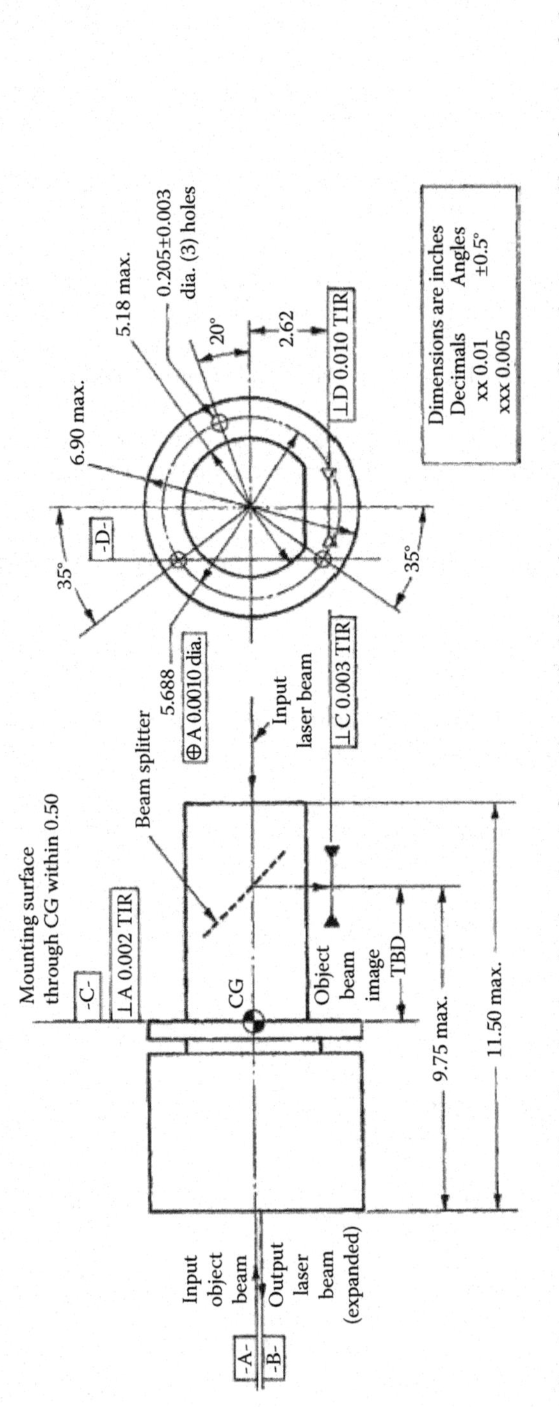

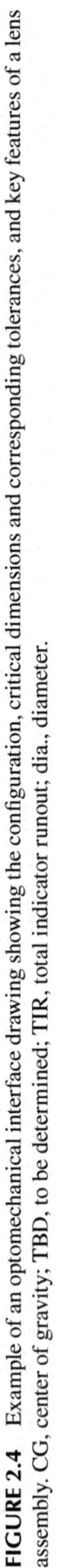

FIGURE 2.4 Example of an optomechanical interface drawing showing the configuration, critical dimensions and corresponding tolerances, and key features of a lens assembly. CG, center of gravity; TBD, to be determined; TIR, total indicator runout; dia., diameter.

(datum -C-), and establishes tolerances for critical dimensions and angles. The associated technical performance specification for this lens defines its optical characteristics (focal length, relative aperture, field of view, image quality, allowable off-axis vignetting, transmission, etc.) as well as constructional features needed for the assembly to accomplish its intended function in a specific environment.

One aspect of optical instrument performance specification preparation worthy of special consideration here is the quantification of what is really needed from the equipment once it has been designed and built. Warren, Smith advises that specifications should ask for just enough to accomplish the intended purpose and no more. Technical requirements should be clear and concise and not overburdened with details, yet not so general as to foster confusion on the part of the designers trying to determine what really is wanted. For example, although it is easy to say that a new photographic lens is to be "diffraction limited," it is not so easy to prove that some lower level of performance would not suffice. It has become a common practice for those wishing a device to be developed to ask first for an analysis of the trade-offs between performance and cost. The time and cost of such analyses, if properly conducted and documented, are usually worthy expenditures. It has often been said that requirements are not absolute and performance is not always the most important attribute of a system. For instance, life cycle cost is sometimes the most vital aspect of new hardware. An affordable system that works adequately may be better in the long run than a new version that offers a performance advantage, but costs more and requires more maintenance.

Strict schedule constraints such as having a new space payload ready to meet a specific launch window that will not occur again for many years also might lead to acceptable compromises in performance because some scientific information from the mission would be better than no information at all. Above all, the project team must understand what the user (or customer) really wants—not just what the initial specification reads! In this case, understanding requires communication and willingness on the part of all parties to examine all aspects of the application to see if the "requirements" are realistic.

Price[4] went a bit further by defining a trade-off as a "balancing of factors or conditions, all of which are not attainable at the same time." He cited and then discussed three useful viewpoints, one or more of which is generally applicable to almost any system:

1. The hardware system including all components from the object to the final output (e.g., a video recording or display system comprising object, illumination, atmosphere, lens, camera, detector, electronics, recorder, tape, player, monitor, and observer's eye)
2. The product–user system, including the interaction between the person and the apparatus (e.g., controls, platforms, handles, switches, eye position, eye–hand coordination requirements, time delays between actions and reactions)
3. The manufacturing system, including raw materials; material handling; part manufacture; assembly; quality control; optics-to-product interfaces and tests; and the attendant costs, schedules, processes, and personnel utilization

Price's paper[4] concluded with the following profound statement: "a well prepared analysis is an essential, but not necessarily sufficient, condition to obtaining acceptance of a proposed system design."

The extent to which the cost of an optical system can be reduced or the product can otherwise be made more attractive to prospective buyers is often intimately related to the allowable degradation from the "perfect" function. Customers faced with the predicted cost of buying state-of-the-art camera systems built to a given specification have been known to ask for a shopping list of alternative designs showing system costs in some quantity such as a function of resolution in line pairs per millimeter. Although a reliable relationship between these factors is quite difficult to derive, its serious consideration would surely help all parties understand the importance of compromise. In 1979, Shannon[5] illustrated this point by pointing out the magnitude of optical distortion introduced by the curved windshields of modern automobiles that is tolerated for style and cost reasons. Walker dealt at length with the compromises appropriate in the design of visual systems such as telescopes, binoculars, or periscopes. Parameters particularly amenable to trade-off in such instruments are image quality, vignetting, and light transmission. To a lesser degree, one might trade magnification, field of view, or aperture against system complexity, size, and cost. At the end of his 1979 paper, Walker[6] provided his version of the dictionary definition of a specification as follows: "a detailed and exact statement prescribing materials, dimensions, workmanship and performance, arrived at after careful and cooperative consideration of the system application and the realistic needs of the end user." This seems to express accurately the viewpoint of many individuals active in optomechanical system design.

Following World War II, most contracts for new optical instruments procured for US government use referred to military specifications, standards, and other government publications. These documents defined general requirements and provided guidance for the selection of materials and the design, inspection, and testing of a variety of equipment items. A shift of official direction occurred in 1994 when the US Armed Services issued a directive stating that all future military procurement contracts would refer to national and international voluntary standards rather than US military specifications. Many existing military specifications relating to optical material were canceled. Others were declared inactive. In some cases, inactive specifications were allowed to apply to existing procurement contracts, but they were not to be used in new contracts. Most other US military specifications were to be reviewed for relevancy to current manufacturing techniques. It was expected that many of these would be rewritten as new voluntary optical standards by voluntary standard bodies, adopted widely, and distributed through the standard organizations of the various countries producing and/or procuring new products.

Work on international optical standards began in 1979 under the auspices of the International Organization for Standards (ISO*), headquartered in Geneva, Switzerland. This effort is conducted within ISO/Technical Committee (TC) 172, "Optics and Optical Instruments." The Deutsches Institut für Normung (DIN) of Germany functions as secretariat of this TC. Currently 12 nations are actively participating in this work through their national standards bodies. See Table 2.2. In addition, 13 nations serve as observers.

ISO/TC 172 was established to promote standardization of terminology, requirements, interfaces, and test methods in the field of optics. These include complete systems, devices, instruments, optical components, auxiliary devices, and accessories, as well as materials. The scope of ISO/TC 172 excludes standardization efforts relative to specific items in the field of cinematography (the responsibility of ISO/TC 36), photography (the responsibility of ISO/TC 42), eye protectors (the responsibility of ISO/TC 94), micrographics (the responsibility of ISO/TC 171), fiber optics for telecommunication (the responsibility of International Electrotechnical Commission [IEC]/TC 86), and electrical safety of optical elements.

To facilitate the development of optical standards and fill the void left by the absence of the US military specifications, a consortium made up of professional societies, trade associations, and companies sponsored the incorporation of the Optics and Electro-optics Standards Council (OEOSC), which acts as the administrator of national optical standards for the United States.† An OEOSC committee called ASC OP, "Optics and Electro-optical Instruments," has been accredited by the ANSI and is authorized to develop US national standards. OEOSC is also responsible for supporting ISO/TC 172 through a US Technical Advisory Group. This group is made up of US optical experts whose primary responsibility is to review drafts of proposed international optical standards so that it can formulate US opinions regarding the suitability of those drafts to become international standards and then to transmit those opinions, through ANSI, to the ISO TC.

* To avoid different acronyms for this organization in different languages, the name ISO is used universally.

† Information regarding the activities, membership, and progress of this council can be found at http://www.optstd.org.

TABLE 2.2
International Organizations Involved in the Development of Voluntary Standards Related to Optics and Optical Instrumentation under ISO/TC 172

- Association Française de Normalisation from France
- American National Standards Institute (ANSI) from the United States
- Asociatia de Standardizare din România from Romania
- British Standards Institution from the United Kingdom
- DIN from Germany (Secretariat)
- Institute of Standards and Industrial Research from Iran
- Japanese Industrial Standards Committee (JISC) from Japan
- Kenya Bureau of Standards from Kenya
- Korean Agency for Technology and Standards from South Korea
- Österreichisches Normungsinstitut from Austria
- State Administration of China for Standardization from China
- State Committee of the Russian Federation for Standardization and Metrology (GOST R) from Russia
- Standards Australia International Ltd. from Australia
- Swiss Association for Standardization (SNV) from Switzerland
- Ente Nazionale Italiano di Unificazione from Italy

Within ISO/TC 172, seven subcommittees (SCs) have been established to address different major topics. Under each active SC, there are several working groups (WGs) that do the actual writing. Table 2.3 depicts the organizational structure down to the WG level as of mid-2011. Draft international standards prepared and adopted by the various ISO TCs are circulated to the international members of ISO for approval before the ISO Council formally approves them. Approval requires at least 75% acceptance by the member bodies voting.

Most optical companies in the United States have long based their engineering drawings for mechanical and optical parts on ANSI Y14.5, *Dimensioning and Tolerancing*,[7] and American Society of Mechanical Engineers (ASME)/ANSI Y14.18M, *Optical Parts*,[8] respectively. These documents were largely based on US military standards including MIL-PRF-13830 (*Performance Specification: Optical Components for Fire Control Instruments; General Specification Governing the Manufacture, Assembly, and Inspection of*),[9] MIL-G-174 (*Military Specification, Glass Optical*),[10] MIL-C-675 (*Military Specification: Coating of Optical Glass*),[11] and MIL-STD-34 (*Military Standard: Preparation of Drawings for Optical Elements and Optical Systems: General Requirements for*).[12] These documents and others of importance to the US optical community are badly out of date and received minimal consideration by ISO during preparation by WG2 of SC 1 of their standard ISO 10110-14, *Optics and Optical Instruments—Preparation of Drawings for Elements and Systems*.[13] The latter standard was based instead on German industry standard DIN 3140, *Dimensions and Tolerance Data for Optical Components*,[14] and differs significantly from the standards used in the United States and some other countries. Notwithstanding this fact, ISO 10110-14 has been adopted by several countries, including Germany, France, Russia, and Japan. ASC OP has voted not to revise ASME/ANSI Y14.18, but to work toward adoption by the United States of ISO 10110-14. One important feature of this standard is that it expresses as many concepts as possible in terms of symbols to minimize the need for notes that would require translation for the drawing to be understood in the languages of non-English-speaking countries. Default tolerances are given in the standard for cases in which a specific tolerance is not required. This simplifies the appearance of drawings in those cases.

ISO 10110-14 has 13 parts as listed in Table 2.4. A few of these parts are worthy of special attention here. The following descriptions are based largely on Parks[15] and Willey and Parks.[16] The first part deals with the mechanical aspects of optical drawings, including lists of items to check for the

TABLE 2.3
Listing of SCs and WGs under ISO/TC 172, "Optics and Optical Instruments"

SC	Title/WG
SC1	Fundamental standards (DIN)
	WG1: "General Optical Test Methods"
	WG2: "Preparation of Drawings for Optical Elements and Systems"
	WG3: "Environmental Test Methods"
SC3	Optical materials and components (JISC)
	WG1: "Raw Optical Glass"
	WG2: "Coatings"
	WG3: "Characterization of IR Materials"
SC4	Telescopic systems (GOST R)
	WG2: "Telescopic Sights"
	WG5: "Night Vision Devices"
SC5	Microscopes and endoscopes (DIN)
	WG3: "Terms and Definitions
	WG6: "Endoscopes"
	WG8: "Immersion Media for Light Microscopy"
	WG9: "Optical Performance of Microscope Components"
SC6	Geodetic and surveying instruments (SNV) (WG not formalized)
SC7	Ophthalmic optics and instruments (DIN)
	WG2: "Spectacle Frames"
	WG3: "Spectacle Lenses"
	WG6: "Ophthalmic Instruments and Test Methods"
	WG7: "Ophthalmic Implants"
	WG8: "Data Interchange"
	WG9: "Contact Lenses"
	WG10: "Devices for Dioptric Power Measurements of Lenses"
SC9	Electro-optical systems (DIN)
	WG1: "Terminology and Test Methods for Lasers"
	WG3: "Safety"
	WG4: "Laser Systems for Medical Applications"
	WG6: "Optical Components and Their Test Methods"
	WG7: "Electro-optical Systems Other than Lasers"

Note: Secretariat for each SC is shown in parentheses (see Table 2.2 for the definitions of acronyms).

completeness of system layouts, subassemblies, and individual optical element drawings. Only such items as are unique to optics are included. All strictly mechanical aspects of optical drawings are covered by ISO standards on technical drawings, as contained in *ISO Standards Handbook 12*[17] and *ISO Standards Handbook 33*.[18]*

The next three parts of ISO 10110-14 deal with optical material specifications and are straightforward adaptations of glass catalog specifications for stress birefringence, bubbles and inclusions, and inhomogeneity (including striae).

Part 5 of ISO 10110-14 deals with optical surface figure errors. Either a visual test plate assessment of figures or computer reduction of interferometric fringe or phase data can be employed.

* Occasionally, the ISO issues groupings of published standards as a handbook. For example, *ISO Standards Handbook 33: Applied Metrology—Limits, Fits, and Surface Properties* was issued in 1988 and included 58 standards developed in seven different TCs, all related to measurement. That handbook includes terminology, indication of mechanical tolerances and surface conditions on technical drawings, limits and fits, properties of surfaces, and measuring instruments.

TABLE 2.4
Subject Matter and Issue/Correction Dates of the 14 Parts of the Standard ISO 10110-14, *Optics and Optical Instruments—Preparation of Drawings for Elements and Systems*

- Part 1: General (2006)
- Part 2: Material imperfections—Stress birefringence (1996)
- Part 3: Material imperfections—Bubbles and inclusions (1996)
- Part 4: Material imperfections—Inhomogeneity and striae (1997)
- Part 5: Surface form tolerances (in preparation)
- Part 6: Centering tolerances (1996/1999)
- Part 7: Surface imperfection tolerances (2008)
- Part 8: Surface texture—Roughness and waviness (2010)
- Part 9: Surface treatment and coating (1996)
- Part 10: Table representing data of optical elements and cemented assemblies (2004)
- Part 11: Non-toleranced data (1996/2006)
- Part 12: Aspheric surfaces (2007)
- Part 14: Wavefront deformation tolerance (2007)
- Part 17: Laser irradiation damage threshold (2004)

Centering tolerances are the subject of Part 6. It shows how to specify centering relative to various datum surfaces. Part 7 covers surface imperfections or cosmetic defects such as those commonly called "scratches and digs." Either of two techniques may be used to evaluate these defects. The defect areas can be measured directly or their visibility assessed against an appropriately illuminated background. Baker[19] described a simple and inexpensive apparatus for quantifying these types of defects. His 2004 book[20] is a definitive reference on this subject.

Part 8 of ISO 10110-14 concerns ground and polished surface texture, while Part 9 tells how to indicate that a surface is to be coated. It does not specify what type of coating is to be applied or what the characteristics and performance of the coating should be. These details are covered in another standard, ISO 9211-1,[21] dealing with optical coatings.

Part 10 of ISO 10110-14 outlines ways to specify simple optical elements in tabular form without preparing a drawing. This is useful, as it facilitates the communication of an optomechanical designer's manufacturing requirements by computer linkage. Part 11 of the ISO standard gives a table of default tolerances applicable to dimensions of manufactured elements for which no tolerances have been given on the drawing. For example, when not otherwise specified, elements from 10 to 30 mm in diameter are expected to have diameters within 0.5 mm of the specified nominal value. If this level of accuracy is adequate for the application, the drawing can be simplified by simply omitting the tolerance.

Part 12 tells us how to specify an aspheric surface in a widely understood and accepted manner. Part 14 gives a default tolerance for wave front deformation. Finally, Part 17 describes how to specify a threshold for laser damage.

To assist designers and engineers in the interpretation and application of ISO 10110-14, the Optical Society of America (OSA) published a user's guide. This guide facilitates the preparation of optical element and system drawings and the inclusion therein of appropriate notations and symbology. It, of course, does not include any recent changes in the standard.

Several other ISO standards are of interest here. Listed in Table 2.5, these cover measurement, inspection, and testing of optics.

Willey and Parks[16] pointed out that the four parts of ISO 9211-1 on optical coatings deal with pertinent subjects in more detail than any other then generally available document. Part 1 clarifies coating terminology and defines 10 coating types by function. It also illustrates many kinds of coating imperfections. Part 2 deals with the optical properties of typical coatings and tells how to specify them. Examples are given to facilitate understanding of this topic. Part 3 covers the environmental

TABLE 2.5
List of ISO Standards Dealing with Measurement, Inspection, and Testing of Optics

- ISO 9022: *Environmental Test Methods* (20 parts)
- ISO 9039: *Determination of Distortion*
- ISO 9211: *Optical Coatings* (4 parts)
- ISO 9335: *OTF, Camera, Copier Lenses, and Telescopes* (3 parts)
- ISO 9336: *Veiling Glare, Definition and Measurement*
- ISO 9802: *Raw Optical Glass, Vocabulary*
- ISO 10109: *Environmental Test Requirements* (7 parts)
- ISO 10934: *Microscopes, Interface Connections*
- ISO 10935: *Microscopes, Interface Connections*
- ISO 10936: *Microscopes, Operations*
- ISO 10937: *Microscopes, Eyepiece Interfaces*
- ISO 11254: *Laser Damage Thresholds*
- ISO 114211: *OTF Measurement Accuracy*
- ISO 11455: *Birefringence Determination*
- ISO 12123: *Bubbles, Inclusions: Test Methods and Classification*

durability of coatings in terms of their intended applications. These range from the relatively benign environment of a sealed instrument to severe outdoor conditions. The consequences of unsupervised (and perhaps improper) cleaning of optical surfaces are also discussed. Part 4 specifies methods for environmental testing.

The reader interested in obtaining copies of published ISO standards should contact the ANSI directly. Activities of ANSI, the SPIE, the OSA, and OEOSC pertinent to standards can be easily accessed through their respective websites.*

Another international organization involved in standardization efforts is IEC. With central offices in Geneva, Switzerland, it is the leading global organization that prepares and publishes international standards for electrical, electronic, and related technologies, including electronics, magnetics and electromagnetics, electroacoustics, multimedia, telecommunication, and energy production and distribution. Some aspects of optomechanical design and development may require inputs from this technology area. Specifics may be accessed in the United States through the ANSI website.

When all inputs to the technical specifications and interface requirements are believed to have been established and documented, it is time for the first design review. During this review, experts in all pertinent technologies critique those documents for adequacy and completeness. Only after approval by this group should the activity proceed into the preliminary design phase. In some cases, approval is granted subject to correction of the technical requirements or constraint documents along specific lines. In other cases, additional trade-off studies and/or confirmation of requirements is needed to resolve perceived problems. Limited approval to proceed may be given for a specific period. Upon resolution of all conflicts, full approval to proceed is issued.

2.5 PRELIMINARY DESIGN

Given the specifications and constraints as well as at least one concept for an optomechanical system, the idealized design process proceeds into the preliminary design phase. Here the optical

* ANSI, 1899 L. St. NW, 11th Floor, Washington, DC 20036, Tel. (202) 293-8020, http://www.ansi.org; SPIE, The International Society for Optical Engineering, PO Box 10, Bellingham, WA 98227-0010, Tel. (360) 676-3290, http://www.spie.org; OSA, 2010 Massachusetts Avenue, NW, Washington, DC 20036-1023, Tel. (800) 762-6960 or (202) 223-1096, http://www.osa.org; The Optical and Electro-optical Standards Council, 5 Sandy Lane, Pittsford, NY 14534, Tel. (585) 387-9913, http://www.optstd.org.

designers, optical engineers, mechanical engineers, and other concerned individuals strive cooperatively to define an approximate assemblage of parts that have a high probability, once finalized, of meeting the design goals and requirements of the system. These individuals must be given sufficient time to sort through design alternatives, scrutinize details, analyze data, and, on occasion, invent new ways to solve perceived problems. Otherwise, under pressure to finish by the quickest route, the design team may produce instruments that reincarnate the weaknesses of earlier designs or embody new unintended weaknesses. At the earliest stage of preliminary design, the optics may be represented as thin lenses or mirrors that possess focal lengths, apertures, and axial separations, but have no specific radii, thicknesses, or material types. The locations, sizes, and orientations of images and pupils should be correct to a first-order approximation in such representations.

Figure 2.5a shows a thin-lens optical schematic for a military periscopic sight with characteristics as listed in the caption. The paths of the marginal rays entering the system parallel to the axis at the rim of the entrance pupil and the principal rays at maximum plus and minus real semifield angles are shown. To provide a vertical or lateral offset in the optical path, flat mirrors or prisms would be inserted into the airspaces to fold the system. It must be remembered, of course, that appropriate space will be needed later to convert the thin lenses into thick ones and, in some cases, into multiple-element groups. For this reason, it is common practice to assume the thin-lens system length to be somewhat shorter than that expected in the final thick-lens system.

The mechanical layout of the housings, cells, and mirror brackets, etc., for an optical instrument known only for thin-lens approximation would not be of much value. Hence, any serious consideration of mountings usually follows completion of the preliminary thick-lens design. At this point, the number and approximate shapes of the optics are known, their separations are nearly final, and all apertures are known approximately. Figure 2.5b illustrates a preliminary thick-lens schematic of the periscope shown in Figure 2.5a. We do not have space here to consider how the lens designer creates the final optical design.

The parameters most responsible for driving the optical design in the telescope system shown in Figure 2.5, if used as an in-line instrument such as a rifle telescope, are overall length, magnification, entrance and exit pupil diameters, allowable vignetting, and field of view. The use of a lens-erecting system instead of prisms to erect the image is appropriate here because folds are not required. All lenses would need longer focal lengths to provide for vertical offsets if the design is intended to be used in a periscope. This would cause the images of a given field of view to grow in diameter. The apparent field of view (in image space), exit pupil diameter, and allowable vignetting combine to determine the diameter of the eyepiece for a given exit pupil distance. The overall length influences the diameters of the erecting lenses. To provide adequate image quality over a large field of view with external pupils, the objective and eyepiece should both be wide-angle types. An Erfle-type eyepiece and an objective styled after a Kellner-type eyepiece are shown. These configurations are described in many optics texts.

Comparison of the thin- and thick-lens designs for this periscope shows the significant change in system length that occurs when real lenses are substituted. The chosen focal lengths of the thin-lens system are preserved in the thick-lens version.

Given a preliminary thick-lens optical design, the mechanical engineer can begin a layout of the metal parts for the instrument. An important input at this time is a preliminary definition of the set of adjustments that should be provided to take care of manufacturing variations in parts at assembly. Knowledge of the predicted sensitivity of the optical design to mispositioned and dimensionally off-nominal components is needed for this determination. These sensitivity data are also needed to assign appropriate tolerances to both optical and mechanical part dimensions and physical properties in the mechanical design.

Another step in the design process that can begin with either version of the optical system in Figure 2.5 is preliminary definition of necessary adjustments such as focusing and component alignment during assembly. Early consideration of these mechanisms will help avoid establishment of an overall system configuration lacking space for these essential features.

Establishing confidence that a proposed preliminary design will really work when finalized entails answering several key questions. Figure 2.6 shows a generic flow diagram for such an evaluation.[22] Starting with the existing requirements and criteria for verifying the capability of the design

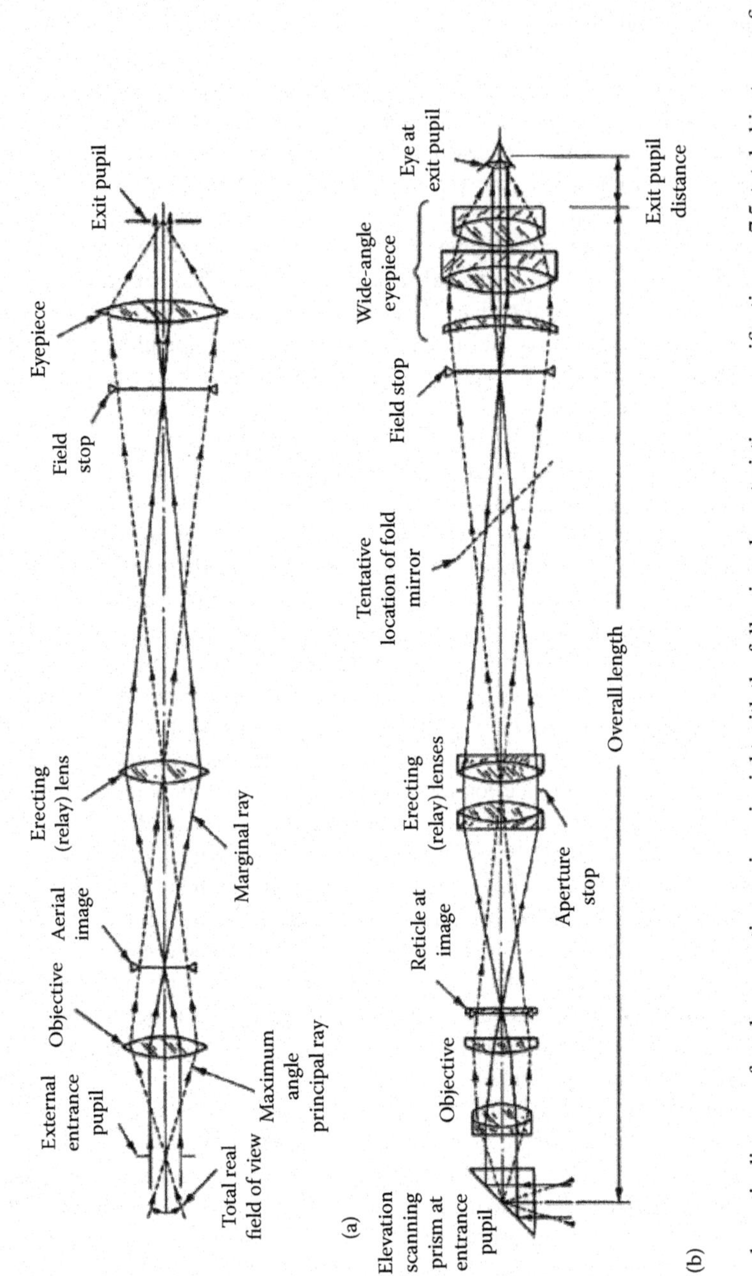

FIGURE 2.5 Optical schematic diagrams for a lens-erecting periscopic sight with the following characteristics: magnification = ×7.5, total object space field of view = 35°, exit pupil diameter = 0.2 in. (5.08 mm), exit pupil distance = 0.68 in. (17.3 mm), and overall length = 23 in. (584 mm). (a) Thin-lens version; (b) preliminary thick-lens version.

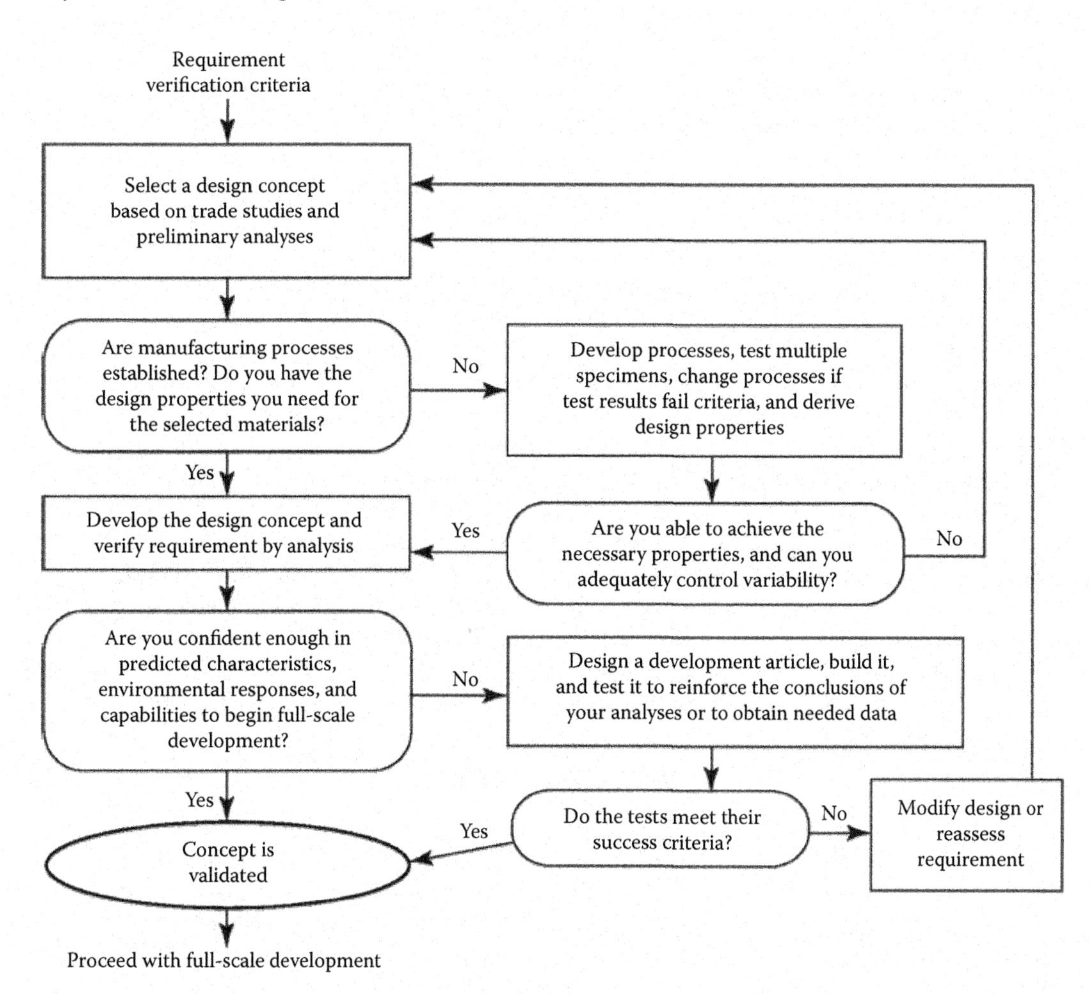

FIGURE 2.6 Flow diagram for design/material/process verification steps during conceptual and preliminary design phases of the project. (From Sarafin, T.P., Developing mechanical requirements and conceptual designs, Chapter 11 in *Spacecraft Structures and Mechanisms*, Sarafin, T.P., and Larson, W.J., Eds., pp. 23–26, Microcosm, Torrance and Kluwer Academic, Boston, MA, 1995.)

to meet them, the design concept is developed. The first set of questions deals with issues on the availability of materials with suitable properties and the adequacy of the manufacturing processes. If these questions are answered "yes," we proceed with the preliminary design. If answered "no," we modify the design accordingly or change materials to ones for which this information is available. It also is appropriate to show that we can tolerate variability in materials and processes. Hopefully, analysis and modeling then confirm that we are ready to begin the final design. If this is not the case, we might design, build, and test models to obtain the needed data. If successful, the project proceeds to the final design step. If not, we modify the preliminary design or, with the concurrence of the customer, revise the requirements and repeat the evaluation process. Eventually, the improved preliminary design for the instrument is accepted and then we proceed into detailed design.

Mirror design is a three-step process comprising predesign, first-order design, and final design analysis. All mirror designs start with performance specifications. In addition to optical performance parameters, such as radius of curvature, diameter, and conic constant, mechanical parameters are also specified. These include self-weight deflection and total weight. Self-weight deflection is normally part of a larger error budget including thermal distortion and fabrication errors. For dynamic applications, self-weight deflection is derived from the relationship between frequency and static deflection.

Scaling from previous designs is a common technique used for initial analysis. Mirror weight can be predicted using scaling laws such as those developed by Valente[23] or Hsu and Johnston.[24] Alternatively, weight is estimated from areal density. A conservative areal density value for lightweight mirrors is 180 kg/m^2, which is represented by the primary mirror of the Hubble Space Telescope. A current state-of-the-art value is about 15 kg/m^2, and this is represented by the primary mirror of the James Webb Space Telescope. Application of simple self-weight deflection equations is another way to estimate mirror performance rapidly. These equations are discussed in Chapter 8.

First-order design is the detailed design of the mirror by using "closed-form" or "hand calculations." In reality, the nonlinear nature of the detailed equations governing mirror self-weight deflection and stiffness makes use of computer aid highly desirable. During this phase of the design, the stiffness and self-weight of the mirror are determined using parametric analysis. It is possible to develop a mirror design by using only first-order analysis. This practice was common before the widespread application of finite-element analysis (FEA).

After first-order design, the candidate mirror configuration is subjected to a detailed analysis by using FEA. There are often large differences in the estimated performance between FEA and first-order analysis. The latter serves as a good check for preliminary FEA. Good agreement is usually defined as agreement between these results within about 20%. The final design is guided by the FEA, although it may be necessary to iterate between the parametric first-order analysis and the FEA to achieve an optimal design. Skipping first-order analysis and going directly to FEA is often unproductive in that it can take a long time to converge to a design solution by using just FEA. The use of FEA is outside the scope of this book. A good introduction to the use of FEA in mirror design analysis is provided in Doyle et al.[25]

2.6 DESIGN ANALYSIS AND COMPUTER MODELING

At this point in the design, the materials for all major parts of the instrument would have been chosen, at least tentatively, so the thermal and dynamic (shock and vibration) characteristics of the design when exposed to the anticipated environment can be analyzed. Any apparent inadequacy of the preliminary design revealed by these analyses should be carefully assessed to determine whether redesign is necessary. The changes might be as simple as substitution of stainless steel for aluminum to reduce the coefficient of thermal expansion of the mechanical parts determining a particularly critical airspace. Glass choices might need reconsideration if, for example, analysis indicates that cemented doublets may not survive thermal shock due to the widely differing expansion properties of the elements or if focal length changes with temperature are excessive. More complex changes in structural design may be found necessary if analytical simulations of shock or vibration loads indicate excessive deformations or even the possibility of structural failure.

In simple instruments, it may be sufficient to estimate the mechanical behavior of the perturbed system by using the classical beam-and-shell theory to quantify component deflections (i.e., strains) and stresses as well as the classical heat transfer theory to quantify temperature effects. In such cases, limited knowledge of temperature distributions, component deformations, and displacements may be adequate to estimate these effects on system performance. Roark's work[26] and later versions of this provided a general set of equations for deflections, internal moments, shears, and stresses of a large variety of geometrical bodies. Roark's equations form the basis for many of the analyses of various types of optomechanical components and assemblies contained in later chapters of this book.

No value calculated with the aid of closed-form equations from reference books can be considered to be exact. The equations are based on certain assumptions regarding the applicability and behavior of a mathematical model of the hardware and the uniformities of the key properties of materials within extended pieces of these materials. Further, they are derived by mathematical procedures that often involve additional approximations. Fortunately, an extremely high accuracy is not always required in engineering design or analysis, so calculations made with these equations are adequate in most cases. More elaborate, but not necessarily more accurate, calculations related to structural design problems are usually accomplished with FEA methods. Even in simple instruments, the calculations may be complex if one has to deal with temperature-dependent material properties, three-dimensional (3D) spatial variations in temperature, temporal variations (such as thermal shock due to rapid temperature changes or development of gradients), and alternative material or configuration trade-offs.

FEA methods have been developed over many years and are the generally accepted tools for design and analysis of mechanical structures. They have great applicability to static, dynamic, and heat transfer analyses of optomechanical instruments. In a typical FEA analysis, a model of the structure is created as a two- or three-dimensional continuum, or mesh, of small elements. It is assumed that deformations (i.e., strains) within these elements are elastic and uniform and are distributed according to some known relationship. The elements usually have triangular, rectangular, or trapezoidal faces and are assumed to be connected by frictionless pins at their vertices or nodes. Elastic-body relationships are utilized to derive polynomial representations of deformations of the structure under applied disturbances. Temperature distributions can be applied to the model to determine thermal effects. Optical components are typically modeled as structures since we are interested in surface deformations as well as stress distributions. Many equations must be solved to describe the behavior of the entire structure, so matrix operations, complex software programs, and high-speed, large-capacity computers are employed. The results are recognized as approximations. They approach truth, i.e., converge, as the assumed model analyzed becomes more complex and greater numbers of smaller elements (i.e., a finer mesh in the model) are considered.

Hatheway in 2004[27] illustrated this convergence property of the FEA method with the following simple example. A 32 in. long square cross section beam with a depth of 16 in. was cantilevered from one end. Its distributed weight caused the free end to droop under gravity. The proportions of the beam were chosen to require consideration of shear effects as well as elastic deformation. The deflection was predicted by linear elastic computations and by FEA assuming different numbers of nodes. Figure 2.7 shows five models having progressively finer meshes and hence more nodes. Table 2.6 lists the characteristics of these models. Figure 2.8 plots the variation in the deflection at the free end of the beam calculated by the linear equation (triangle) and by FEA (solid circles) for different numbers of nodes. The curve shows the FEA approximation approaching the linear value, as the elements grow smaller, i.e., as the number of nodes increases. This characteristic can be used to estimate the degree of accuracy of an FEA analysis by testing a given calculation with an increased number of nodes. If the calculation result changes by only a very small amount, convergence is occurring and the result can be considered reasonably accurate.

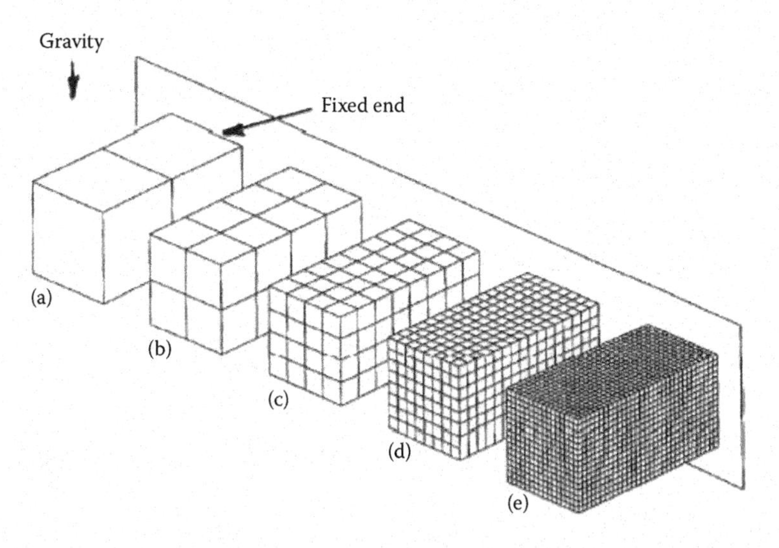

FIGURE 2.7 Five FEA models used to illustrate the effect of complexity on convergence; properties (a)–(e) of the models are given in Table 2.6. (Adapted from Yoder, P.R., Jr. and Vukobratovich, D., *Opto-Mechanical Systems Design*, 4th ed., CRC Press, 2015.)

TABLE 2.6
Characteristics of FEA Models of Beams Shown in Figure 2.7

View	(a)	(b)	(c)	(d)	(e)
Element size, in.	16	8	4	2	1
Number of nodes	12	45	225	1377	9537
End deflection, in. × 10^{-5}	6.63	6.95	7.16	7.26	7.30

Source: Adapted from Yoder, P.R., Jr. and Vukobratovich, D., *Opto-Mechanical Systems Design*, 4th ed., CRC Press, 2015.

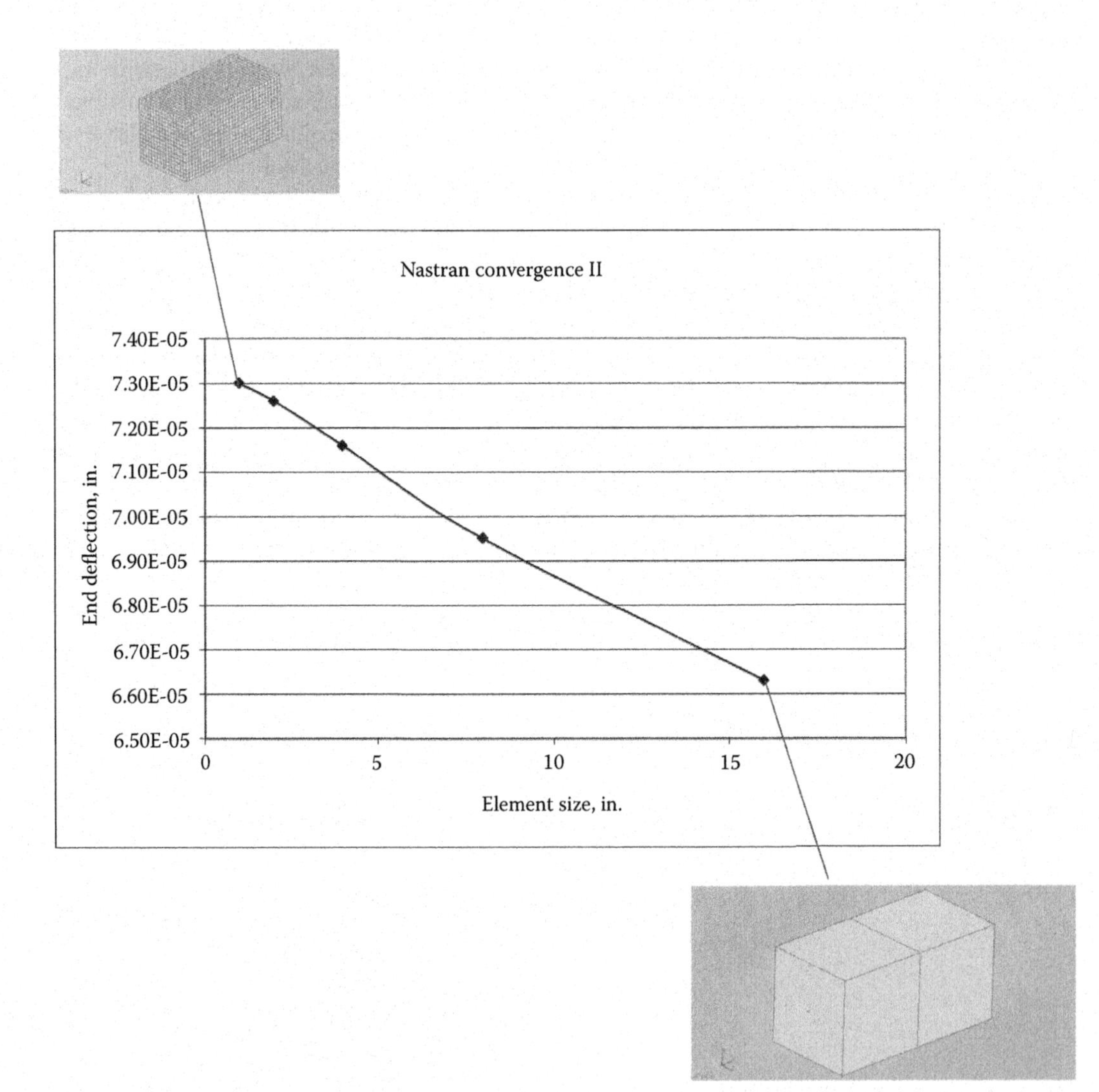

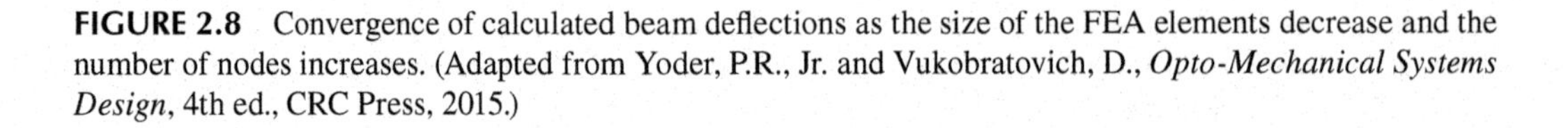

FIGURE 2.8 Convergence of calculated beam deflections as the size of the FEA elements decrease and the number of nodes increases. (Adapted from Yoder, P.R., Jr. and Vukobratovich, D., *Opto-Mechanical Systems Design*, 4th ed., CRC Press, 2015.)

Computerized structural analysis and related pre- and postprocessing codes such as ANSYS, NASTRAN, PATRAN, and STARDYNE have been developed to facilitate structural design and analysis. Among their other capabilities, these codes can compute the elastic deflections of structures under an assumed set of imposed loads, either static (steady state) or dynamic (time varying), as well as the localized stresses in those structures. Stress computations are frequently used to estimate the potential for structural damage under extreme environmental conditions. Doyle et al.[25] explained the basic considerations involved in the use of FEA methods to model and analyze optical instruments.

FEA codes are very effective from the viewpoint of optomechanic systems if used in combination with other codes. Mechanical and thermal analyses require the use of many different types of codes. When optics are involved, we must add optical analysis models and codes. The most powerful of these are general-purpose lens design and analysis codes such as Code V, OSLO, and ZEMAX. These codes calculate multiple-ray trajectories and intercepts by using the laws of reflection, refraction, and diffraction. Optical performance is usually evaluated in terms of geometrical aberrations, modulation transfer functions, point and line spread functions, Zernike polynomial representations of aberrations, and/or optical surface distortions. Optical system performance degradation usually results from rigid-body component tilts, from displacements from nominal orientation and location, and from surface deformations. Frequently, the last ones are the most significant causes of performance degradation.

Most sophisticated applications, such as the design and analysis of complex instruments for space exploration or large ground-based astronomical telescopes, involve disciplines other than optics and mechanics. Control systems, fluid mechanics, electromagnetics, electronic signal processing, communications, etc., may need to be considered. All the disciplines that comprise the total system may need to share data to permit analysis of the performance of the instrument.

Contemporary CAD packages (such as AutoCad, Pro/Engineer, and SolidWorks) are very powerful tools for formatting analytical models of optomechanical systems and for graphically portraying computational results. Optical design codes and structural or thermal analysis codes use different techniques for solving their equations. Their input/output data formats are not generally compatible, so the computational routines may not be directly linked. To evaluate the optical effects of mechanical or thermal disturbances, it has become common practice to evaluate the optical performance of the unperturbed system, compute the elastic deformations due to external influences such as vibration or temperature change, and then to input those results into the optical design code where the optical performance is recomputed. Coronato and Juergens[28] described a technique for transferring data by using Zernike circular polynomials.

In an *integrated analysis method* (see Figure 2.9) described by Hatheway in 2004,[27] each technical discipline uses its own software and a database manager moves data from code to code and reformats (or translates) the output of each code to serve as the input to the next code. Data values are interpolated or extrapolated to fit the unique requirements of the various codes as they are transferred. Errors introduced in the data transfer steps may be large. They then are difficult to quantify. Validation of results may require checking with a more rigorous procedure or by experimentation. In some cases, solving problems for which the results are already known from prior closed-form calculations or tests can validate the calculation routines.

This method has a series of individual software codes available for database managing and translating software (DBM/TS). The latter uses a central computer or system of computers programmed to interface and control all the computational steps. Once required input files (models) for each code have been prepared, the DBM/TS can conduct the appropriate calculations and direct the output from each code to the proper next recipient. If the indicated recipient is another code, the DBM/TS automatically processes the data into the proper format for input to that code. For example, data-processing algorithms may convert data from a cylindrical to a rectangular coordinate system or interpolate temperature distribution data to a finer grid than that originally computed. This represents a very sophisticated operating system for the software codes that it is designed to integrate.

Figure 2.10 shows a representation of another integrated analysis method with potential pathways for complex flow of design data between various software programs during the design or

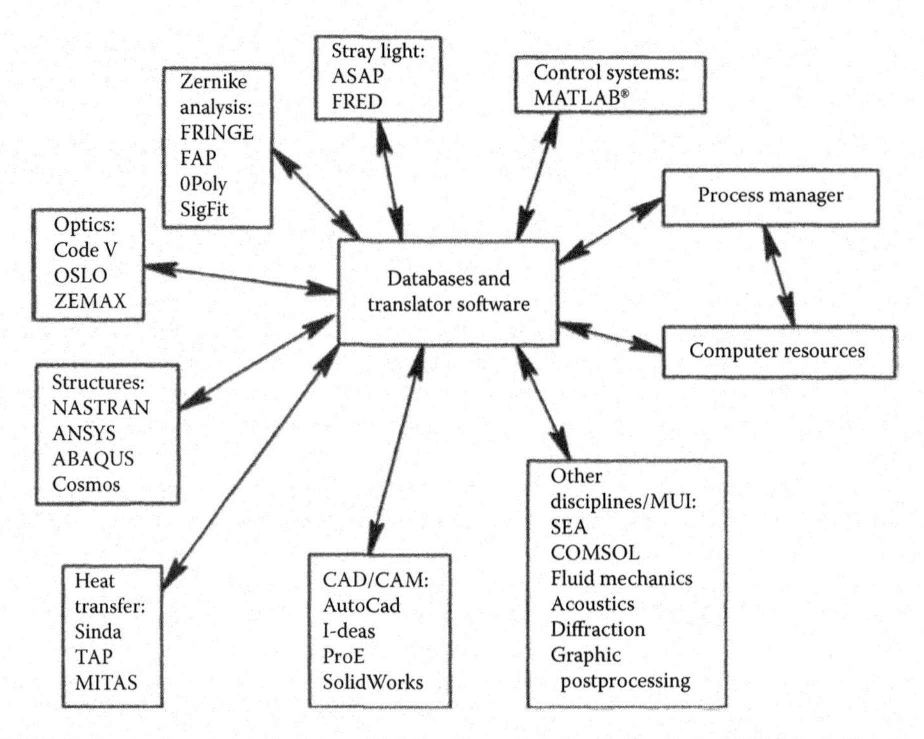

FIGURE 2.9 Interactions between various disciplines in optomechanical analysis. MUI, Magic User Interface. (From Hatheway, A.E., Error budgets for optomechanical modeling, *Proc. SPIE*, 5178, 1, 2004.)

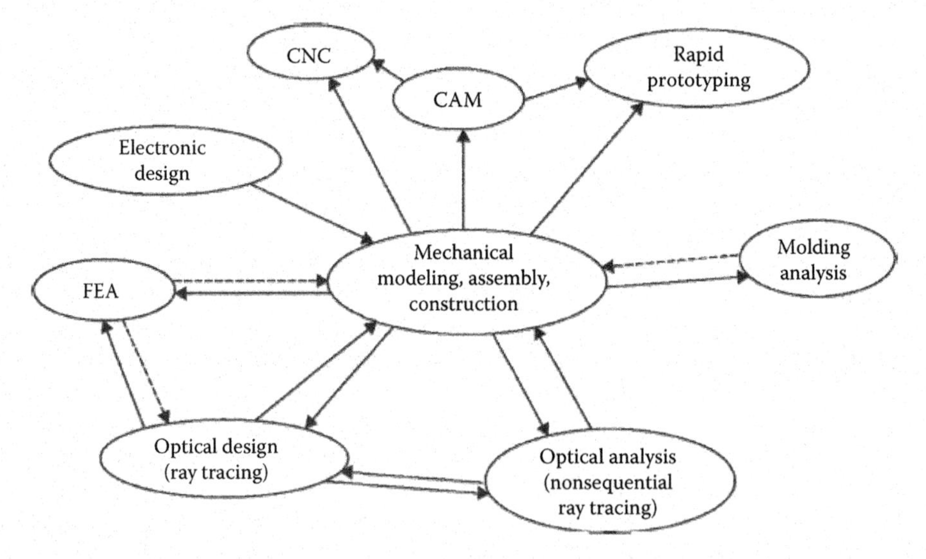

FIGURE 2.10 Potential pathways for design data flow between software packages. The dashed-line arrows represent paths wherein analysis data can be used as a basis for design changes, but the analysis program does not modify the design data directly. CNC, computer numerical control. (From Shackelford, C.J., and Chinnock, R.B., Making software get along: Integrating optical and mechanical design programs, *Proc. SPIE*, 4198, 148, 2000.)

development process.[29] Solid lines imply direct influences, while dashed lines indicate data flow that may form the basis for manual design changes. Data exchange between programs is facilitated if they all use standard file formats. Examples of existing data exchange formats are listed in Table 2.7. The reader's attention is drawn especially to the last entry in that table (STEP), which is a program from ISO that produces and applies an international standard for product data representation

TABLE 2.7
Some CAD and Graphics File Formats for Electronic Data Exchange

Format	Name	Maintained by	Comments
ACIS	Alan, Charles Ian's System	Spatial Corp.	Three-dimensional modeling engine
BMP	BitMaP	Microsoft Corp.	Graphics file format used by Windows File with array of RBG graphics data for each image pixel
DXF	Drawing eXchange Format	Autodesk	Vector-based 3D format
IGES	International Graphics Exchange Specification	National Computer Graphics Association	Set of protocols for transfer display of graphical data
JPEG	Joint Photographic Experts Group	C-Cube Microsystems	Compression algorithm for encoding bitmap data and not a file format
JFIF	JPEG File Interchange Format	C-Cube Microsystems	De facto standard Internet JPEG format
STL	STereo Lithography Interface Format	3D Systems	ASCII or binary files allow CAD data to be read by stereolithography apparatus.
VDA-FS	Verband Der Automobileindustrie Flachen Snittstelle	Verband Automobileindustrie	German international standard
STEP	STandard for the Exchange of Product	ISO	ISO 10303-21:1994, *Industrial Automation Systems—Product Representation and Exchange*
PNG	Portable Network Graphic		Raster graphic file format supporting lossless data compression

Source: Shackelford, C.J., and Chinnock, R.B., Making software get along: Integrating optical and mechanical design programs, *Proc. SPIE*, 4198, 148, 2000.
Note: Solid modeling uses STEP, ACIS, and IGES. Two-dimensional drawings use BMP, JPEG, and DXF.

and exchange. It is intended, in part, to address the need for data exchange throughout the life cycle of a product. This period may extend well beyond the lifetime of the computer program used to design the product. During this period, proprietary programs may become obsolete or lose their ability to communicate with other needed programs. STEP also provides a neutral data model that allows data to be stored on any database platform and to be accessed from any application through a standard interface. A different approach to solving multidisciplinary computational problems related to optomechanics is to develop a mathematical analogy between the discipline of interest and the FEA code being used. Hatheway[30] defined this as a *unified analysis method.* Linearized versions of more complex equations produce acceptable results for analysis of thermal, elastic, and optical response in most cases. By linearizing the equations used in the analogy, the solution often requires only one software code, thus avoiding the interpolations, extrapolations, format changes, truncations, data expansions, and contractions that might otherwise be required to move the problem solution back and forth among software codes en route to the desired result.

To illustrate this technique, Figure 2.11a shows an FEA model of a structure supporting the primary and secondary mirrors of an afocal telescope.[31] This telescope was not performing well during the test. It was suggested that mechanical strains introduced into the primary mirror by deformation of its structure due to particulates trapped in the mounting flange interface might be the cause. Direct evaluation was complicated by the smallness of the distortions relative to rigid-body motions of the mirror. In an FEA simulation, an optical analog of an interferometer was combined with a model of the mirror surface. This interferometer was constrained mathematically to move with the mirror surface when the structure was deformed as shown in Figure 2.11b. Then, only relative motions (deformations)

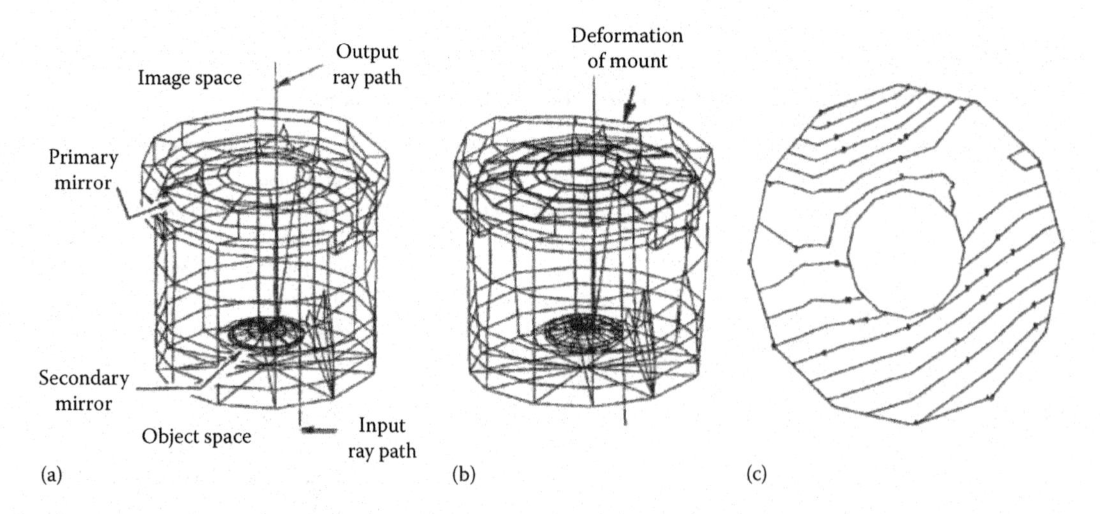

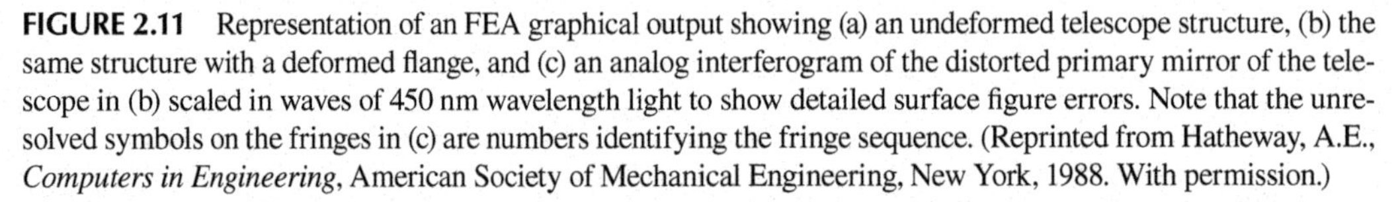

FIGURE 2.11 Representation of an FEA graphical output showing (a) an undeformed telescope structure, (b) the same structure with a deformed flange, and (c) an analog interferogram of the distorted primary mirror of the telescope in (b) scaled in waves of 450 nm wavelength light to show detailed surface figure errors. Note that the unresolved symbols on the fringes in (c) are numbers identifying the fringe sequence. (Reprinted from Hatheway, A.E., *Computers in Engineering*, American Society of Mechanical Engineering, New York, 1988. With permission.)

appeared in the output data. Plots of simulated interferograms at appropriate scales (in wavelengths) allowed the determination of the disturbing effects of various-sized particles trapped under the mounting flange of the mirror. In Figure 2.11c, mirror deformations resulting from a single, hard, 125 μm (0.005 in.) diameter particle located under the mounting flange for the primary mirror at a particular grid location were represented by contour lines separated by 0.5 wave at 450 nm wavelength. The total deflection range equaled 0.71 wave at 633 nm wavelength and explains why the system did not perform properly. Disassembly, cleaning, and reassembly corrected the problem in the hardware.

The capability of properly applied FEA programs to perform analyses such as that just described may lead some inexperienced analysts to believe computed results without adequately questioning the validity of the assumptions and modeling inputs to the program or the accuracy limits of the selected analytical model. The design engineer must never lose sight of the fact that the FEA model may, under certain circumstances, neglect important characteristics of the structure analyzed or may be improperly applied. It will then give misleading results. As mentioned earlier, by carefully applying selected classical methods of elastic structure behavior analysis such as Roark's formulas as presented by Young[32] or Timoshenko and Goodier,[33] along with an FEA program, as synergistic tools of the trade, greater confidence in the results can be generated. Genberg et al.[34] wisely indicated that FEA results should be considered "guilty until proven innocent." Other authors advised that the engineer must accept the burden of understanding the underlying theory of structural/thermal/FEA; understand the working details of the FEA program, including its pre- and postprocessor features; make and verify modeling decisions and assumptions; interpret the results and draw the appropriate conclusions; and properly document the analysis.

No matter what form the analysis of a design may take, it is imperative that records be kept as the design progresses. Items such as the reasons for choosing particular materials, the basis for concluding that a design (or a portion thereof) will or will not work reliably, and the logic behind the choice of specific commercial parts for incorporation into the design are important enough to document. These records serve as valuable backup information for design reviews. Experience has shown that these records also are well worth the trouble of preparation if they are needed for future reference in the solution of unanticipated problems, for support of patent applications or disputes, or for protection from product liability claims. These documents should become part of the formal design files and not reside in the individual designer's or engineer's files, where they are likely to become lost with time.

If, as is generally the case, the production cost and maintainability of the instrument through its life cycle are critical to the intended application, analyses of these aspects of the design would be appropriate. Trade-off studies of alternative versions of cost-driving features should help indicate the most cost-effective design. Maintainability analyses may lead to design improvements that reduce the number of spare parts to be inventoried, eliminate needs for special tooling, or facilitate (or eliminate needs for) manual adjustments by highly trained personnel. Willey,[35] Fischer,[36] Willey and Durham,[37] Smith,[38] and Fischer et al.[39] contributed many examples of good and bad designs as well as excellent technical guidelines for avoiding production or testing problems by proper instrument design and thoughtful selection of materials and processes before finalizing the drawings.

2.7 ERROR BUDGETS AND TOLERANCES

Closely related to the performance specifications and constraint definitions considered in Section 2.4 are the multilevel budgets on allowable deviations from perfection of component dimensions and alignments relative to other components in the instrument. Tolerances strongly influence how an optomechanical system will perform and the life cycle cost of that instrument. For example, let us consider an electro-optical star sensor system intended for use as a precision attitude reference on a spaceborne platform as described by Cassidy.[40] The achievement of pointing accuracies of 0.5 arcsec over an 8° field of view required extremely uniform symmetry and encircled energy consistency of all-star images over the full field of view. System performance analysis showed that for proper function, the actual spot diameter at, for instance, the 75% encircled energy level should be considerably larger than the diffraction limit corresponding to the chosen fast (*f*/1.5) system relative aperture. For the designer of this lens system to do his or her job effectively, the target energy distribution in the axial image produced by the lens and a budget on permitted perturbation of that distribution due to aberrations as a function of semifield angle were specified as design parameters. Once it was determined that this performance was easily achievable in a perfectly built and aligned system, an error analysis indicated how tilt, decentration, and despace of the individual optical components were related to image degradation. A portion of the total error budget was then assigned to individual and ensemble internal component misalignments, and the detailed mechanical design was allowed to proceed. Since the ability of technicians to assemble the lens to meet component decentration, tilt, and despace budgets depends in part on their ability to detect small errors, portions of the error budget were assigned to instrumental and random errors in the measurement processes used. Inherent in this budgeting process is the assumption that the focus of the lens system remains perfect during operation. Obviously, temperature changes could affect focus, so a portion of the mechanical design error budget was allocated to uniformly distribute thermal effects. Thermal gradients across the lens affect symmetry of the image, so they also received due attention and were assigned another portion of the budget. As a result of careful manufacture and the assembly being closely monitored by quality control inspectors, who ensured that the design was accurately represented in the hardware, the system met all requirements for its application.

Smith[41] pointed out that since many potential sources of error unique to any particular design situation need consideration, the allocation of error budgets should be systematized to ensure a successful design. Ginsberg[42] outlined a technique used successfully for this purpose. He stated that the purpose of the process was "to determine the loosest tolerances that can be specified for optical and mechanical parts and assemblies which will still provide adequate performance." A more recent summary of the basic error budget/tolerancing process was given by Fischer et al.[39] This basic process occurs in almost all optomechanical hardware development projects.

A block diagram relating various steps in the process is shown in Figure 2.12. It begins with the nominal system optomechanical design in block 1. In block 2, tentative tolerances are assigned. These are usually based on experience or common manufacturing practices (see Table 2.8 for a listing of typical published values for optical components[39,43]). In block 3, certain adjustments are defined. These might be small lateral movements of selected lenses to minimize off-axis aberrations

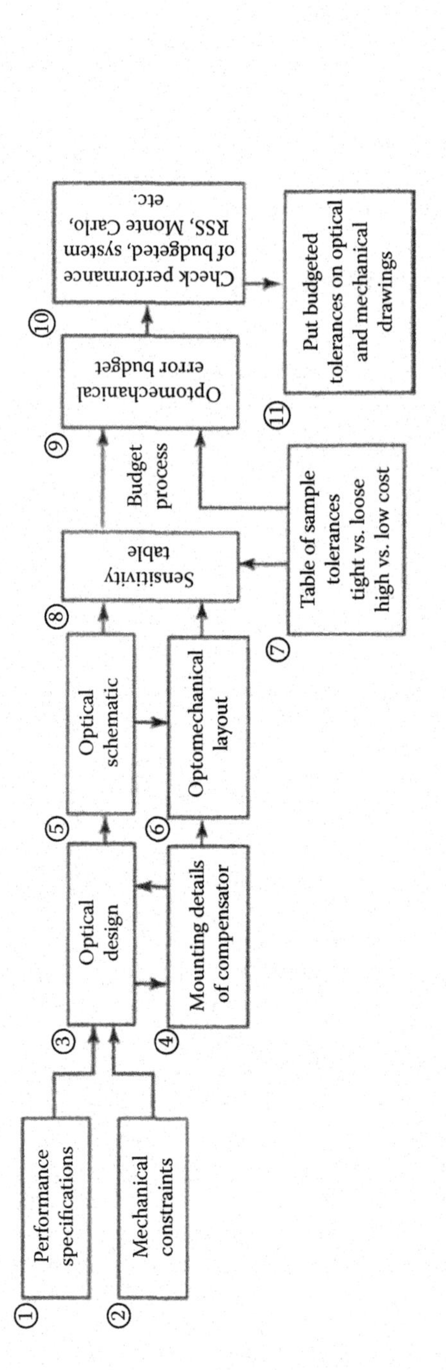

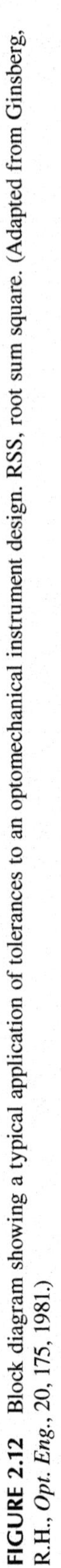

FIGURE 2.12 Block diagram showing a typical application of tolerances to an optomechanical instrument design. RSS, root sum square. (Adapted from Ginsberg, R.H., *Opt. Eng.*, 20, 175, 1981.)

TABLE 2.8
Typical Tolerances on Optomechanical Parameters for Optics

Parameter	Units	Tolerance: Loose	Tolerance: Moderate	Tolerance: Tights	Cost Impact (%)
Index of refraction[a]	–	±0.0005	±0.0003	±0.0002	N/A[b]
Diameter (D)[b]	mm	±0.1	±0.025	±0.005	>125
Radius departure from test (50 mm diameter lens)	Fringes[c]	±5	±2	±0.125	~250
Departure from spherical or flat test plate: power (irregularity)	Fringes	±5 (±2)	±3 (±0.5)	±1 (±0.1)	~250[d]
Element thickness (t)	mm	±0.2	±0.05	±0.01	~200
Physical wedge	arcmin	3	1	0.25	~150
Aspect ratio (D/t)	–	10/1	20/1	50/1	~350
Decentration, physical	mm	0.10	0.010	0.005	N/A
Tilt, physical	arcmin	3	0.3	0.1	N/A
Dimensional errors for prisms	mm	0.25	0.010	0.005	N/A
Angle errors: prisms and windows	arcmin	5	0.5	0.1	N/A
Scratch/dig (per MIL-PRF-13830)	–	80–50	60–40	10–5	N/A

Source: Plummer, J., and Lagger, W., *Photonics Spectra*, 65, December 1982; Fischer, R.E. et al., *Optical System Design*, McGraw-Hill, New York, 2009.

Note: N/A, not available. Table includes miscellaneous manufacturers' advertized capabilities.

[a] Depends upon element size.

[b] Assumes close fit to cell ID.

[c] One fringe equals 0.5 wavelength at 0.546 μm (mercury green). Fringes are specified over the maximum dimension of the clear aperture.

[d] Depends on the manufacturing process.

(coma, astigmatism, and distortion) or adjustment of the axial position of one or more lenses to minimize spherical aberration and/or optimize focus. These adjustments would be accomplished at final assembly in a test fixture that allows measurement of the pertinent aberrations. In block 4, the lens design program is utilized to determine the sensitivities of the system performance to small variations in each parameter. In block 5, the parameter tolerances are adjusted so very sensitive values are assigned relatively tight tolerances, while less sensitive parameters are given looser tolerances.

Figure 2.13 shows an example of an afocal telescope used as a laser beam expander from the earlier referenced paper by Ginsberg[42] to illustrate aberration compensation and final alignment of a typical simple assembly. The telescope is to be flange-mounted at datum -A- and located laterally with respect to a pilot diameter (datum -B-). It is assumed that the laser beam will enter perpendicular to datum -A- and will be centered with respect to the pilot diameter. The larger lens is to be used as a compensator for focusing the output beam and also for aligning that beam normal to datum -A- (by sliding its mount on surface -C-). Because the first lens registers to its first polished surface, an assembly clearance between its outer diameter and the metal inner diameter will allow the lens to tilt slightly about the center of curvature of that surface. The second lens, on the other hand, is referenced to a plane surface so it cannot tilt, but can only decenter. The first lens will tilt if the shoulder against which it is mounted is tilted with respect to surface -A-. Because the cell of the larger lens rotates in its threads for focus adjustment, it should be centered after focusing. Other design features requiring tolerances include the parallelism of surfaces -A- and -C- and the fits of the threads that seat retainers against curved lens surfaces.

To prepare the sensitivity data of block 4 in Figure 2.12, the maximum reasonable magnitudes for all potential errors are approximated and then each parameter is changed by a convenient small

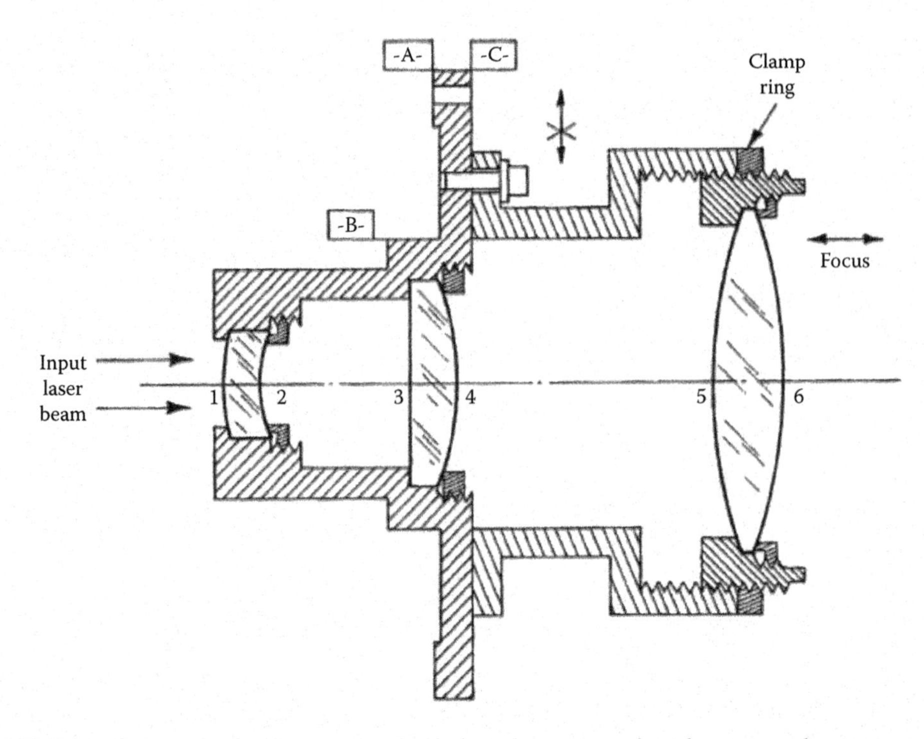

FIGURE 2.13 Optomechanical layout of a simple laser beam expander telescope used as an example in considerations of a suggested process for budgeting error tolerances. (From Ginsberg, R.H., *Opt. Eng.*, 20, 175, 1981.)

step. The corresponding change in performance in terms of some merit function or aberration that has previously been agreed upon is computed and entered into the blank spaces on a table such as that shown in Figure 2.14. Linearity with small parameter changes is assumed.

The sensitivity data in Figure 2.14 apply to the hardware example of Figure 2.13. The performance characteristic of importance here is the output beam divergence, Δdiv (in microradians), due to the changes listed in the column "Change." Footnote a in Figure 2.14 states that axial adjustment of the third lens element corrects defocus, while lateral displacement of that lens corrects error in the direction of the output beam before the change in divergence is calculated. The magnitudes of these adjustments are entered into the final two columns. Additional columns could be added to the table to record sensitivities of other performance criteria as appropriate.

The error budget (block 9 of Figure 2.12) is developed from the sensitivities and the approximated maximum errors. Errors introduced by outside factors during operation are considered in block 6. These might include predicted atmospheric turbulence or vibration.

Figure 2.15 shows a budget applicable to the example considered by Ginsberg.[42] All parameters of interest for each component are listed together to facilitate transferring the information to the optical and mechanical drawings. The applicable tolerances must, of course, be considered as an ensemble. If the errors are reasonably independent, we may estimate their overall effect as the root sum square of those errors. This should be compared with the total allowable system error. A worst-case budget would allow the errors to add directly. This is not a reasonable representation of items to be produced in quantity because statistically the parameters will not all be at the maximum tolerance.

A better error budget results if the effects of a group of toleranced parameters are combined by the Monte Carlo method. Here, a group of at least 25 versions of the basic optomechanical design is created. Each parameter of each design is varied randomly within its tolerance range in accordance with a normal (Gaussian) distribution. The designs then represent a group of manufactured lenses.

Sensitivity table

1 2 3 4 5 6

Surface element or group	Change	Parameter and comments	Δdiv (μrad)[a]	Required refocus 5–6 in.	Required decenter 5–6 in.
1–2	0.001	Index of refraction			
3–4	0.001	-do-			
5–6	0.001	-do-			
1–2	0.00001	Homogeneity			
3–4	0.00001	-do-			
5–6	0.00001	-do-			
1–2	0.001″	Thickness or airspace			
2–3	0.001″	-do-			
3–4	0.001″	-do-			
4–5	0.001″	-do- (without compensation)			
5–6	0.001″	-do-			
1	0.1%	Radius			
2	0.1%	-do-			
4	0.1%	-do-			
5	0.1%	-do-			
6	0.1%	-do-			
3	1 Frng	Nonflat over ″ϕ			
1	1 Frng	Irregty over ″ϕ			
2	1 Frng	-do- ″ϕ			
3	1 Frng	-do- ″ϕ			
4	1 Frng	-do- ″ϕ			
5	1 Frng	-do- ″ϕ			
6	1 Frng	-do- ″ϕ			
1–2	1 Frng	Wedge at 2			
3–4	1 mr	Wedge at 4			
5–6	1 mr	Wedge at 6			
1–2	0.001″	Roll at 1			
5–6	0.001″	Roll at 5			
1–2	0.001″	Decenter			

[a] After refocusing output beam with lenses 5–6, or correcting output beam direction with lenses 5–6.

FIGURE 2.14 Typical sensitivity table applicable to the optomechanical assembly shown in Figure 2.13. Clear aperture, C.A.; FR, fringe; Frng, fringe; Irregty, irregularity; mr, milli-radian; over, over diameter of optic. (From Ginsberg, R.H., *Opt. Eng.*, 20, 175, 1981.) *(Continued)*

Note that some parameter distributions may be skewed from a Gaussian one because of the way the parts are produced. For example, lens elements are usually made with thicknesses on the high side of nominal. This provides material to be removed if a surface is scratched during processing and the lens has to be reground and repolished.

Because application of this process to even the simplest of optical instruments can result in an unacceptable "first-cut" error budget, it may be necessary to iterate the process until a satisfactory distribution is achieved. This is represented in Figure 2.12 by block 8. If no acceptable budget can be achieved, the optical design or the mechanical design may have to be revised and a new budget developed.

One aspect of optomechanical design that is sometimes forgotten (or ignored until problems are discovered while building hardware) is the producibility of the optical and mechanical subsystems. Figure 2.16 shows additional loops that should be inserted into the design process to make sure that critical producibility factors are adequately analyzed.[44] Early consultation with those individuals who will later

Sensitivity table

Surface element or group	Change	Parameter and comments	Δdiv (μrad)[a]	Required refocus 5–6 in.	Required decenter 5–6 in.
3–4	0.001″	Decenter			
5–6	0.001″	-do- (without compensation)			
1–2	1 mr	Tilt at 1 C.A.			
3–4	1 mr	Tilt at 3			
5–6	1 mr	Tilt at 5 C.A.			
		Axial displacement[b]			
	0.010″	Laser decenter			
	1 mr	Laser tilt at 1			
	10 mr	Laser tilt at 1			

[a] After refocusing output beam with lenses 5 and 6, or correcting output beam direction with lenses 5 and 6.
[b] Information available from air space changes.

FIGURE 2.14 (CONTINUED) Typical sensitivity table applicable to the optomechanical assembly shown in Figure 2.13. Clear aperture, C.A.; FR, fringe; Frng, fringe; Irregty, irregularity; mr, milli-radian. (From Ginsberg, R.H., *Opt. Eng.*, 20, 175, 1981.)

be asked to fabricate and assemble the optics and mechanical parts, test the instrument, and maintain it during use will allow their feedback to be incorporated into the design while it is still relatively fluid.

Optical systems with required performance higher than that expected of the simple laser beam expander example considered earlier should be treated differently. We can specify optical surfaces to be close fits to calibrated test plates. In such cases, the actual radii in a particular instrument are then well established. We can obtain measured index of refraction data for each melt of glass used in the optics from the manufacturer. We can also measure actual axial thicknesses of lenses, prisms, etc. With this information, we can then reoptimize the nominal optical design (usually by adjusting airspaces) and maximize performance. This generally is cost-effective for high-performance systems.

In complex systems, it may be appropriate to consider error budgets for individual optical subassemblies and then higher-level error budgets for those subassemblies acting as rigid bodies relative

Error budget

1 2 3 4 5 6

	Surface element or group	Change	Parameter and comments	Δdiv (μrad)[a]	Required refocus in.	Required decenter in.
	1–2		Index of refraction			
			Homogeneity			
		"	Thickness			
	1	"	Radius error FR. %			
	2	"	Radius error FR. %			
	1	FR	Irregty over. " %			
	2	FR	Irregty over. " %			
	at 2	mr	Wedge			
	at 1	"	Roll			
			Decenter, RSS of all causes			
		"	Axial displacement, RSS			
	at 1 C.A.	mr	Tilt, RSS			
	3–4		Index of refraction			
			Homogeneity			
		"	Thickness			
	3	FR	Nonflat over. "ϕ			
	4	"	Radius error FR. %			
	3	FR	Irregty over. "ϕ			
	4	FR	Irregty over. "ϕ			
	at 4	mr	Wedge			
		"	Roll			
		"	Decenter. RSS of all causes			
		"	Axial displacement, RSS			
	at 4	mr	Tilt, RSS			
		"	Laser decenter			
		mm	Laser tilt			

[a] After refocusing output beam with lenses 5 and 6, or correcting output beam direction with lenses 5 and 6.

FIGURE 2.15 Typical error budget derived from the sensitivity table in Figure 2.14. Clear aperture, C.A.; FR, fringe; Frng, fringe; Irregty, irregularity; mr, milliradian; over, over diameter of optic. (From Ginsberg, R.H., *Opt. Eng.*, 20, 175, 1981.) *(Continued)*

to the balance of the system. This just adds complexity to the process. The basic process at each level remains as considered here.

The costs, fabrication time, and quality of optics have been reduced since the early 1990s by introduction and wide adoption of computer numerically controlled grinding and polishing and precision machining of critical interfacing surfaces. More recently, the use of magnetorheological finishing techniques in which the stiffness of the abrasive/polishing fluid (slurry) is adjusted within the contact region and that slurry is constantly renewed. This promotes high surface accuracy and lessens the adverse effects of heating; see Pollicove and Golini.[45] The inherently high degree of surface figure, position, and orientation achievable by those techniques make near perfection possible.

A vital part of optomechanical design is to achieve a good balance between the tightness of tolerances (and hence increased costs) for all types of components and the need for adjustments

Error budget

	Surface element or group	Change	Parameter and comments	Δdiv (μrad)[a]	Required refocus in.	Required decenter in.
	5–6		Index of refraction			
			Homogeneity			
		"	Thickness			
	5	"	Radius error FR. %			
	6	"	Radius error FR. %			
	5	FR	Irregty over. "ϕ			
	6	FR	Irregty over. "ϕ			
	at 6	mr	Wedge			
	at 5	"	Roll			
		"	~~Decenter, RSS of all causes~~			
		"	Axial displacement, RSS			
	at 5 C.A.	mr	Tilt. RSS			
			Index of refraction			
			Homogeneity			
		"	Thickness			
		FR	Irregty over. "ϕ			
		FR	Irregty over. "ϕ			
		mr	Wedge			
		"	Roll			
		"	Decenter, RSS of all causes			
		"	Axial displacement, RSS			
		mr	Tilt, RSS			
			Σ			
			RSS			

[a] After refocusing output beam with lenses 5 and 6, or correcting output beam direction with lenses 5 and 6.

FIGURE 2.15 (CONTINUED) Typical error budget derived from the sensitivity table in Figure 2.14. Clear aperture, C.A.; FR, fringe; Frng, fringe; Irregty, irregularity; mr, milliradian; over, over diameter of optic. (From Ginsberg, R.H., *Opt. Eng.*, 20, 175, 1981.)

(with the associated costs of instrumentation for determining when the adjustments are adequately achieved as well as for labor to accomplish the adjustments). The tolerances discussed in this section apply primarily to errors in the dimensions, locations, and orientations of optical components. Mechanical part designs also need tolerances to fabricate, inspect, and assemble those parts. In general, loose tolerances mean lower production costs because less expensive fabrication methods can be used and fewer inspections are required.

Some frequently overlooked instances in which tight tolerances on mechanical part designs are advisable include dimensions of holes for shear-carrying fasteners (to maintain in-plane stiffness and distribute loads), optomechanical interfaces which allow the application of proper preloads, minimization of the entry of contaminants, dimensional stability of structural parts that determine optical alignment and focus in high-performance instruments, and fits between optical and mechanical parts in applications involving very high accelerations (such as sensors in gun-fired projectiles).

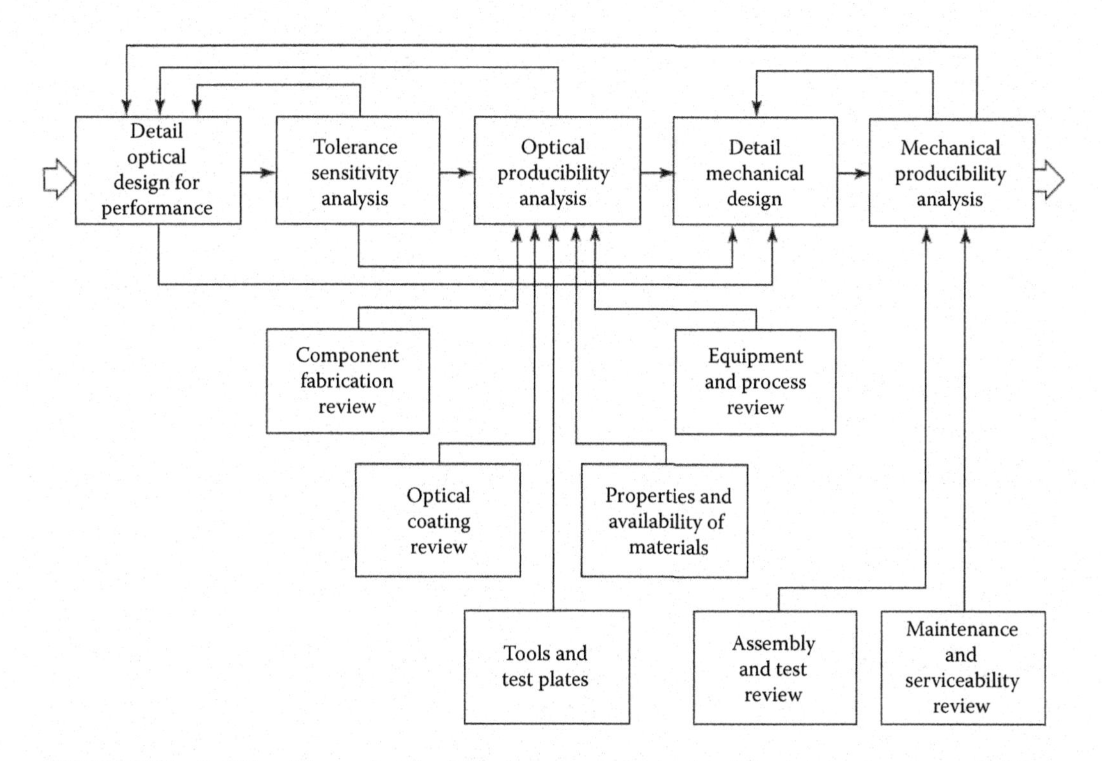

FIGURE 2.16 Additional loops that should be incorporated into the design/tolerancing process in Figure 2.12 to ensure producibility of the optical and mechanical systems. (From Willey, R.R. Economics in optical design, analysis and production, *Proc. SPIE*, 399, 371, 1983.)

REFERENCES

1. Krim, M., Athermalization of optical structures, *SPIE Short Course Notes SC2*, SPIE Press, Bellingham, WA, 1990.
2. Petroski, H., *The Evolution of Useful Things*, Vintage Press, New York, 1994, 231.
3. Sarafin, T.P., Developing mechanical requirements and conceptual designs, Chapter 2 in *Spacecraft Structures and Mechanisms*, Sarafin, T.P., and Larson, W.J., Eds., Microcosm, Torrance and Kluwer Academic, Boston, MA, 1995.
4. Price, W.H., Trade-offs in optical system design, *Proc. SPIE*, 531, 148, 1985.
5. Shannon, R.R., Making the qualitative quantitative—A discussion of the specification of visual systems, *Proc. SPIE*, 181, 42, 1979.
6. Walker, B.H., Specifying the visual optical system, *Proc. SPIE*, 181, 48, 1979.
7. ANSI (American National Standards Institute), *Dimensioning and Tolerancing*, ANSI Y14.5, ANSI, New York, 1982.
8. ANSI, *Optical Parts*, ASME/ANSI Y14.18M, ANSI, New York, 1987.
9. *Performance Specification: Optical Components for Fire Control Instruments; General Specification Governing the Manufacture, Assembly, and Inspection of*, MIL-PRF-13830.
10. *Military Specification: Glass Optical*, MIL-G-174.
11. *Military Specification: Coating of Optical Glass*, MIL-C-675.
12. *Military Standard: Preparation of Drawings for Optical Elements and Optical Systems: General Requirements for*, MIL-STD-34.
13. ISO (International Organization for Standardization), *Optics and Optical Instruments—Preparation of Drawings for Optical Elements and Systems*, ISO 10110-14, ISO Central Secretariat, Geneva.
14. DIN (Deutsches Institut für Normung), *Inscription of Dimensions and Tolerances for Optical Components—Form Errors*, DIN 3140, DIN, Berlin, 1978.
15. Parks, R.E., private communication, 1991.
16. Willey, R.R., and Parks, R.E., Optical fundamentals, Chapter 1 in *Handbook of Optomechanical Engineering*, Ahmad, A., Ed., pp. 1–38, CRC Press, Boca Raton, FL, 1997.

17. ISO, *ISO Standards Handbook 12: Technical Drawings*, ISO, Geneva, 1991.
18. ISO, *ISO Standards Handbook 33: Applied Metrology—Limits, Fits, and Surface Properties*, ISO, Geneva, 1988.
19. Baker, L., Surface damage metrology: Precision at low cost, *Proc. SPIE*, 4779, 41, 2002.
20. Baker, L., *Metrics for High-Quality Specular Surfaces, Tutorial Text TT65*, SPIE Press, Bellingham, WA, 2004.
21. ISO, *Optics and Optical Instruments—Optical Coatings*, ISO 9211-1, ISO Central Secretariat, Geneva.
22. Sarafin, T.P., Developing confidence in mechanical designs and products, Chapter 11 in *Spacecraft Structures and Mechanisms*, Sarafin, T.P., and Larson, W.J., Eds., pp. 309–312, Microcosm, Torrance and Kluwer Academic, Boston, MA, 1995.
23. Valente, T.M., Scaling laws for light-weight optics, *Proc. SPIE*, 1340, 47, 1990.
24. Hsu, Y.W., and Johnston, R.A., Design and analysis of one meter beryllium space telescope, *Proc. SPIE*, 2542, 244, 1995.
25. Doyle, K.B., Genberg, V.L., and Michels, G.J., *Integrated Optomechanical Analysis, TT58*, SPIE Press, Bellingham, WA, 2002.
26. Roark, R.J., *Formulas for Stress and Strain*, McGraw-Hill, New York, 1954.
27. Hatheway, A.E., Error budgets for optomechanical modeling, *Proc. SPIE*, 5178, 1, 2004.
28. Coronato, P.A., and Juergens, R.C., Transferring FEA results to optics codes with Zernikes: A review of techniques, *Proc. SPIE*, 5176, 1, 2003.
29. Shackelford, C.J., and Chinnock, R.B., Making software get along: Integrating optical and mechanical design programs, *Proc. SPIE*, 4198, 148, 2000.
30. Hatheway, A.E., Optics in the finite element domain, in *Computers in Engineering*, American Society of Mechanical Engineering, New York, 3, 1988.
31. Hatheway, A.E., *Computers in Engineering*, American Society of Mechanical Engineering, New York, 1988.
32. Young, W.C., *Roark's Formulas for Stress and Strain*, McGraw-Hill, New York, 1989.
33. Timoshenko, S.P., and Goodier, J.N., *Theory of Elasticity*, 3rd ed., McGraw-Hill, New York, 1950.
34. Genberg, V., Michels, G., and Doyle, K., Integrated opto-mechanical analysis, *SPIE Short Course Notes SC254*, SPIE, Bellingham, WA, 2002.
35. Willey, R.R., Optical design for manufacture, *Proc. SPIE*, 1049, 96, 1989.
36. Fischer, R.E., Optimization of lens designer to manufacturer communications, *Proc. SPIE*, 1354, 506, 1990.
37. Willey, R.R., and Durham, M.E., Ways that designers and fabricators can help each other, *Proc. SPIE*, 1354, 501, 1990.
38. Smith, W.J., *Modern Optical Engineering*, 3rd ed., McGraw-Hill, New York, 2000.
39. Fischer, R.E., Tadick-Galeb, B., and Yoder, P.R. Jr., *Optical System Design*, McGraw-Hill, New York, 2008.
40. Cassidy, L.W., Advanced stellar sensors—A new generation, *Proc. AIAA/SPIE/OSA Technology for Space Astrophysics Conference: The Next 30 Years*, 164, 1982.
41. Smith, W.J., Fundamentals of establishing an optical tolerance budget, *Proc. SPIE*, 531, 196, 1985.
42. Ginsberg, R.H., Outline of tolerancing (from performance specification to toleranced drawings), *Opt. Eng.*, 20, 175, 1981.
43. Plummer, J., and Lagger, W., Cost-effective design—A prudent approach to the design of optics, *Photon. Spectra*, Dec., 65, 1982.
44. Willey, R.R. Economics in optical design, analysis and production, *Proc. SPIE*, 399, 371, 1983.
45. Pollicove, H., and Golini D., Deterministic manufacturing processes for precision optics, *Key Engineering Materials,* Vol. 238, 53 (2003).

3 Material Selection

3.1 INTRODUCTION AND SUMMARY

Material properties for selection in optomechanics are those that influence the shape and position of the optical surfaces of a system. Optical, mechanical or structural, and thermal properties are of interest. Cost is also important and is usually related to the ease of fabrication. Typically, materials are chosen based on figures of merit (FOMs), which are combinations of properties related to certain kinds of performance. Care is necessary to avoid the "single property fallacy," in which a material is chosen based on a single performance parameter without consideration of other possibly adverse properties.

3.2 OPTICAL FIGURES OF MERIT

Optical FOMs for refractive materials include index of refraction, dispersion, transmission, and stress optic effects. For reflective materials, reflectivity is important. The index of refraction determines the bending of a ray of light when incident on a boundary between two different refractive materials. The change in the index of refraction with wavelength is the dispersion of the refractive material. Transmission determines the amount of light that passes through a refractive material of some thickness. The stress in a refractive material changes the index of refraction, inducing an optical path difference (OPD). Reflectivity is the ratio of light incident on a reflective surface to light reflected from the surface.

The index of refraction n is the ratio of the speed of light v in a refractive medium to the speed of light c in a vacuum, or $n = c/v$. Snell's law gives the bending of a light ray passing through the interface between two refractive materials with different indices of refraction:

$$n_1 \cos\theta_1 = n_2 \cos\theta_2, \tag{3.1}$$

where n_1 and n_2 are the indices of refraction and θ_1 and θ_2 are the angles of incidence.

Normally, the index of refraction of a material is measured relative to that of air, which is considered to have an index of refraction of unity at standard conditions (temperature = 295 K and pressure = 1013 mbar). For glass in the visible region, the index of refraction is 1.5–2.0. An index of $n = 1.5$ is a common approximation for glass in the visible region. The refractive index of plastic is 1.31–1.65. The index of refraction of infrared (IR) materials is from about 1.2 for lithium fluoride to just over 4.00 for germanium.

The index of refraction changes nonlinearly with wavelength as shown in Figure 3.1 for visible glass. This variation in the index of refraction with wavelength is called the dispersion of the material. Over a range of wavelengths, the index of refraction for optical glass is given by the Sellmeier formula[1]:

$$n^2 - 1 = \sum_{i=1}^{3} \frac{B_i \lambda^2}{\lambda^2 - C_i}, \tag{3.2}$$

where λ is the wavelength in micrometers and B_i and C_i are the Sellmeier coefficients.

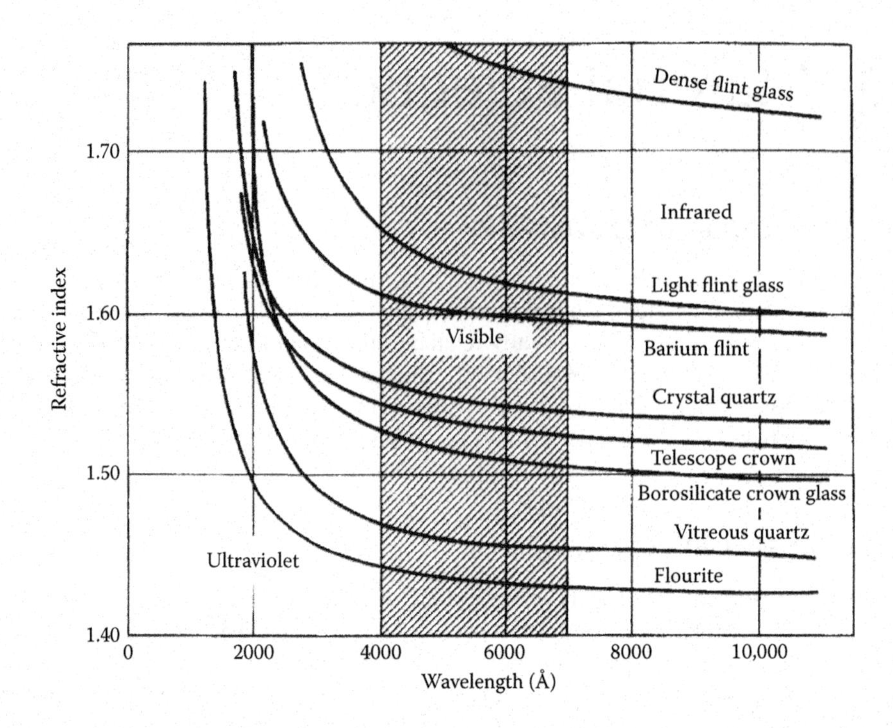

FIGURE 3.1 Dispersion curves for several materials used for refracting optical components. (Adapted from Jenkins, F.A., and White, H.E., *Fundamentals of Optics*, McGraw-Hill, New York, 1952.)

Example 3.1

Find the index of refraction of Schott F2, code 620364, at a wavelength of 550 nm. The constants of the Sellmeier dispersion formula for F2 are

$B_1 = 1.34533359,$
$B_2 = 0.209073176,$
$B_3 = 0.937357162,$
$C_1 = 0.00997743871,$
$C_2 = 0.0470450767,$
$C_3 = 111.886,$

$$n = \left[\frac{B_1(0.55)^2}{(0.55)^2 - C_1} + \frac{B_2(0.55)^2}{(0.55)^2 - C_2} + \frac{B_3(0.55)^2}{(0.55)^2 - C_3} + 1\right]^{\frac{1}{2}}$$

$n = 1.62366.$

Another measure of the variation in the index of refraction with wavelength is the principal dispersion, which is the difference in index between blue ($\lambda = 486$ nm) and red ($\lambda = 656$ nm) wavelengths. These wavelengths correspond to the F and C spectral lines, so the principal dispersion is $n_F - n_C$. Dispersion is most commonly given for glass in the visible region by the Abbe number ν_d, which is the ratio of the index of refraction n_d at the wavelength of the d line ($\lambda = 588$ nm) to the principal dispersion:

$$\nu_d = \frac{n_d - 1}{n_F - n_C}. \tag{3.3}$$

In an Abbe diagram, glasses are plotted by the index of refraction versus the Abbe number. Figure 3.2 is an example of an Abbe diagram or glass map for Schott glasses for indices of refraction at a wavelength of 588 nm. Normal glasses lie along a line connecting $n_d = 1.511$, $\nu_d = 60.5$ and $n_d = 1.620$, $\nu_d = 36.3$ in the Abbe diagram. The Abbe number of optical glass is 20–91. The Abbe number of optical plastics is 18–57.

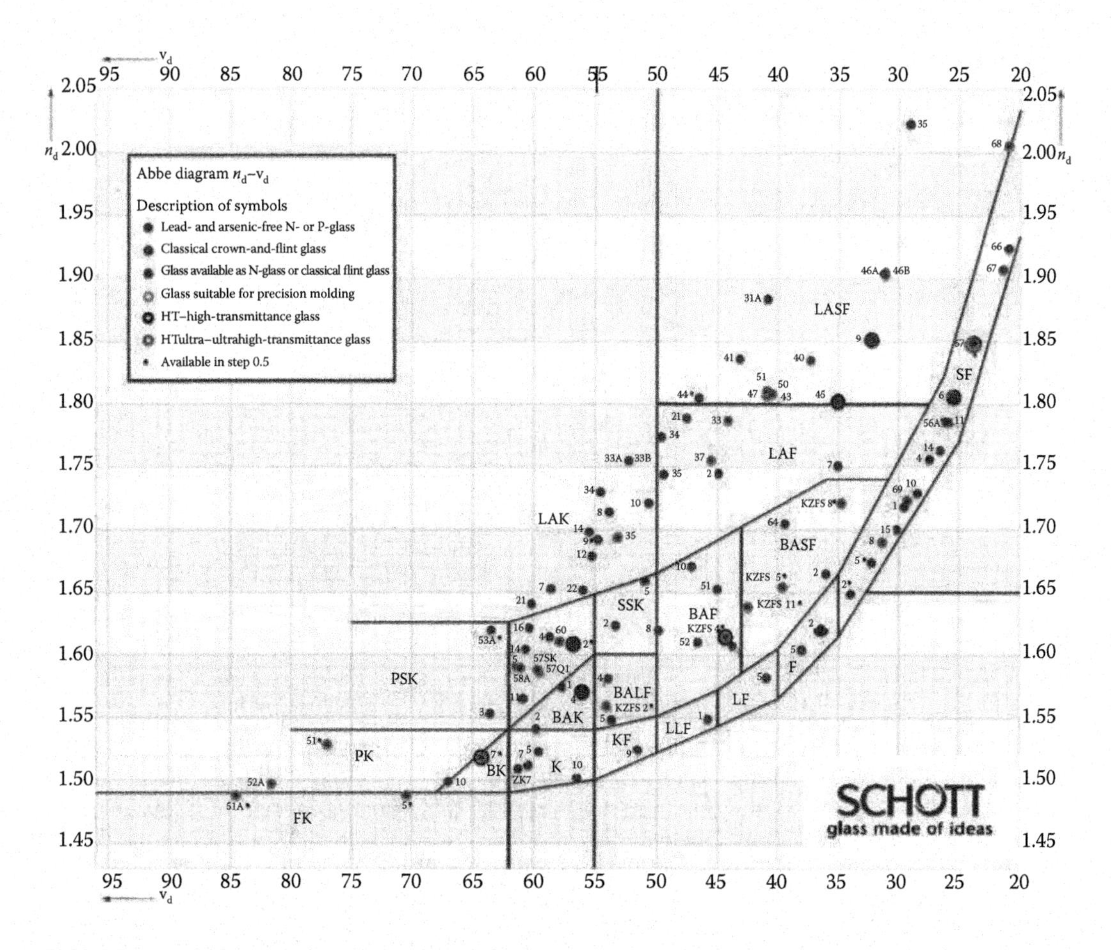

FIGURE 3.2 Current glass map showing dispersion ν_d versus index n for glasses available from Schott in January 2014. (Courtesy of Schott North America, Inc., Duryea, PA.)

Light is absorbed by refractive materials. Absorption is nonlinear with the thickness of the material. Internal transmittance τ is defined as the ratio of the light intensity I_0 before entering the transmissive medium and the light intensity I_1 after exiting the medium, or $\tau = I_1/I_0$. The absorption coefficient β is related to the material thickness h and internal transmittance by

$$\tau = I_1 e^{-\beta h}. \tag{3.4}$$

Typically, the absorption coefficient β is given in units of inverse centimeters (cm^{-1}). It is not possible to measure the absorption coefficient directly, and a correction is necessary for reflection losses at the interface between two mediums of different refractive indices. The Fresnel reflection P loss at an air-to-glass interface is

$$P = \left(\frac{n-1}{n+1}\right)^2. \tag{3.5}$$

Then, the total transmission through a plane-parallel plate is[2]

$$\tau_\infty = \frac{(1-P)^2\tau}{1-P^2\tau^2}. \tag{3.6}$$

For $n \approx 1.5$, $P = 0.04$; if the absorption loss is small ($\beta \approx 0$), $\tau_\infty \approx 0.92$.

Example 3.2

Transmission through a 15 mm thick ZnS window at a wavelength of 1.06 μm:
For ZnS,

$n = 2.289$ (at 1.06 μm),
$\beta = 0.02\ \text{cm}^{-1}$ (at 1.06 μm).

From Equations 3.4 through 3.6,

$$\tau = \exp\left[-(0.02\text{cm}^{-1})(1.5\text{cm})\right]$$
$$\tau = 0.970,$$

$$P = \left(\frac{2.289-1}{2.289+1}\right)^2$$
$$P = 0.154,$$

$$\tau_\infty = \frac{\left[1-(0.154)^2\right](0.970)}{1-(0.154)^2(0.970)^2}$$
$$\tau_\infty = 0.969.$$

Optical glass transmits from 400 nm to about 1.8 μm, as shown in Figure 3.3.[3] Specialized ultraviolet (UV) glasses transmit down to a wavelength of 200 nm, while IR glasses transmit up to 3.5 μm. Optical plastics transmit from 400 nm to 1.1 μm. The transmission of IR materials varies with the wavelength and material. For example, at a wavelength of 4 μm, the middle of the midwave infrared (MWIR) band, the absorption coefficient of calcium fluoride (CaF_2) is less than 0.001 cm^{-1}, while at the same wavelength, the absorption coefficient of fused silica is 1.2 cm^{-1}.

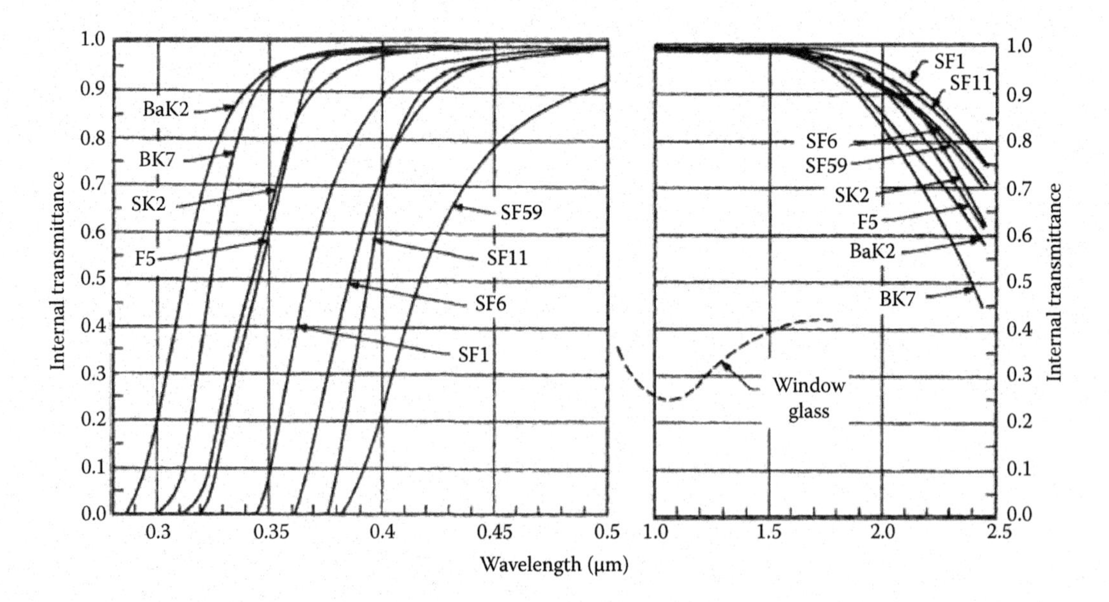

FIGURE 3.3 Internal transmittance of several representative optical glasses and of common window glass, all for a thickness of 25 mm. (From Smith, W.J., *Modern Optical Engineering*, 3rd ed., McGraw-Hill, New York, 2008.)

When light rays of different polarizations pass through a stressed refractive medium, the length of the path for each ray is different. This OPD induces optical aberration in the image. For two perpendicular polarized light rays, the OPD for a thickness h is given by[4]

$$\mathrm{OPD} = K_S \sigma h, \tag{3.7}$$

where K_S is the stress optic coefficient (units of TPa^{-1}) and σ is the stress.

The stress optic coefficient of most optical glasses is 1.5–3.5 TPa^{-1}. One of the lowest stress optic coefficients for glass is 0.02 TPa^{-1} for Schott SF57, while one of the highest is 4.21 TPa^{-1} for N-KZFS11. Stress optic coefficients for plastics are relatively high; for Lexan polycarbonate, K_S = 93 TPa^{-1}. Stress optic effects limit the amount of stress in transmissive mediums.

Reflecting materials are typically metal films, although bare copper mirrors are used in some high-energy laser applications. Reflectance R is the ratio of reflected to incident radiant intensity. In Figure 3.4, reflectance is plotted versus wavelength from 0.2 to 10 μm for freshly deposited aluminum, silver, gold, and copper films.[5]

The reflectance of metal films degrades with time. At 550 nm, the reflectance of silver is 0.983, which is superior to those of many other metal films, including that of aluminum, which is 0.915 at the same wavelength. Unfortunately, bare silver tarnishes, with reflectance falling to 0.77 or less. Aluminum films produced by vacuum deposition replaced silver since the reflectance of the aluminum did not degrade as rapidly. However, the surface of an aluminum film oxidizes with time, the reflectance dropping by 10% or more. A transparent overcoat can protect the metal film from chemical attacks. With a silicon monoxide overcoat, aluminum is protected from oxidation. The overcoat drops the reflectance by about 1% while protecting the underlying film from much greater oxidation losses. Today special overcoated silver films are competitive with aluminum, offering better protection from tarnishing than in the past.

3.3 STRUCTURAL FIGURES OF MERIT

Resistance to deflection, primarily from self-weight, is the most important structural property of optomechanical materials. Normally, optical systems will carry only loads associated with the weight of the optical components. Self-weight is the remaining loading condition in the absence of external loads. Lens barrels support the weight of the lenses of a system, but again deflection, or the position of the centers of the elements relative to the optical axis, is critical.

The self-weight deflection δ_b of a uniformly loaded beam with a constant cross section, with gravity acting normal to the long axis of the beam, is given by

$$\delta_b = C_b \left(\frac{\rho}{E}\right)\left(\frac{A}{I}\right) L^4, \tag{3.8}$$

where C_b is a dimensionless beam support parameter, ρ is the density of the beam material, E is the elastic modulus of the beam material, A is the cross section of the beam, I is the moment of inertia of the cross section of the beam, and L is the beam length.

An optical component can be approximated as a plate for estimating self-weight deflection. Maximum self-weight deflection δ_p occurs when gravity is normal to the surface of a right circular plate, and it is given by[6]

$$\delta_p = C_p \frac{qr^4}{D} \tag{3.9}$$

where C_p is the dimensionless plate support parameter, q is the load per unit area on the plate, r is the plate radius, and D is the flexural rigidity.

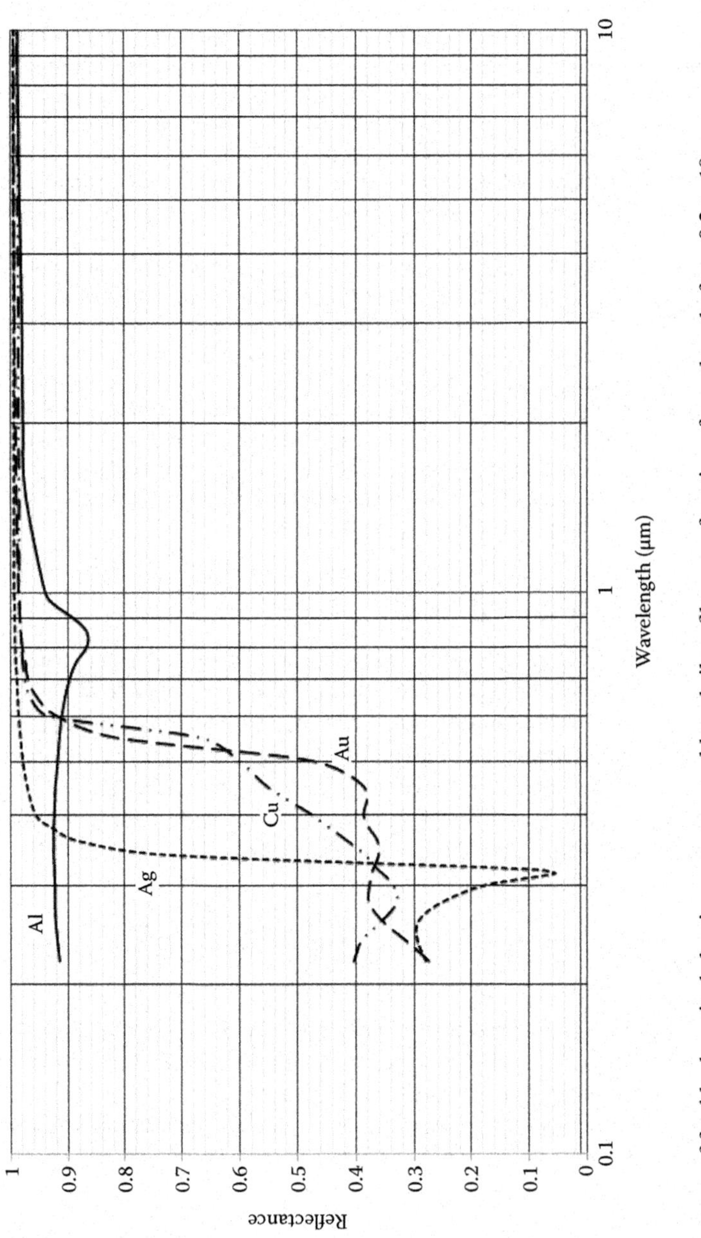

FIGURE 3.4 Reflectance of freshly deposited aluminum, copper, gold, and silver films as a function of wavelength, from 0.2 to 10 µm.

For a circular plate of constant thickness h,

$$q = \rho h,$$
$$D = \frac{Eh^3}{12(1-\nu^2)}. \qquad (3.10)$$

Substituting this into Equation 3.9,

$$\delta_p = C_p \left[\frac{\rho}{E}(1-\nu^2) \right] \left(\frac{r}{h} \right)^2 r^2. \qquad (3.11)$$

Equation 3.11 shows that the material parameter determining the self-weight deflection of an optical component is $(p/E)(1 - \nu^2)$; this parameter is the inverse specific stiffness. For most common materials used in optomechanics, Poisson's ratio is 0.17–0.33. Over this range of values, the effect of changing Poisson's ratio is less than about 10%. Hence, Poisson's ratio is often neglected in discussions of specific stiffness, and the simplified ratio ρ/E used as a basis for material selection. Table 3.1 gives the specific stiffness values of some materials used in optomechanics.

TABLE 3.1
Values of Inverse Specific Stiffness ρ/E for Selected Materials

Material	ρ/E (× 10^{-9} m^{-1})
Beryllium, wrought S-200FH	59.8
Silicon carbide CVD beta	67.7
Silicon	174.4
Graphite–epoxy composite GY-70×30	187.8
Schott Zerodur™	273.9
Corning fused silica 7940	296.3
Schott N-BK7	300.3
Diamond	301.3
Molybdenum	309.1
Corning Pyrex 7740	347.2
Aluminum, cast A356	374.0
Stainless steel, 440C	377.7
Titanium, 6Al–4V	381.2
Aluminum, wrought 7075	381.7
Steel, 4340	385.5
Magnesium AZ31B	387.6
Aluminum, wrought 6061	389.8
Germanium	503.6
Invar 36	537.2
Zinc sulfide	537.2
Schott F2	619.6
Copper C10100	749.6
Brass, 70/30	825.0
Sapphire	973.6
Glass reinforced polycarbonate, Makrolon 8035	2,943,000.0

Note: CVD, chemically vapor deposited.

Force units, rather than mass units, are used in evaluating structural FOMs. Mass under acceleration results in force, and force induces deflection. In Système International (SI) kilogram-force units, rather than kilogram-mass units are appropriate.

The inverse specific stiffness values of most common optomechanical materials range from 257×10^{-9} m^{-1} for Schott Zerodur to 377×10^{-9} m^{-1} for titanium. Excluding the anomalously low inverse specific stiffness of near-zero thermal expansion mirror substrate materials, for most glasses and metals, $\rho/E \approx 350 \times 10^{-9}$ m^{-1} to within 10%. When using ordinary materials, self-weight deflection does not change significantly when substituting materials. Figure 3.5 is a plot of the inverse specific stiffness values of materials.

Example 3.3

Compare the thickness of mirrors made out of silicon and Corning 7740 Pyrex and monocrystalline silicon assuming that the diameter, mounting geometry, and deflection tolerance are held constant.

For silicon,

$$\rho_{Si} = 2330 \text{ kg/m}^3,$$
$$E_{Si} = 13.1 \text{ GPa},$$
$$\nu_{Si} = 0.42.$$

For Corning 7740 Pyrex,

$$\rho_{Pyrex} = 2230 \text{ kg/m}^3,$$
$$E_{Pyrex} = 6.30 \text{ GPa}.$$
$$\nu_{Pyrex} = 0.2.$$

From Equation 3.11,

$$\frac{h_{Si}}{h_{Pyrex}} = \left[\frac{\dfrac{\rho_{Si}\left(1-\nu_{Si}^2\right)}{E_{Si}}}{\dfrac{\rho_{Pyrex}\left(1-\nu_{Pyrex}^2\right)}{E_{Pyrex}}} \right]^{\frac{1}{2}}$$

$$\frac{h_{Si}}{h_{Pyrex}} = 0.657.$$

The silicon mirror is about two-thirds as thick as the mirror made of Corning 7740 Pyrex.

Exceptions to the rule that most materials have about the same inverse specific stiffness are beryllium, silicon carbide, and composites (both polymer and metal matrix). The density of beryllium is comparable to that of magnesium, while its elastic modulus is 1.5 times higher than that of steel. The inverse specific stiffness of beryllium is less than 20% of that of most materials, making this exotic metal of considerable interest in lightweight structures. Unfortunately, the cost of beryllium is relatively high.

The inverse specific stiffness of silicon carbide is about 25% that of common structural materials. Like beryllium, silicon carbide is attractive for lightweight systems and for some applications is a substitute for the more expensive and complex to fabricate beryllium. Polymer matrix composites typically have a density comparable to that of magnesium with stiffness similar to that of titanium, again offering a superior stiffness-to-weight ratio for lightweight structures. Finally, metal matrix composites have a density near that of aluminum, with an elastic modulus that is comparable to that of titanium.

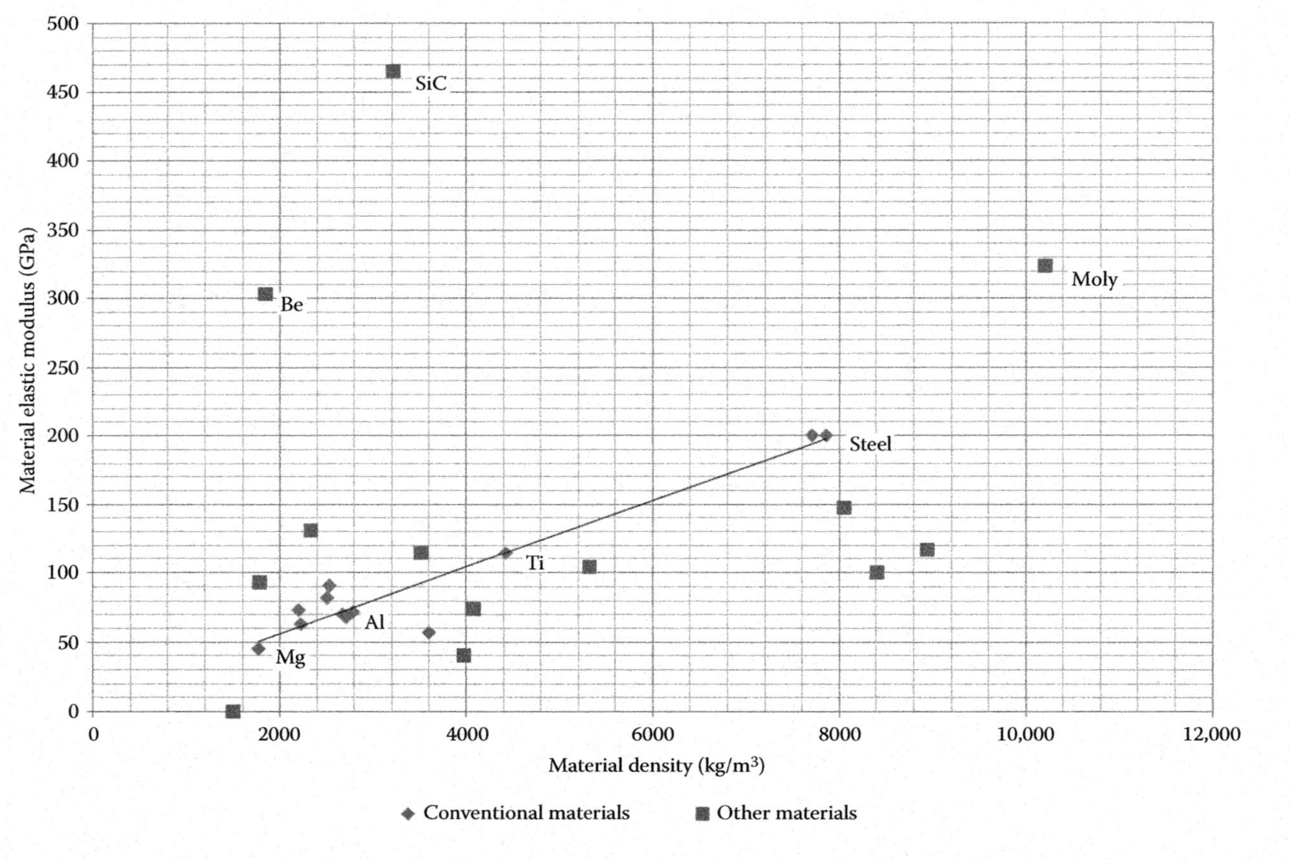

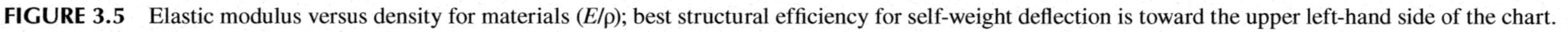

FIGURE 3.5 Elastic modulus versus density for materials (E/ρ); best structural efficiency for self-weight deflection is toward the upper left-hand side of the chart.

TABLE 3.2
Representative MYSs of Common Optomechanical Materials

Material	MYS (MPa)
Copper, OFC	12
Beryllium, S-200FH	21
Magnesium, AZ-31	33
Invar 36	41
Titanium, 6AL–4V	50
Stainless steel, 304	65
Aluminum, cast, A356	70
Steel, 1040	138
Aluminum, wrought, 6061	140
Molybdenum	280

Stress does not appear in the analysis of deflection, and inverse specific stiffness is independent of the strength of the material. Hence, for most optomechanical applications, there is no need for high-strength materials. For example, cast alloys with lower strength than that of more expensive wrought materials are acceptable choices for lens barrels and optical housings since the inverse specific stiffness values of the two types of materials are identical.

Dimensional instability is the change in the size and shape of a material with time. One source of dimensional instability is the release of stress over a long period. Another is microyielding, which is a permanent deformation from a stress well below the nominal 2×10^{-3} strain yield stress of the material.

The release of stress induces changes in glasses and metals. The change in dimension is proportional to the ratio of the residual stress σ_R to elastic modulus, or σ_R/E. Although this ratio favors high-stiffness materials, good engineering practice is to reduce the amount of residual stress to insure long-term stability.

The FOM for the dimensional stability of metals is the microyield strength (MYS), which is defined as the amount of stress needed to produce a permanent plastic strain of 10^{-6}. When the strain ε is between $10^{-7} \le \varepsilon \le 10^{-4}$, the stress–strain curve is not linear and follows a power law relationship:

$$\sigma = K'\varepsilon^{n'}, \tag{3.12}$$

where σ is the applied stress and K' and n' are material parameters.

Microyield is influenced by the prior history of the metal; heat treatment and mechanical strain can change the MYS. Microyield is not associated with the yield strength of the material, and testing is necessary to determine this material property. Table 3.2 gives the MYSs of some metals common in optomechanics.

A rule of thumb is that dimensional stability will not be a problem if the applied stress is kept at about one-half of microyield. For common metals, n' is from about 0.03 for a 5000 series aluminum alloy to about 0.34 for beryllium. At microyield, the deformation strain is 10^{-6}, so at half of microyield, the worst-case strain is about 10^{-7}. For a 250 mm diameter mirror, this would be a deformation of 25 nm, or about 0.04 wave (1 wave = 633 nm), which is well below the tolerance of most systems.

3.4 THERMAL FIGURES OF MERIT

Resistance to thermal deformation is the most important thermal property of optomechanical materials. Thermal FOMs include the thermal coefficient of expansion, change in refractive index with temperature, thermal distortion index, thermal diffusivity, and thermal shock resistance. Many thermal properties are nonlinear with temperature; over restricted temperature ranges (military standard or from –50°C to +70°C), linear approximations are acceptably accurate.

The thermal coefficient of expansion, symbolized by α, is also called the coefficient of thermal expansion. Although nonlinear over a wide range of temperatures, α is assumed to be constant where the temperature range is between –50°C to +50°C. The thermal coefficient of expansion α is related to the size change ΔL of a body of length L by

$$\Delta L = \alpha L \Delta T, \tag{3.13}$$

where ΔT is the change in temperature.

For optical glasses, the thermal coefficient of expansion is from 4×10^{-6} to 16×10^{-6} K^{-1}. There is a larger range of thermal coefficients of expansion among IR materials, with silicon at $\alpha = 2.7 \times 10^{-6}$ K^{-1}, while KRS5 has $\alpha = 58 \times 10^{-6}$ K^{-1}. Typically, the thermal coefficient of expansion of metals is much higher than that of optical glass, which can induce thermal stress in glass-to-metal interfaces. For example, the thermal coefficient of expansion of 6061 aluminum is about 23×10^{-6} K^{-1}.

"Zero-coefficient-of-expansion" materials minimize changes with temperature. In reality, no material has a zero thermal coefficient of expansion. Some materials have a thermal expansion curve that crosses zero, but this is of no utility since the slope of the curve determines change with temperature. Invar is the oldest low-thermal-coefficient-of-expansion material, with $\alpha = 1.26 \times 10^{-6}$ K^{-1}, while the α of Super Invar is even lower, with $\alpha = 0.31 \times 10^{-6}$ K^{-1}. For glasses and glass ceramics in mirror substrates, the range of thermal coefficients of expansion is from $\alpha = 3.3 \times 10^{-6}$ K^{-1} for Corning Pyrex 7740 to $\alpha = 0.05 \times 10^{-6}$ K^{-1} for Schott Zerodur.

The refractive index n of glass changes with temperature. The variation of the refractive index with temperature and wavelength is nonlinear. Over a restricted range of temperatures, from –50°C to +50°C, and for visible wavelengths, the departure from linearity is less than 10% for most optical glasses and can be neglected for preliminary analysis. The change in the refractive index is absolute when measured in a vacuum since there is no change in the index of the surrounding medium with temperature. More common is a relative measurement made in air; the index of refraction of the air changes with temperature. The values of the change in index, dn/dT, for optical glass range from -6.7×10^{-6} K^{-1} for N-PK51 to $+12.5 \times 10^{-6}$ K^{-1} for SF57 (relative to air). There is a large variation in dn/dT for IR materials, from -254×10^{-6} K^{-1} for KRS5 in the visible region to $+424 \times 10^{-6}$ K^{-1} for germanium.

When subjected to a temperature gradient, a material deforms. An optical component can be approximated as a plane-parallel plate. When a plane-parallel plate is subjected to a linear transverse temperature gradient (normal to the surface), the surface becomes spherical with a radius R given by

$$R = \frac{h}{\alpha \Delta T} = \frac{\kappa}{\alpha}\frac{1}{q}, \tag{3.14}$$

where h is the plate thickness, ΔT is the temperature difference from top to bottom in the plate, κ is the thermal conductivity of the plate material, and q is the heat flux per unit area in the plate.

The ratio α/κ is the thermal distortion index of the material. To keep the distortion from a temperature gradient small, the thermal distortion index should also be small. The thermal conductivity of glasses is relatively low, 0.6–1.5 W/m K; hence, a low coefficient of thermal expansion is necessary to minimize the thermal distortion index. Although the thermal coefficient of expansion of metals is larger than that of glasses, this is offset by the high thermal conductivity of most metals. For example, the thermal distortion index of 6061 aluminum is about 5% of that of Corning Pyrex 7740.

Thermal gradients often arise after a sudden change in the surface temperature of an optical component. The rate of change in the temperature of the center of the component after a surface temperature change is determined by the heat capacity and thermal conductivity of the material. After a temperature change T, at the surface, the difference in temperature ΔT as a function of time t after the change is given by

$$\Delta T = T \exp\left(\frac{-t}{\tau}\right),$$

$$\tau = \frac{h^2}{\pi^2} \frac{1}{\left(\frac{\kappa}{\rho c_p}\right)}, \tag{3.15}$$

where ρ is the density of the material and c_p is the specific heat of the material.

The parameter τ with units of time is the thermal time constant and is the time required for the interior to change its temperature by a factor of e (≈ 2.72). As a rough approximation, thermal equilibrium is reached after five thermal time constants, since ΔT is then about 1% of T. The thermal diffusivity a is the ratio $\kappa/(\rho c_p)$; for a minimum thermal time constant, the thermal diffusivity should be large. The thermal diffusivity of Corning Pyrex 7740 is about 0.65×10^{-6} m²/s, while that of 6061 aluminum is about 69×10^{-6} m²/s.

Example 3.4

To minimize errors from thermal gradients, time is allowed before testing an optic for the mirror to reach thermal equilibrium. Assuming that thermal equilibrium is at a temperature difference of 1% between the surface and interior, how much time must be allowed before testing a 50 mm thick Corning Pyrex 7740 mirror?

For Corning Pyrex 7740,

$\kappa = 1.13$ W/m K,
$\rho = 2230$ kg/m³,
$c_p = 1050$ J/kg K.

From Equation 3.15,

$$\tau = \frac{(50 \times 10^{-3}\,\text{m})^2}{\pi^2} \frac{1}{\frac{1.13\,\text{W/mK}}{(2230\,\text{kg/m}^2)(1050\,\text{J/kgK})}}$$

$\tau = 525$ s,

$0.01 = \exp(x\tau)$,
$x = 4.605\ \tau$,
$t = 2417$ s.

About 40 minutes must be allowed for the mirror to reach thermal equilibrium before testing. In this example, a more exact value of the thermal time constant, 4.605, is used rather than the rule of thumb of "five thermal time constants to reach equilibrium."

A temperature gradient induces stress in the material; under certain conditions, this stress may be large enough to cause structural failure. The severity of the thermal shock is determined by the dimensionless Biot number Bi, which is the ratio of heat transfer at the surface to internal heat transfer. If Bi $\leq$ 0.1, the thermal shock is mild, while if Bi $>$ 1.0, the thermal shock is severe. Assuming a worst-case or severe thermal shock (Bi $\gg$ 1.0) and using the plane-parallel plate approximation for an optical component, the thermal stress σ induced by a temperature gradient ΔT is

$$\sigma = \frac{\alpha E \Delta T}{1-\nu}. \tag{3.16}$$

If σ_{YS} is the yield stress of the material, the maximum instantaneous temperature change ΔT_{max} that the material can survive is given by

$$\Delta T_{max} = \frac{\sigma_{YS}(1-\nu)}{\alpha E}. \tag{3.17}$$

The right side of Equation 3.17 defines the FOMs for thermal shock, or $\sigma_{YS}(1 - \nu)/(\alpha E)$. This is again for a worst-case thermal shock (Bi $\gg$ 1.0); mild thermal shocks require a modification for the actual value of the Biot number. The yield stress of the material is required for evaluation of this FOM. The next section discusses the strength of glass and similar brittle optical materials.

3.5 STRENGTH OF BRITTLE OPTICAL MATERIALS

Glass fails at a relatively low stress in comparison with structural metals. Glasses and other brittle optical materials fail by fracture from preexisting surface flaws initiated by tensile stress. Unlike metals, there is no plastic deformation in glasses prior to failure.

If sufficient tensile stress is applied to a surface flaw, a crack is initiated, and the crack will move with some velocity v_c. If the velocity of crack propagation is above 1 m/s, failure is by fast fracture. At crack velocities below 1 m/s, failure is by static fatigue, and v_c is influenced by contaminants, notably water vapor. The critical stress σ_{CR} necessary to initiate a crack is given by the Griffith theory and is

$$\sigma_{CR} = \frac{K_{IC}}{\pi\sqrt{h_c}}, \tag{3.18}$$

where K_{IC} is the material critical stress intensity or fracture toughness and h_c is the surface flaw depth.

The fracture toughness of a material is an FOM of strength, with units (SI) of Pa $m^{1/2}$. Typical values of K_{IC} of glass are from 0.5 to 1.0 MPa $m^{1/2}$. For the IR materials, K_{IC} is larger, up to 2.0 MPa $m^{1/2}$ for sapphire.

The surface flaw depth h_c in Equation 3.18 includes not only the visible depth, but also the depth of subsurface damage, or cracks extending into the glass from the flaw. Currently there is no means of determining the depth of the subsurface damage with nondestructive testing. There is a large range of values for empirical relationships between visible flaw sizes and subsurface damage, leading to strength estimates that may vary by an order of magnitude or more. Hence, the Griffith relationship is not commonly used to find glass strength. Strength is dependent on flaw depth according to the Griffith equation; increasing flaw depth lowers strength. Surface finish is highly important in determining glass strength.

An empirical probability relationship based on sample testing predicts fast fracture failure. Although Gaussian and other distributions can be used, the two-parameter Weibull probability distribution is the most common. As rule of thumb, about 30 samples are necessary for good statistics. The stress associated with the failure of each sample is the basis for the statistical distribution.

Samples are preferably tested by the ring-on-ring method, described by ASTM 1499-09, *Standard Test Method for Monotonic Equibiaxial Flexural Strength of Advanced Ceramics at*

Ambient Temperature.[7] Four-point bending testing does not stress much of the surface area of the sample, and the stress distribution is not uniform. A disk of glass is clamped between two different diameters in ring-on-ring testing, providing a larger, more uniform stress area in comparison. Ideally, the surface finish of the test samples is the same as that of the optical components. In reporting Weibull statistics, the surface finish and test area of the sample should be given; the time rate of the change of the test loading is also useful.

Weibull statistics are fitted to the test data by using ASTM C1239-07, *Standard Practice for Reporting Uniaxial Strength Data and Estimating Weibull Distribution Parameters for Advanced Ceramics*.[8] The two Weibull parameters are the scale factor σ_0 and the Weibull modulus m. The scale factor is in units of strength and is an indication of the mean strength of the material. The dimensionless Weibull modulus is the slope of the failure curve and is a measure of variation in the statistics. The failure probability P_f associated with an applied stress σ for the two-parameter Weibull distribution is given by

$$P_f = 1 - \exp\left(-\frac{\sigma}{\sigma_0}\right)^m. \tag{3.19}$$

The probability of failure is associated with the surface area; larger areas are more likely to have bigger flaws, with reduced strength. Since test sample areas are relatively small, it is necessary to scale results from the sample area A_0 to the actual area subjected to stress A. A modified form of the Weibull distribution incorporating scaling for stress area is

$$P_f = 1 - \exp\left[-\frac{A}{A_0}\left(\frac{\sigma}{\sigma_0}\right)^m\right]. \tag{3.20}$$

From the Griffith equation, at constant failure probability, strength is scaled for flaw depth according to

$$\frac{\sigma_1}{\sigma_2} = \sqrt{\frac{h_2}{h_1}}, \tag{3.21}$$

where σ_1 and σ_2 are the respective strengths and h_1 and h_2 are the respective flaw depths.

The effect of surface finish on Zerodur strength is shown in Figure 3.6.

Since glass strength is dependent on a variety of factors, including surface finish, surface area, environment, and stress rate, the use of a single value for yield is inadvisable. The two-parameter Weibull distribution is extremely conservative at small values of failure probability $P_f \leq 10^{-3}$ which are required for optical systems. An additional factor of safety is obtained from this conservatism. Different distributions such as a three-parameter Weibull or Weibull distribution normalized to a unit area are suggested to overcome some of the limitations associated with the classic two-parameter Weibull distribution. Currently the lack of standards for these alternatives limits their utility. Table 3.3 gives representative Weibull statistics for materials.

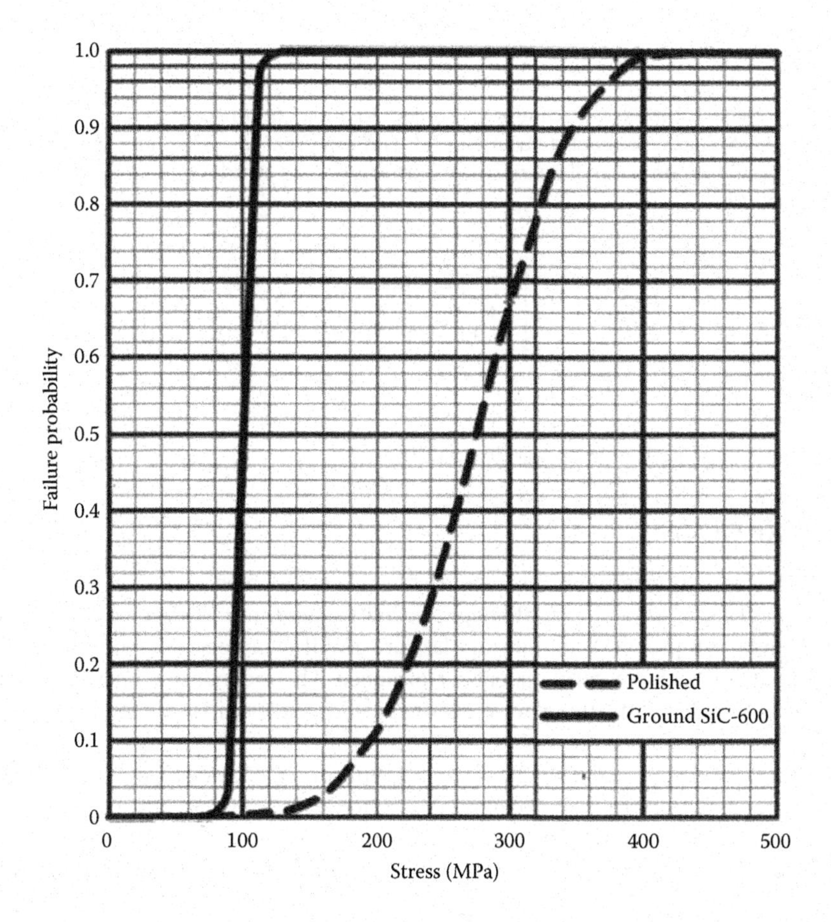

FIGURE 3.6 Weibull probability curves for Schott Zerodur with ground (dashed line) and polished (solid line) surfaces. The higher strength of the polished sample is apparent.

TABLE 3.3
Representative Weibull Statistics for Optical Materials

Material	Weibull Modulus m	Scale Factor σ_0 (MPa)	Test Area A (mm^2)	Surface Flaw Depth h (μm)
Schott Zerodur	16.0	108	113	38 (SiC 600)
Schott Zerodur	5.3	294	113	Polished
Corning fused silica	9.89	116	100	Polished
Schott BK-7	30.4	70.6	113	38 (SiC 600)
Schott F-2	25.0	57.1	113	38 (SiC 600)
Multispectral ZnS	9.99	81.5	100	Polished
CVD ZnS	15.7	56.7	100	Polished
Monocrystalline silicon	4.54	347	–	Polished
Germanium	34	120	–	Polished
Calcium fluoride	6.47	72.1	100	Polished
Sapphire (r-plane)	4.10	600	100	Polished
Oxyfluoride glass	3.71	132	100	Polished
Spinel	19.5	255	100	–

Example 3.5

A 300 mm circular Schott N-BK7 window is subject to a maximum stress of 14 MPa. If the scale factor σ_0 = 50.3 MPa, Weibull modulus m = 13.3, and test area A_0 = 113 mm², what is the probability of failure P_f? Assume that the stress is uniform across the window.

From Equation 3.20,

$$P_f = 1 - \exp\left[-\frac{\frac{\pi}{4}(300\ \text{mm})^2}{113\ \text{mm}^2} \left(\frac{14\ \text{MPa}}{50.3\ \text{MPa}} \right)^{13.3} \right]$$

$$P_f = 2.56 \times 10^{-5}.$$

The probability of failure is about 1 part in 40,000.

3.6 OPTICAL GLASS

Optical glass is used in transmissive optics for the visible range. Nominally, the useful wavelength range for optical glass is 380–780 nm, although specialty glasses are used in the near-UV and near-IR. The most important properties of optical glass are a well-characterized index of refraction and dispersion, which are used to control optical performance. The secondary properties of interest are density, variation in index with temperature, hardness, and susceptibility to staining from chemical exposure.

Optical glasses are identified using glass codes. Typically, the glass code is a six-digit number, with the first three digits representing the index of fraction n_d (λ = 587 nm) and the second three digits the Abbe number ν_d. The glass code gives the first three digits after the period dropping the first number, which is always 1 for optical glass. For the conventional six-digit code, the code is given by

$$\text{code} = 1000(n_d - 1) + 10\nu_d \tag{3.22}$$

Occasionally a nine-digit class code is used, with a period after the first six digits, and then three digits for the density of the glass in grams per cubic meter. Density increases with the index of refraction. Caution is necessary in working with equivalent glass codes from different suppliers, since other properties may vary despite identical codes.

Glasses are either crown or flint types. For crown glasses, $n_d < 1.60$, $\nu_d > 55$ and $n_d > 1.60$, $\nu_d > 50$. Other glasses are flints. These names identify glasses of low and high dispersion types.

In the past, a large number of optical glasses were available; for example, the Schott catalog in 1967 listed 273 different glasses. High cost, limited utility, and environmental concerns led to a decrease in the number of glass types. In 2016, there were over 100 glass types available. An even smaller number of glasses are preferred for most optical designs. Appendix 1 gives the properties of preferred optical glasses based on works by Walker[9] and Zhang and Shannon.[10] Glass types go out of production with time, and it is necessary to contact the supplier during design to insure availability.

Glass hardness is normally given as Knoop hardness, in accordance with ISO 9385.[11] The Knoop hardness of optical glass varies from 335 to 810. Hardness usually correlates with the elastic modulus of the glass. Hardness determines the ability of the glass to resist scratching; higher-hardness glasses are usually more durable.

When exposed to various chemicals, the surface of optical glass may stain and become hazy. Two types of chemicals in service environments are moisture in the air and cleaning solutions. Stain resistance is a factor in the cost of fabrication since glasses that are easily stained are more expensive to grind and polish.

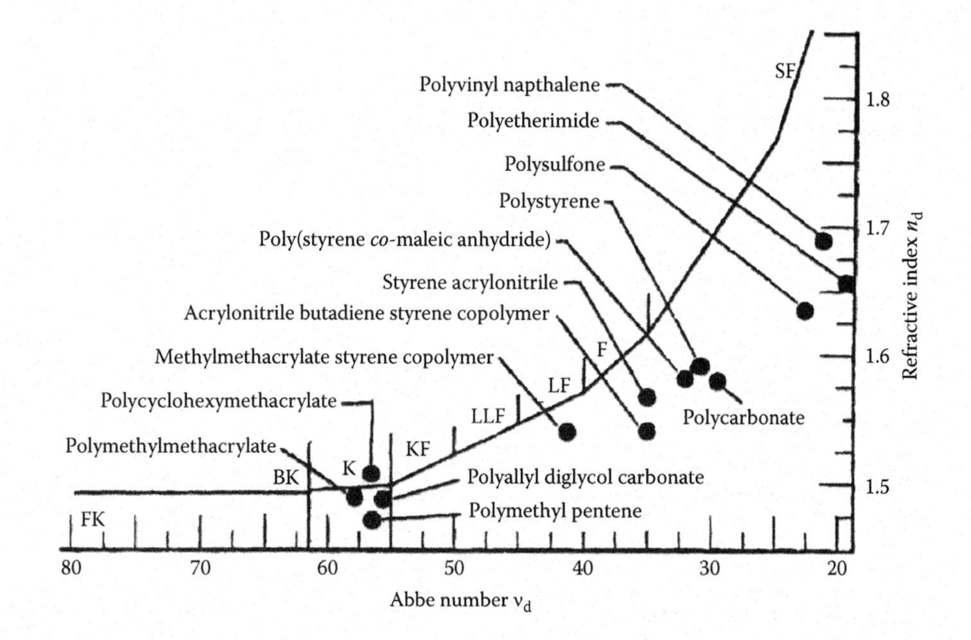

FIGURE 3.7 Optical polymer map showing plastic optic materials referenced to adjacent glass types. (Adapted from Lytle, J.D., Polymeric optics, Chap. 34, in *OSA Handbook of Optics*, 2nd ed., Bass, M. et al. Eds., Vol. II, pp. 34.1–34.20, McGraw-Hill, New York, 1995.)

There are no standards for the stain resistance of optical glasses; each glass supplier uses their own test methods. Common tests are exposure to acid and alkali solutions; with haze measured afterward. Weathering is tested by exposure to neutral aqueous solutions.

3.7 OPTICAL PLASTICS

Low density compared to that of optical glass and reduced costs are the advantages of optical plastics compared to traditional optical glasses. The density of optical plastics is 0.83–1.46 g/cm^3 compared to 2.3–6.0 g/cm^3 for optical glass. The cost of plastic optics in volume production can be 6–24% of the cost of comparable glass optics. An additional advantage is the ability to mold plastic in complex geometries. In some applications, it is possible to mold the mount with the optic, with further reduction in system cost and complexity. Also, aspheric and off-axis geometries can be produced, increasing optical design opportunities. Figure 3.7 gives the glass map for plastics. Table 3.4 gives the properties of common optical plastics.

Offsetting these advantages are drawbacks including high index changes with temperature, poor thermal properties, low hardness, and index changes with exposure to moisture. Additionally, there are not as many different types of optical plastics compared to optical glass. The index change with temperature dn/dT for optical plastics is −143 to −100 × 10^{-6} K^{-1}, which is an order of magnitude greater than that of optical glass. The expansion coefficients of optical plastics are 25–130 × 10^{-6} K^{-1}, which are higher than that of optical glass. The thermal conductivity of optical plastics is lower, between 0.1 and 0.3 W/m K, than that of optical glass. Some optical plastics absorb as much as 0.3% of water after 24 hours of immersion, inducing a change in the index of refraction.

3.8 INFRARED MATERIALS

Selection of IR materials is based primarily on transmission since this property varies substantially with the material and wavelength. The importance of other properties depends on the application.

TABLE 3.4
Properties of Optical Plastics

		Crownlike		Flintlike			
Material		**Polymethyl Methacrylate Acrylic**	**Cyclic Olefin Copolymers**	**Polystyrene**	**Polycarbonate**	**Methyl Methacrylate Styrene**	**Styrene Acrylonitrile**
Supplier		AtoHaas	Topas	Monsanto	GE	Richardson	Monsanto
Name/Code		V920	6015	Lustrex	Lexan 121	3070	
Property	**Units**						
Indices of refraction	F (486.1 nm)	1.4978	1.5398	1.6041	1.5994	1.5755	1.579
	d (587.6 nm)	1.4918	1.5332	1.5905	1.5855	1.567	1.5674
	C (656.3 nm)	1.4892	1.5304	1.5849	1.5799	1.5592	1.5627
Abbe value	v_d	57.19	56.72	30.76	30.03	34.79	34.81
dN/dT	$\times 10^{-6}$/K	−105	−101	−140	−107	−115	−110
Thermal coefficient of expansion	$\times 10^{-6}$ m/m K	60	60	50	68	58	50
Thermal conductivity	$\times 10^{-3}$ W/m K	170–250	120–150	100–140	200	190	120
Density	kg/m^3	1180	1020	1050	1250	1090	1070
H_2O absorption	% after 24 hour immersion	0.3	0.01	0.1	0.2	0.15	0.28
Transmission (visible)	3 mm thick	0.92	0.92	0.9	0.88	0.9	0.9
Upper limit service temperature	C	85	150	75	120	80	80

In transmission, IR materials are used for optical systems and windows. For optical systems, the same properties as those of optical glass are important, including index of refraction, dispersion, and change of index with wavelength. For windows, strength and hardness are important, as are thermal properties.

The IR spectrum extends from about 750 to 1 mm in wavelength and is divided into a series of wavelength regions or bands. ISO 20473:2007[12] defines three IR bands: near infrared (NIR), from 0.78 to 3 μm; mid infrared (MWIR), from 3 to 50 μm; and far infrared (FIR), from 30 μm to 1 mm. Unfortunately, the use of the ISO definitions is uncommon in the IR community, and there is a wide variety of definitions for the different bands. The common practice in science and aerospace engineering for the different wavelength regions is given in Table 3.5. Also shown in Table 3.5 are different wave band definitions based on a combination of atmospheric transmission and sensor response. The common practice is used throughout this text, not the ISO- or sensor-and-transmission-based definitions.

Refractive IR materials include alkali and alkaline halides, glasses and oxides, semiconductors, and chalcogenides. Most IR materials are suitable for temperatures up to about 600 K. Refractory window materials for high-temperature applications up to 2300 K include aluminum oxynitride, spinel (magnesium aluminate), sapphire (aluminum oxide), magnesium oxide, and diamond.

Alkali and alkaline halides are crystalline materials with reasonably good transmission and are relatively inexpensive. Poor mechanical properties, including low strength and hardness, are disadvantages of these materials. These materials are often used for relatively low-temperature applications ($T < 600$ K) where high strength and hardness are not required. For example, calcium fluoride is used in terrestrial scientific instruments, where its susceptibility to failure from thermal shock is not a concern. Appendix 2 gives the properties of selected alkali and alkaline earth halides.

The upper limit for the transmission of amorphous optical glasses containing silicon is about 3.7 μm. Replacing silicon with other materials extends the upper limit of transmission. For example, the upper limit of transmission of calcium aluminate glass is about 5.5 μm. Glasses are relatively inexpensive and are the easiest of the IR materials to fabricate. Glasses are not suitable for most window applications since the thermal shock resistance and hardness of glasses are low when compared to those of other IR materials. Fused silica is an exception, and some high-energy-laser windows are made of this material. Appendix 3 gives the properties of some selected IR glasses.

Semiconductors have high refractive indices and reasonably good thermal and mechanical properties. Transmission is generally lower than that of the IR crystals and glasses. Transmission also falls off at high temperatures, typically above 500 K. At temperatures above 500 K, germanium may experience structural failure through a phenomenon known as thermal runaway. Germanium and silicon are the most common of these materials in IR systems. Diamond is also classed as a semiconductor, with outstanding hardness and thermal properties. Diamond transmits over an enormous spectral range, from 0.22 to 80 μm. These exceptional properties are offset by an equally exceptional high cost, although techniques for making synthetic diamond are reducing prices. Appendix 4 gives the properties of some selected semiconductors.

TABLE 3.5
IR Regions

IR Region	Wavelength Range (μm)	
	Common	Sensor/Transmission
NIR	0.75–1.4	0.7–1.0
Short-wave infrared	1.4–3	1.0–3
MWIR	3–8	3–5
Long-wave infrared	8–15	7–14
FIR	15–1000	12–30

Chalcogenides are materials containing a chalcogen such as oxygen, sulfur, selenium, or tellurium. These materials have high refractive indices, moderately low absorption coefficients, and fair mechanical and thermal properties. The transmission of zinc sulfide (ZnS) and zinc selenide (ZnSe) is unusually wide, extending from the upper end of the visible at 600 nm to around 13 μm or higher for some grades of ZnSe. Although the transmission cutoff for ZnSe is around 20 μm, this property is offset by lower hardness and greater susceptibility to surface damage or erosion than ZnS. The properties of selected chalcogenides are given in Appendix 5.

3.9 MIRROR SUBSTRATE MATERIALS

Mirror substrates are selected based on the stiffness-to-weight ratio, thermal properties, and smoothness of the finished optical surface. Dimensional stability is also important, particularly for metal mirrors. Cost is an important parameter for mirror substrates. Many mirrors are larger than typical refractive components, requiring reductions in cost per volume and cost per area during fabrication compared to traditional optical glasses. Appendix 6 gives the properties of glassy mirror substrate materials, and Appendix 7 gives the properties of metallic and nonmetallic mirror substrate materials.

Since mirrors reflect light, the character of the reflecting surface of the substrate material is an index of performance. Errors in the surface of a mirror surface are optical figure and surface finish, or surface roughness. Surface errors are described by spatial frequency, normally in units of cycles per millimeter or cycles over the mirror aperture. Optical figure is the deviation from the ideal or desired shape. Errors in figure cause optical aberrations. Figure errors are low, mid, and high frequency. At the low end, there are about eight cycles over the clear aperture, described in wavelength units (1 wave = 633 nm is the convention). High spatial frequency errors are also called surface roughness, with a frequency of up to 50 cycles/mm. Surface roughness is averaged over the entire mirror aperture and is given in units of angstroms root-mean-square (RMS). Surface roughness scatters light, reducing the intensity of the central maximum of the diffraction disk and lowering image contrast. For digital detectors, surface roughness affects the energy on detector.

Common types of mirror materials are glasses, metals, and ceramics. Glass ceramics are considered to be glasses for material evaluation. Exotic materials sometimes used for mirror substrates include composites, both polymer and metal matrix, and silicon. Appendices 6 and 7 give the properties of glass and metallic mirror substrate materials, respectively. Table 3.6 compares structural and thermal FOMs for different mirror substrate materials.

There is about a 30% variation in the inverse specific stiffness between different glass types from a high of 334×10^{-9} m^{-1} for Corning Pyrex 7740 to a low of 257×10^{-9} m^{-1} for Schott Zerodur. The thermal conductivity of glass substrates varies between 1.13 W/m K for Corning Pyrex 7740 and 1.64 W/m K for Schott Zerodur. Thermal diffusivity is 482×10^{-9} m^2/s for Corning Pyrex 7740 to 776×10^{-9} m^2/s for Corning ULE 7972. Since structural and most thermal properties do not change much with glass type, selection is based on thermal coefficient of expansion and cost.

Crown glasses are the lowest in cost, with the highest thermal expansion coefficient, and are used in environments where thermal performance is not important. The thermal expansion coefficient of borosilicate crown glasses, including Corning Pyrex 7740, Schott Duran 50, and Ohara E-6, is between 2.8 and 3.3×10^{-6} K^{-1}, which is about 40% of that of the crown glasses. The thermal coefficient of expansion curves of the zero-coefficient-of-expansion materials, such as Schott Zerodur and Sitall 115, are zero crossing near room temperature. This zero crossing occurs only at one temperature and is not useful. Over a more reasonable temperature range of ±60°C, the effective thermal coefficient of expansion of these materials can be as low as 30×10^{-9} K^{-1} for Corning ULE 7972.

Metals are an alternative to glassy substrates and are used when high thermal conductivity, high specific stiffness, and low manufacturing cost are desirable. Drawbacks of metals are increased surface roughness compared to glass substrates and dimensional instability. Common metals used for mirrors are aluminum, copper, and beryllium.

TABLE 3.6
FOMs for Some Selected Mirror Substrate Materials

Material	FOM		
	Inverse Specific Stiffness ρ/E ($\times 10^{-9}$ m^{-1})	Thermal Distortion Index α/κ ($\times 10^{-9}$ m/W)	Thermal Diffusivity $\kappa/(\rho - c_p)$ ($\times 10^{-6}$ m^2/s)
Aluminum, 6061	384.7	141.0	69.0
Beryllium, S-200FH	59.8	52.8	0.605
Copper, OFHC	760.5	43.4	114.0
Corning fused silica 7980	297.3	400.0	0.788
Corning Pyrex 7740	346.6	2920.0	0.483
Corning ULE 7972	319.5	22.9	0.776
Graphite epoxy composite GY70/x30	187.9	0.571	21.3
Invar 36	560.6	1620.0	2.53
Magnesium, AZ-31	386.2	271.0	54.2
Monocrystalline silicon	174.6	16.0	92.9
Schott Zerodur	274.8	68.5	0.721
Silicon carbide, cast, reaction bonded	79.1	12.3	411.0
Stainless steel, 304	398.8	1070.0	4.11
Titanium, 6Al–4V	381.7	1280.0	2.88

Note: OFHC, oxygen-free high thermal conductivity.

The thermal conductivity of metal mirrors is much higher than that of glass substrates; for aluminum, $\kappa = 167$ W/m K. IR systems are often cooled to reduce thermal background radiation. The high thermal conductivity of metal substrates reduces cooling time and the power required for cooling. The thermal distortion index of 6061 aluminum alloy is 141×10^{-9} m/W, which is about 5% of that of Corning Pyrex. Aluminum is also used to maintain focus and alignment through same material athermalization. Copper is even better than aluminum in thermal performance, with a thermal distortion index of about 43×10^{-9} m/W. For high-energy laser mirrors, where thermal distortion under laser irradiation is an important performance index, copper is the preferred material.

The specific stiffness of beryllium is about six times higher than that of aluminum or glass. Beryllium is used in high-speed scanner mirrors, where high specific stiffness minimizes surface distortion at high angular accelerations. Weight-critical applications such as satellite optics often employ beryllium since its density is about three quarters of that of aluminum or glass.

The surface roughness of metal mirrors limits performance, usually restricting applications to longer-wavelength IR systems. The grain structure of metals is large in comparison with the wavelength of light in the visible, increasing surface scatter. Diamond turning is a low-cost way of producing optical surfaces in metal mirrors, but leaves a residual regular pattern of grooves that increases surface scatter. As a rule of thumb, the surface roughness of a diamond turned aluminum mirror is 60–100 Å RMS. The surface roughness of diamond turned copper and amorphous nickel plating is better, at around 40 Å RMS. Postpolishing of electroless nickel after diamond turning can reduce surface roughness to 20 Å RMS or less. New methods for postpolishing aluminum after diamond turning can reduce surface roughness to between 10 and 20 Å.

A surface roughness of about 5 Å RMS is possible with glassy substrates. More common is a surface roughness between 10 and 20 Å RMS. Surface roughness is an important parameter in selecting glassy substrates for demanding applications, particularly in the visible.

The inverse specific stiffness of chemically vapor deposited (CVD) silicon carbide is 68×10^{-9} m^{-1}, which is about 20% of that of conventional glassy or metal mirror substrates. The thermal distortion parameter of silicon carbide is 16×10^{-9} m/W, which is also much better than those of most mirror substrate materials. These two properties make silicon carbide a competitive alternative to conventional mirror substrates, particularly for lightweight applications. Silicon carbide is also an alternative to copper for high-energy laser mirrors.

Silicon carbide is made by chemical vapor deposition or a reaction bonding process. Cast substrates are made using the latter process. Silicon carbide is difficult to polish, so a thin silicon cladding is placed atop the substrate for optical finishing. The silicon cladding is diamond turned or polished using ordinary methods. Silicon carbide mirrors must be mounted using the same methods as for glass mirrors since the material is too brittle for the types of mounts used for metal mirrors. Currently, costs of silicon carbide mirrors are high, comparable to those of beryllium ones. With increased use, this cost is expected to become lower in the near future.

Monocrystalline or single-crystal silicon is another alternative mirror substrate material, with an inverse specific stiffness of 174×10^{-9} m^{-1}, which is about half that of conventional mirror substrate materials. Silicon is a low-cost alternative to more exotic substrate materials such as beryllium or silicon carbide for lightweight applications. The thermal distortion index of silicon is 19×10^{-9} m/W, which is comparable to that of silicon carbide. The low thermal distortion index of silicon makes it suitable as a high-energy laser mirror substrate. Silicon mirrors are diamond turned or conventionally polished. Since silicon is a brittle material, it is mounted the same way as glassy substrates. One technique frit bonds Invar inserts into the back of a silicon mirror substrate for mounting.

3.10 MATERIAL COST

The cost of an optical material is the sum of the bulk cost and processing cost. The cost of materials varies considerably and changes with time. The cost of fabrication to the desired final shape is influenced by the type of material. Cost is an important consideration when selecting materials. Use of a less than optimum material is often due to cost factors.

Bulk cost is determined by the type of material, volume purchased, and shape of the bulk material. More common materials cost less than exotic materials. For example, in metals, aluminum is lower in cost than titanium, which is in turn lower in cost than beryllium. "Preferred glasses" are lower in cost than "standard glasses," which are lower in cost than special-order glasses. For most materials, purchasing in larger volumes lowers cost. Finally, shape influences cost; extruded or rolled sections cost more than simple slabs or rounds.

The cost of optical materials is influenced by the difficulty of fabrication. For glasses, the two material properties influencing cost are stain resistance and grindability. According to Willey and Parks,[13] the cost of fabrication varies in accordance with stain resistance and grindability as

$$\text{Cost} = \left[\frac{1+0.01(\text{SC})^3}{\text{HG}}\right] \times \text{Base}_{\text{GP}}, \tag{3.23}$$

where Cost is the total cost associated with fabrication of the material, HG is the grindability factor, Base_{GP} is the base cost, and SC is the stain code of the material.

This equation assumes that there is a minimum or base cost associated with the setup for fabrication. The base cost is material independent and is constant for all types of materials. The grindability factor is proportional to the material removal rate during grinding and polishing. Finally, the stain code is an integer varying from 0 to 4 for different glass types.

ISO 12844, *Raw Optical Glass—Grindability with Diamond Pellets—Test Method and Classification*,[14] quantifies relative removal rates during grinding for various types of optical glass. Surface removal rates are in units of volume per unit time and are referenced to a standard glass type during a standardized grinding test. Relative removal rates are given in terms of an integer

TABLE 3.7
Grindability Classifications from ISO 12844

Grindability Class	Limiting Values Referenced to Schott N-SK16
HG 1	≤30
HG 2	>30 and ≤60
HG 3	>60 and ≤90
HG 4	>90 and ≤120
HG 5	>120 and ≤150
HG 6	>150

from 1 to 6, with larger numbers representing faster removal rates or more volume per unit time. Since grinding cost is proportional to time, faster removal rates imply lower cost.

According to ISO 12844,[14] the grindability HG rounded to the nearest integer is given by

$$HG = \frac{w_X/\rho}{w_0/\rho_0} \times 100, \tag{3.24}$$

where w_X is the mean mass removal of the test specimens, ρ is the density of test glass, w_0 is the mean mass removal of the reference glass, and ρ_0 is the density of the reference glass.

The reference glass for ISO 12844 is a fine annealed dense barium crown glass, with a glass code of 620603 and density of 3.58 g/cm^3. Schott N-SK16 is a commercial glass meeting the reference standard. Table 3.7 from ISO 12844 gives the HG classes. The grindability of Schott glass varies from HG 1 for SF57 to HG 6 for N-FK51A. The grindability of common glass types such as F2 and N-BK7 is HG 2 or HG 3.

PROBLEMS

1. What is the index of refraction of sapphire at a wavelength of 1.06 μm, using the following Sellmeier coefficients: $B_1 = 1.4313493$, $C_1 = 0.0726631$, $B_2 = 0.65054713$, $C_2 = 0.1193242$, $B_3 = 5.3414021$, and $C_3 = 18.028251$. Give the result to six decimal places. *Answer:* $n = 1.754558$
2. The index of refraction of spinel is given in an abbreviated Sellmeier equation. What is the index of refraction at a wavelength of 1.55 μm when using the following coefficients: $B_1 = 1.8938$, $C_1 = 0.09942$, $B_2 = 3.0755$, and $C_2 = 15.826$? Give the result to four decimal places. *Answer:* $n = 1.6946$
3. What is the transmission of a 15 mm thick spinel window at a wavelength of 4 μm if the absorption coefficient is 0.018 cm^{-1}? Hint: use the dispersion equation from problem 2. *Answer:* $\tau_\infty = 0.865$
4. If the glass code of Schott F2 is 620364.360, what is the transmission at a wavelength of 587 nm, neglecting absorption losses? *Answer:* $\tau_\infty = 0.894$
5. The critical angle is the angle of incidence at which the angle of refraction is 90°. What is the critical angle for optical glass if $n = 1.5$? *Answer:* $\theta_1 = 41.81°$
6. The maximum tolerable OPD for a window made of Schott F2 is 0.10 wave, at a wavelength of 550 nm. If the stress optic coefficient for F2 is 2.81×10^{-6} MPa^{-1}, how much stress is tolerable in a 20 mm thick window? *Answer:* $\sigma = 979$ kPa
7. If $n' = 0.029$ and $k' = 221$ MPa, what is the MYS of 5456 aluminum? What is the strain at half of microyield? Hint: see Equation 3.12. *Answer:* $\sigma_{MYS} = 148$ MPa and strain at $\sigma_{MYS}/2$ is 41×10^{-18}

8. At constant deflection, compare the relative weight of mirrors made of Corning 7740 Pyrex and S-200FH beryllium alloy. Assume that the mirror support is the same for mirrors made of both materials. Use material properties from Appendices 6 and 7. *Answer:* relative weight beryllium/Pyrex = 0.351
9. At constant thickness, compare the deflections of mirrors made of Schott Zerodur and 6061 aluminum: Assume that mirror size and support are identical. Use the properties from Appendices 6 and 7. *Answer:* relative thickness Zerodur/aluminum = 0.753
10. A flat Corning 7740 Pyrex mirror is 40 mm thick with a diameter of 250 mm. What is the deviation from flatness if there is a 10 K temperature gradient in the mirror? Use properties from Appendix 6. *Answer:* $\delta = 0.165\ \mu m$
11. A flat OFHC copper mirror is used in a high-energy laser cutting system. The mirror is 15 mm thick with a diameter of 50 mm. The mirror absorbs 1% of the 1 kW laser energy on its surface, producing a temperature gradient from front to back. Assuming that all of the absorbed energy flows through the mirror, what is the deviation of the mirror from flatness? Use properties from Appendix 7. *Answer:* $\delta = 6.102 \times 10^{-9}$ m
12. In 1919 there were problems with the thermal behavior of the 2.5 m Hooker telescope primary mirror in the Mount Wilson observatory. Air temperature changes about 10 K in the first hour after sunset. The mirror is 300 mm thick. What is the time required for the mirror to reach thermal equilibrium (1% difference from front to back)? What is the maximum surface deformation, assuming that the 10 K temperature change is instantaneous? Use the following properties for the mirror material: density = 2500 kg/m^3, thermal conductivity = 1.1 W/m K, specific heat = 860 J/kg K, and thermal coefficient of expansion = 7×10^{-6} K^{-1}. *Answer:* t = 22.8 hours and $\delta = 182\ \mu m$
13. The Russian BTA telescope primary mirror is 600 mm thick. If the mirror material properties are the same as those of Corning Pyrex 7740, what is time required for this mirror to reach thermal equilibrium, where thermal equilibrium is defined as a 1% temperature difference from front to back? Use properties from Appendix 6. *Answer:* t = 96 hours
14. A missile dome made of sapphire is subjected to an instantaneous increase in temperature of 300 K from air friction when fired. If the breaking strength of sapphire is 300 MPa, will the dome survive this thermal shock? Use properties from Appendix 3. *Answer:* The dome will fail since ΔT_{max} = 124 K is less than the 300 K instantaneous temperature change.
15. Ignoring area scaling, what is the failure stress associated with a probability of failure of 0.001 for polished CVD ZnS? Use the data in Table 3.3. *Answer:* σ = 36.5 MPa
16. A giant right circular cylinder Zerodur telescope mirror is 8 m in diameter. During fabrication, the mirror is subjected to a handling stress of 12 MPa when the mirror surface condition is ground, equivalent to a SiC 600 condition. Using data from Table 3.3, what is the probability of failure during this handling operation? Assume that the stress is distributed uniformly on the lower surface of the mirror, with the mirror axis vertical. *Answer:* $P_f = 2.4 \times 10^{-10}$
17. Schott BK7 is substituted for BaK4 in making prisms. What is the estimated change in fabrication cost? The grindability of BK7 is HG = 3, and its stain code is SC = 2, while the grindability factor of BaK4 is HG = 2 and its stain code is SC = 1. Assume that the base cost is the same for both materials. *Answer:* The cost decreases by a factor of 0.714.
18. A 300 mm diameter, 40 mm thick calcium fluoride lens blank is subjected to a 50 K thermal shock during handling. What is the probability of breaking the lens blank? Use data from Table 3.3 for the Weibull statistics and assume that the stress is uniform on the entire surface area of the right cylindrical cylinder lens blank. Use data from Appendix 4 for other calcium fluoride properties. *Answer:* $P_f = 2.85 \times 10^{-9}$

REFERENCES

1. Lentes, F.-T., Refractive index and dispersion, Section 2.1 in *The Properties of Optical Glass*, Bach, H., and Neuroth, N., Eds., pp. 19–29, Springer, Berlin, 1995.
2. Wolfe, W., *Introduction to Infrared System Design*, SPIE Press, Bellingham, WA, 1994.
3. Smith, W.J., *Modern Optical Engineering*, 3rd ed., McGraw-Hill, New York, 2008.
4. Hofman, H.-J., Differential changes of the refractive index, Section 2.4 in *The Properties of Optical Glass*, Bach, H., and Neuroth, N., Eds., pp. 96–120, Springer, Berlin, 1995.
5. Levi, L., *Applied Optics*, John Wiley & Sons, Hoboken, NJ, 1968.
6. Timoshenko, S., and Woinowsky-Krieger, S., *Theory of Plates and Shells*, 2nd ed., McGraw-Hill, New York, 1959.
7. ASTM International, *Standard Test Method for Monotonic Equibiaxial Flexural Strength of Advanced Ceramics at Ambient Temperature*, ASTM 1499-09, ASTM International, West Conshohocken, PA, 2013.
8. ASTM International, *Standard Practice for Reporting Uniaxial Strength Data and Estimating Weibull Distribution Parameters for Advanced Ceramics*, ASTM C1239-07, ASTM International, West Conshohocken, PA, 1995.
9. Walker, B.H., Select optical glasses, in *The Photonics Design and Applications Handbook*, p. H-356, Lauren, Pittsfield, MA, 1993.
10. Zhang, S., and Shannon, R.R., Lens design using a minimum number of glasses, *Opt. Eng.*, 34, 3536, 1995.
11. ISO (International Organization for Standardization), *Glass and Glass-Ceramics—Knoop Hardness Test*, ISO 9385, ISO, Geneva, 1990.
12. ISO, *Optics and Photonics—Spectral Bands*, ISO 20473:2007, ISO, Geneva, 2007.
13. Willey, R.R., and Parks, R.E., Chapter 1, Optical fundamentals, in *Handbook of Optomechanical Engineering*, Ahmad, A., Ed., pp. 1–44, CRC Press, Boca Raton, FL, 1997.
14. ISO, *Raw Optical Glass—Grindability with Diamond Pellets—Test Method and Classification*, ISO 12844, ISO, Geneva, 1999.

FURTHER READING

Ashby, M., *Materials Selection in Mechanical Engineering*, 4th ed., Butterworth-Heinemann, Oxford, 2010.

Harris, D.C., *Materials for Infrared Windows and Domes: Properties and Performance*, SPIE Press, Bellingham, WA, 1999.

Hartmann, P., *Optical Glass*, SPIE Press, Bellingham, WA, 2014.

Schaub, M.P., *The Design of Plastic Optical Systems*, SPIE Press, Bellingham, WA, 2009.

4 Window Design and Mounting

4.1 INTRODUCTION AND SUMMARY

A window is used in an optical instrument primarily as a transparent interface between the internal components and the outside environment. Usually, it is a plane-parallel plate of optical glass, fused silica, plastic, or crystalline material that allows the desired radiation to pass through with minimal effect on intensity and image quality, but excludes dirt, moisture, and other contaminants and, in some cases, maintains a positive or negative pressure differential between the internal and external atmospheres. For infrared (IR) applications, the window must not radiate due to its temperature in a manner that interferes with the function of the system. Some windows take the form of a deep spherical shell. A deep shell is called a dome.

Critical aspects of window-mounting design generally include freedom from mechanically and thermally induced distortions, strength, sealing provisions, and location in the optical system. If at or near the aperture stop or a pupil, a prime consideration is deformation introduced into the transmitted wave front. Cosmetic defects such as scratches, digs, and dirt are less significant. If the window is near an image, it can have little effect on the wave front, but defects or dirt on the surfaces of the window may appear superimposed on that image. In high-energy or high-power laser applications, cleanliness and freedom from minute scratches or other physical defects are vital to prevent or, at least, minimize laser-induced damage.

Because plane-parallel plates have essentially no effect on collimated light beams or ones with very slight convergence or divergence, those plates may be considered relatively benign elements of the optical system from an alignment viewpoint. If thick plane-parallel windows are located in beams with significant convergence or divergence, they will contribute aberrations in the same manner as prisms and are alignment sensitive. Usually, windows located inside an optical system are thin enough for this not to be a serious problem.

The optical performance of each of the window types discussed in this chapter is usually specified in terms of the maximum allowable deterioration of a transmitted plane wave front over the full window aperture or selected subapertures. The requirements depend, of course, on the specific application and the wavelength used. For example, a visual viewing system may allow a wave front error as large as 1 wave peak to valley (p-v) at 0.63 μm. A window for a conventional forward-looking IR sensor operating at 10.6 μm may tolerate 0.1 wave p-v wave front error at that wavelength (equivalent to 1.7 waves p-v at 0.63 μm). A window or dome for a high-velocity missile might need less than 0.1 wave root-mean-square (RMS) for any subaperture as defined by the line-of-sight direction and overall sensor aperture. A high-performance photographic camera application may require the p-v error of a protective window in front of its lens to contribute less than 0.05 wave error at 0.63 μm.

Window design is a multistep process, beginning with estimates of performance, refined analysis if necessary, and finally detailed design, including mounting. Simple closed-form analyses coupled with worst-case assumptions are desirable for initial design estimates. If this analysis indicates that there is sufficient performance margin, detailed design is next. When simple analysis indicates insufficient margin, more accurate methods, such as finite-element analysis (FEA), are necessary before beginning the development of the final configuration.

Window deformation and strength are affected by the type of edge constraint. The edge of the window is free to rotate when simply supported. Deflection under load and stress are at the maximum for simple support. Deflection and stress are minimized when the window edge is clamped or fixed. Defects in the mounting surfaces can induce significant deformations, and thermal stress is a concern in windows with clamped edges. An intermediate case is the "held but not fixed" edge constraint with limited rotation possible at the window edge. For the held but not fixed edge condition, either FEA or more advanced numerical methods are required for window design; hence, this edge condition is not considered.

Design and mounting configurations for typical types of windows and domes are considered in the following sections. Optical effects in plane-parallel windows arising from environmental conditions such as temperature and pressure differentials are first discussed. The next topic is the strength of plane-parallel windows. Following are design guides for mounting flat windows. Methods of analyzing dome performance are next. Finally, guidelines for dome mounting are given.

4.2 WINDOW DESIGN

A window provides protection to an optical system from an adverse exterior environment. Window thickness is determined by a combination of optical performance and strength. In some applications, window thickness is set by optical aberrations of the wave front passing through a window distorted by pressure or thermal effects. In others, window strength decides thickness. A final consideration is the type of mounting for the window. Performance is often neglected in optical design since windows are zero-power elements. As shown in the next section, pressure and temperature can induce significant effects in the wave front passing through a window.

4.2.1 Optical Performance of Windows

A wave front may have optical aberration introduced after passing through a window that is distorted from an ideal plane-parallel, zero-power form. Causes of window distortion include pressure differentials, temperature effects, and improper mounting. A pressure differential deforms a plane-parallel window into a shallow shell which acts as a weak meniscus lens. Temperature gradients through the thickness of or on the plane of the window introduce optical path differences (OPDs) into the wave front. Distortion from improper mounting may also introduce optical aberration. In this section, pressure and temperature effects on optical performance are considered. Improper mounting effects are difficult to analyze and should be minimized by the mounting techniques discussed later.

Common window shapes are circular, elliptical, and rectangular. More exotic shapes, such as trapezoids, are relatively uncommon. Simple closed-form analysis of optical performance is limited to circular or right cylindrical windows. Axial symmetry and zero power (for a plane-parallel cross section) simplify analysis. More complex methods such as FEA are required for noncircular geometries such as elliptical or rectangular windows. As a worst case, the window edge is assumed to be simply supported.

A pressure differential through the thickness of a circular or axisymmetric plane-parallel window distorts its shape. After passing through the distorted window, an OPD or optical aberration is introduced into the wave front. Since both sides of the window bend almost equally, the window becomes a weak lens. If the focus of system behind the window is adjusted, the residual OPD is usually negligible.

For a window that is plane-parallel and circular, Sparks and Cottis[1] provided equations for the minimum thickness required to reduce the OPD from pressure-induced distortion to $\lambda/8$, where λ is the wavelength. An OPD of $\lambda/8$ is below the diffraction limit for an optical component, and for most applications, it is negligible. The OPD estimate includes focus and hence is conservative. Two different equations are necessary, one for the fixed edge and one for the simply supported edge.

In the following equations, Poisson's ratio is $\nu = 0.3$, and the surrounding medium is air, with an index of refraction of unity:

$$h_{\min} = 0.842\left[(n - n_0)\left(\frac{P}{E}\right)^2\left(\frac{d}{\lambda}\right)\right]^{\frac{1}{5}} d;\ \text{fixed}, \tag{4.1}$$

$$h_{\min} = 1.01\left[(n - n_0)\left(\frac{P}{E}\right)^2\left(\frac{d}{\lambda}\right)\right]^{\frac{1}{5}} d,\ \text{simply supported}, \tag{4.2}$$

where ν is the Poisson's ratio of the window material, n is the index of refraction of the window material, n_0 is the index of refraction of the surrounding medium ($n_0 \approx 1$ for air), P is the pressure differential, E is the elastic modulus of the window material, d is the window diameter, $h_{\min}$ is the minimum window thickness, and λ is the wavelength.

In many applications, focus can be adjusted after the wave front passes through the window. If focus is removed, a less conservative estimate of the OPD induced by a pressure differential in a plane-parallel circular window is given in Equations 4.3 and 4.4. Again, there are two edge constraint equations, fixed and simply supported. These equations include the actual value of Poisson's ratio for the window material:

$$\text{OPD} = \frac{9}{2048}(n-1)\left(\frac{P}{E}\right)^2\left(\frac{d}{h}\right)^5 d\,(1-\nu^2)^2,\ \text{fixed}, \tag{4.3}$$

$$\text{OPD} = \frac{9}{2048}(n-1)\left(\frac{P}{E}\right)^2\left(\frac{d}{h}\right)^5 d\,(\nu-1)^2(\nu+1)(\nu+5),\ \text{simply supported}. \tag{4.4}$$

Temperature gradients deform an initially plane-parallel circular window. The temperature distortion of the window introduces an OPD into the wave front. An axial gradient through the window thickness distorts a simply supported circular window into a spherical shell, or meniscus. There is no bending stress from the temperature gradient in a simply supported window. The radius of curvature of the resulting shell is given by

$$\frac{1}{R} = \frac{\alpha \Delta T_{\text{h}}}{h}, \tag{4.5}$$

where R is the radius of curvature of the spherical shell, α is the window material thermal coefficient of expansion, and ΔT_{h} is the temperature gradient through the window thickness.

The axial gradient turns the window into a weak meniscus lens. The power, defined as the reciprocal of the equivalent focal length (FL) is given by[2]

$$\frac{1}{\text{FL}} = \left(\frac{n-1}{n}\right)\alpha^2 \Delta T_{\text{h}}^2 \frac{d}{h^2} = \left(\frac{n-1}{n}\right)\left(\frac{\alpha}{k}\right)^2 dq^2, \tag{4.6}$$

where k is the window thermal conductivity and q is the heat flux per unit area in the window.

An in-plane or radial temperature gradient changes the index of refraction of the window material. The optical effects of a radial gradient are more complex than those of an axial gradient. A rough-order-of-magnitude estimate of the OPD from an in-plane gradient is made by considering the maximum temperature difference. Caution is necessary in evaluating this estimate, since some of the OPD may be in the form of focus. In many applications, focus can be adjusted in the system, with a negligible residual optical aberration. The OPD from a maximum in-plane temperature gradient ΔT_{max} is given by[3]

$$\text{OPD} = nh\,\Delta T_{max}\left(\alpha + \frac{1}{n}\frac{\mathrm{d}n}{\mathrm{d}T}\right), \tag{4.7}$$

where $\mathrm{d}n/\mathrm{d}T$ is the change in window index of refraction with temperature.

For many common thermal loading conditions, the radial gradient ΔT_r can be approximated as being parabolic, with the form $\Delta T_r = C_r r^2$, where C_r is a constant and r is the distance from the window center. A focus shift is induced with this temperature gradient, and the residual OPD is negligible. If the edge is simply supported and the window is thin ($h \ll d$), the window becomes a weak lens with an effective FL given by

$$\frac{1}{\text{FL}} = 8\frac{h}{d^2}\Delta T_r\left[(n-1)(1+\nu)\alpha + \frac{\mathrm{d}n}{\mathrm{d}T}\right]. \tag{4.8}$$

Most of the effect on window performance from a radial temperature gradient occurs where the gradient is strongest, which is at the edge. In many cases, the optical effect of a radial temperature gradient becomes negligible if the window is made sufficiently oversized with respect to the required clear aperture. As a rough rule of thumb, increasing the diameter by 20–25% over the clear aperture suppresses radial gradient effects.

4.2.2 Window Strength

The strength of windows is evaluated using classic plate equations. Handbook equations for plate stress often assume that Poisson's ratio for the material is about 0.3. This assumption is incorrect for most window materials, leading to errors in calculating stress. For a circular window, there is a change in stress of about 4% between a material with a Poisson's ratio of 0.3 and one of 0.17 (fused silica). Simple equations without corrections for Poisson's ratio are suitable for conceptual design. Detailed or final design analysis should incorporate the actual Poisson's ratio of the material unless the design margin is large. Unsupported, rather than overall, dimensions should be used in stress analysis.

For brittle optical materials, a Weibull analysis should be used to find the design-allowable stress. Weibull statistics are not available for many window materials, requiring use of a "rupture modulus" or fracture strength. There is considerable uncertainty in design-allowable stress from published values of strength since factors such as surface finish and area are not known. This uncertainty leads to large safety factors (SFs). Table 4.1 provides values of fracture strength for some selected IR window materials at room temperature.

The selection of the SF depends upon the window application. An SF of 10× is required for untempered glass in the American Society of Mechanical Engineers pressure vessel code.[4] NASA-STD-5001 requires a minimum SF of 3×.[5] For reasonable values of the Weibull parameters, $m > 5$ and $\sigma_0 \approx 100$ MPa, an SF of 2–4 provides better than 0.99 confidence. An SF of $2 \le \text{SF} \le 4$ is hence suggested for most applications.

TABLE 4.1
Approximate Fracture Strengths of IR Window Materials

Material	σ_{YS} (MPa)	Material	σ_{YS} (MPa)
Silicon carbide	600	CaF_2 (single crystal)	100–150
Silicon nitride	600	Gallium phosphide	100–120
Sapphire	300–1000	ZnS (standard)	100
Diamond (CVD)	100–800	ZnS (multispectral)	70
ALON	300	Geranium	90
Spinel	190	SrF_2 (single crystal)	70–110
Yttria (doped/undoped)	160	Fused silica	60
Silicon	120	Zinc selenide	50
MgF_2 (hot-pressed)	100–150		

Source: Harris, D.C., *Materials for Infrared Windows and Domes: Properties and Performance*, SPIE Press, Bellingham, WA, 1999.

Note: Material strength should always be considered approximate. It varies with the quality of surface finish, fabrication method, material purity, type of test, and size of sample. Values shown are at room temperature. CVD, chemically vapor deposited; ALON, aluminum oxynitride.

Further increase in the SF is necessary to include the effect of possible window damage, especially near the end of life. A window on an aircraft may be scratched or pitted from sand and rain impacts. These surface defects lower the window strength. Improper handling may scratch the window surface, again reducing strength. As a rough rule of thumb, for aircraft windows exposed to rain and dust erosion, an additional SF of about 2x is desirable. The SF associated with scratches depends on the scratch depth. Window strength varies as $h_s^{-1/2}$, where h_s is the scratch depth. When the scratch depth increases by a factor of 2, the strength of the window material is about 0.7 when compared to its initial condition.

The minimum thickness h_{min} for a circular window exposed to a pressure differential P is given approximately by Equation 4.9. In this equation, $C_1 = 1.25$ for simply supported edges and 0.75 for a clamped edge. Although there is no correction for Poisson's ratio, the equation is conservative, providing a greater minimum thickness than produced by a more accurate analysis. Note that the diameter is the unsupported dimension, not the total window size:

$$h_{min} = \frac{d}{2}\left(\frac{C_1 \, \mathrm{SF} \, P}{\sigma_{YS}}\right)^{\frac{1}{2}}, \tag{4.9}$$

where SF is the safety factor and σ_{YS} is the window material strength.

Less conservative equations for minimum window thickness are given in the following, including Poisson's ratio for the material[6]:

$$h_{min} = \frac{d}{4}\left[\frac{3(3+\vartheta)\,\mathrm{SF}\,P}{2\sigma_{YS}}\right]^{\frac{1}{2}}, \text{ simply supported,} \tag{4.10}$$

$$h_{min} = \frac{d}{4}\left[\frac{3(1+\vartheta)\,\mathrm{SF}\,P}{2\sigma_{YS}}\right]^{\frac{1}{2}}, \text{ clamped.} \tag{4.11}$$

Oblique beams and scanning systems are examples of applications of elliptical windows. If the ellipse length or major axis is a and the width or minor axis is b, the minimum thickness h_{min} for a pressure differential P, including Poisson's ratio, assuming simply supported edges is given by[7]

$$h_{min} = \frac{b}{2}\left\{\left[2.816+1.581\vartheta-(1.691+1.206\vartheta)\frac{b}{a}\right]\frac{\text{SF}\,P}{\sigma_{YS}}\right\}^{\frac{1}{2}},\ 0.2<\frac{b}{a}<1.0. \tag{4.12}$$

If the edge of the elliptical window is clamped, the minimum thickness is given by

$$h_{min} = \frac{b}{2}\left\{\frac{6\,\text{SF}\,P}{\sigma_{YS}\left[3+2\left(\frac{b}{a}\right)^2+3\left(\frac{b}{a}\right)^4\right]}\right\}^{\frac{1}{2}}. \tag{4.13}$$

Rectangular windows are a lower-cost alternative to elliptical or more complex window geometries. The minimum window thickness h_{min} for a pressure differential P is estimated with the following equations. A value of Poisson's ratio of 0.3 is assumed. The equations are conservative in that the estimated minimum thickness is greater than that for smaller values of Poisson's ratio. In these equations, the unsupported dimensions are the length L and width w. For simply supported window edges,

$$h_{min} = L\left\{\frac{0.75\,\text{SF}\,P}{\sigma_{YS}\left[1.61\left(\frac{w}{L}\right)^3+1\right]}\right\}^{\frac{1}{2}}. \tag{4.14}$$

For clamped window edges,

$$h_{min} = w\left\{\frac{\text{SF}\ P}{2\sigma_{YS}\left[0.623\left(\frac{w}{L}\right)^6+1\right]}\right\}^{\frac{1}{2}}. \tag{4.15}$$

Example 4.1

A plane-parallel circular window made of ZnS on an aircraft is subjected to an axial temperature gradient of 70 K and a pressure differential of 45 kPa. The unsupported window diameter is 300 mm. An SF = 4 is required, and an additional factor of 2 is required to allow for damage in service. A probability of failure of 10 parts in a million ($P = 10 \times 10^{-6}$) is required. The window is used in the midwave IR, where $n = 2.252$ and $dn/dT = 43 \times 10^{-6}\ \text{K}^{-1}$ at $\lambda = 4\ \mu\text{m}$. What is the minimum thickness of the window using the Weibull data from Table 3.3 assuming a polished surface? What is the FL of the equivalent lens due to the axial temperature gradient? For maximum accuracy, include Poisson's ratio $\nu = 0.29$ in the thickness calculation. Use $\alpha = 7.0 \times 10^{-6}\ \text{K}^{-1}$ for ZnS.

From Table 3.3, $m = 9.99$, $\sigma_0 = 81.5$ MPa, and $A_0 = 100\ \text{mm}^2$. In bending, one side of the window is in tension and one is in compression; the probability of failure for the compressive side is much lower than that for the tensile side. Hence, the stress area is taken to be just the side in tension, or $A = (\pi/4)(300\text{ mm})^2 = 70.7 \times 10^3\text{ mm}^2$. The Weibull equation (Equation 3.20) is

$$10\times10^{-6} = 1-\exp\left[-\frac{70.7\times10^3\,\text{mm}^2}{100\,\text{mm}^2}\left(\frac{\sigma_{YS}}{81.5\text{MPa}}\right)^{9.99}\right].$$

Solving for the stress, $\sigma_{YS} = 13.4$ MPa.

The minimum window thickness h_{min} is

$$h_{min} = \frac{300\,\text{mm}}{4}\left[\frac{3(3+0.29)(4)(2)(45\text{kPa})}{2(13.4\text{MPa})}\right]^{\frac{1}{2}} = 27.4\,\text{mm}.$$

The window FL from the axial temperature gradient is

$$\text{FL} = \left[\frac{(2.252-1)}{2.252}(7.0\times10^{-6}\text{K}^{-1})^2(70\text{K})^2\frac{300\,\text{mm}}{(27.4\,\text{mm})^2}\right]^{-1} = 18.7\times10^3\,\text{m}.$$

4.2.3 Mounting Flat Windows

Figure 4.1 is an example of a simply supported mounting design for small round aperture windows used to seal the interior of an optical system from the outside world. This window is a 20 mm (0.79 in.) diameter by 4 mm (0.16 in.) thick disk of BK7 glass. It is to be used in the *f*/10 beam of the illumination path in a military telescope reticle projection subsystem. The surfaces need be flat only to ±10 waves p-v of visible light and parallel to 30 arcmin. The window is bonded into a stainless steel (corrosion-resistant steel [CRES] type 303) cell with a polysulfide base sealing compound according to US military specification MIL-S-11031.[8] A room-temperature–vulcanization (RTV)-type

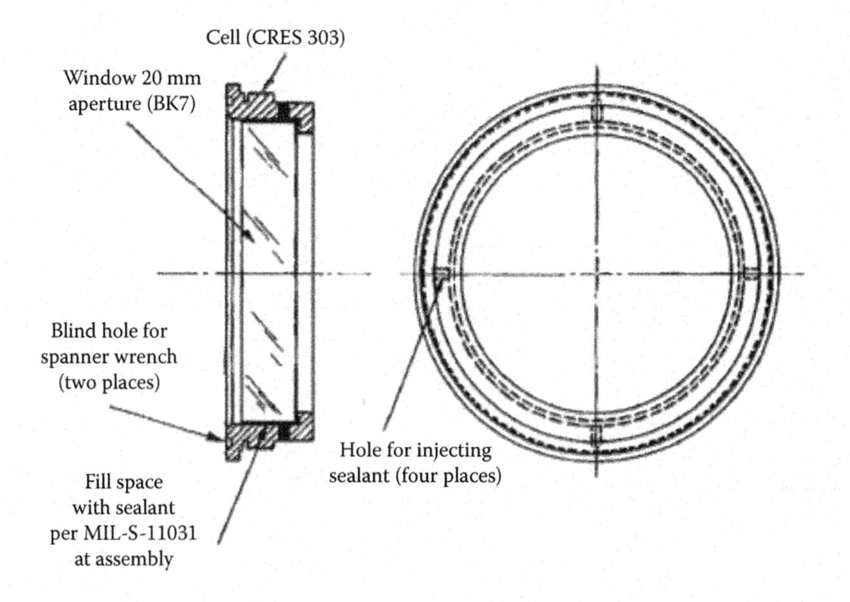

FIGURE 4.1 Instrument window subassembly featuring a bonded-in-place glass element. (Courtesy of US Army, Washington, DC.)

elastomer would serve as well. This secures the window and forms an effective seal. Note that the glass is positioned axially against a flat annular shoulder inside the cell and that the adhesive fills the annular space created by undercutting that shoulder. The uniformity of the radial thickness of the encapsulating adhesive layer can easily be achieved by inserting shims or gauge wires between the glass and metal before the adhesive is injected. These items can be removed after the sealant has cured. For best sealing, the spaces left after removing the shims or wires should be filled with sealant and cured.

Slightly simpler construction is provided with a little less reliability if the cell of Figure 4.1 does not have sealant injection holes, and the sealant is either carefully applied to the window rim and cell inside diameter (ID) before inserting the window into the cell or inserted into the narrow groove between the rim of the window and the cell ID with a hypodermic syringe. In either case, any excess sealant should be cleaned away before it sets. The suitability of the seal may be inferred from observation of sealant bead continuity around the window rim or, preferably, checked by pressure testing. An external thread on the cell mates with a threaded hole in the instrument housing. A flat gasket or O-ring would typically be used between the flange of the cell and the housing to seal the interface.

Figure 4.2 shows an example of a fixed-edge window mount. This subassembly has a BK7 window of 50.80 mm (2.0 in.) diameter and 8.8 mm (0.346 in.) thickness sealed with military specification sealant into a stainless steel (CRES 416) cell and secured with a threaded retainer that also is made of stainless steel. This window is intended to be used as an environmental seal in front of the objective of a high-power telescope. Since the light beam transmitted through this window is collimated and fills the clear aperture of the window at all times, the critical optical specifications are transmitted wave front error (±5 waves of spherical power and 0.05 wave p-v of irregularity for green light) and wedge angle (30 arcsec maximum).

The maximum and minimum allowable diametric clearances between the glass and the metal at ambient temperature are 0.53 mm (0.020 in.) and 0.22 mm (0.009 in.), respectively. The cell is

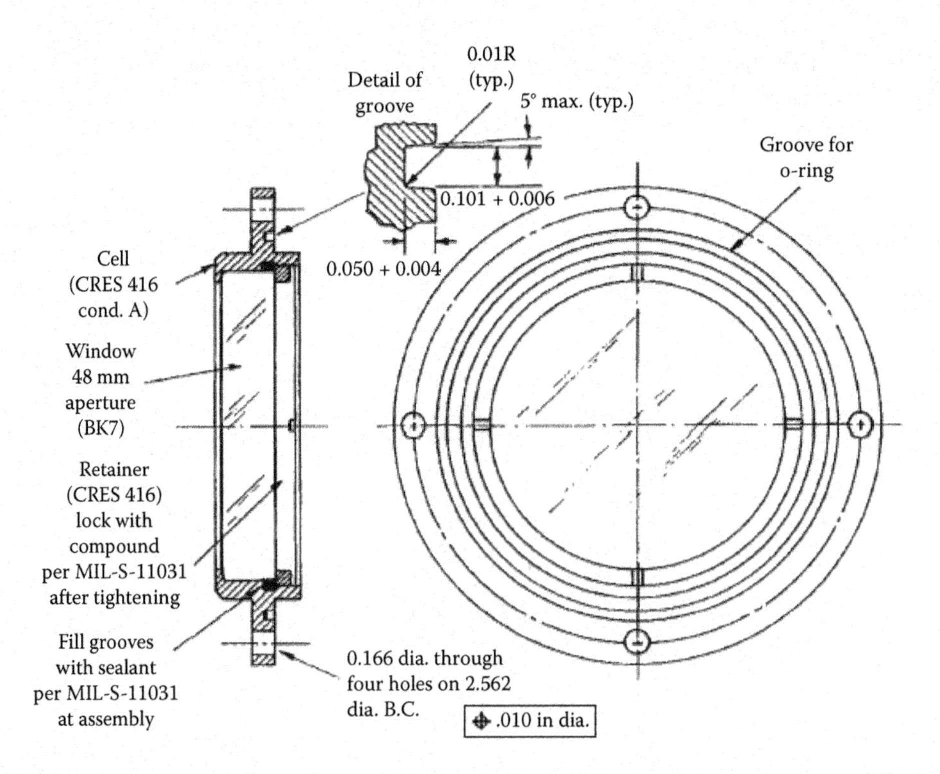

FIGURE 4.2 Instrument window subassembly with window held in place by a retaining ring. Dimensions are in inches. (Courtesy of US Army, Washington, DC.)

provided with an annular groove for an O-ring that is used to seal the cell to the instrument housing at the next level of assembly. Dimensions of the groove are shown in the detail view. Note that the mounting holes are outside this seal. Screws used to attach the subassembly should thread into blind holes in the instrument housing. The seals are designed to hold 34.5 kPa (5 lb/in.^{2}) positive pressure inside the telescope for an extended period.

Figure 4.3 shows an elliptical window for a low-light-level television (LLLTV) system utilizing light in the spectral region of 0.45–0.9 μm and mounted on a pod under the wing of an aircraft. Two plane-parallel plates of zinc crown glass are laminated together to form the single 19 mm (0.75 in.) thick glazing of elliptical aperture approximately 25 cm × 38 cm (9.8 in. × 15.0 in.). It is mounted into a cast aluminum frame. The frame interfaces with the curved surface of the camera pod and is held in place by several screws through the recessed holes visible in the figure. The internal construction of this subassembly is shown schematically in the exploded view of Figure 4.4. The wires connect to an electrically conductive coating applied to one glass plate. The second plate is attached with optical cement over this coating to protect it. This coating provides heat for anti-icing and defogging purposes during a military mission. It also attenuates electromagnetic radiation. The design is such that the optic may be replaced if damaged. The window is sealed in place with an RTV-type sealant. The assembled window is capable of withstanding, without damage, a proof pressure differential of 76 kPa (11 lb/in.^{2}) in either direction. Both exposed surfaces of the window are broadband antireflection coated. The frame exterior is painted white.

A window assembly developed for vacuum cryogenic applications on a double-walled dewar is illustrated in Figure 4.5.[9] The window is germanium and has a racetrack-shaped aperture of about 133.3 mm × 33.0 mm (5.25 in. × 1.30 in.). Since a vacuum-tight seal is required, a gasket of indium is compressed by a spring-loaded piston onto the heavily beveled rim of the window as shown in Figure 4.6. Deflection of the spring plate provides sufficient total preload of 2.35 kN (530 lbf) to hold the window in place at all temperatures between 77 and 373 K and to create a peak compressive stress in the indium of about 8.3 MPa (1200 lb/in.^{2}).

The spring plate is slit radially at its inner boundary to distribute the force evenly around the edge of the window. Titanium is used for the spring because of its low α, high Young's modulus, and high yield strength. The window frame is Nilo 42 ($Ni_{42}Fe_{58}$), which approximates the α of germanium. The piston is aluminum for ease of fabrication to its unusual shape.

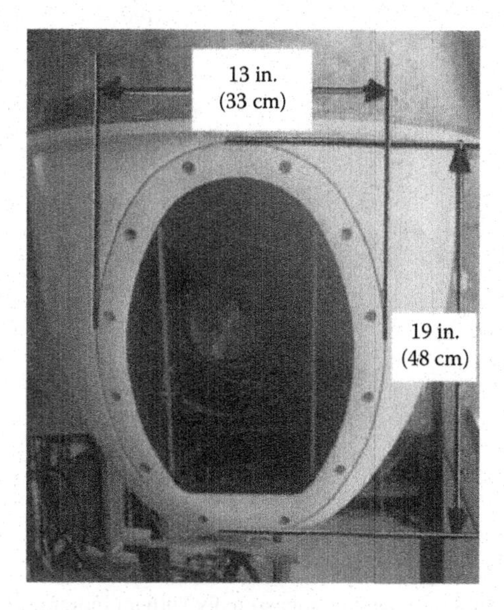

FIGURE 4.3 Elliptically shaped laminated glass window used in an LLLTV system mounted on an aircraft pod. (Courtesy of Goodrich Corporation, Danbury, CT.)

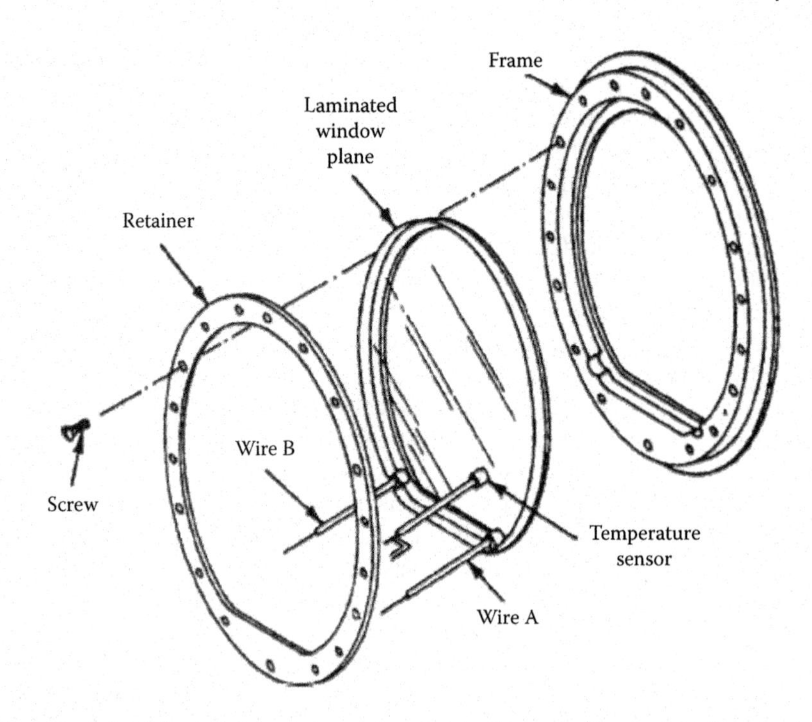

FIGURE 4.4 Exploded view of the window subassembly shown in Figure 4.3. (Courtesy of Goodrich Corporation, Danbury, CT.)

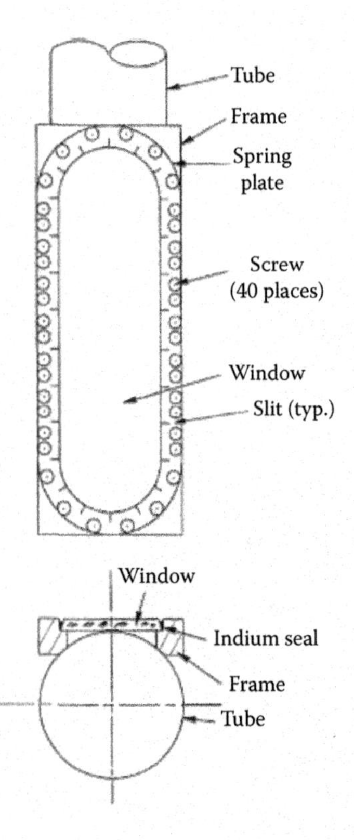

FIGURE 4.5 Plan and end views of a window subassembly with an indium seal to withstand vacuum cryogenic conditions. (Adapted from Haycock, R.H. et al., A compact indium seal for cryogenic optical windows, *Proc. SPIE*, 1340, 165, 1990.)

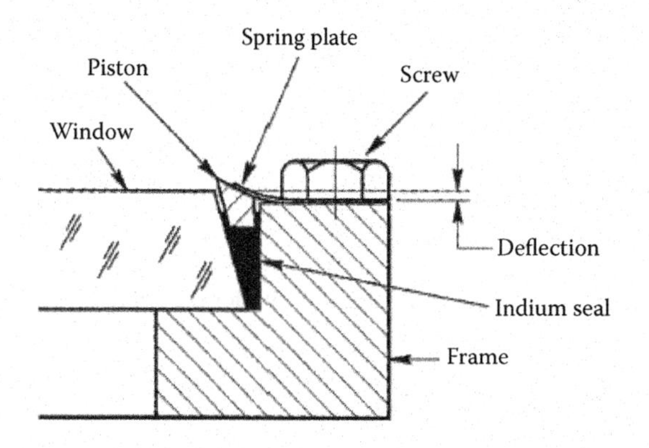

FIGURE 4.6 Detail view of the pressure-loaded indium seal for the window subassembly shown in Figure 4.5. (Adapted from Haycock, R.H. et al., A compact indium seal for cryogenic optical windows, *Proc. SPIE*, 1340, 165, 1990.)

One important design parameter for this window is the width at the narrower end (bottom) of the triangular gap into which the indium is pressed. A small dimension is needed to maintain the required pressure within the seal, but if too small, it would be difficult to assemble the seal with complete packing of the indium into the available volume. A dimension of 254 μm (0.010 in.) was satisfactory for this application. Another critical dimension is the gap on either side of the piston under spring load. A value of 25 μm (0.001 in.) was found to be appropriate to sufficiently minimize extrusion of the indium at higher temperatures so that the seal would remain intact over periods on the order of 1 week. Testing of the assembly at cryogenic temperature indicated that it was leak-proof to the accuracy of the test apparatus (approximately 10^{-10} standard atm cm^3/s) during and after repeated cycling (>200 cycles) throughout the temperature range of 293–77 K.

4.3 DOMES

4.3.1 Dome Strength

Meniscus-shaped shells are used as windows for applications requiring wide fields of view, those requiring an ability to scan a line of sight over a large conical space, or protective windows for underwater systems. Domes are deep shells and are defined by the angle between the axis and dome edge. For a spherical segment, the angle is less than 180°; for a hemisphere, the angle is exactly 180°; and hyperhemispheres are domes that extend beyond 180°.

The two most common dome loading conditions are uniform external pressure and uniform pressure loading on the projected area of the dome. Underwater applications subject a dome to uniform external pressure. Air drag on a missile dome is approximated by uniform pressure loading on the projected area of the dome. Very high accelerations (11,000*g*) and shocks are associated with military applications such as gun-launched projectiles. High stagnation temperatures (1500 K) produced by air friction at supersonic speeds are encountered in missile domes. These latter two loading conditions are uncommon, requiring advanced methods for design analysis and are not discussed here.

To minimize stress, the dome mounting must provide tangential support. This requires a mount with a conical surface or socket, with the apex of the conical mounting surface coincident with the center of curvature of the dome. The dome edge is made to the same conical shape. For the design equations given in the next paragraph, this type of edge mount is assumed. In high-acceleration applications, it is common to make the dome edge flat and perpendicular to the axis.

Finding the minimum thickness for a dome under uniform external pressure loading depends on the ratio of dome radius r to dome thickness h, or r/h.[10] If $r/h > 10$, the dome is considered thin, while if $r/h \leq 10$, the dome is thick. In preliminary analysis, the dome is considered thin if

$$\frac{2\sigma_{YS}}{SF\ P} > 10. \tag{4.16}$$

For a thin dome, the minimum wall thickness h_{min} under external pressure P is given by

$$h_{min} = \frac{SF\ Pr}{2\sigma_{YS}}. \tag{4.17}$$

Usually the inner radius of the dome r_i is defined by the swept volume of the optical system. If the dome is thick, $r/h \leq 10$, the minimum thickness h_{min} is found by solving for r_o:

$$\sigma_{YS} = \frac{3SF\ Pr_o^3}{2\left(r_o^3 - r_i^3\right)}, \tag{4.18}$$

where : r_o is the outer radius of the dome and r_i is the inner radius of the dome.
Then, $h_{min} = r_o - r_i$.

Example 4.2

A fused silica dome with radius $r = 75$ mm is used underwater, at a depth of 200 m, where $P =$ 2.1 MPa. What is the minimum dome thickness h_{min}, using the fracture strength of 60 MPa from Table 4.1, if SF = 4?

Determine if the dome is thin, using Equation 4.17:

$$\frac{2(60\ \text{MPa})}{4(2.1\ \text{MPa})} = 14.2 > 10, \text{ dome is thin.}$$

The minimum dome thickness h_{min} is given using Equation 4.18:

$$h_{min} = \frac{4(2.1\ \text{MPa})(75\ \text{mm})}{2(60\ \text{MPa})} = 5.3\ \text{mm}.$$

Figure 4.7 shows the geometry associated with air drag on a dome, assuming that the flight direction is coincident with the dome axis. The pressure loading is also assumed to be uniform on the projected area. The approximate uniform pressure as a worst case is assumed to be equivalent to the stagnation pressure P_{ST} at the front of the dome. Stagnation pressure is a function of the ambient pressure P_∞ and dimensionless flight Mach number M. The Mach number is the ratio of flight velocity v to the speed of sound c_a. Note that the Mach number at constant flight velocity varies with the change in speed of sound at altitude. The stagnation pressure is given by

$$\frac{P_{ST}}{P_\infty} = \left(\frac{\gamma+1}{2}M^2\right)^{\frac{\gamma}{\gamma-1}} \left(\frac{2\gamma}{\gamma+1}M^2 - \frac{\gamma-1}{\gamma+1}\right)^{-\frac{1}{\gamma-1}}, \tag{4.19}$$

where γ is the specific heat ratio ($\gamma \approx 1.4$ for air).

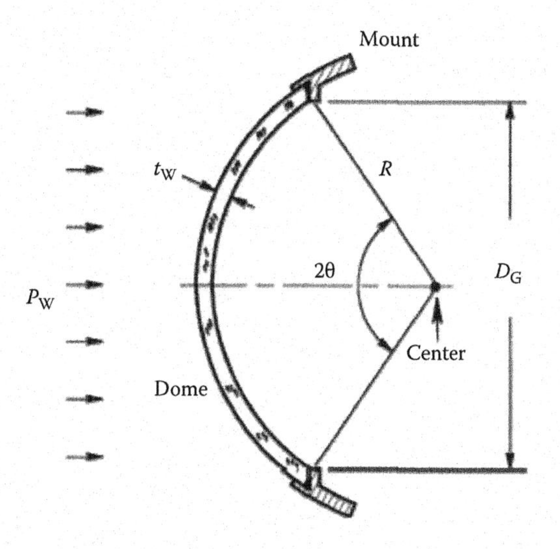

FIGURE 4.7 Geometry of a dome with a simply supported base subjected to a pressure differential. (Adapted from Harris, D.C., *Materials for Infrared Windows and Domes, Properties and Performance*, SPIE Press, Bellingham, WA, 1999.)

Equation 4.19 is plotted in Figure 4.8 as a function of altitude from sea level to 20 km and for flight velocities $1 \le M \le 5$.

Dome stress depends on the edge constraint. If the dome is simply supported, the dome edge is free to rotate and displaces in the radial direction. Tensile stress is associated with a simply supported dome edge. If the dome edge is constrained against rotation and radial displacement, the dome edge is fixed and the stress is compressive. For brittle optical materials, failure is much more likely with tensile stress than compressive stress. Fixing the dome edge, typically with a raised lip around the circumference, is one way of decreasing failure probability under load. A minimum SF = 4 is recommended for dome stress.

From Figure 4.7, θ is the half angle of the dome. If P_i is the interior pressure of the dome, the minimum dome thickness h_{min} for the simply supported edge is found by solving

$$\frac{\sigma_{YS}}{SF(P_{ST}-P_i)} = \frac{r}{2h_{min}}\left\{\left[1.6+2.44\sqrt{\frac{r}{h_{min}}}\sin(\theta)\right]\cos(\theta)-1\right\}. \tag{4.20}$$

For Equation 4.20 to hold, the following condition must be satisfied:

$$\frac{\sin^2(\theta)}{12} \le \frac{h_{min}}{r} \le \frac{\sin^2(\theta)}{1.2}. \tag{4.21}$$

For preliminary estimates, assuming agreement with the condition given by Equation 4.21, the minimum thickness h_{min} is given for the simply supported edge approximately by

$$\frac{\sigma_{YS}}{SF(P_{ST}-P_i)} = 0.581\left(\frac{h_{min}}{r}\right)^{-\frac{3}{2}}. \tag{4.22}$$

When the dome edge is fixed, with radial constraint, the minimum thickness h_{min} is given by

$$h_{min} = 1.2\frac{SF(P_{ST}-P_i)r}{\sigma_{YS}}. \tag{4.23}$$

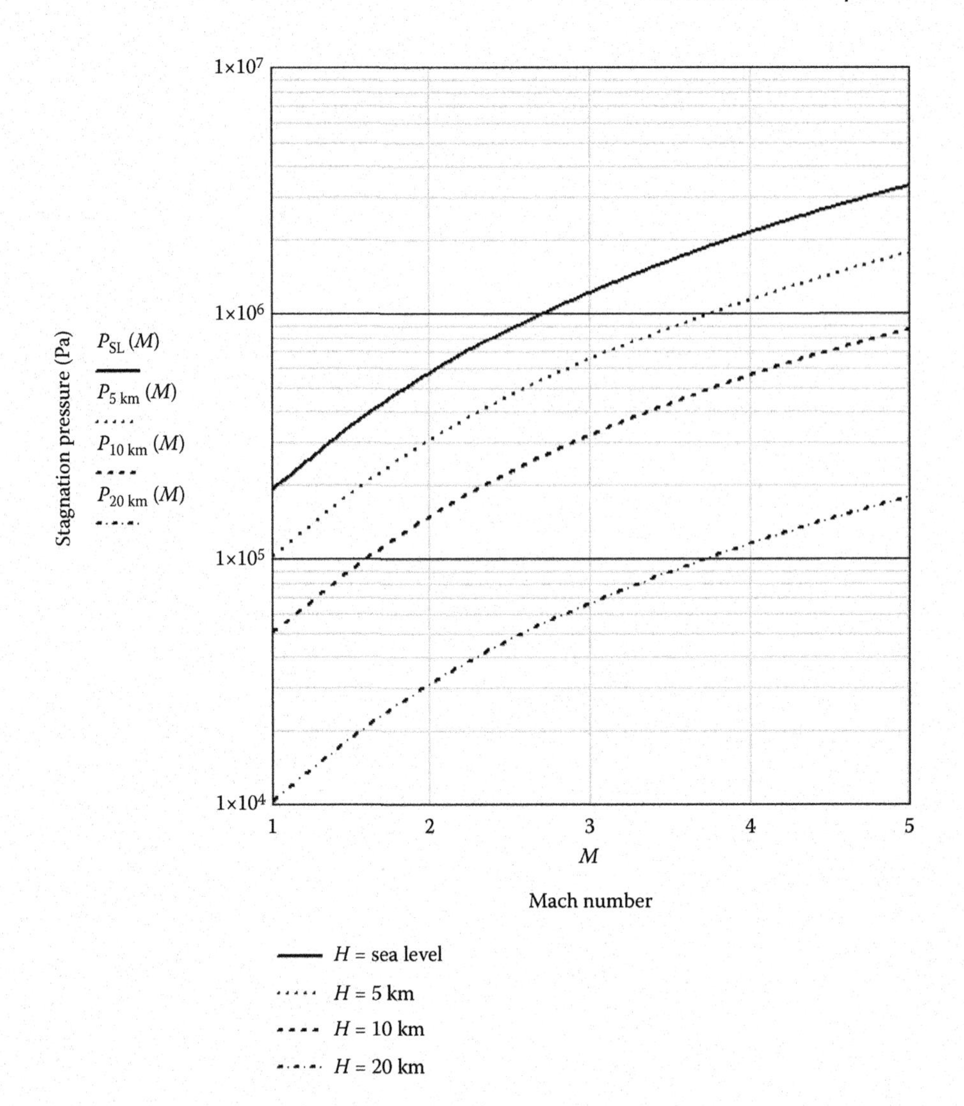

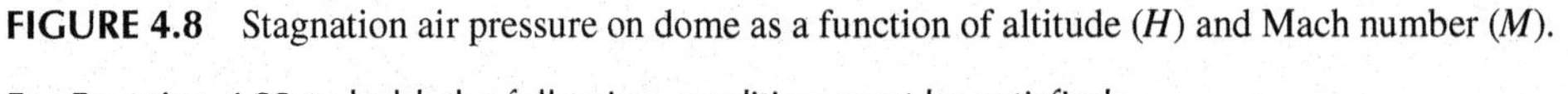

FIGURE 4.8 Stagnation air pressure on dome as a function of altitude (H) and Mach number (M).

For Equation 4.23 to hold, the following condition must be satisfied:

$$\frac{\sin^2(\theta)}{12} \leq \frac{h_{min}}{r} \leq \frac{\sin^2(\theta)}{3}. \tag{4.24}$$

Example 4.3

A simply supported ZnSe dome with radius $r = 50$ mm is on the front of a missile. The dome apex angle $2\theta = 60°$ and the missile interior is vented so that $P_i = P_\infty$. The missile flight velocity is $M = 3.0$ at an altitude of 10 km, where $P_\infty = 26.4$ kPa. Using the fracture strength of $\sigma_{YS} = 50$ MPa from Table 4.1 and SF = 4, what is the approximate minimum dome thickness h_{min}?

From Equation 4.19, the stagnation pressure P_{ST} on the dome is

$$P_{ST} = 26.4\text{ kPa}\left[\frac{1.4+1}{1.4}(3)^2\right]^{\frac{1.4}{1.4-1}}\left[\frac{2(1.4)}{1.4+1}(3)^2 - \frac{1.4-1}{1.4+1}\right]^{\frac{1}{1.4-1}} = 318\text{ kPa}.$$

Solving Equation 4.22 for h_{min},

$$h_{min} = \frac{50\,\text{mm}}{\left[\dfrac{50\,\text{MPa}}{0.581(4)(318\,\text{kPa} - 26.4\,\text{kPa})}\right]^{\frac{2}{3}}} = 2.85\ \text{mm}.$$

Checking for applicability of Equation 4.22, h_{min}/r = 2.85 mm/50 mm = 0.057 and

$$\frac{\sin(30°)^2}{12} = 0.021 \quad \text{and} \quad \frac{\sin(30°)^2}{1.2} = 0.208.$$

Since $0.057 \leq h_{min}/r = 0.057 \leq 0.208$, Equation 4.23 is valid. Note that the numerical solution of Equation 4.20 gives a value of h_{min} = 2.82 mm, suggesting that the approximation is conservative.

4.3.2 Dome Mounting

A photograph of a mounted hyperhemisphere is shown in Figure 4.9. The outside diameter of this dome is 127 mm (5.0 in.), the dome thickness is 5 mm (0.2 in.), and the angular aperture is approximately 210°. The subassembly shown in this figure is made of crown glass, but many domes used for military applications are made of IR-transmitting materials.

In mounting domes, especially for underwater applications, the angles of the conical mounting surface and dome edge should be matched to a tolerance of about 1 arcmin (≈ 300 μrad). In addition, the apex location of the conical socket should be within $0.001r_0$ of the center of curvature of the dome, where r_0 is the dome outer radius. A minimum 32 RMS surface finish is required on both dome and socket surfaces.

Mounting tolerances are relaxed by bonding the dome in place. A typical bond line thickness is 125 μm (0.005 in.), and both epoxy and RTV adhesives are common. Placing the bond around the outer dome circumference, between the mount and dome, provides isolation from the stress induced by differences in thermal expansion between the dome and mount. Extending the bond line below the dome permits an increase in mounting tolerance. However, for high-acceleration applications such as gun-launched projectiles and missiles, direct contact between the dome and mount is desirable to minimize rebound from the compressibility of the adhesive. A flat dome edge and mounting

FIGURE 4.9 Photograph of a crown glass hyperhemispherical dome potted with elastomer into a metal flange.

surface, normally to a flatness tolerance ≈ 2.5 μm (0.0001 in.) or better, is necessary to reduce contact stress during acceleration.

Mountings for shells and domes usually involve (1) attaching them to a metallic ring-shaped flange or directly to an instrument housing by potting them in place with elastomers, (2) mechanically clamping them in place and sealing them with a gasket or elastomer, and (3) brazing the optic to the housing. Figure 4.10 illustrates three typical mounting configurations of types 1 and 2. The view in Figure 4.10a shows a dome constrained by a ring-shaped flange that acts through a soft Neoprene gasket to seal the interface. The view in Figure 4.10b shows a dome bonded to a flange or bezel by using epoxy and backed up axially by a retaining ring. The view in Figure 4.10c shows a hyperhemisphere potted with elastomer into a ring-shaped mount in the manner illustrated by Figure 4.9.

Figure 4.11 shows schematically a military missile with a deep dome attached to the front end to protect a radar transceiver or an IR or visible light sensor used for target acquisition and homing.

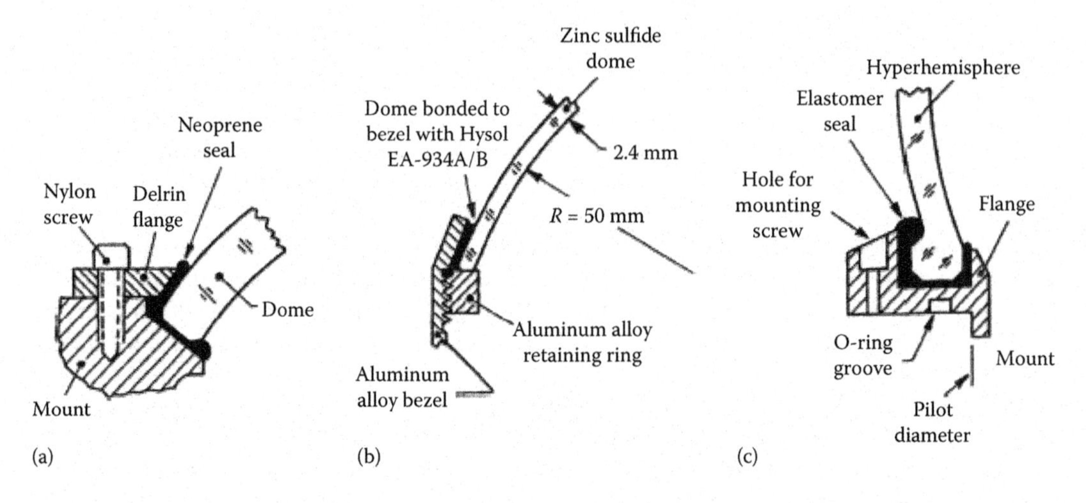

FIGURE 4.10 Three configurations for dome mountings. (a) Dome clamped through a soft gasket with a flange. (Adapted from Vukobratovich, D., Introduction to opto-mechanical design, SPIE short course, 2014.) (b) Dome constrained by an internal retainer. (Adapted from Speare, J., and Belioli, A., *Proc. SPIE*, 450, 182, 1983.) (c) Hyperhemisphere secured and sealed with elastomer.

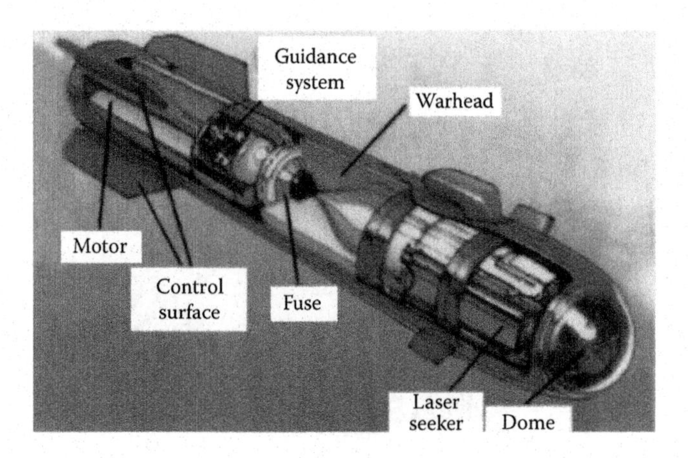

FIGURE 4.11 Schematic side view of a generic rocket-propelled missile with a dome-shaped window at the forward end that protects an internal homing sensor.

At the rear of the missile are the rocket motor and fins for stabilizing and guiding the weapon. The dome serves as a window transmitting the appropriate radiation to the sensor and as an aerodynamic structure while the missile is in flight. The missile body is essentially cylindrical. It contains the sensor and its electronics, a guidance controller, motors for driving the fins, a warhead, and fuel for the rocket motor. When the velocity of flight is subsonic, the dome may be made of optical plastic, glass, germanium, or some other crystals, depending on the wavelength used. For applications involving supersonic velocities, airstream friction raises the temperature of the dome to a very high value. This temperature (called stagnation temperature) occurs at relatively low altitudes because the transfer of heat from the air to the dome material becomes smaller as the air density decreases with altitude. IR radiation from this hot material is determined, in part, by its emittance, which measures the fraction of the radiation that would radiate from a blackbody if it were at the temperature of the dome. When the dome surface becomes hot in an IR sensor, the emission from that hot body will reduce the signal-to-noise ratio of the sensor or, in extreme cases, saturate the sensor so it becomes useless. Materials frequently used for such applications involving wavelengths of 3–5 μm are ALON, sapphire, and spinel.

PROBLEMS

1. Assuming a maximum OPD tolerance of $\lambda/8$, $\lambda = 550$ nm, what is the minimum thickness of a simply supported right circular cylinder fused silica window with a 150 mm unsupported diameter subjected to a pressure differential of 100 KPa? Consider optical deformation only. For fused silica at $\lambda = 550$ nm, $n = 1.460$ and $E = 73$ GPa. *Answer:* $h_{min} = 11.8$ mm
2. A simply supported plane-parallel circular window made of standard ZnS is used at a wavelength of 10.6 μm. The unsupported window diameter is 200 mm, and there is a 75 kPa pressure differential through the window. What is the minimum window thickness, using a SF of 4, including a Poisson's ratio of 0.29 for ZnS, if the fracture strength is 100 MPa? What is the OPD for a window of the minimum thickness required for strength, ignoring focus changes? For ZnS at 10.6 μm, $n = 2.192$. Use E = 74.5 GPa. *Answer:* $h_{min} =$ 6.08 mm, OPD = 0.137 μm
3. A plane-parallel circular ZnSe window with an unsupported clear aperture of 25 mm transmits a high-power laser beam for cutting; the beam power is 1 kW at a wavelength of 10.6 μm. If the window absorbs 0.5% of the transmitted laser energy, what is the FL of the distorted window? For ZnSe at $\lambda = 10.6$ μm, $n = 2.403$, $\alpha = 7 \times 10^{-6}$ K^{-1} and $\kappa = 18$ W/m K. Assume that the laser energy absorption is uniform across the diameter. *Answer:* FL = 4.37×10^6 m
4. Consider a simply supported right circular cylinder window with an unsupported diameter of 300 mm, subjected to a pressure differential of 100 KPa. Is the minimum thickness determined by strength or optical deformation? Assume an SF of 4 and a maximum OPD of $\lambda/8$, with focus included. Use $n = 1.5$ at $\lambda = 0.55$ μm, $E = 82$ GPa, and a fracture strength of 10 MPa and neglect Poisson's ratio effects. *Answer:* The minimum window thickness set by strength is 33.5 mm, while for optical performance, it is 16.0 mm; hence, the minimum thickness is set by strength.
5. If focus is removed, what is the ratio of thickness for simply supported and fixed windows, assuming identical materials and the same OPD and pressure differential? The answer should be in equation form, with a correction for Poisson's ratio. For a Poisson's ratio of 0.3, what is the thickness ratio? *Answer:* $h_{simply\ supported} = h_{fixed}((\nu + 5)/(\nu + 1))^{1/5}$; for $\nu = 0.3$, $h_{simply\ supported} = 1.325 h_{fixed}$
6. A 25 mm thick elliptical plane-parallel window is subjected to a fourth-order in-plane temperature gradient with a maximum temperature differential from center to edge of 30 K. The window material is fused silica with $n = 1.460$, $\alpha = 580 \times 10^{-9}$ K^{-1}, and $dn/dT = 10 \times 10^{-6}$ K^{-1}. What is the estimated worst-case OPD? *Answer:* OPD = 8.14 μm
7. Consider a plane-parallel circular window subjected to a parabolic in-plane temperature gradient. Comparing Schott BK-7 and fused silica, which material gives the longer

equivalent FL from the temperature gradient? For fused silica, use the properties given in problem 6, with $\nu = 0.17$. For BK-7, $n = 1.518$, $\nu = 0.21$, $dn/dT = 3.0 \times 10^{-6}$ K^{-1} and $\alpha = 8.2 \times 10^{-6}$ K^{-1}. *Answer:* The equivalent FL of the BK-7 window is 1.27 times longer than that of the fused silica window.

8. A 12 mm thick simply supported sapphire window is subjected to a pressure differential of 500 kPa. The window is elliptical, with a 100 mm minor axis and 150 mm major axis. The fracture strength of the window is 300 MPa and $\nu = 0.27$. Is the window thick enough to withstand the pressure differential, with an SF of 10? *Answer:* With SF = 10, h_{min} = 8.89 mm < 12 mm, so the window is safe.
9. A polished plane-parallel rectangular fused silica window is 25 mm thick, with unsupported length of 150 mm and width of 100 mm. The edge condition is fixed, and the window is subjected to a 500 kPa pressure differential. With an SF of 10, what is the probability of failure of the window under load? Use the Weibull data in Table 3.3, and assume that only one side is stressed. *Answer:* $P_F = 2.365 \times 10^{-3}$
10. A germanium dome with radius r = 100 mm is used underwater, where P = 3 MPa. What is the minimum dome thickness, with fracture strength of 90 MPa and SF of 4? *Answer:* h_{min} = 6.67 mm
11. A germanium dome is used underwater at a depth where the pressure is 5 MPa. The internal swept radius of the dome is 100 mm. Is the dome thick or thin? What is the minimum dome thickness with an SF of 4 if the fracture strength of germanium is 90 MPa? *Answer:* The dome is thick, and h_{min} = 14.5 mm.
12. A sapphire dome is on the front end of a missile flying at M = 5 at an altitude of about 10 km, where the pressure is approximately 26.4 kPa. The included angle of the apex of the dome $2\theta = 60°$, and its radius is 75 mm. The dome edge is fixed and the missile interior is vented to the exterior. The fracture strength is 300 MPa, and an SF of 8 is required. What is the minimum dome thickness? *Answer:* h_{min} = 2 mm

REFERENCES

1. Sparks, M., and Cottis, M., Pressure-induced optical distortion in laser windows, *J. Appl. Phys.*, 44, 787, 1973.
2. Barnes, W.P. Jr., Some effects of aerospace thermal environments on high-acuity optical systems, *Appl. Opt.*, 5, 701, 1966.
3. Harris, D.C., *Materials for Infrared Windows and Domes*, SPIE Press, Bellingham, WA, 1999.
4. ASME (American Society of Mechanical Engineers), *Safety Standard for Pressure Vessels for Human Occupancy*, ASME PVHO-1-2007, ASME, New York, 2007.
5. NASA (National Aeronautics and Space Administration), *Structural Design and Test Factors of Safety for Spaceflight Hardware*, NASA-STD-5001B, NASA, Washington, DC, 2014.
6. Timoshenko, S., and Woinowsky-Krieger, *Theory of Plates and Shells*, 2nd ed., McGraw-Hill, New York, 1959.
7. Young, W.C., and Budynas, R.G., *Roark's Formulas for Stress and Strain*, 7th ed., McGraw-Hill, New York, 2002.
8. *Sealing Compound, Adhesive: Curing (Polysulfide Base)*, MIL-S-1103 I B, US Government Printing Office, Washington, DC, 1998.
9. Haycock, R.H., Tritchew, S., and Jennsion, P., A compact indium seal for cryogenic optical windows, *Proc. SPIE*, 1340, 165, 1990.
10. Klein, C.A., Infrared missile domes: Is there a figure of merit for thermal shock? *Proc. SPIE, 1760,* 338, 1992.

FURTHER READING

Harris, D.C., *Materials for Infrared Windows and Domes: Properties and Performance*, SPIE Press, Bellingham, WA, 1999.

5 Mounting Individual Lenses

5.1 INTRODUCTION AND SUMMARY

This chapter covers ways of mounting individual axisymmetric lenses ~10 to ~250 mm (~0.4 to ~10 in.) in diameter. Mounting techniques for multiple-lens assemblies are discussed in Chapter 6.

The preferred design is one in which the effects of manufacturing and assembly tolerances are minimal and the location and orientation of each optic are maximally benign to adverse environments. These conditions are hard to achieve in many designs—increasingly so as performance demands increase.

First discussed is what is meant by "centered" optics and how centration is achieved. Next are methods of estimating the weight of typical lens elements and locating their centers of gravity (CGs). Techniques for preloading individual refracting elements axially are then addressed. Some designs provide direct metal-to-glass interfaces at or near the rim of the polished surfaces of the lens, while others encapsulate the lens rim locally or continuously with resilient elastomeric material. Different mechanical contours for the interfaces with the lens surfaces are considered along with the localized contact stresses in the glass and metal parts resulting from preloads. Some typical means of aligning individual lenses in their mount are also summarized.

Development is from simple, low-precision designs to more complex, high-precision mounting techniques. The emphasis in this chapter is on optics made of glass-type materials, although many of the mounting techniques discussed are also applicable to crystalline and plastic lenses. Mountings specific to plastic optics are described at the end of this chapter.

5.2 CENTERING OPTICS

Because of the inherent rotational symmetry of spherical surfaces and their aberrations, lens design begins by defining a straight line in space and locating all surfaces with optical power symmetrically about that line. If the centers of curvature of an ensemble of such surfaces lie on this same line, the line is the optical axis and the system is centered. Figure 5.1a illustrates a perfectly centered biconvex lens element. The centers of curvature C_1 and C_2 of the first and second lens surfaces, respectively, define the optical axis of the lens. The rim of the lens is nominally cylindrical with its axis on the system axis.

A lens with a flat (i.e., plano) surface is shown in Figure 5.1b. In Figure 5.1b, the plano-convex lens is tilted with respect to an arbitrarily oriented line A–A' (dashed line). This line may be the mechanical axis of a cell into which the lens is to be mounted. The optical axis of the lens is defined as the line passing through the center C_1 of surface 1 and perpendicular to surface 2. Symmetry exists only about this axis. Systems with intentionally tilted surfaces such as wedges or those that need asymmetry for aberrational correction reasons cannot be considered to be centered or rotationally symmetric. These asymmetric systems are not discussed here.

The preceding definition of a centered system also applies to folded assemblies and systems in which the optical axis is deviated through a constant angle by reflection from one or more intentionally tilted reflecting surface(s). An example of an optical instrument with such a fold is a visual astronomical telescope of the Newtonian form. The folding mirror is generally located a distance X inside the focus of the primary mirror, as indicated in Figure 5.2. The optical axis of the parabolic primary passes through the center C of its vertex sphere and the optical vertex A of the mirror, as defined by the symmetry of its aspheric contour. The optical axis after reflection is defined as the line connecting the reflected location A' of the primary vertex (serving as a virtual object) and the

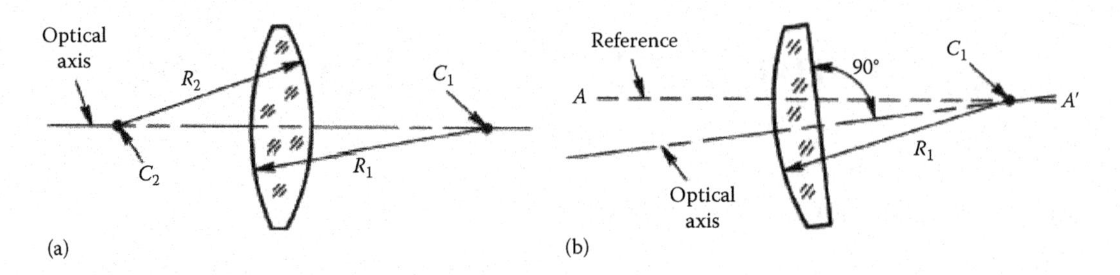

FIGURE 5.1 (a) Perfectly centered biconvex lens. The centers C_1 and C_2 of surfaces with radii R_1 and R_2 define the optical axis of the lens. (b) Plano-convex lens tilted with respect to the mechanical reference line A–A' passing through the convex surface center C_1. The optical axis of this lens passes through C_1 and is perpendicular to the plano surface.

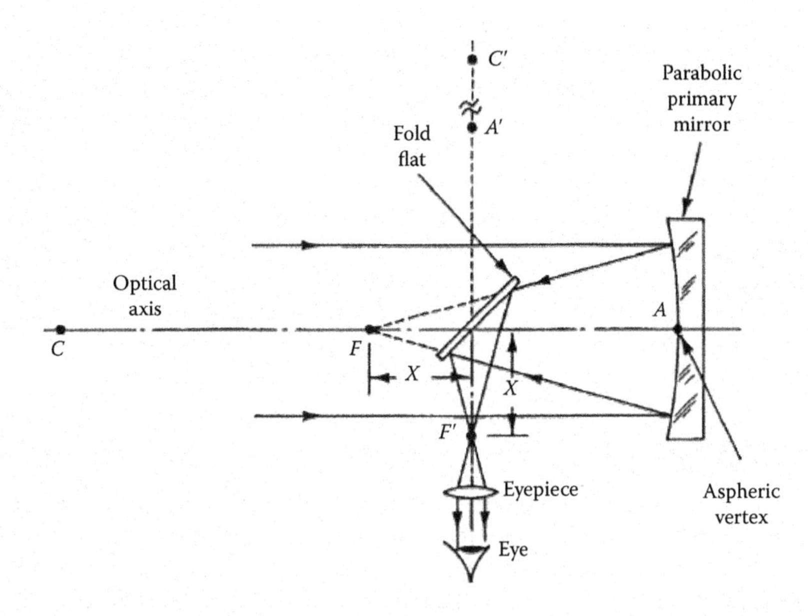

FIGURE 5.2 The Newtonian telescope is an example of a folded optical system with centered optics.

reflected location C' of the center of curvature of the primary. If the optical surfaces of the eyepiece are centered about this reflected optical axis, the entire optical system is centered.

Figure 5.3 shows an unobscured-aperture Cassegrain telescope in which the optical surfaces of the mirrors are made from aspheric parent surfaces (dashed lines). The latter surfaces are centered to the optical axis so that the system is centered. The entrance pupil of the system is located off the optical axis, so the design is called a decentered-pupil system, or an eccentric-pupil system.

Practically all lens elements have cylindrical rims produced by edge-grinding the element after the surfaces are polished. This cylinder defines a mechanical axis of the lens that may or may not coincide with the lens' optical axis, depending on how the latter axis is oriented while the rim is being ground. Figure 5.4 shows a biconvex lens mounted on the bell (or chuck) of a typical precision lens-edging machine. The lip of the bell is rounded to a smooth toroidal shape, and the lens is generally secured to the bell with an adhesive such as wax or pitch, as shown in the detail view. In other types of machines, the lens is held with a vacuum chuck. Warming the adhesive to soften it or partially releasing the vacuum and judiciously pushing the lens sideways until it is centered aligns the lens to the rotational axis of the machine. Care must be exercised to ensure that the glass remains in contact with the bell throughout the edging process. This ensures that the center of curvature of

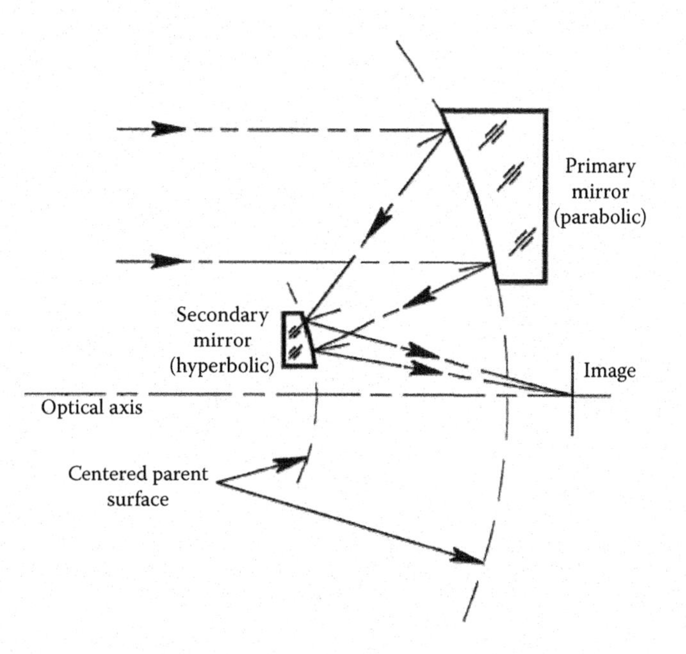

FIGURE 5.3 Unobstructed-aperture Cassegrain telescope with mirror segments made from centered aspheric parent surfaces. This is sometimes called an off-axis (or preferably, an eccentric-pupil) system.

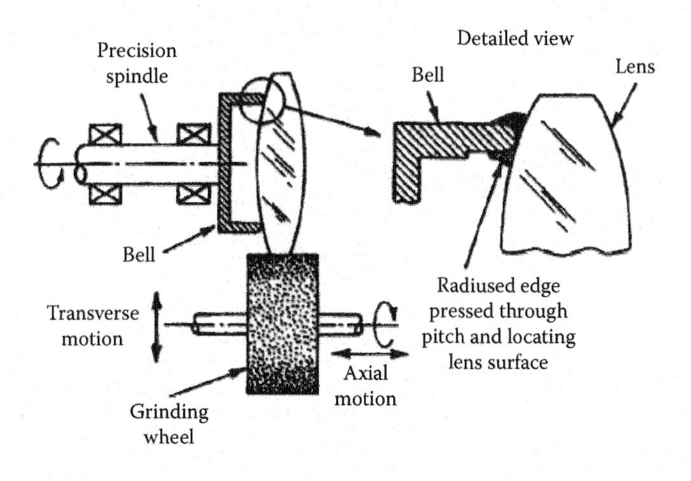

FIGURE 5.4 Typical setup for edging a lens element. The detail view shows one means for attaching the lens to the bell of the machine. (Adapted from Yoder, P.R. Jr., *Mounting Optics in Optical Instruments*, 2nd ed., SPIE Press, Bellingham, WA, 2008.)

that lens surface lies on the rotational axis of the machine. The exposed surface will wobble as the bell rotates if its center of curvature is not on the rotational axis.

Figure 5.5a illustrates a lens decentered on an edging machine bell. If the lens is edged under this condition, it may be considered to have an integral geometric wedge of angle $\theta = w/2h$, where w is the full indicator movement (runout) of the surface at the height h. This wedge will deviate a transmitted beam approximately by an angle $\delta = (n - 1)\theta$. Here, n is the glass refractive index. Both θ and δ are in radians. To convert these angles into minutes of arc, multiply by 3438. This optical wedge is frequently given a tight tolerance in the specification of the lens.

Various methods can be utilized for determining when the exposed surface of the lens is properly centered so that the edging process can begin. The simplest mechanical means is illustrated in

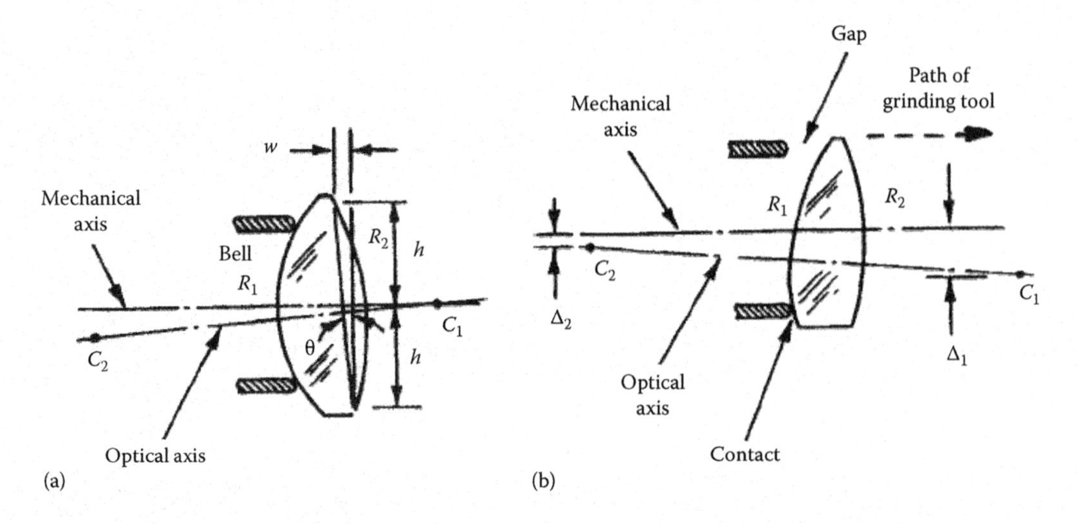

FIGURE 5.5 Typical (exaggerated) misalignments of the lens during edging. (a) C_1 on and off C_2 off the rotational axis; (b) both C_1 and C_2 off the rotational axis by unequal amounts on the plane of the figure.

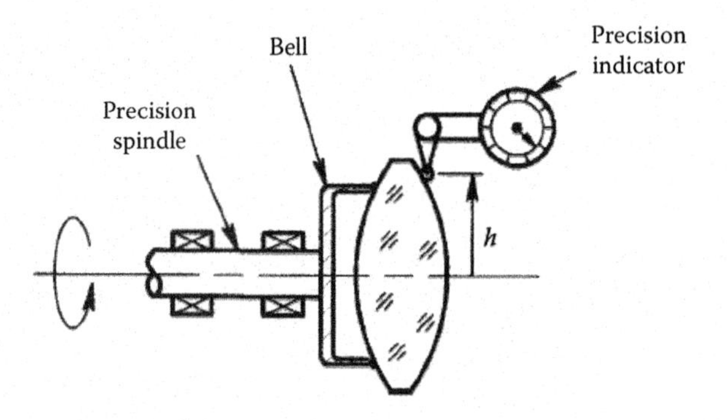

FIGURE 5.6 Technique for measuring the centering error of a lens element on a spindle. The dial indicator shows the full indicator movement (runout) of the lens for each revolution.

Figure 5.6. The tip of a precision dial gauge gently touches the exposed polished surface at radius *h* outside its clear aperture. As the bell is rotated slowly, the gauge needle will oscillate, showing the wobble of surface. What is measured after one complete revolution is the total mechanical wobble *w* of the surface. An electronic indicator can be used in a similar manner.*

In the schematic test arrangement of Figure 5.7,[2] a point light source is set up at some convenient distance from the exposed lens surface and the real or virtual image of the source formed by reflection from that lens surface is observed by the unaided eye or with optical aid (loupe, microscope, or telescope, as appropriate for the existing image location). If the image appears to move as the bell is rotated slowly, the lens is decentered.

Another technique for measuring the runout of the decentered lens surface is sketched in Figure 5.8. A collimated light beam from a crosshair reticle pattern is directed through a beam splitter and focused by an auxiliary lens at the center of curvature of the decentered surface. A portion of the beam reflects from the decentered surface back through the auxiliary lens where it is recollimated. Within the cube, a portion of the light reflects into an observation telescope containing another crosshair reticle. If the reflected crosshair image does not coincide with the telescope reticle, the

* Bayar[1] reported that a runout of 5 μin. (0.13 μm) peak to peak can be measured with a high-quality capacitance gauge.

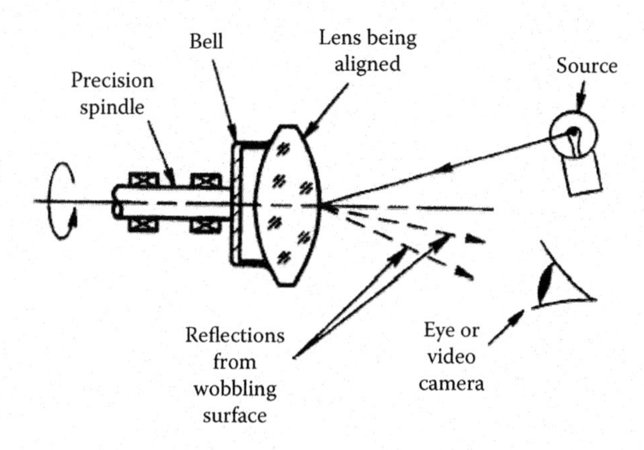

FIGURE 5.7 Means for observing centering errors by observing movements of a reflection from the exposed surface of the rotating element. (From Yoder, P.R. Jr., *Mounting Optics in Optical Instruments*, 2nd ed., SPIE Press, Bellingham, WA, 2008.)

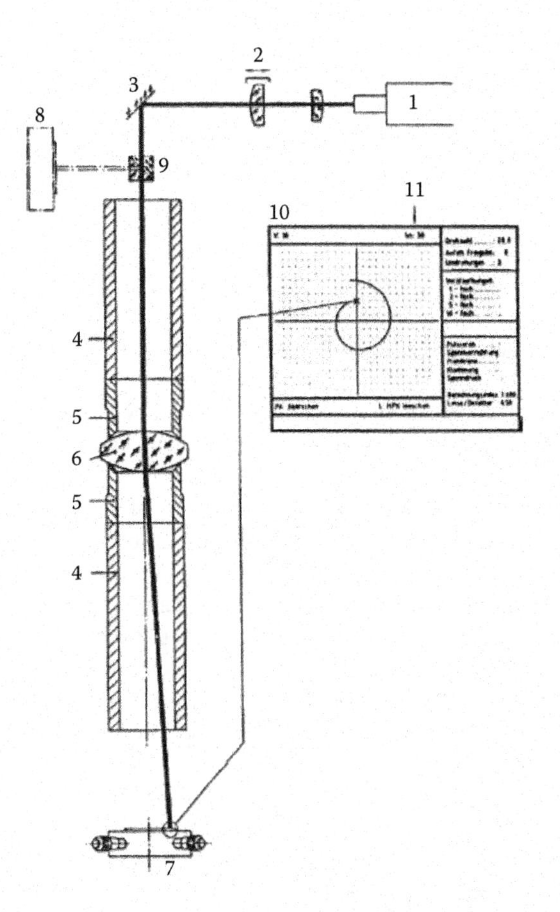

FIGURE 5.8 Optomechanical configuration of a commercial sensor used to measure lens centration errors by reflection or transmission during edging. 1, laser; 2, focusing system; 3, reflecting mirror; 4, centering spindle; 5, centering mandrel; 6, lens; 7, detector for transmission; 8, detector for reflection; 9, reflecting prism (reflection method); 10, monitor, laser mask; 11, surface angel of tilt(s). (Courtesy of LOH Optical Machinery, Inc., Milwaukee, WI.)

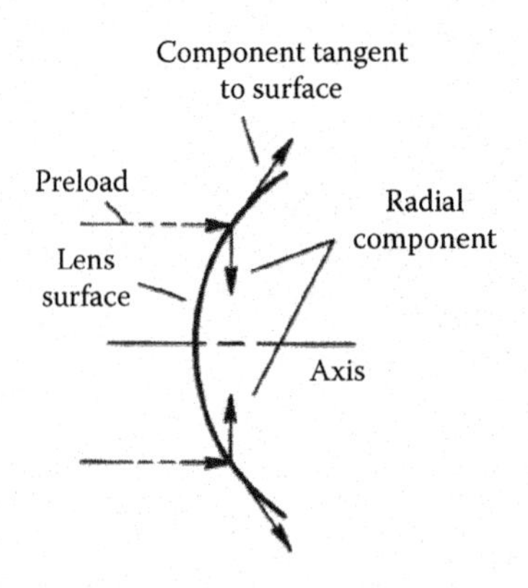

FIGURE 5.9 Schematic showing equal radial force components created by preload against a curved lens surface when the lens is perfectly aligned with the axis. If the lens becomes tilted or decentered, the radial components on the plane of the misalignment become unequal. This tends to realign the lens.

exposed lens surface is decentered. In this figure, the lens is clamped between two opposing centering bells on coaxial spindles, so no adhesive or vacuum is needed to hold it in place. If the lens is tilted, an axial compressive force exerted on the curved surfaces of the lens around its periphery creates unequal radial force components that tend to squeeze the lens toward its centered position, which is illustrated schematically in Figure 5.9.

When centering a lens with this type of apparatus, the factor Z in the following equation should not exceed 0.56:

$$Z = \frac{2y_{C1}}{R_1} - \frac{2y_{C2}}{R_2}, \tag{5.1}$$

where y_{C1} and y_{C2} are the bell contact heights from the axis and R_1 and R_2 are the optical radii of the lens.

Lenses with long radii (i.e., those with $Z > 0.56$) that cannot be adequately centered by the opposing bell process must be aligned manually by external mechanisms applying radial force(s) to slide the lens on the bell.

An interferometric technique for detecting decentration of the lens on the centering bell is sketched in Figure 5.10. A laser beam is directed through a beam splitter and through a plano-concave test plate located near the decentered surface. The curved surface of the plate must have a radius approximating that of the decentered surface. Two beams, one from each of the adjacent surfaces, are reflected back to the beam splitter, which reflects portions of both beams into the eye or, preferably, a video camera.* Interference effects within the cavity formed by the adjacent surfaces cause fringes to appear on a monitor. As the bell rotates, these fringes move on the monitor. A mark on the surface of the monitor can serve as a fixed reference. The lens centration is adjusted until the fringes appear to remain stationary.

Once the lens is aligned and secured in place on the bell, its cylindrical rim is fine ground to the appropriate outside diameter (OD). As indicated in Figure 5.4, the mechanisms driving the grinding wheel provide means for the linear travel of the cutting edge parallel to the axis of rotation, as well as carefully controlled transverse (radial) travel of that edge to remove material from the lens.

Figure 5.11 illustrates a typical setup, the tools used, and the sequence of operations in edging, centering, and beveling a typical double-concave lens that has multiple secondary ground

---

* Direct observation requires appropriate filtering to protect the eye.

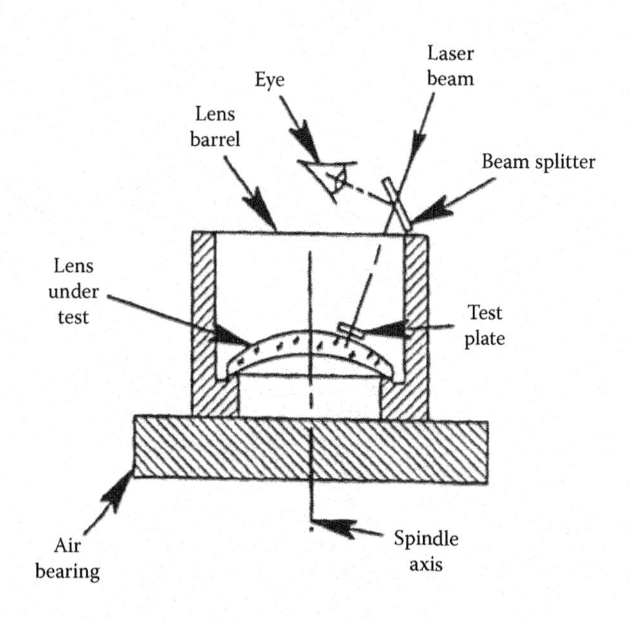

FIGURE 5.10 Technique for testing alignment of a lens on a rotating air-bearing spindle by observing the movement of Fizeau interference fringes due to surface runout. (Adapted from Carnell, K.H. et al., *Optica Acta*, 21, 615, 1974.)

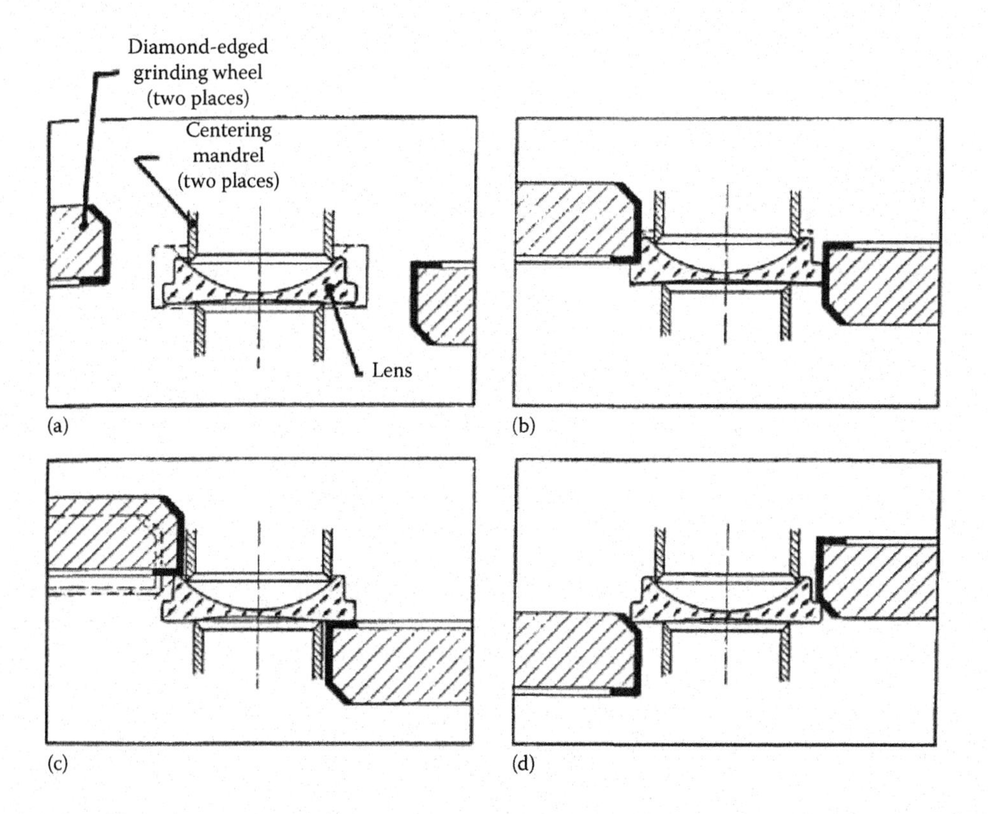

FIGURE 5.11 Lens installed in an opposing-bell centering machine for CNC centering, edging, and beveling. Sequence of operations: (a) Start position. (*Note:* Excess material to be removed is indicated by the dashed outline.) (b) Machining two cylinders and one flat bevel. (c) Machining two flat bevels and one protective bevel. (d) Machining two additional protective bevels. (Courtesy of LOH Optical Machinery, Inc., Milwaukee, WI.)

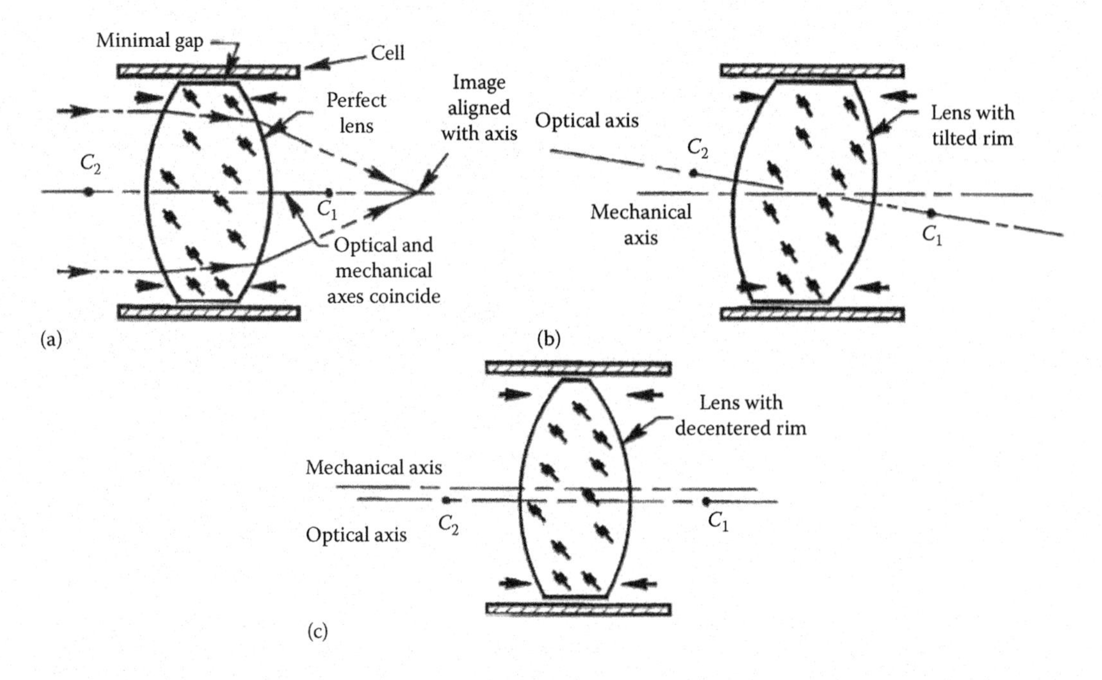

FIGURE 5.12 Holding a tilted-rim lens (view (b)) or a decentered lens (view (c)) in a rim-contact mount will deviate the image of a distant on-axis point object from the ideal condition shown in view (a) and will tend to introduce unsymmetrical aberrations in that image. Axial preload is indicated by the symmetrical set of horizontal arrows in each view.

surfaces (three flat bevels and two cylindrical surfaces), as well as three protective bevels. The machine represented here is a precision, automatic, computer numerically controlled (CNC) centering machine. Two multiple-surface, diamond-coated grinding wheels are used to speed operations.

A wedge built into a lens element during the edging process is a very undesirable attribute for a *rim-contact*-mounted lens. This type of mounting, which has minimal clearance* between the OD of the lens and the inside diameter (ID) of the cell, is illustrated in Figure 5.12. Three conditions are shown. In the view in Figure 5.12a, a perfect lens is centered in a perfect cylindrical cell with essentially zero radial clearance between the glass rim and the cell ID. The lens optical axis is then closely aligned with the mechanical axis of the cell. In the view in Figure 5.12b, the lens rim is edged in a tilted condition with respect to the optical axis, so the surface centers of curvature are both off the mechanical axis. The transmitted beam is then tilted. In the view in Figure 5.12c, the optical axis of the lens is decentered from the cell ID mechanical axis, but is parallel to that axis. Once again, the transmitted beam is tilted.

In the case of a lens with a mechanical wedge after edging, using currently available sensors and applying good optical workshop practices, the edge thickness difference ΔET of moderate-size optics can be controlled to <5 μm (0.0002 in.). High-precision techniques can measure ΔET values as small as 1.3 μm (50×10^{-6} in.). For a lens of diameter 60 mm (2.36 in.), these ΔET values correspond to geometric wedge angles of 17.2 and 4.5 arcsec, respectively. Larger lenses with the same ΔET values would have smaller wedge angles.

Lens alignment problems would be expected if the lens is perfectly centered and edged, but the associated mechanical parts for a rim-contact mount are not correctly made. Figure 5.13 illustrates this situation for three common machining errors. In the view in Figure 5.13a, the lens is tilted

* From experience, the smallest diametric clearance allowing a lens to be inserted (carefully) into a cell is ~0.0002 in. (~5 μm).

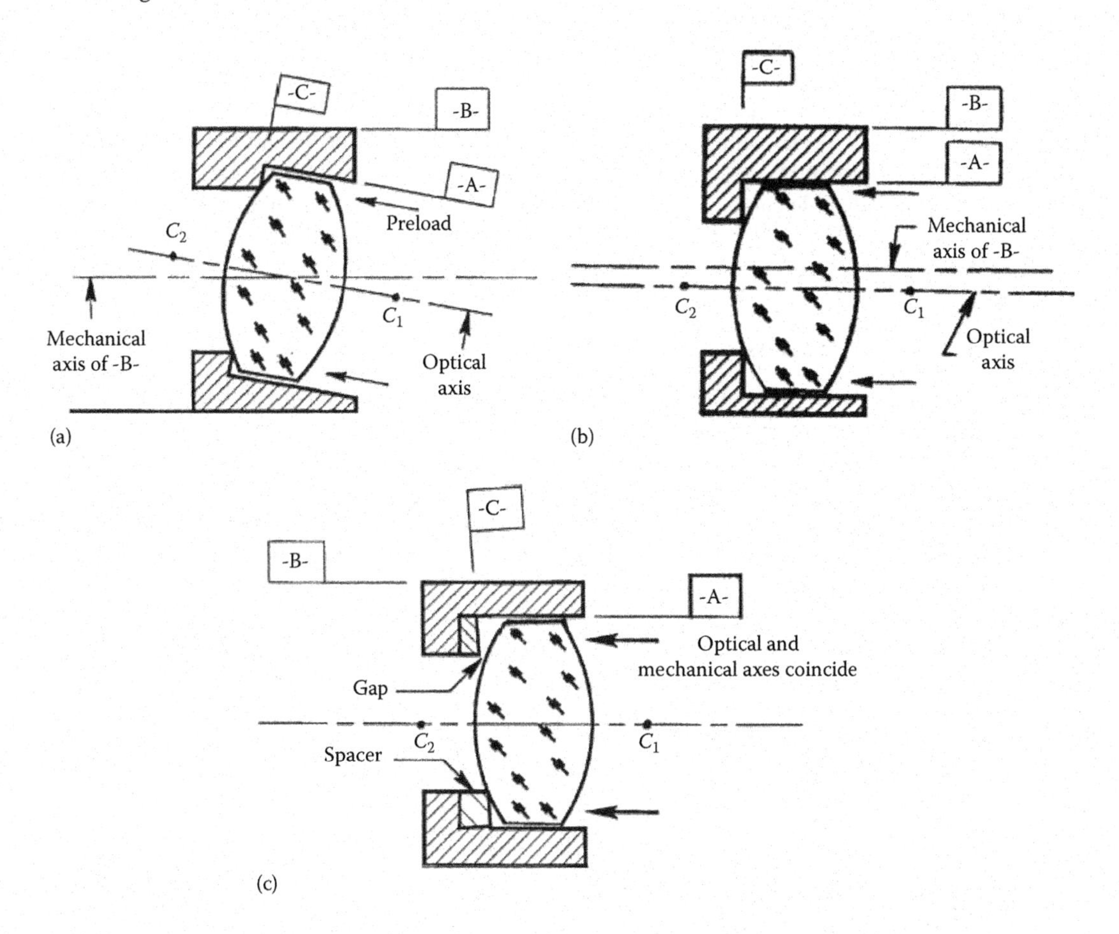

FIGURE 5.13 A perfect lens in an imperfect rim contact will introduce alignment errors relative to a mechanical reference such as datum -B- as shown in (a) for a tilted bore and shoulder or in (b) for a decentered bore. A wedged spacer, as in (c), will cause nonuniform distribution of preload and unsymmetrical stress buildup. (Adapted from Yoder, P.R. Jr., *Mounting Optics in Optical Instruments*, 2nd ed., SPIE Press, Bellingham, WA, 2008.)

with respect to the cell OD (datum -B-) because the cylindrical bore (datum -A-) and shoulder face (datum -C-) are tilted. In the view in Figure 5.13b, the cylindrical bore (datum -A-) is decentered with respect to datum -B-. The lens is therefore decentered as well. In the view in Figure 5.13c, a spacer used to locate the lens axially is wedged. The presence of a localized gap between a lens surface and the interfacing mechanical part as in the view in Figure 5.13c would allow unsymmetrical localized stress buildup at the contact owing to axial preload. Obviously, control of dimensions and surface relationships in both optical and mechanical parts is needed.

It is frequently possible to configure the lens-to-mount interface so that alignment of the lens with its mount is determined by contacting the polished optical surfaces themselves and not any secondary, i.e., ground, reference surface such as the rim, an annular flat, or a bevel. This is *surface contact mounting*. This type of mounting is discussed later in this chapter.

There is no basic difference in the use of the edging and mounting techniques described in this chapter if the element to be mounted is a flat or curved (i.e., image-forming) mirror rather than a lens as long as its external configuration has rotational symmetry. The optical consequences of centering errors are, however, significantly more pronounced in the case of a mirror since the deviation of a reflected beam due to a mirror surface being tilted through the angle θ is 2θ rather than $(n - 1)\theta$ as it would be for refraction through a lens element wedged by the angle θ.

5.3 LENS MASS AND CENTER OF GRAVITY LOCATION

Part of the design process of an optical instrument is the estimation of the mass of its optical, mechanical, and other components and the locations of the CGs of those components. With this and other design information, the approximate weight of the complete instrument can be predicted and the CG of the ensemble can be located before any of the parts are available to be measured. In this section, the procedures for making these calculations for lenses or simple mirrors with shapes similar to the common forms of lenses are given.

5.3.1 Estimating Lens Mass

Any lens can be divided into a combination of any of three basic parts: a spherical segment, a right circular cylinder, and a truncated cone. These are the *disk*, *cap*, and *cone* shapes, respectively, as shown in Figure 5.14. The volume of a convex cap is considered to be positive, while that of a concave cap is taken to be negative. For a biconvex lens, the masses of the two caps are added to that of a disk. For a meniscus lens, the mass of the convex cap is added to that of the disk and the mass of concave cap is subtracted. For simplicity, lens shapes are considered to extend to sharp corners, so weight reductions from bevels are ignored. For this reason, such calculations may tend to slightly overestimate mass. Masses for cemented doublets and more complex configurations are obtained by adding the contributions from the individual elements. Figure 5.15 shows section views of nine basic lens configurations.

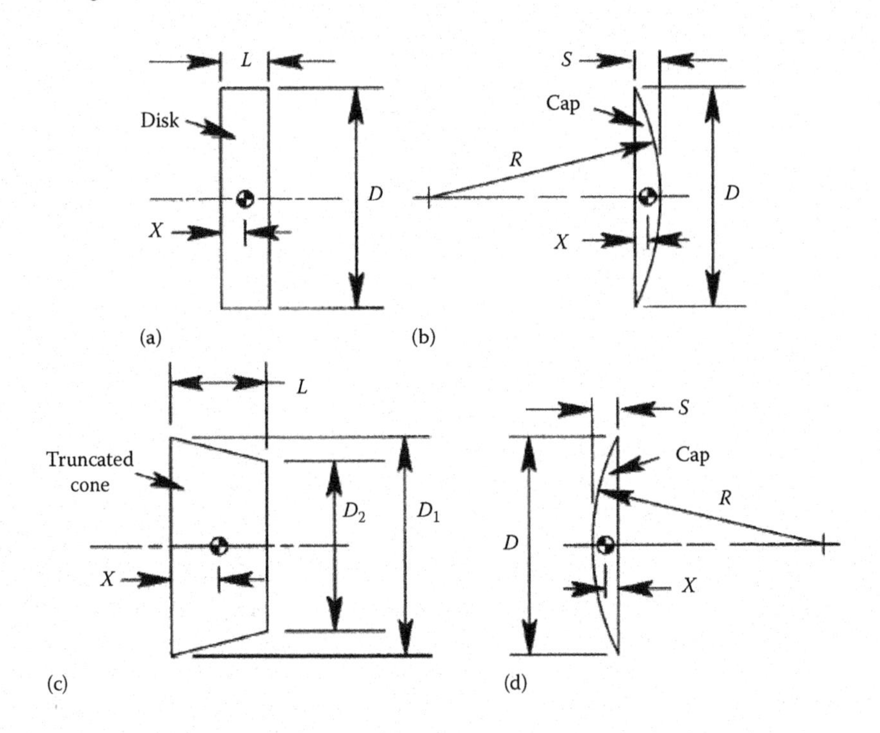

FIGURE 5.14 Schematic section views of basic solid shapes from which lenses are formed. The approximate location of the CG for each shape is indicated. (a) Disk, (b) and (d) spherical caps, and (c) tapered disk. (Adapted from Yoder, P.R. Jr., *Mounting Optics in Optical Instruments*, 2nd ed., SPIE Press, Bellingham, WA, 2008.)

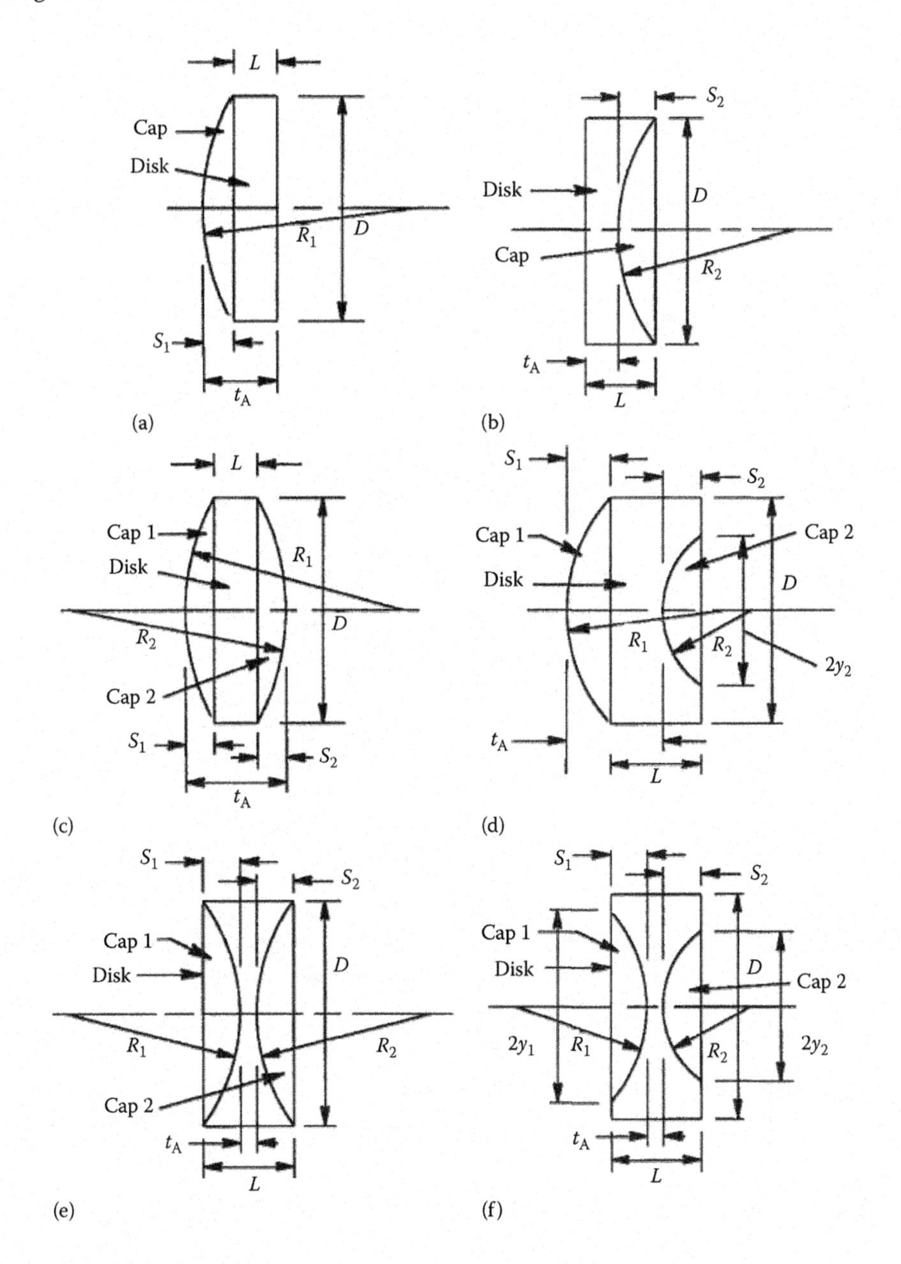

FIGURE 5.15 Schematic section views of nine lens configurations dimensioned as required to estimate lens weights: (a) plano-convex, (b) plano-concave, (c) biconvex, (d) meniscus, (e) biconcave, (f) biconcave with dual flat bevels. *(Continued)*

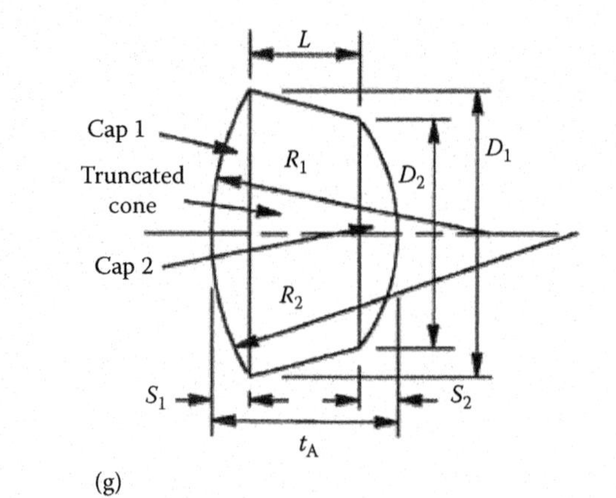

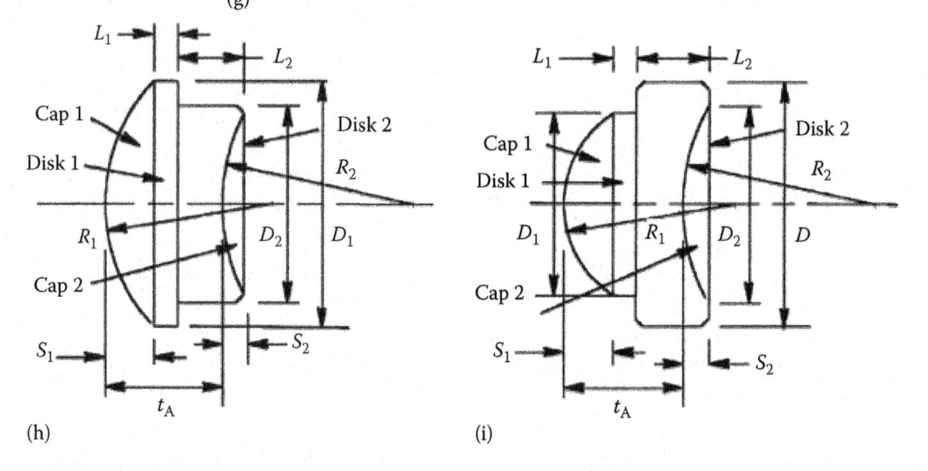

FIGURE 5.15 (CONTINUED) Schematic section views of nine lens configurations dimensioned as required to estimate lens weights: (g) biconvex with conical section, (h) cemented meniscus with larger plano-convex element, and (i) cemented meniscus with larger plano-concave element.

The sagittal depth S and mass W_{CAP} of a cap are given by

$$S = R - \left(R^2 - \frac{D^2}{4} \right)^{\frac{1}{2}}, \tag{5.2}$$

$$W_{CAP} = \pi \rho S^2 \left(R - \frac{S}{3} \right), \tag{5.3}$$

where S is the sagittal thickness, R is the radius, D is the diameter, W_{CAP} is the mass of the shape, and ρ is the mass density of the lens material.

The mass W_{DISK} of a disk of axial length L and diameter D is given by

$$W_{DISK} = \frac{\pi \rho L D^2}{4}. \tag{5.4}$$

The mass W_{CONE} of a truncated cone is given by

$$W_{CONE} = \frac{\pi \rho L\left(D_1^2 + D_1 D_2 + D_2^2\right)}{12}, \tag{5.5}$$

where L is the axial length of the truncated cone, D_1 is the diameter of the smaller end of the truncated cone, and D_2 is the diameter of the bigger end of the truncated cone.

In general, caps with aspheric surfaces are treated as spherical ones unless the asphericity is strong, as in the case of a very deep paraboloid. If this is the case, the volume (and mass) is determined by calculating the cross-sectional area of the element and multiplying that by 2π times the height from the axis of symmetry of the centroid of the area. Most general aspherics can be approximated by conics. The pertinent equations for all conic sections are in standard solid analytic geometry texts.

For quick estimates, if $S \leq 0.1R$, S and the mass of a spherical cap are given approximately (error $\varepsilon \leq 10\%$) by

$$S \cong \frac{D^2}{8R}, \tag{5.6}$$

$$W_{CAP} \cong \pi \rho R S^2. \tag{5.7}$$

Example 5.1

A biconvex lens as shown in Figure 5.15c has dimensions $D = 50$ mm, $t_A = 13$ mm, $R_1 = 76$ mm, and $R_2 = 114$ mm. The lens is made of Schott NBK7 glass with $\rho = 2519$ kg/m^3. What is the mass of the lens? Compare the exact answer with the mass given by the approximations in Equations 5.6 and 5.7.

From Equation 5.2,

$$S_1 = (76\times10^{-3}\,\text{m}) - \left[(76\times10^{-3}\,\text{m})^2 - \frac{(50\times10^{-3}\,\text{m})^2}{4}\right]^{\frac{1}{2}} = 4.23\times10^{-3}\ \text{m},$$

$$S_2 = (114\times10^{-3}\,\text{m}) - \left[(114\times10^{-3}\ \text{m})^2 - \frac{(50\times10^{-3}\,\text{m})^2}{4}\right]^{\frac{1}{2}} = 2.78\times10^{-3}\,\text{m}.$$

From the geometry of the figure,

$$L = 13\times10^{-3}\,\text{m} - 4.23\times10^{-3}\,\text{m} - 2.78\times10^{-3}\,\text{m} = 5.99\times10^{-3}\,\text{m}.$$

From Equation 5.3,

$$W_{CAP1} = \pi(2.519\times10^3\,kg/m^3)(4.23\times10^{-3}\,m)^2\left(76\times10^{-3}\,m - \frac{4.23\times10^{-3}\,m}{3}\right)$$
$$= 10.6\times10^{-3}\,kg,$$

$$W_{CAP2} = \pi(2.519\times10^3\,kg/m^3)(2.78\times10^{-3}\,m)^2\left(114\times10^{-3}\,m - \frac{2.78\times10^{-3}\,m}{3}\right)$$
$$= 6.92\times10^{-3}\,kg.$$

From Equation 5.4,

$$W_{DISK} = \frac{\pi(2.620\times10^3\,kg/m^3)(5.99\times10^{-3}\,m)(50\times10^{-3}\,m)}{4} = 29.6\times10^{-3}\,kg.$$

Summing, the total lens mass is

$$W_{LENS} = 10.6\times10^{-3}\,kg + 6.92\times10^{-3}\,kg + 29.6\times10^{-3}\,kg = 47.1\times10^{-3}\,kg.$$

Comparing this result to the approximations, from Equation 5.6,

$$S_1 \cong \frac{(50\times10^{-3}\,m)^2}{8(76\times10^{-3}\,m)} = 4.11\times10^{-3}\,m.$$

$$S_2 \cong \frac{(50\times10^{-3}\,m)^2}{8(114\times10^{-3}\,m)} = 2.74\times10^{-3}\,m.$$

The approximation given in Equation 5.6 is valid since $S_1/R_1 < 0.1$ and $S_2/R_2 < 0.1$:

$$L = 13\times10^{-3}\,m - 4.11\times10^{-3}\,m - 2.74\times10^{-3}\,m = 6.15\times10^{-3}\,m.$$

From Equation 5.7,

$$W_{CAP1} = \pi(2.519\times10^3)(76\times10^{-3}\,m)(4.11\times10^{-3}\,m)^2 = 10.2\times10^{-3}\,kg,$$

$$W_{CAP2} = \pi(2.519\times10^3)(114\times10^{-3}\,m)(2.74\times10^{-3}\,m)^2 = 6.77\times10^{-3}\,kg.$$

Then, the disk mass is

$$W_{DISK} = \pi(2.519\times10^3\,kg/m^3)(6.15\times10^{-3}\,m)\,\frac{(50\times10^{-3}\,m)^2}{4}$$
$$= 30.4\times10^{-3}\,kg.$$

The lens mass is

$$W_{LENS} = 10.2 \times 10^{-3}\ \text{kg} + 6.77 \times 10^{-3}\ \text{kg} + 30.4 \times 10^{-3}\text{kg} \\ = 47.4 \times 10^{-3}\text{kg}.$$

The error is

$$\varepsilon = \frac{47.4\times10^{-3}\,\text{m} - 47.1\times10^{-3}\,\text{m}}{47.1\times10^{-3}\,\text{m}} = 6.37 \times 10^{-3} \quad (\approx 6.4\%).$$

Example 5.2

A biconcave lens as shown in Figure 5.15e has dimensions D = 50 mm, t_A = 7.1 mm, R_1 = 76 mm, and R_2 = 114 mm. The lens is made of Schott NF2 glass with ρ = 2650 kg/m^3. What is the mass of the lens?

From Equation 5.2,

$$S_1 = (76\times10^{-3}\,\text{m}) - \left[(76\times10^{-3}\,\text{m})^2 - \frac{(50\times10^{-3}\,\text{m})^2}{4}\right]^{\frac{1}{2}} = 4.23\times10^{-3}\ \text{m},$$

$$S_2 = (114\times10^{-3}\,\text{m}) - \left[(114\times10^{-3}\ \text{m})^2 - \frac{(50\times10^{-3}\,\text{m})^2}{4}\right]^{\frac{1}{2}} = 2.78\times10^{-3}\,\text{m}.$$

From the geometry of the figure,

$$L = 7.1\times10^{-3}\,\text{m} + 4.23\times10^{-3}\ \text{m} + 2.78\times10^{-3}\ \text{m} = 14.1\times10^{-3}\ \text{m}.$$

From Equation 5.3,

$$W_{CAP1} = \pi(2.650\times10^{3}\,\text{kg/m}^3)(4.23\times10^{-3}\,\text{m})^2\left(76\times10^{-3}\,\text{m} - \frac{4.23\times10^{-3}\,\text{m}}{3}\right) \\ = 11.1\times10^{-3}\,\text{kg},$$

$$W_{CAP2} = \pi(2.650\times10^{3}\,\text{kg/m}^3)(2.78\times10^{-3}\,\text{m})^2\left(114\times10^{-3}\,\text{m} - \frac{2.78\times10^{-3}\,\text{m}}{3}\right) \\ = 7.28\times10^{-3}\,\text{kg}.$$

From Equation 5.4,

$$W_{DISK} = \pi(2.650\times10^{3}\,\text{kg/m}^3)(14.1\times10^{-3}\,\text{m})\frac{(50\times10^{-3}\,\text{m})^2}{4} = 72.8\times10^{-3}\,\text{kg}.$$

Summing, the total lens mass is

$$W_{LENS} = -11.1\times10^{-3}\,\text{kg} + 72.8\times10^{-3}\,\text{kg} - 7.28\times10^{-3}\,\text{kg} = 54.4\times10^{-3}\,\text{kg}.$$

Example 5.3

A biconvex lens with a conical rim as shown in Figure 5.15g has radii and dimensions R_1 = 150 mm, R_2 = 100 mm, D_1 = 50 mm, D_2 = 44.5 mm, and t_A = 22 mm. The lens is made of Schott N-SK16 glass with ρ = 3580 kg/m³. What is the mass of the lens?

From Equation 5.2,

$$S_1 = (150\times10^{-3}\,\text{m}) - \left[(150\times10^{-3}\,\text{m})^2 - \frac{(50\times10^{-3}\,\text{m})^2}{4}\right]^{\frac{1}{2}} = 2.10\times10^{-3}\,\text{m}.$$

$$S_2 = (100\times10^{-3}\,\text{m}) - \left[(100\times10^{-3}\,\text{m})^2 - \frac{(44.5\times10^{-3}\,\text{m})^2}{4}\right]^{\frac{1}{2}} = 2.51\times10^{-3}\,\text{m}.$$

From the geometry of the figure,

$$L = 22\times10^{-3}\,\text{m} - 2.10\times10^{-3}\,\text{m} - 2.51\times10^{-3}\,\text{m} = 17.4\times10^{-3}\,\text{m}$$

From Equation 5.3,

$$W_{CAP1} = \pi(2.620\times10^3\,\text{kg/m}^3)(2.10\times10^{-3}\,\text{m})^2\left(150\times10^{-3}\,\text{m} - \frac{2.10\times10^{-3}\,\text{m}}{3}\right)$$
$$= 7.41\times10^{-3}\,\text{kg},$$

$$W_{CAP2} = \pi(3.580\times10^3\,\text{kg/m}^3)(2.51\times10^{-3}\,\text{m})^2\left(100\times10^{-3}\,\text{m} - \frac{2.51\times10^{-3}\,\text{m}}{3}\right)$$
$$= 7.01\times10^{-3}\,\text{kg}.$$

From Equation 5.5,

$$W_{CONE} = \frac{\pi(5.580\times10^3\,\text{kg/m}^3)(17.4\times10^{-3}\,\text{m})}{12}\Big[(50\times10^{-3}\,\text{m})^2$$
$$+ (50\times10^{-3}\,\text{m})(44.5\times10^{-3}\,\text{m}) + (44.5\times10^{-3}\,\text{m})^2\Big] = 109\times10^{-3}\,\text{kg}.$$

Then, the lens mass is

$$W_{LENS} = 7.14\times10^{-3}\,\text{kg} + 109\times10^{-3}\,\text{kg} + 7.41\times10^{-3}\,\text{kg} = 124\times10^{-3}\,\text{kg}.$$

5.3.2 Lens Center of Gravity Location

In Figure 5.14, which shows the cross sections of the four basic shapes that make up lenses, the dimensions X indicate the locations of their CGs relative to their left surfaces. The following equations give X for each section using the dimensions shown in the figure:

$$X_{DISK} = \frac{L}{2}, \tag{5.8}$$

$$X_{\text{CAP}} = \frac{S(4R-S)}{4(3R-S)}, \tag{5.9}$$

$$X_{\text{CONE}} = \frac{2L\left(\frac{D_1}{2}+D_2\right)}{3(D_1+D_2)}. \tag{5.10}$$

For quick estimates, if $S \le 0.1R$, X_{CAP} is given approximately (error $\varepsilon \le 10\%$) by

$$X_{\text{CAP}} \cong \frac{S}{3}. \tag{5.11}$$

The location of the CG of a lens comprising N parts and weighing W_{LENS} can be estimated from

$$X_{\text{LENS}} = \sum_{i=1}^{i=N} \frac{X_i' W_i}{W_{\text{LENS}}}. \tag{5.12}$$

In Equation 5.12, all distances X_{LENS} and all values of the moment arms X_i' for the individual parts are measured from the same point along the axis of the lens. For example, in view (c) of Figure 5.14, choose the left vertex as the reference and remember that X_{CAP} is always measured from the plano side. Then, the moment arms X_i' are $X'_{\text{CAP1}} = S_1 - X_{\text{CAP1}}$, $X'_{\text{DISK1}} = S_1 + X_{\text{DISK}}$, and $X'_{\text{CAP2}} = S_1 + L + X_{\text{CAP2}}$.

Example 5.4

Find the CG for a meniscus lens of the shape shown in Figure 5.15d, with $D = 75$ mm, $t_A = 40$ mm, $R_1 = 150$ mm, $R_2 = 75$ mm, and $2y_2 = 60$ mm. The lens is made of Schott F2 glass with $\rho = 3.6 \times 10^3$ kg/m^3. Compare exact and approximate results.

First, find S_1 and S_2 by using Equation 5.2:

$$S_1 = (150\times10^{-3}\,\text{m}) - \left[(150\times10^{-3}\,\text{m})^2 - \frac{(75\times10^{-3}\,\text{m})^2}{4}\right]^{\frac{1}{2}} = 4.763\times10^{-3}\ \text{m},$$

$$S_2 = (75\times10^{-3}\,\text{m}) - \left[(75\times10^{-3}\,\text{m})^2 - \frac{(60\times10^{-3}\,\text{m})^2}{4}\right]^{\frac{1}{2}} = 6.261\times10^{-3}\ \text{m},$$

$$L = 40\times10^{-3}\,\text{m} + 6.261\times10^{-3}\,\text{m} - 4.763\times10^{-3}\,\text{m} = 41.49\times10^{-3}\,\text{m}.$$

Find the volume and mass of the lens sections by using Equations 5.3 and 5.4:

$$W_{\text{CAP1}} = \pi(3.6\times10^{-3}\,\text{kg/m}^3)(4.763\times10^{-3}\,\text{m})^2\left(150\times10^{-3}\,\text{m} - \frac{4.763\times10^{-3}\,\text{m}}{3}\right)$$
$$= 38.08\times10^{-3}\,\text{kg},$$

$$W_{CAP2} = \pi(3.6\times10^{-3}\,\text{kg/m}^3)(6.261\times10^{-3}\,\text{m})^2\left(75\times10^{-3}\,\text{m}-\frac{6.2612\times10^{-3}\,\text{m}}{3}\right)$$
$$= 32.33\times10^{-3}\,\text{kg},$$

$$W_{DISK} = \pi(3.6\times10^{-3}\,\text{kg/m}^3)(41.49\times10^{-3}\,\text{m})\frac{(75\times10^{-3}\,\text{m})^2}{4} = 660.0\times10^{-3}\,\text{kg},$$

$$W_{LENS} = 660.0\times10^{-3}\,\text{kg}+38.08\times10^{-3}\,\text{kg}-32.33\times10^{-3}\,\text{kg} = 665.7\times10^{-3}\,\text{kg}.$$

Find the locations of the CGs of the lens sections by using Equations 5.8 and 5.9:

$$X_{DISK} = \frac{41.49\times10^{-3}\,\text{m}}{2} = 20.75\times10^{-3}\,\text{m},$$

$$X_{CAP1} = \frac{(4.763\times10^{-3}\,\text{m})\left[4(150\times10^{-3}\,\text{m})-4.763\times10^{-3}\,\text{m}\right]}{4\left[3(150\times10^{-3}\,\text{m})-4.763\times10^{-3}\,\text{m}\right]} = 1.592\times10^{-3}\,\text{m},$$

$$X_{CAP1} = \frac{(6.261\times10^{-3}\,\text{m})\left[4(75\times10^{-3}\,\text{m})-6.261\times10^{-3}\,\text{m}\right]}{4\left[3(75\times10^{-3}\,\text{m})-6.261\times10^{-3}\,\text{m}\right]} = 2.102\times10^{-3}\,\text{m},$$

Finally, find the CG location using Equation 5.12:

$$\begin{aligned}CG_{LENS} = \frac{1}{665.7\times10^{-3}\text{kg}}&\left[(4.763\times10^{-3}\,\text{m}-1.592\times10^{-3}\,\text{m})(38.08\times10^{-3}\,\text{kg})\right.\\&+(660.0\times10^{-3}\,\text{kg})(20.75\times10^{-3}\,\text{m}+4.763\times10^{-3}\,\text{m})\\&-(32.33\times10^{-3}\,\text{kg})(20.75\times10^{-3}\,\text{m}+4.763\times10^{-3}\,\text{m}\\&\left.-6.261\times10^{-3}\,\text{m}+2.102\times10^{-3}\,\text{m})\right] = 24.44\times10^{-3}\,\text{m}.\end{aligned}$$

For comparison, use Equation 5.6 to find the approximate values of S_1 and S_2:

$$S_1 \cong \frac{(75\times10^{-3}\,\text{m})^2}{8(150\times10^{-3}\,\text{m})} = 4.69\times10^{-3}\,\text{m},$$

$$S_2 \cong \frac{(60\times10^{-3}\,\text{m})^2}{8(75\times10^{-3}\,\text{m})} = 6.00\times10^{-3}\,\text{m}.$$

Finding L,

$$L = 40\times10^{-3}\,\text{m}+6.00\times10^{-3}\,\text{m}-4.69\times10^{-3}\,\text{m}.$$

Use Equation 5.7 to find the approximate masses of the spherical sections; the disk mass is found by using Equation 5.4 as before:

$$W_{CAP1} \cong \pi(3.6\times10^{-3}\,\text{kg/m}^3)(150\times10^{-3}\,\text{m})(4.69\times10^{-3}\,\text{m})^2 = 37.3\times10^{-3}\,\text{kg},$$

$$W_{CAP2} \cong \pi(3.6\times10^{-3}\,\text{kg/m}^3)(75\times10^{-3}\,\text{m})(6.00\times10^{-3}\,\text{m})^2 = 30.5\times10^{-3}\,\text{kg},$$

$$W_{LENS} \cong 660.0\times10^{-3}\,\text{kg} + 37.3\times10^{-3}\,\text{kg} - 30.5\times10^{-3}\,\text{kg} = 667\times10^{-3}\,\text{kg}.$$

The error in lens mass when using the approximation is about 0.15%!

The approximate CG locations for the spherical sections are given by Equation 5.11; the disk is the same as before:

$$X_{CAP1} \cong \frac{4.69\times10^{-3}\,\text{m}}{3} = 1.56\times10^{-3}\,\text{m},$$

$$X_{CAP2} \cong \frac{6.00\times10^{-3}\,\text{m}}{3} = 2.00\times10^{-3}\,\text{m}.$$

Finally, Equation 5.12 gives the CG location:

$$\begin{aligned} CG_{LENS} \cong \frac{1}{667\times10^{-3}\text{kg}}\Big[&(1.56\times10^{-3}\,\text{m})(37.3\times10^{-3}\,\text{kg}) \\ &+(660.0\times10^{-3}\,\text{kg})(20.75\times10^{-3}\,\text{m}+4.69\times10^{-3}\,\text{m}) \\ &-(30.5\times10^{-3}\,\text{kg})(20.75\times10^{-3}\,\text{m}+4.69\times10^{-3}\,\text{m} \\ &-6.00\times10^{-3}\,\text{m}+2.00\times10^{-3}\,\text{m})\Big] = 24.3\times10^{-3}\,\text{m}. \end{aligned}$$

The error in CG location is about 0.6%. This example shows that for many estimates, the approximate methods are acceptably accurate.

5.4 LENS AXIAL PRELOAD

5.4.1 General Considerations

Mounting lens elements on their optical polished (generally spherical) surfaces provides a number of advantages. The optical surfaces are the most accurate, and using these surfaces for mounting minimizes errors. Self-centering by clamping the lens between two circular surfaces eliminates errors from improper centering of the lens edge.

In optical surface mounting, sufficient radial clearance is provided between the lens rim and the mount ID to ensure that the lens rim does not touch the ID of the mount. Axial preload that holds the lens against a mechanical reference surface, such as a shoulder in the mount or a spacer, is provided by tightening a threaded retaining ring, attaching a continuous flange, or applying some other mechanical clamping technique. A generic version of the threaded retaining ring constraint is illustrated in Figure 5.16.[3] In Figure 5.16, a single biconvex lens is held against a cell shoulder by a threaded retainer. The chief variables in such designs are the magnitude of the applied preload and

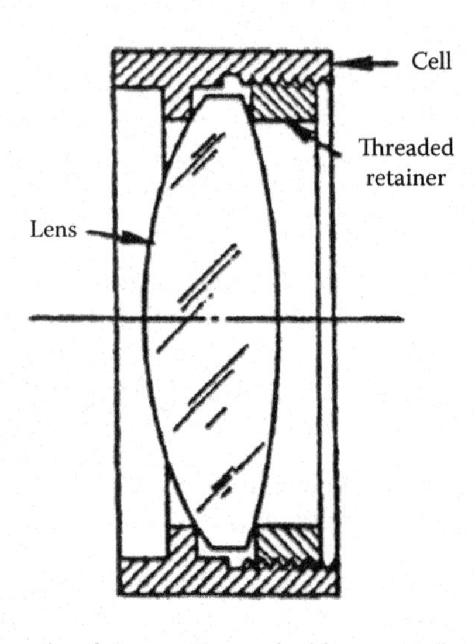

FIGURE 5.16 Common configuration for mounting a double-convex lens with a threaded retaining ring. Interfaces with the glass are generic 90° sharp corner interfaces. (From Yoder, P.R. Jr., Lens mounting techniques, *Proc. SPIE*, 389, 2, 1983.)

the shapes of the optomechanical interfaces between the glass and the metal parts. This mounting can be assembled and disassembled relatively easily; it accommodates axial thickness variations of the element; it lends itself easily to environmental sealing with a formed-in-place elastomeric seal or with an O-ring; and it is compatible with mounting multiple elements in the same cell or housing.*

Another distinct advantage of this so-called surface contact lens-mounting design is that tolerances on the location and orientation of secondary references surfaces such as the lens rim, bevels, and chamfers can be relaxed since they are not contacted by the cell. Figure 5.17a shows a case in which the lens rim is tilted considerably, while Figure 5.17b shows a decentered rim. These errors would have no effect on the lens alignment. The diameter of the lens is not critical with this type of mounting.

In applications allowing decentering or residual deviation due to a geometric wedge of 25–75 μm (0.001–0.003 in.) and 0.75–3 arcmin, it is common practice to premachine the optic and cell to radial dimensions and tolerances that will preclude damage to the glass under the most extreme temperature conditions. The lens is just inserted and clamped in place. This is sometimes called a *drop-in* assembly. Advantages include simplicity and ease of assembly and disassembly.

Neglecting the effects of temperature changes, the total nominal axial force (preload) F_A that should be exerted on the lens at assembly to hold it in place by any means of constraint, is given by

$$F_A = W_{LENS} \sum a_G, \tag{5.13}$$

where W_{LENS} is the lens mass and a_G is the acceleration.

The acceleration terms include constant acceleration, random vibration (3σ), amplified resonant vibration (sinusoidal), acoustic loading, and shock. For ease of defining accelerations in equations, these accelerations are here expressed as multiples, of ambient gravity a_G. Since all types of external accelerations do not generally occur simultaneously, the summation term in Equation 5.13 does not need to be taken literally. In some cases, it may be appropriate to root-sum-square the accelerations occurring in orthogonal directions. For simplicity, consider a_G to be a single-valued, worst-case number. Hence,

$$F_A = W_{LENS}\, SF a_G. \tag{5.14}$$

* As a rule of thumb, it is unwise to hold more than three lenses with a single retainer.

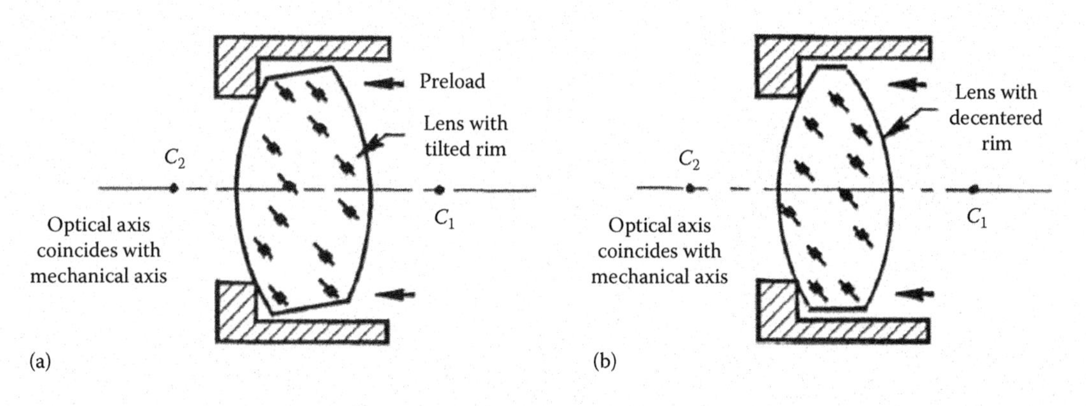

FIGURE 5.17 With interfaces directly on the polished refracting surfaces of a lens, errors in edging the lens and in lens diameter are not critical. (a) Tilted rim and (b) decentered rim.

In Equation 5.14, friction and moments imposed at the interfaces are ignored and sealing compound is assumed to be not present. The term SF is a safety factor that tends to compensate for approximations in the calculation. Typically, SF is 1.5–2.0.

An axial preload applied to a curved surface of a lens tends to center the lens to the mechanical axis of the interfacing mount. This effect is illustrated by Figure 5.9 in conjunction with self-centering of a lens in preparation for edging. If the curvature of the surface and the preload are sufficiently large, the difference between the opposing radial components of the axial force pressing against a decentered lens will overcome friction, and the lens will tend to self-center to the extent possible, given the geometry.

Axial preload also can help resist decentering of a centered lens with curved surfaces in surface contact mounting if the subassembly is subjected to radial acceleration. In the presence of such a preload, the centered lens is nominally clamped symmetrically between the mechanical interfaces on opposite sides of the optic. See, for example, Figure 5.18. The lens is shown as plano-convex to simplify the geometry. Radial clearance between the lens rim and the mount ID is assumed to exist. Radial acceleration of magnitude a_G acting through the CG of a lens with mass W will exert a downward radial force of $-Wa_G$ that tends to force the mechanical interfaces apart by the wedging effect of the curved lens surface or surfaces trying to move in the direction of the acceleration. On the plane of the figure, the total wedge angle is θ. If the mechanical mount and the lens are sufficiently stiff that they do not deform and the interfaces do not separate axially, the mount is seen to exert an upward force on the lens equal to $+Wa_G$ to balance the downward force. Radial motion of the lens is prevented.

As shown in the detail view of Figure 5.18, the upward force has a component tangential to the curved surface equal to ($Wa_G \cos\theta$). Associated with this component is another component normal to the lens surface equal to $(Wa_G/\mu)\cos\theta$, where μ is the coefficient of friction at the glass-to-metal interface. This component has yet another component in the axial direction that equals the axial preload F_R required to overcome the radial acceleration-induced force. This preload is

$$F_R = \frac{Wa_G \cos^2(\theta)}{\mu}. \tag{5.15}$$

Figure 5.19 shows four general lens configurations and the angle θ for each. Note that interfaces that oppose downward motion of the lens in the view in Figure 5.19b under downward

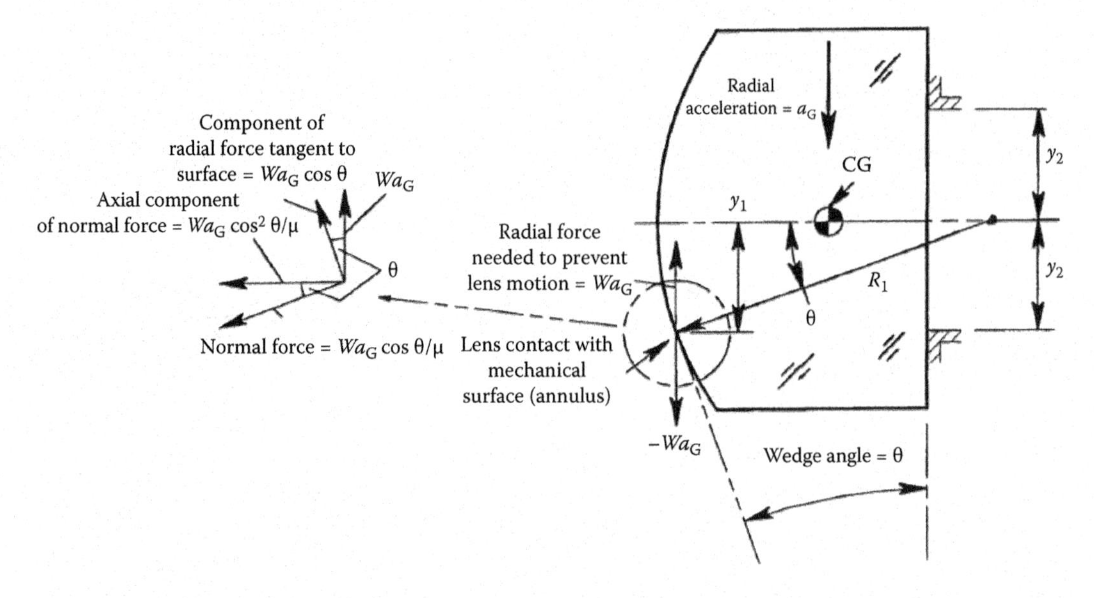

FIGURE 5.18 Geometry for estimating preload required to prevent a surface contact lens from decentering under radial acceleration.

acceleration are on opposite sides of the axis. Other general cases (not shown) would be a plane-parallel plate (window, filter, or reticle) or a lens with flat bevels registering on both sides against flat mechanical surfaces oriented perpendicular to the axis. Such configurations would have nominal angles θ of zero, and the preloads that just prevent radial motion would equal (Wa_G/μ).

The view in Figure 5.19a shows a biconvex lens with interfaces on the spherical surfaces at different heights from the axis. The angle θ is seen to be the sum of two parts relative to a plane normal to the axis and passing through the apex of the angle θ. These angles are

$$\theta_1 = \arcsin\left(\frac{y_1}{R_1}\right), \tag{5.16}$$

$$\theta_2 = \arcsin\left(\frac{y_2}{R_2}\right), \tag{5.17}$$

$$\theta = \theta_2 - \theta_1, \tag{5.18}$$

where y_1 and y_2 are the radial distances to the line of contact with the lens, R_1 and R_2 are the optical radii of the lens, and θ_1 and θ_2 are the contact angles with the lens surfaces.

Convention dictates that radii of surfaces with centers to the left of the surface are negative, while those with centers to the right of the surface are positive. A common value for the friction coefficient between uncoated optical glass and metal is $\mu \approx 0.15$. Optical coatings, and the plating or coating of the mounting surfaces, can increase the friction coefficient substantially. For example, the friction coefficient for an MgF optical coated glass lens against a black anodized mounting surface is $\mu \approx 0.375$. Table 5.1 gives the friction coefficients for coated glass surfaces against a variety of mounting materials.

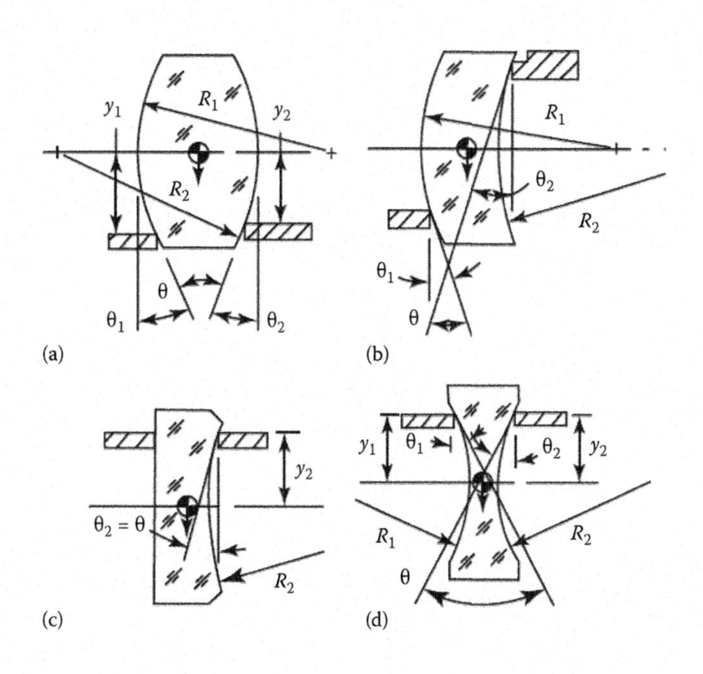

FIGURE 5.19 Determining the angle θ of four types of lenses: (a) biconvex, (b) meniscus, (c) plano-concave, and (d) biconcave.

Example 5.5

A biconvex lens as shown in Figure 5.19a has R_1 = 76.2 mm and R_2 = −114.3 mm. Surface contacts are provided at $y_1 = y_2$ = −22.86 mm, and the friction coefficient between lens and contacts is μ = 0.20. Radial clearance is provided around the lens rim. The lens mass is 46 × 10^{-3} kg. The lens is subjected to a downward (radial direction) acceleration of a_G = 20 g. What axial preload force F_R is necessary to prevent radial motion (decentration) of the lens?

From Equations 5.16 through 5.18,

$$\theta_1 = \sin^{-1}\left(\frac{-\ 22.86\ \text{mm}}{76.2\ \text{mm}}\right) = -17.458°,$$

$$\theta_2 = \sin^{-1}\left(\frac{-22.86\ \text{mm}}{-114.3\ \text{mm}}\right) = 11.537°,$$

$$\theta = -(-17.458°) + 11.537° = 28.995°.$$

From Equation 5.15, the required axial preload force F_R is

$$F_R = \frac{20(9.81\ \text{m/s}^2)(46\times10^{-3}\ \text{kg})\left[\cos(28.995°)\right]^2}{0.20} = 34.52\ \text{N}.$$

A potential design problem related to axial acceleration occurs when the temperature rises above the assembly temperature and, as is usually the case, the mount expands more than the lens does. The axial preload then decreases from that applied at assembly and may disappear completely.

TABLE 5.1
Friction Coefficients, Glass-to-Metal, for Various Surface Coatings on Glass and Metal

Lens Optical Coating	Aluminum Surface Coating	Static Friction Coefficient Metal-to-Glass (μ_G)
MgF_2	Black anodize[a]	0.375
MgF_2	Pioneer Optical Black[b]	0.294
MgF_2	Anodize, Bodycoat[c]	0.344
SiO_2	Black anodize[a]	0.205
SiO_2	Pioneer Optical Black[b]	0.218
SiO_2	Anodize, Bodycoat[c]	0.212
Uncoated	Black anodize[a]	0.165
Uncoated	Pioneer Optical Black[b]	0.146
Uncoated	Anodize, Bodycoat[c]	0.169

[a] Black anodize per MIL-A-8625F type II, class 2, 15 μm.
[b] Optical Black from Pioneer Metal.
[c] Low-reflection special anodizing from Bodycote.

5.4.2 Threaded Retaining Ring Mounting

Figure 5.20a shows the functional aspects of the threaded retaining ring design depicted in Figure 5.16. The torque applied to the ring is converted into axial force (preload) against the lens by the thread. Sets of holes or transverse slots are usually machined into the exposed face of the retainer to accept pins or rectangular lugs on the end of a cylindrical wrench that is used to tighten the retainer (see Figure 5.20b). Adjustable spanner wrenches may also be used (see Figure 5.20c and d). Alternatively, a flat, plate-type tool (not shown) that spans the retainer can be used as the wrench. All tools should be shaped so as always to clear the polished surface of the lens to prevent damage to that surface. The cylindrical wrench is easiest to use and is most conducive to measurement of torque applied to the ring. Figure 5.21 is a partial drawing of an actual retainer.

The axial preload clamps the lens against the cell shoulder shown to the left of Figure 5.20a. To minimize deformation of the lens surfaces if the required axial force is large or the lens is thin, the heights of contact y_C on both sides of the lens should be essentially the same. Means for estimating bending stress in the lens if this is not the case are described in Section 5.5.

The fit of the ring threads into the cell threads should be loose: class 1 or 2 per the American Society of Mechanical Engineers' publication B1.1-1989 (R2001) so that the ring can tilt slightly, if necessary, to accommodate the residual wedge angle in the lens or in the retainer when the lens axis is properly centered. This helps to ensure that the preload is distributed uniformly around the lens periphery. A rule-of-thumb criterion for suitability of the fit of the threaded parts is to assemble the ring and the retainer in the mount without the optic in place, hold the subassembly to the ear, and shake it gently. You should be able to hear the ring rattle in the mount.

Equation 5.19 gives the axial preload F_A produced by tightening a threaded retainer with thread pitch diameter d_T with a torque M against a lens surface. The first term in the denominator results from the classical equation for a body sliding slowly with frictional constraint on an inclined plane (i.e., the thread). The second term represents the friction effects at the circular interface between the lens surface and the end of the rotating retainer:

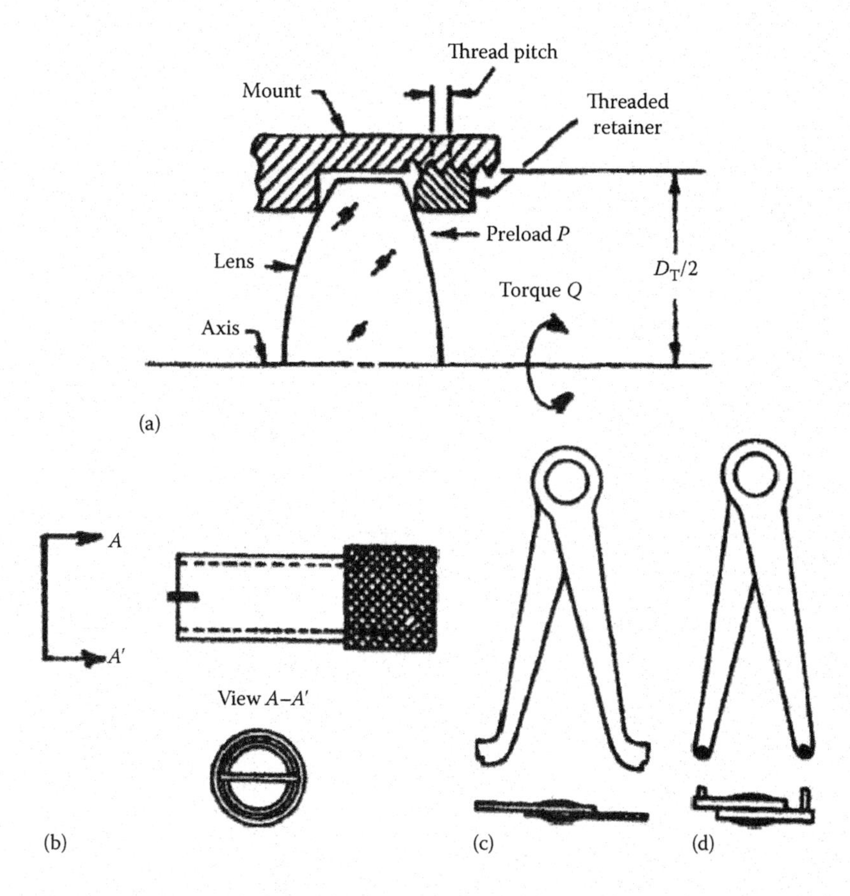

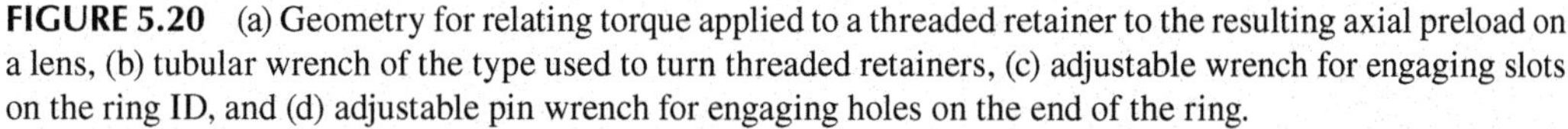
FIGURE 5.20 (a) Geometry for relating torque applied to a threaded retainer to the resulting axial preload on a lens, (b) tubular wrench of the type used to turn threaded retainers, (c) adjustable wrench for engaging slots on the ring ID, and (d) adjustable pin wrench for engaging holes on the end of the ring.

$$F_A = \frac{M}{d_T(0.577\mu_M + 0.500\mu_G)}, \tag{5.19}$$

where μ_M is the coefficient of sliding friction of the metal-to-metal interface and μ_G is the coefficient of friction of the glass-to-metal interface.

Equation 5.19 is an approximation because of small factors neglected in the derivation and significantly larger uncertainties in the values of μ_M and μ_G. The latter values depend strongly on the smoothness of the metal surfaces (which depends, in part, upon the machined finish as well as how many times the thread has been tightened) and whether the surfaces are dry or moistened by water or a lubricant. For black anodized aluminum, μ_M is about 0.19, while μ_G for uncoated glass on metal is about 0.15. Unfortunately, friction coefficients are often not known with accuracy, leading to the use of the following approximation:

$$F_A = \frac{5M}{d_T}. \tag{5.20}$$

Similar metals, such as aluminum and aluminum, should never be in contact in a threaded joint without some form of lubrication or a hard coating or plating on the mating parts because the rubbing surfaces may gall and seize. The surface preparations can significantly alter the coefficient of friction, making Equation 5.20 less accurate.

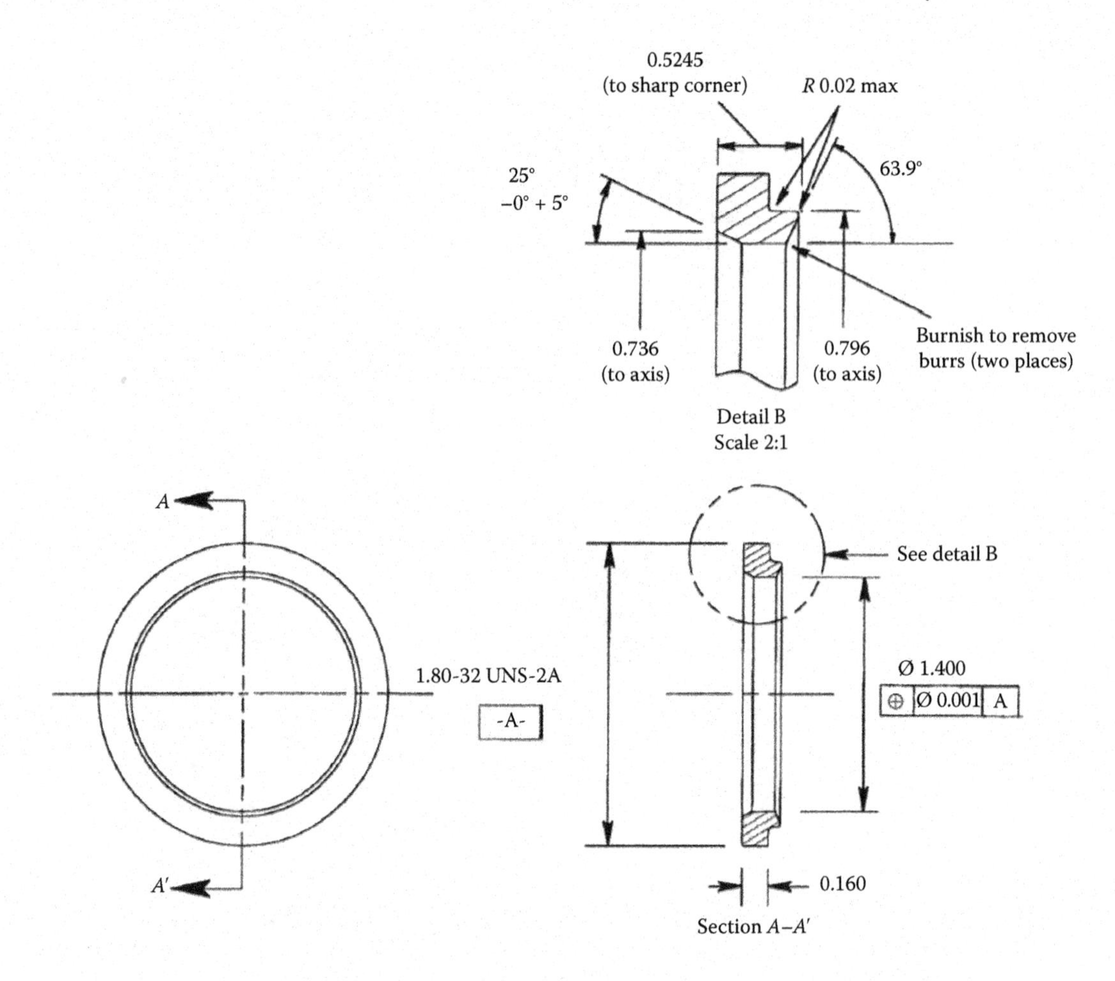

FIGURE 5.21 Representations of a typical threaded retaining ring as shown on an engineering drawing. The right end of this part has a conical surface for contacting a convex spherical lens surface with a radius of 1.671 in. (42.44 mm) at a height of 0.748 in. (19.00 mm). *Notes:* (1) material, stainless steel round, type 416 H per ASTM A582; (2) finish: passivate per SAE-AMS-QQ-P-35, type 2; (3) bag and tag with part number; (4) surface finish: 32; (5) dimensions are in inches; (6) tolerances: .xxx ± 0.005, .xxxx ± 0.00010, ±0.5°.

Example 5.6

A 53.34 mm diameter lens is clamped with a total axial preload of 55.6 N by using a threaded retaining ring with a pitch diameter of 55.88 mm. Using the approximate Equation 5.20, what torque should be applied to the retainer?

From Equation 5.20,

$$M = \frac{(55.88 \times 10^{-3}\,\text{m})(55.6\,\text{N})}{5} = 621 \times 10^{-3}\,\text{N m}.$$

Dimensional or "packaging" constraints may require the use of fine threads to minimize the wall thickness and overall diameter of the mount. Extreme care is necessary assembling a fine-threaded retainer to prevent "crossing" the threads and rendering the parts unusable. An additional concern is possible thread yielding induced by the preload force.

Figure 5.22a shows the commonly used terms for screw threads, while Figure 5.22b shows the basic profile of a thread. The dimension designations apply to a bolt (here, the threads of the retained)

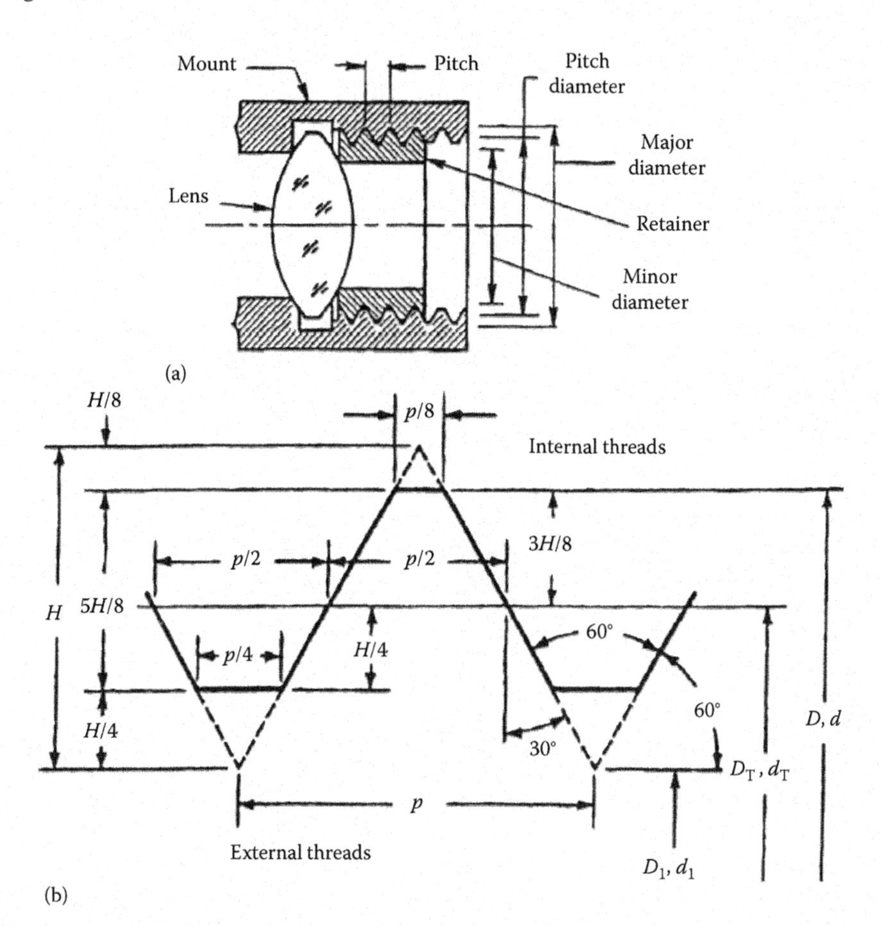

FIGURE 5.22 (a) Schematic showing terminology for retainer screw threads. (b) Basic thread profile where D (d) is the major diameter, D_1 (d_1) is the minor diameter, D_T (d_T) is the pitch diameter, and p is the thread pitch. Capital and lowercase letters represent external and internal threads, respectively. (Adapted from Shigley, J.E., and Mischke, C.R., The design of screws, fasteners, and connections, Chapter 8 in *Mechanical Engineering Design*, 5th ed., pp. 409–458, McGraw-Hill, New York, 1989.)

and its matching nut (here, the internal thread in the mount). These apply to the unified thread system, with two major series called "UNC" and "UNF" for coarse and fine pitches, respectively.

The average stress in the threads is the total axial preload divided by the total annular area over which that force is distributed. From the geometry of Figure 5.22b, the crest-to-root thread height H is related to the thread pitch p by the following equation:

$$H = 0.5\, p\, \sqrt{3}. \tag{5.21}$$

From the figure, the annular contact area has a radial dimension of (5/8)H. Hence, the annular area A_T per thread is given by

$$A_T = \frac{5\pi d_2 H}{8} = 1.700\, p d_T, \tag{5.22}$$

where d_T is the pitch diameter of the thread.

The first few (typically three) threads on a machine screw carry most of the tensile load developed when the screw is tightened. Assuming that this is also the case for a threaded lens retainer, the total annular area in contact is $3A_T$. Hence, the stress in the threads, σ_T, is given approximately by

$$\sigma_T = \frac{F_A}{3A_T} = \frac{0.196F_A}{pd_T}. \tag{5.23}$$

For this stress not to exceed the yield stress of the metal σ_{ys}, divided by an appropriate SF, the thread pitch p should be no smaller than that given by the following equation:

$$p = \frac{0.196\ SF\,F_A}{d_T \sigma_{ys}}. \tag{5.24}$$

It is common engineering practice to define a thread in terms of the number of threads per unit length rather than the linear dimension from crest to crest. This parameter is $1/p$. For the usual designs in which the mount has a coefficient of thermal expansion (CTE) higher than that of the lens, the thread stress should be estimated at the lowest survival temperature since the preload is the greatest then and represents a worst-case situation.

Example 5.7

A 53.34 mm diameter lens is clamped with an axial preload F_A = 55.6 N with a threaded retainer of thread diameter D_T = 55.88 mm. At the lowest design temperature, an additional axial preload of 1.112 kN is added. The retainer is made of 6061 aluminum with a yield stress of σ_{ys} = 55 MPa. Using SF = 4, what is the minimum thread pitch p?

From Equation 5.24,

$$p = \frac{0.196(4)(55.6\text{N} + 1.112\times10^3\,\text{N})}{(53.34\times10^{-3}\,\text{m})(55\times10^6\,\text{Pa})} = 312\times10^{-6}\,\text{m}.$$

This is about 3.2 threads/mm or 81 threads/in. This fine a thread pitch is difficult to machine, hard to assemble, and liable to damage during handling. Although a thread pitch of 250 μm (100 threads/in.) can be produced, a practical limit for most applications is about 625 μm, or 40 threads/in.

5.4.3 Continuous Flange Mounting

Flange retainers are most frequently used with large-aperture lenses, i.e., those with diameters exceeding about 6 in. (152 mm), where the manufacture and assembly of a truly round threaded retaining ring would be difficult. Their functions are essentially the same as that of the threaded retaining ring described in the last section.

Typical designs for a lens mounting with a continuous flange-type retainer are shown in Figure 5.23a and b.[4] These differ only in that the flange in the view in Figure 5.23a is constrained by multiple screws threaded into the cell, while in the view in Figure 5.23b, it is held with a threaded cap. With screws, a backup ring might be needed to stiffen the rim of the thin flange and minimize its bending between the screws. This is not necessary with the cap design.

The magnitude of the force produced by a given axial deflection Δ of the flange shown in either of the figures can be approximated by considering the flange to be a perforated circular plate with its

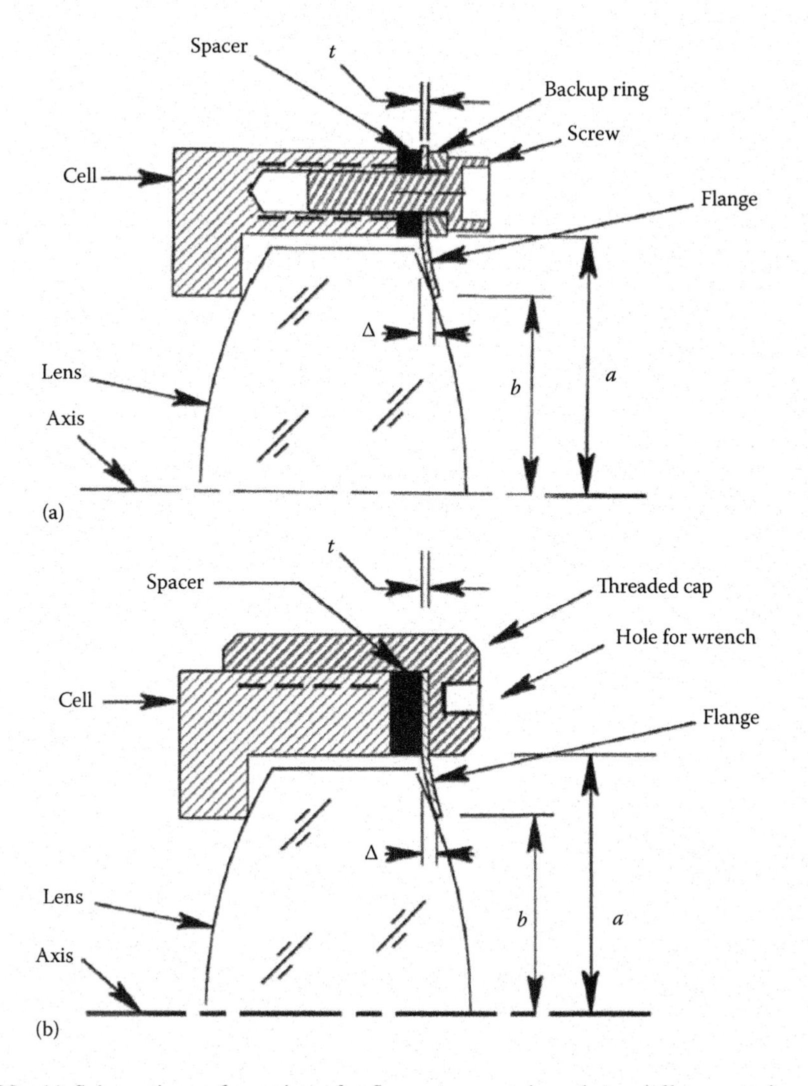

FIGURE 5.23 (a) Schematic configuration of a flange-type retainer that axially constrains a lens in a cell. The flange is stiffened with a backup ring and secured with multiple screws. (b) Schematic of an alternative design for the flange-type lens retainer that uses a threaded cap to secure the flange to the cell. (From Yoder, P.R. Jr., *Mounting Optics in Optical Instruments*, 2nd ed., SPIE Press, Bellingham, WA, 2008.)

outer edge fixed. An axially directed load is applied uniformly along the inner edge of the flange to deflect that edge. The relationship between the flange deflection Δ to the total preload F_A is given by

$$\Delta = (K_A - K_B)\frac{F_A}{t^3}, \tag{5.25}$$

$$K_A = \frac{3(m^2-1)\left[a^4 - b^4 - 4a^2b^2 \ln\left(\frac{a}{b}\right)\right]}{4\pi m^2 E_M a^2}, \tag{5.26}$$

$$K_{\mathrm{B}} = \frac{3(m^2-1)(m+1)\left[2\ln\left(\frac{a}{b}\right)+\frac{b^2}{a^2}-1\right]\left[b^4+2a^2b^2\ln\left(\frac{a}{b}\right)-a^2b^2\right]}{4\pi m^2 E_{\mathrm{M}}\left[b^2(m+1)+a^2(m-1)\right]}, \tag{5.27}$$

where t is the thickness of the cantilevered section of the flange, a is the outer radius of the cantilevered section, b is the inner radius of the cantilevered section, m is the reciprocal of the Poisson's ratio (ν_{m}) of the flange material, E_{m} is the Young's modulus of the flange material.

The spacer between the cell and the flange can be ground at assembly to the particular thickness that produces the predetermined flange deflection after firm metal-to-metal contact is achieved by tightening the clamping screws or cap. Customizing the spacer accommodates variations in as-manufactured lens thicknesses. The flange material and thickness are the main design variables. The dimensions a and b, and hence the annular width $(a - b)$, can also be varied, but these are usually set primarily by the lens aperture, mount wall thickness, and overall dimensional requirements.

The design of a flange retainer is not complete until the stress σ_{B} built up in the bent portion of the flange is determined. The flange bending stress σ_{B} must not exceed the yield stress σ_{ys} of the material. An SF = 2 with regard to the yield stress is advisable. The following equations apply:

$$\sigma_{\mathrm{B}} = \frac{K_{\mathrm{C}} F_{\mathrm{A}}}{t^2} = \frac{\sigma_{\mathrm{ys}}}{\mathrm{SF}}, \tag{5.28}$$

$$K_{\mathrm{C}} = \frac{3}{2\pi}\left[1 - \frac{2mb^2 - 2b^2(m+1)\ln\left(\frac{a}{b}\right)}{a^2(m-1)+b^2(m+1)}\right]. \tag{5.29}$$

A major advantage of the flange-type constraint over the threaded retainer is that it can be calibrated. After calibration, the preload will be known precisely when the flange is deflected by a particular amount. Calibration of the flange can be done off-line using a load cell or other means to measure the force produced by various deflections. This procedure refines the flange deflection versus force prediction made during design using the preceding equations. Because the test is nondestructive, it can be safely assumed that during actual use, the hardware will behave as measured. As a rule of thumb for reliable measurement of Δ by using conventional micrometers or height gauges, Δ should be at least 10 times the resolution capability of the measuring device. Typically, these devices can resolve ±13 μm (±0.0005 in.), so the deflection should be at least 130 μm (0.005 in.).

Example 5.8

A 400 mm diameter corrector plate is held in place on an astronomical telescope with an axial preload force F_{A} = 534 N distributed uniformly around an annulus near the edge by a titanium (Ti–6Al–4V) flange. The inner diameter and outer diameter of the flange are a = 200.3 mm and b = 197 mm. Assume that ν_{m} = 0.34, E_{M} = 114 GPa, and σ_{ys} = 114 GPa. What should be the flange thickness if SF = 2, and how much is it deflected?

From Equation 5.29,

$$K_{\mathrm{C}} = \frac{3}{2\pi}\left[1 - \frac{2\left(\frac{1}{0.34}\right)(197\ \mathrm{mm})^2 - 2(197\ \mathrm{mm})^2\left(\frac{1}{0.34}+1\right)\ln\left(\frac{200.3\ \mathrm{mm}}{197\ \mathrm{mm}}\right)}{(200.3\ \mathrm{mm})^2\left(\frac{1}{0.34}-1\right)+(197\ \mathrm{mm})^2\left(\frac{1}{0.34}+1\right)}\right] = 0.0158.$$

Solving for the flange thickness, using Equation 5.28,

$$t = \left[\frac{2(0.0158)534\,\text{N}}{827\times10^{6}\,\text{Pa}}\right]^{\frac{1}{2}} = 143\times10^{-6}\,\text{m}.$$

Using Equations 5.25 through 5.27,

$$\begin{aligned} K_A = {} & 3\left[\left(\frac{1}{0.34}\right)^2 - 1\right]\Bigg[(200.3\times10^{-3}\,\text{m})^4 - (197\times10^{-3}\,\text{m})^4 \\ & - 4(200.3\times10^{-3}\,\text{m})^2(197\times10^{-3}\,\text{m})^2 \ln\left(\frac{200.3\times10^{-3}\,\text{m}}{(197\times10^{-3}\,\text{m})}\right)\Bigg] \\ & \times\left[4\pi\left(\frac{1}{0.34}\right)^2(114\times10^{9}\,\text{Pa})(200.3\times10^{-3}\,\text{m})^2\right]^{-1} \\ = {} & 879\times10^{-21}\,\text{m}^2/\text{Pa}, \end{aligned}$$

$$\begin{aligned} K_B = {} & 3\left[\left(\frac{1}{0.34}\right)^2 - 1\right]\left(\frac{1}{0.34}+1\right) \\ & \times\left[2\ln\left(\frac{200.3\times10^{-3}\,\text{m}}{197\times10^{-3}\,\text{m}}\right) + \frac{(197\times10^{-3}\,\text{m})^2}{(200.3\times10^{-3}\,\text{m})^2} - 1\right] \\ & \times\Bigg[(197\times10^{-3}\,\text{m})^4 + 2(200.3\times10^{-3}\,\text{m})^2(197\times10^{-3}\,\text{m})^2 \ln\left(\frac{200.3\times10^{-3}\,\text{m}}{197\times10^{-3}\,\text{m}}\right) \\ & - (200.3\times10^{-3}\,\text{m})^2(197\times10^{-3}\,\text{m})^2\Bigg] \times \Bigg\{4\pi\left(\frac{1}{0.34}\right)^2(114\times10^{9}\,\text{Pa}) \\ & \left[(197\times10^{-3}\,\text{m})^2\left(\frac{1}{0.34}+1\right) + (200.3\times10^{-3}\,\text{m})^2\left(\frac{1}{0.34}-1\right)\right]\Bigg\}^{-1} \\ = {} & 724\times10^{-24}\,\text{m}^2/\text{Pa}, \end{aligned}$$

$$\Delta = \left[(879\times10^{-21}\,\text{m}^2/\text{Pa} - 724\times10^{-24}\,\text{m}^2/\text{Pa})\frac{554\,\text{N}}{(143\times10^{-6}\,\text{m})^3}\right] = 166\times10^{-6}\,\text{m}.$$

5.4.4 Multiple Cantilever Spring Clip Mounting

A simple way to clamp a lens into its mount is illustrated by Figure 5.24. In the figure, the lens rests against three thin Mylar pads (shown with exaggerated thickness) attached to a shoulder in the cell. The pads are located at ~120° intervals and serve as semikinematic registration surfaces. Three metal clips that act as cantilevered springs apply preload. The outermost ends of the clips are attached to the cell with screws and washers. Spacers between the clips and the cell are machined to produce specific deflections of the clips, thereby providing axial preload. The clips are located such that preload passes through the lens directly toward the Mylar pads. These locations are chosen

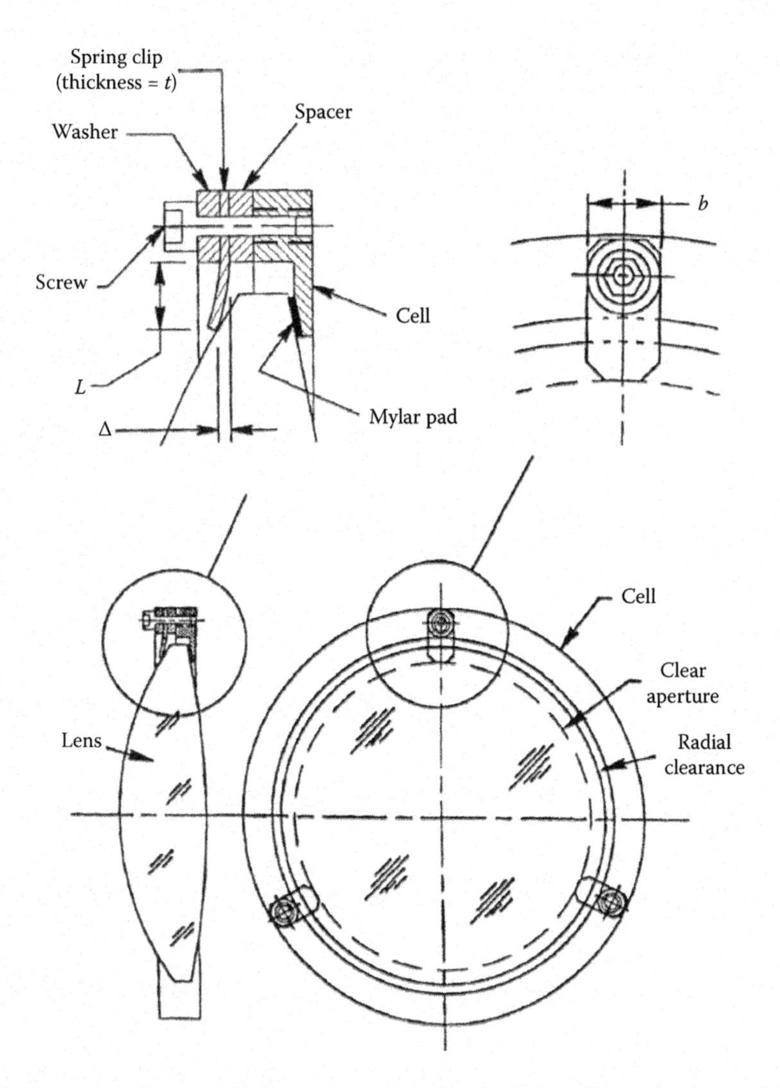

FIGURE 5.24 Concept of lens mounting using three radially oriented cantilevered springs to locally preload a lens against pads on a cell shoulder. (From Yoder, P.R. Jr., *Mounting Optics in Optical Instruments*, 2nd ed., SPIE Press, Bellingham, WA, 2008.)

because they minimize bending moments that could otherwise be applied to the optic. The use of the slightly resilient Mylar pads also helps reduce the need to tightly tolerance the shoulder as a geometrically accurate and smooth surface.

The following equations are used to calculate the deflection of each of N identical spring clips from its relaxed or undeflected condition to provide a specific total preload F_A and the bending stress σ_B in each bent clip:

$$\Delta = \frac{4F_A L^3 \left(1 - \nu_M^2\right)}{E_M b t^3 N}, \tag{5.30}$$

$$\sigma_B = \frac{6F_A L}{bt^2 N}, \tag{5.31}$$

where ν_M is the Poisson's ratio of the clip material, L is the cantilevered length of the clip, E_M is the elastic Young's modulus of the clip material, b is the width of the clip, t is the uniform thickness of each clip, and N is the number of clips.

In applications such as laser diode beam collimators, optical correlators, anamorphic projectors, and some scanning systems, the natural aperture shapes of lenses, windows, prisms, and mirrors are rectangular, racetrack, trapezoidal, etc., because their beams or fields of view are different in the vertical and horizontal meridians. Cylindrical, toroidal, and nonrotationally symmetrical aspheric optical surfaces are frequently used in such optical systems to create other desired beam shapes or to introduce different magnifications in orthogonal directions, i.e., anamorphism. These lenses may have nonrotationally symmetrical surface shapes and/or noncircular apertures, so they cannot be mounted conventionally in circular cells or held in place by threaded retainers since the surface sagittas at a given distance from the axis are not equal. Mounts for these optics are usually customized for the particular needs of the component under consideration.

A simple example of such a mounting design is shown in Figure 5.25. The lens is a plano-concave cylindrical lens with a 2:1 aperture aspect ratio. The lens is clamped with four spring clips into a rectangular recess machined into a flat metal plate. The plate is circular, so it can be attached conventionally to the structure of the instrument. Note that a slot is provided to align the cylindrical axis of the lens rotationally to the system coordinate system as represented by a pin or key (not shown) on the instrument housing. The clips provide localized preload to hold the lens in the recess under the anticipated acceleration forces.

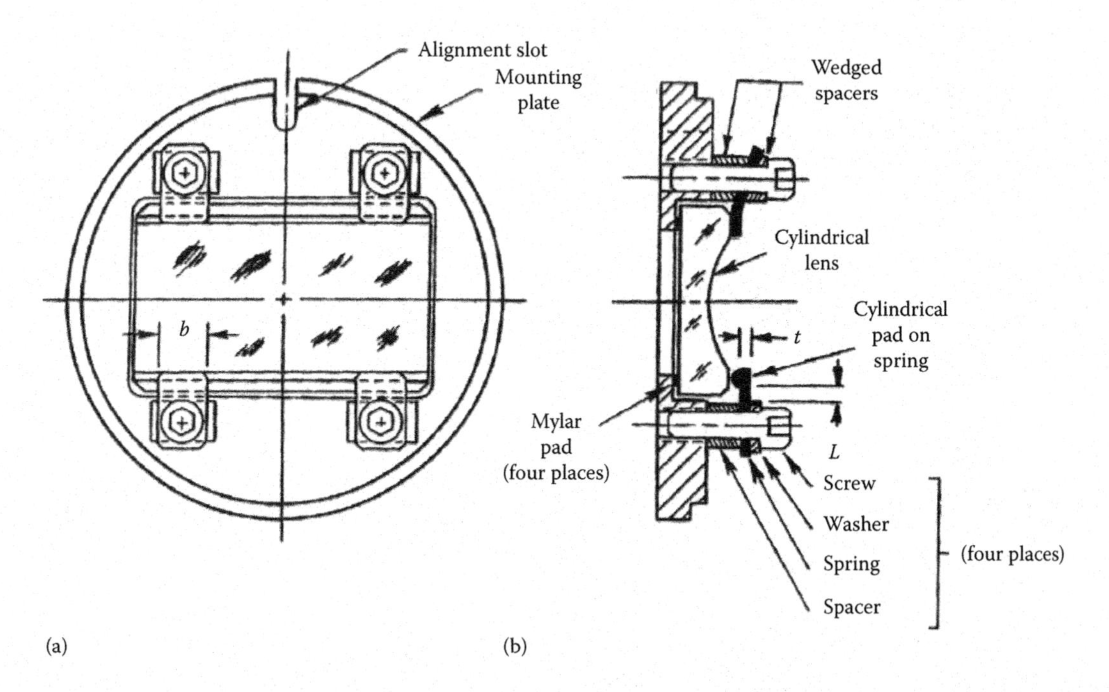

FIGURE 5.25 Schematic diagram of mounting for a rectangular aperture lens. (a) Frontal view showing how preload is provided by four cantilevered spring clips. (b) Sectional view showing means for attaching the clips. *Note:* Two types of clips are shown. The upper clip is a flat spring tilted so as to prevent sharp corner contact with the lens. In (b), a cylindrical pad on the clip provides a more controlled interface with the bevel. (From Yoder, P.R. Jr., *Mounting Optics in Optical Instruments*, 2nd ed., SPIE Press, Bellingham, WA, 2008.)

Example 5.9

A circular lens is constrained axially by an 8.5 N preload by three titanium spring clips with L = 8 mm, b = 5 mm, and t = 250 μm. Assume that $\nu_m = 0.34$ and E_m = 114 GPa. How much is each clip deflected, what is the stress in each clip, and what is the SF against yield if σ_{ys} = 827 MPa?

From Equation 5.30, the deflection is

$$\Delta = \frac{\left[1-(0.34)^2\right]4(8.5\text{N})(8\times10^{-3}\,\text{m})^3}{3(114\times10^9\,\text{Pa})(5\times10^{-3}\,\text{m})(250\times10^{-6}\,\text{m})^3} = 576\times10^{-6}\,\text{m}.$$

From Equation 5.31, the bending stress is

$$\sigma_B = \frac{6(8.5\text{N})(8\times10^{-3}\,\text{m})}{3(5\times10^{-3}\,\text{m})(250\times10^{-6}\,\text{m})^2} = 435\times10^6\,\text{Pa}.$$

The SF is the ratio of the yield stress to bending stress, or SF = σ_{ys}/σ_B:

$$\text{SF} = \frac{827\times10^6\,\text{Pa}}{435\times10^6\,\text{Pa}} = 1.90.$$

5.5 LENS BENDING

5.5.1 Bending Stress

If the axial clamping force by the retainer and the constraint or seat are not on a common diameter (at the same radial distance from the axis on both sides), a bending moment is created in the optic. This moment deforms the optic so that one side becomes more convex and the other side becomes more concave, as shown in Figure 5.26. This bending may degrade the optical performance of the

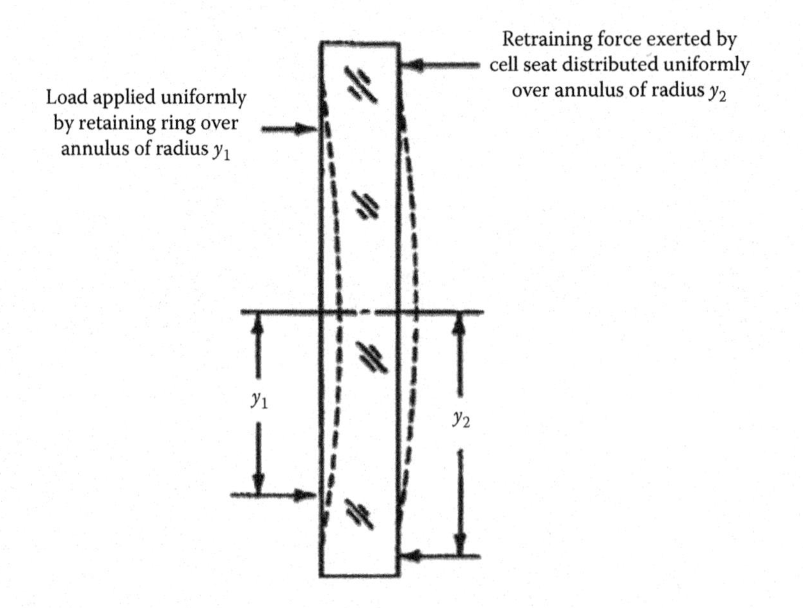

FIGURE 5.26 Geometry allowing the estimation of the effects of bending moments from axial preload and constraining force applied at different radial heights on opposite sides of a plane-parallel plate. (Adapted from Bayar, M., *Opt. Eng.*, 20, 181, 1981.)

optics. The side of the optic that becomes more convex is placed in tension, while the other surface is in compression. Brittle optical materials are much weaker in tension than in compression, so there is concern about possible structural failure.

Bending stress is estimated using an equation from plate theory. Since most optical elements are not plane-parallel plates, Equation 5.32 is an approximation. For a curved lens with variable thickness, accuracy is improved by selecting a lens thickness midway between the edge and center, typically at about one-half to two-thirds the lens diameter. Alternately, the average of the edge and center thickness can be used. As a worst case, the minimum thickness of the lens gives the highest bending stress. The bending stress σ_B is given by

$$\sigma_B = \frac{3F_A}{2\pi t^2}\frac{1}{m}\left[\frac{m-1}{2} + (m+1)\ln\left(\frac{y_1^2}{y_2^2}\right) - (m-1)\left(\frac{y_1^2}{2y_2^2}\right)\right], \tag{5.32}$$

where F_A is the axial preload force; t is the lens thickness (see the preceding discussion for selection); m is the inverse Poisson's ratio of the lens, $m = 1/\nu_G$; y_1 is the smaller contact radius, and y_2 is the larger contact radius.

5.5.2 Self-Weight Deflection of Lenses

When gravity acts normal to the optical surface of a lens or parallel to the optical axis, the lens is subjected to a load depending on the local lens thickness and material density. This load bends the lens out of shape, with the upper side becoming more concave, while the lower side is more convex. Bending can degrade the optical performance of the optic.

Bending from gravity loads is estimated by an equation from plate theory. This method is approximate for the same reasons given in the preceding section: lenses are not plane-parallel plates. An optical path difference (OPD) is induced in a wave front passing through a pressure deformed plane-parallel circular window; this topic is discussed in Section 4.2.1. If the OPD is $\lambda/8$, where λ is the wavelength, the effect is negligible; the window is "diffraction limited." Equations 4.1 and 4.2 give the minimum thickness necessary for a clamped and simply supported window respectively to limit the pressure-induced OPD in the wave front to $\lambda/8$. Replacing the pressure term P with the weight per unit area ρh, where ρ is the material density and h is the thickness, Equation 4.2 for a simply supported window becomes

$$h = 1.01d\left[(n-1)\left(\frac{\rho h}{E}\right)^2\frac{d}{\lambda}\right]^{\frac{1}{5}}, \tag{5.33}$$

where d is the window (lens) diameter, n is the index of refraction of the material, and E is the elastic modulus of the material.

Solving for the thickness,

$$h = 1.02d^2\left[(n-1)\left(\frac{\rho}{E}\right)^2\frac{1}{\lambda}\right]^{\frac{1}{3}}. \tag{5.34}$$

For glass in the visible, $n \approx 1.5$, $\rho \approx 2500$ kg/m^3, $E \approx 70$ GPa, and $\lambda \approx 550$ nm. Putting these values into Equation 5.34, for a diameter of 1 m, the required thickness is about 5 mm or $d/h = 200$! Since $d/h < 20$ even for very large lenses, the preceding example shows that gravity deflection is negligible in lenses. This rough-order-of-magnitude estimate is conservative since focus is included. If the focus is adjusted, the lens can be even thinner.

5.6 SURFACE CONTACT OPTOMECHANICAL INTERFACES

5.6.1 Sharp Corner Interface

A sharp corner interface is created by the intersection of a cylindrical hole in a lens mount and a flat surface machined perpendicular to the axis of that hole. It is the interface easiest to produce and is used in a vast majority of optical instruments. In reality, the sharp corner is not actually a knife edge. For typical machine shop practice, there is a small radius on sharp corner, so the actual contact geometry is toroidal. Normally, the sharp corner radius is 50–250 μm (0.002–0.010 in.), although smaller values are more common. This small radius surface contacts the glass at the height y_C. Figure 5.27a illustrates typical interfaces on convex spherical lens surfaces, while Figure 5.27b shows typical interfaces with concave lens surfaces.

The accuracy of the sharp corner interface is degraded by the presence of burrs left over from machining. Removing the burrs often produces an irregular edge, again lowering accuracy. Plating tends to build up preferentially and irregularly on a sharp corner, with resulting loss of accuracy. A contact angle of 45° as shown in Figure 5.27b is easier to produce without burrs or other irregularities and is a more accurate form of sharp corner contact. Disadvantages of the 45° are higher production cost and requirement for more room around the lens for the mount.

The preload force on the lens induces contact stress in the glass. Contact stress is determined by Hertz mechanics. The toroidal contact against the spherical surface of the lens is equivalent to two cylinders of finite length in contact. The length of the contact is the circumference of the contact; one cylindrical radius is that of the sharp corner or toroidal contact, and the other is the optical radius of curvature of the lens. Figure 5.28 shows the geometry of the equivalent contact, while Figure 5.29 shows a cross section through the lens mount. The maximum contact stress σ_C is compressive and is given by

$$\sigma_C = 0.798\left[\frac{\dfrac{F_A(D_1+D_2)}{2\pi y_C D_1 D_2}}{\dfrac{1-\nu_G^2}{E_G}+\dfrac{1-\nu_M^2}{E_M}}\right]^{\frac{1}{2}}, \tag{5.35}$$

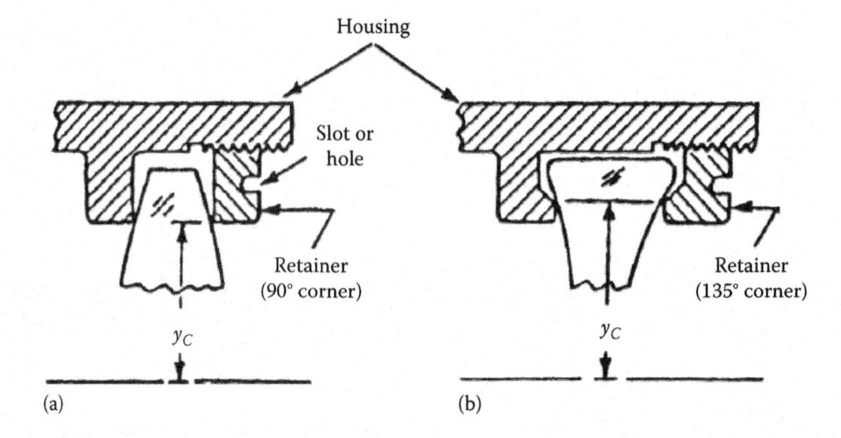

FIGURE 5.27 Typical sharp corner interfaces (a) with convex lens surfaces and (b) with concave lens surfaces.

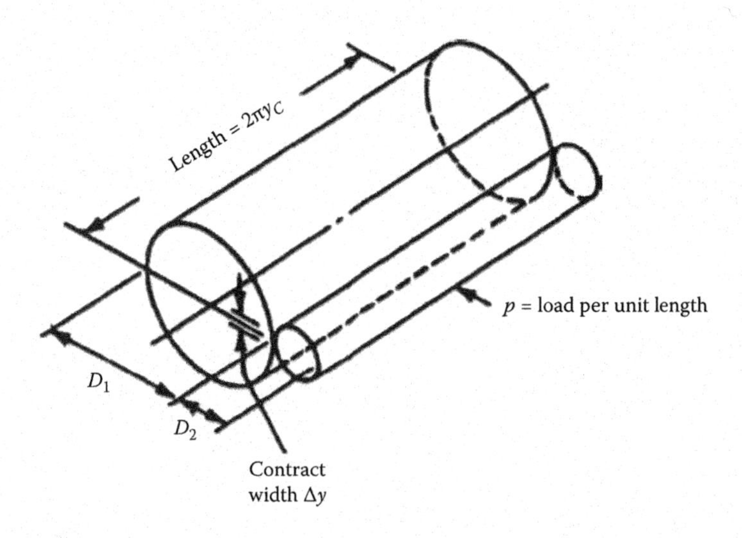

FIGURE 5.28 General analytical model of the annular interface between a convex mechanical constraint (small cylinder) and a convex surface of a lens or mirror (large cylinder).

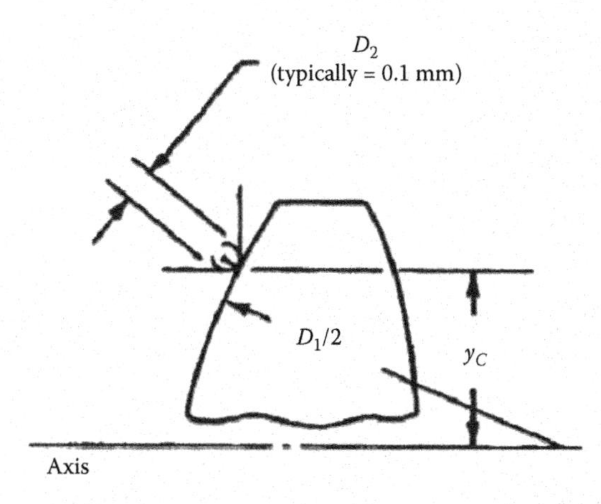

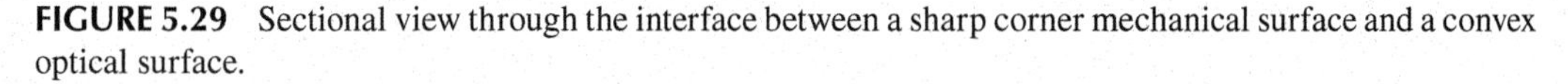

FIGURE 5.29 Sectional view through the interface between a sharp corner mechanical surface and a convex optical surface.

where F_A is the axial preload, D_1 is twice the optical radius R_1 of the lens ($D_1 = 2R_1$), D_2 is twice the radius R_2 of the mounting surface ($D_2 = 2R_2$), y_C is the radial distance from the optical axis to the point of contact, ν_G is the Poisson's ratio of the lens material, E_G is the elastic modulus of the lens material, ν_M is the Poisson's ratio of the lens mount material, and E_M is the elastic modulus of the lens mount material.

For maximum accuracy, y_C should be the distance from the optical axis to the center of curvature of the toroidal surface in contact with the lens. This improvement in accuracy is offset by the uncertainty in the radius of sharp corner contact. For most practical applications, the radial distance y_C can be assumed to be the distance to the contact.

When D_1 is about 10 times larger than D_2 ($D_1 > 10D_2$), D_1 can be ignored and Equation 5.37 from the next section is used. The error in this approximation is less than 5% for $D_1 = 10D_2$ and is usually less than 2% for most lens mounts. If $D_2 \approx 50$ μm, the minimum optical radius for this approximation is 250 μm ($R_1 = 10D_1/2$), which is a relatively small lens.

Example 5.10

A biconvex germanium lens with optical radii of 457 and 1829 mm is mounted in a 6061 aluminum cell with a sharp corner interface and $y_C = 38$ mm. The preload force $F_A = 90$ N. Assume that the sharp corner interface is $D_2 = 100$ μm. What is the contact compressive stress?

From Appendices 4 and 7,

$E_G = 103.7$ GPa,
$\nu_G = 0.278$,
$E_M = 68.2$ GPa,
$\nu_M = 0.332$.

From Equation 5.35 for the 457 mm radius surface ($D_1 = 2 \times 457$ mm = 904 mm),

$$\sigma_C = 0.798\left[\frac{\dfrac{(90\,\text{N})(904\times10^{-3}\,\text{m}+100\times10^{-6}\,\text{m})}{2\pi(38\times10^{-3}\,\text{m})(904\times10^{-3}\,\text{m})(100\times10^{-6}\,\text{m})}}{\dfrac{1-(0.278)^2}{103.7\times10^9\,\text{Pa}}+\dfrac{1-(0.332)^2}{68.2\times10^9\,\text{Pa}}}\right]^{\frac{1}{2}} = 330.8\times10^6\,\text{Pa}.$$

For the 1829 mm radius surface ($D_1 = 2 \times 1829$ mm = 3654 mm),

$$\sigma_C = 0.798\left[\frac{\dfrac{(90\,\text{N})(3.654\,\text{m}+100\times10^{-6}\,\text{m})}{2\pi(38\times10^{-3}\,\text{m})(3.654\,\text{m})(100\times10^{-6}\,\text{m})}}{\dfrac{1-(0.278)^2}{103.7\times10^9\,\text{Pa}}+\dfrac{1-(0.332)^2}{68.2\times10^9\,\text{Pa}}}\right]^{\frac{1}{2}} = 330.7\times10^6\,\text{Pa}.$$

Since $D_1 > 10D_2$ (457 mm > 1.00 mm), Equation 5.37 could also be used; for both surfaces, the contact stress is

$$\sigma_C = 0.798\left[\frac{\dfrac{90\,\text{N}}{2\pi(38\times10^{-3}\,\text{m})(100\times10^{-6}\,\text{m})}}{\dfrac{1-(0.278)^2}{103.7\times10^9\,\text{Pa}}+\dfrac{1-(0.332)^2}{68.2\times10^9\,\text{Pa}}}\right]^{\frac{1}{2}} = 330.7\times10^6\,\text{Pa}.$$

The error associated with this approximation is about $\varepsilon \approx 302 \times 10^{-6}$, or 0.03%.

5.6.2 Tangential Interface

If the mechanical surface contacting a convex spherical lens surface is conical, the design is said to have a tangent cone interface or, more simply, a tangential interface as shown in Figure 5.30. The cone half angle ψ is determined by the following equation, where R is the optical radius of curvature of the lens:

$$\psi = 90° - \arcsin\left(\frac{y_C}{R}\right). \tag{5.36}$$

The tangential interface is not feasible with a concave lens surface, but it is generally regarded as the nearly ideal interface for convex lens surfaces. Easily and economically made by modern

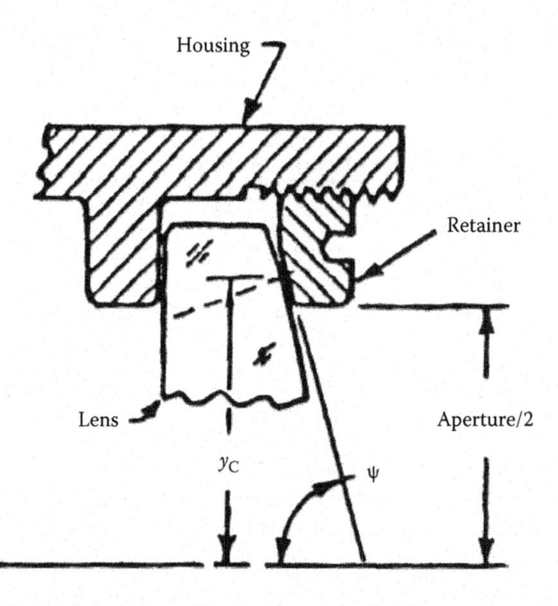

FIGURE 5.30 Tangential (conical) interface with a convex spherical lens surface. Angle ψ is the half angle of the cone.

machining technology, the conical interface tends to produce smaller contact stress in the lens for a given preload than the sharp corner interface. Manufacturing defects such as burrs or buildup from plating are not concerns with the tangential interface. Disadvantages of the tangential interface are the need for more room around the lens for mounting and lower accuracy in lens location. A retaining ring with a conical surface is shown in Figure 5.30.

The contact stress between the lens and mount is determined by a modified form of Equation 5.35. For the tangential contact, $D_2 \gg D_1$, so the contact stress is given by

$$\sigma_C = 0.798\left[\frac{\dfrac{F_A}{2\pi y_C D}}{\dfrac{1-\nu_G^2}{E_G}+\dfrac{1-\nu_M^2}{E_M}}\right]^{\frac{1}{2}}, \tag{5.37}$$

where D is twice the lens optical radius R ($D = 2R$).

Example 5.11

A biconvex germanium lens with optical radii of 457 and 1829 mm is mounted in a 6061 aluminum cell with a tangential interface and y_C = 38 mm. The preload force F_A = 90 N. What is the contact compressive stress? This is the same as Example 5.10, but with a tangential contact.

From Equation 5.37, for the 457 mm radius surface,

$$\sigma_C = 0.798\left[\frac{\dfrac{90\text{N}}{2\pi(38\times10^{-3}\text{ m})(2)(457\times10^{-3}\text{ m})}}{\dfrac{1-(0.278)^2}{103.7\times10^9\text{ Pa}}+\dfrac{1-(0.332)^2}{68.2\times10^9\text{ Pa}}}\right]^{\frac{1}{2}} = 3.459\times10^6\text{ Pa}.$$

For the 1829 mm radius surface,

$$\sigma_C = 0.798\left[\frac{\dfrac{90\,\text{N}}{2\pi(38\times10^{-3}\,\text{m})(2)(1.829\,\text{m})}}{\dfrac{1-(0.278)^2}{103.7\times10^9\,\text{Pa}}+\dfrac{1-(0.332)^2}{68.2\times10^9\,\text{Pa}}}\right]^{\frac{1}{2}} = 1.729\times10^6\,\text{Pa}.$$

For the 457 mm surface, the tangential stress is 96 times smaller, and for the 1829 mm surface, 256 times smaller, than the sharp corner contact.

5.6.3 Toroidal Interface

Figure 5.31a shows a toroidal or donut-shaped mechanical surface contacting a convex spherical lens surface of radius R, while Figure 5.31b shows a similar interface on a concave lens surface. Contact nominally occurs at the center of the toroidal land in both cases. The radial distance from the optical axis to the point of contact is y_C. Figure 5.32a and b shows families of toroids interfacing with convex and concave lens surfaces of radii R.[5] The toroid radii vary from $-R/2$ to $-R/32$ in each case. This type of interface is particularly useful on concave lens surfaces where the tangent interface cannot be used. Contact stress is reduced with the toroidal interface.

Contact stress for the toroidal interface is evaluated by Equation 5.35, with D_2 replaced with $2R_T$, where R_T is the radius of the toroidal surface. Toroidal interfaces are almost always convex. When $2R_T \ll D_1$, the interface is a sharp corner. When $R_T \approx \infty$, the interface is tangential. Finally, when $2R_T = D_1$, the interface is spherical, and this is discussed in the next section.

5.6.4 Spherical Interface

Figure 5.33a and b shows, respectively, spherical lands interfacing with convex and concave spherical lens surfaces. Contacts occur over the entire spherical lands. Since the preload is distributed over a relatively large area, the contact stress is very low. If the spherical radius of the contact accurately matches the optical radius of curvature to within a few wavelengths of light, the contact stress is given by dividing the preload by the annular contact area. If the surfaces do not match closely, the contact degenerates to a sharp corner or tangential interface.

The spherical interface surfaces of the mechanical parts must be accurately ground and lapped to match the radii of the lenses within a few wavelengths of light. Each mechanical part that is to

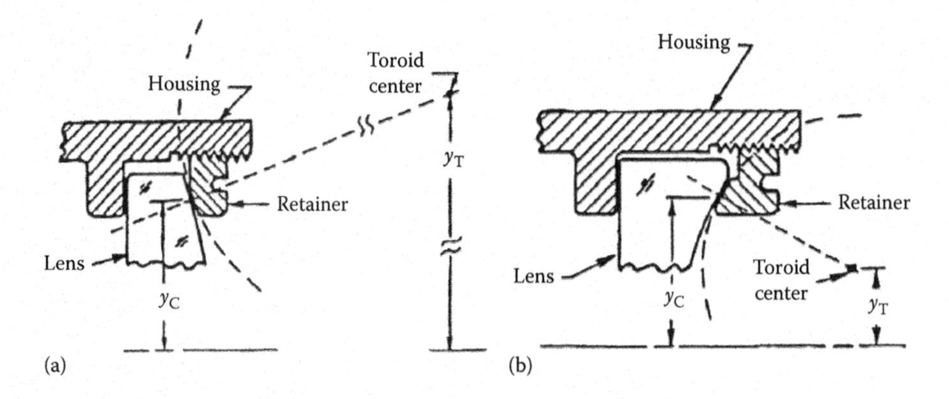

FIGURE 5.31 Toroidal (donut) interface: (a) contacting a convex spherical lens surface and (b) contacting a concave lens surface.

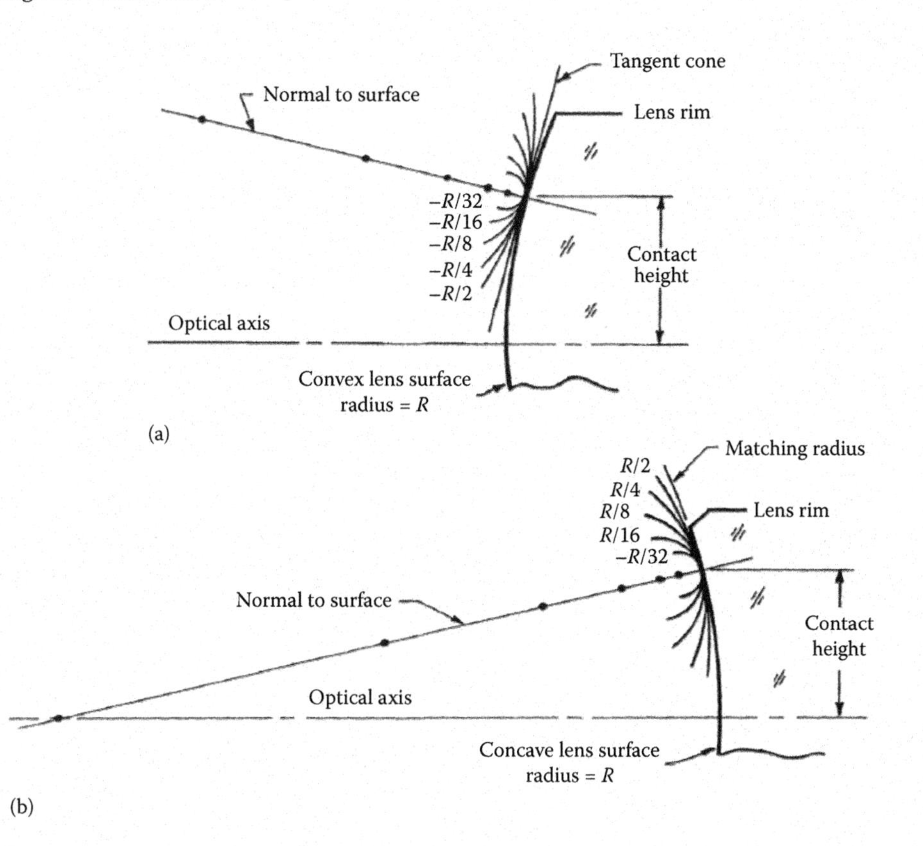

FIGURE 5.32 (a) Convex lens surface of radius R interfacing with convex toroidal surfaces of differing radii. The tangent cone is a limiting case. (b) Concave lens surface of radius R interfacing with convex toroidal surfaces of differing radii. The matching radius or spherical radius is a limiting case. (From Yoder, P.R. Jr., Axial stresses with toroidal lens-to-mount interfaces, *Proc. SPIE*, 1533, 2, 1991.)

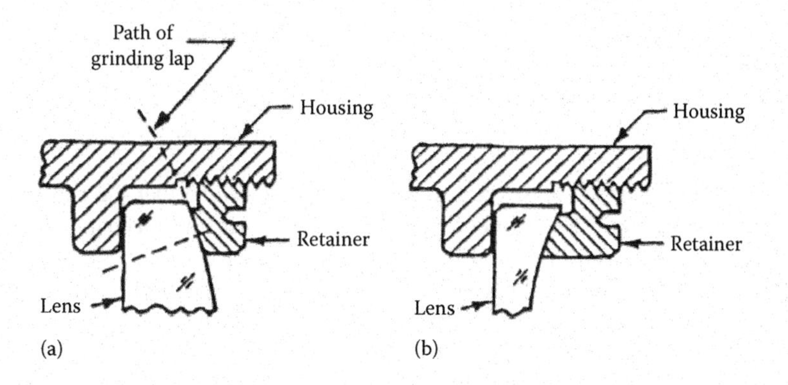

FIGURE 5.33 Spherical interface with (a) a convex spherical surface and (b) a concave surface.

touch a lens surface must be designed with access for lapping. The final stages of manufacture of these surfaces are usually conducted in the optical shop by using tools equivalent to those used to make the corresponding glass surfaces. Because the manufacture and testing of these surfaces are expensive, the spherical interface mounting technique is seldom used. The spherical interface does have the distinct advantage that axial forces are distributed over large areas, so they cause relatively low contact stresses under very high preloads. Hence, high accelerations can be survived. With a relatively large contact area, the spherical contact also facilitates heat transfer through the interface.

5.6.5 Interfaces on Bevels

It is standard optical shop practice to lightly bevel all sharp edges of optics. This minimizes the danger of chipping, so such bevels are called *protective bevels.* Larger bevels are used to remove unneeded material when weight is critical or packaging constraints are tight and to provide mounting surfaces. Usually, all these secondary surfaces are ground with several steps of progressively finer abrasives. If the lenses are likely to have to endure severe stress, the bevels and the lens rims may also be given a crude polish by buffing with polishing compound on a cloth- or felt-covered tool. These grinding and polishing procedures tend to strengthen the lens material by removing subsurface damage resulting from the grinding operations.

Figure 5.34a through c show three lenses with bevels. The plano-convex element in the view in Figure 5.34a has minimum protective bevels that traditionally might be specified as "0.5 mm maximum face width at 45°" or " 0.4 ± 0.2 mm face width symmetric to surfaces." Each surface of the biconcave lens in the view in Figure 5.34b has two wider annular flat bevels oriented perpendicular to the optical axis of the lens. Applying axial preload cannot center this lens; some external means must be utilized. Tight tolerances on perpendicularity must be specified for such bevels if both centers of curvature of the lens surfaces are to be brought to the mechanical axis of the mount simultaneously by lateral translation of the lens.

Typically, the accuracy of location with the bevel interface is the lowest of all contact geometries since it is difficult for the optics shop to locate the mounting surfaces relative to the optical axis. Mounting cost is also highest, since additional shop operations are required to produce the bevels. For these reasons, the bevel interface is uncommon and is not a preferred design.

Figure 5.34c shows a meniscus lens with a wide 45° bevel on the concave side and a step bevel ground into the rim on the other (convex) side to form a flat surface perpendicular to the axis and recessed into the lens. A conventional retainer or a spacer can be brought to bear against the latter surface as shown in Figure 5.34d. A 45° bevel or a radius should be built into the inner leading edge of the retainer so that it does not interfere with a rounded inside corner that usually is created in a step bevel during manufacture.

As a general rule, the only types of bevels used as mechanical references in mounting lenses are the flat bevel or the equivalent surface of a step bevel. This is because dimensional and angular tolerances on all other bevels are usually much more lenient and their positions may not be predictable.

Cemented doublet and triplet lenses are sometimes configured so that the rim of one element extends radially beyond the other element(s). The interface with the mechanical surround then occurs on just that larger element. Two designs with this type construction are shown in Figures 5.35 and 5.36. There are at least two advantages to these designs: the weight of one element is reduced and small geometrical wedges in the cantilevered element or in the cement joint will not affect the symmetry of the mounting interface. As mentioned earlier, the thread fits for the retainers are intentionally slightly loose so that they can register against the lens surface despite small residual wedges in the clamped elements.

In Figure 5.35, the crown element is the larger element. The mechanical interface on the convex surface is a conical surface on the retainer, while that on the concave surface is a toroidal surface machined into a shoulder in the cell. For clarity, the toroid radius in the figure is shown shorter than the true scale. The OD of that element is slightly smaller than the local ID of the cell so that its rim does not contact the cell wall.

In Figure 5.36, the flint element is the larger element. It has precision flat bevels on both sides that interface with the flat sides of a shoulder in the cell and a threaded retainer. Any small wedge in the

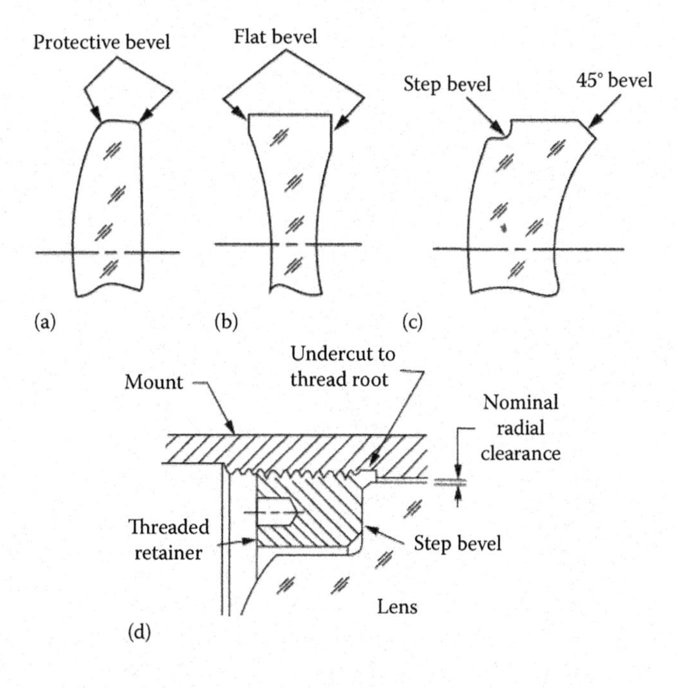

FIGURE 5.34 Types of flat bevels as applied to lenses: (a) protective bevels, (b) flat bevels, and (c) step and 45° bevels; (d) details of a step bevel on a convex surface.

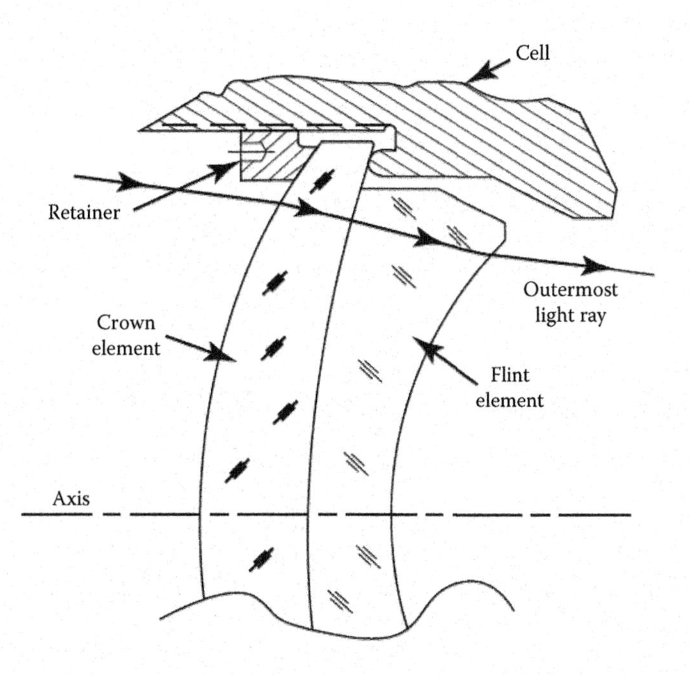

FIGURE 5.35 Mounting for a cemented doublet lens with the crown element larger in diameter than the flint element. (Adapted from Yoder, P.R. Jr., *Mounting Optics in Optical Instruments*, 2nd ed., SPIE Press, Bellingham, WA, 2008.)

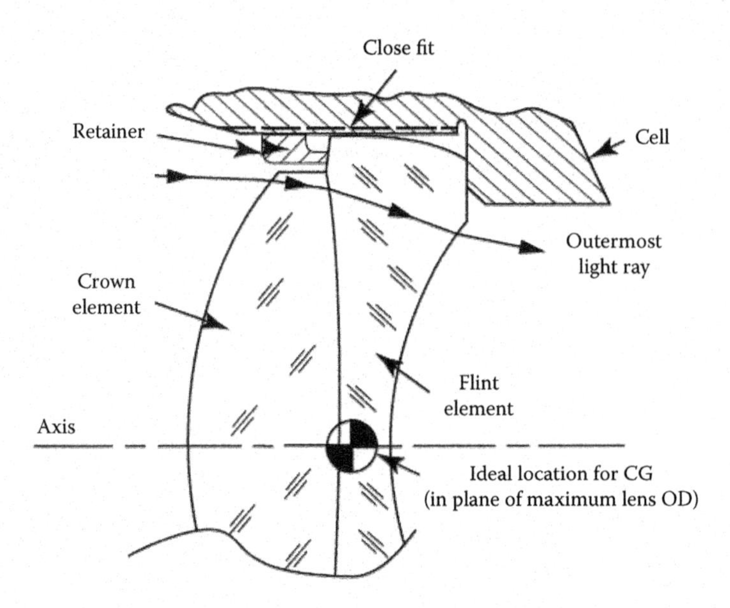

FIGURE 5.36 Mounting for a cemented doublet lens with the flint element larger in diameter than the crown element. In this rim-contact type of mounting, the rim of the flint element is spherical. (Adapted from Yoder, P.R. Jr., *Mounting Optics in Optical Instruments*, 2nd ed., SPIE Press, Bellingham, WA, 2008.)

cement joint has no effect on the mounting. The shoulder face should be toleranced to be accurately perpendicular to the mechanical axis of the cell. The rim of the flint element is spherical (shown exaggerated) so that it can fit closely into the ID of the lens barrel. The high point on the rim lies approximately on the plane of the CG of the lens for maximum stability under shock and vibration.

5.6.6 Parametric Evaluation of Contact Stress

Figure 5.37a shows the variation in contact stress with interface geometry for a convex lens surface with optical radius $R = 457$ mm (18 in.), radial contact distance $y_C = 39.4$ mm (1.505 in.), and radial preload $F_A = 91.9$ N (20.7 lb), where the lens material is germanium and the cell material is aluminum. When the toroidal interface radius is about 10 times bigger than the surface radius ($D_2 = 10D_1$), the contact stress is equivalent to that of a tangential interface. When the interface is spherical, the limiting contact stress occurs when the interface radius matches the optical surface radius and is equivalent to the preload divided by the contact area. Figure 5.37b shows a cross section through the interfaces.

5.7 ELASTOMERIC RING MOUNTINGS FOR LENSES

A simple technique for mounting lenses in cells is illustrated schematically in Figure 5.38. This figure shows a typical configuration of a lens constrained by an annular ring of a resilient elastomeric material within a cell. Typically, room-temperature-vulcanizing (RTV) sealing compounds such as Dow Corning RTV732, Dow Corning RTV93-500, General Electric RTV88, and General Electric RTV8112 have been used successfully. EC2216B/A epoxy made by 3M has also been used for this purpose and is representative of that type of elastomer. Unfortunately, important parameters, such as Poisson's ratio and Young's modulus, are not always specified by the manufacturers of the materials. If they are given, the numbers generally represent averages of many production lots, so they may not accurately apply to a specific lot. In especially critical applications, it is advisable for the parameters of the selected material lot to be measured before using.

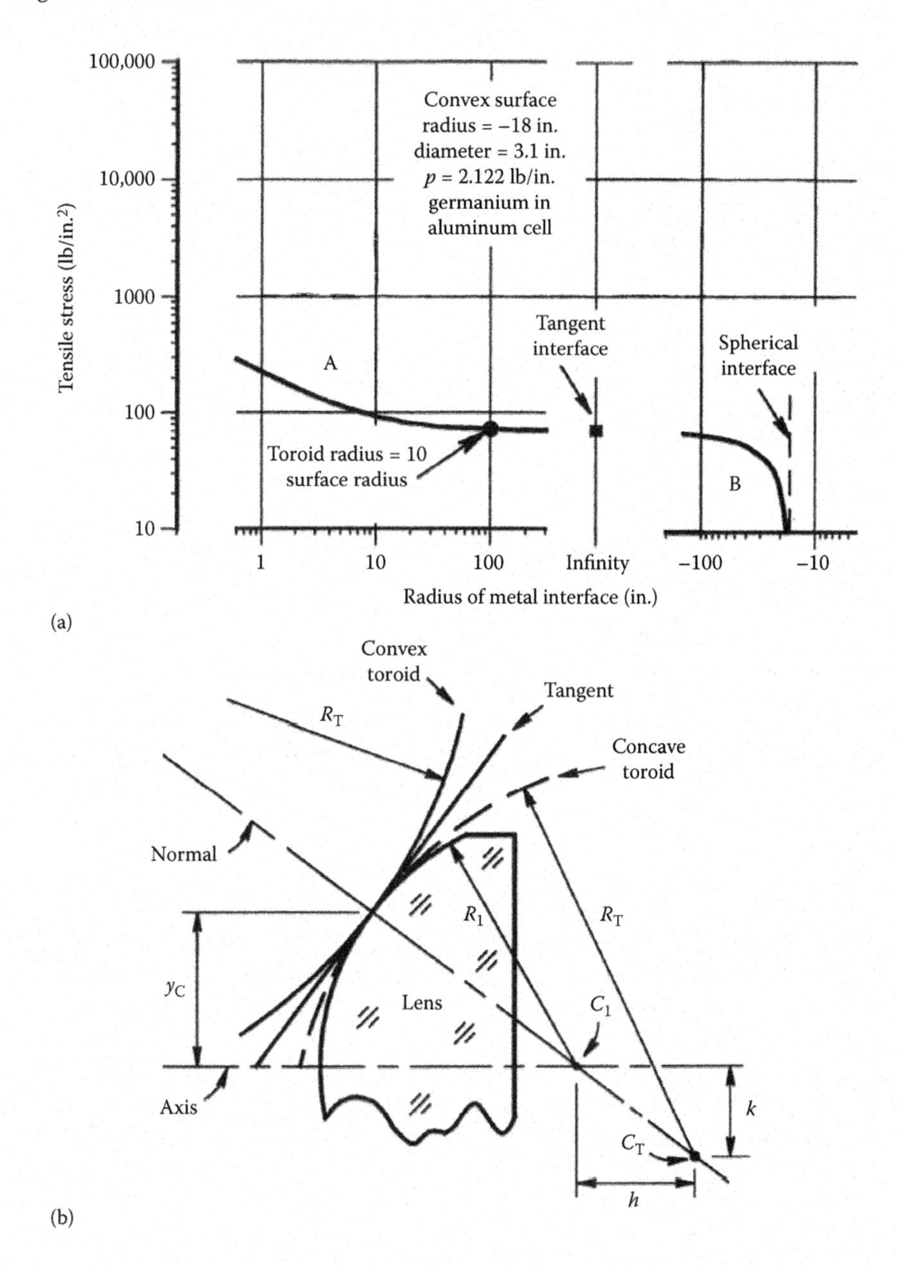

FIGURE 5.37 (a) Variations in tensile contact stress as a function of the sectional radius of the mechanical contacting surface for a convex lens surface. Curve A represents a convex toroidal interface, and curve B represents a concave toroidal interface. That curve ends at the spherical interface. The tangential interface is shown as an infinite toroidal interface radius. The design dimensions and characteristics are indicated. (b) Shows that interfaces for all three contacting surfaces

One (or sometimes both) sides of the elastomeric ring is (are) intentionally left exposed so that the material surface can deform under radial compression or tension caused by curing and/or temperature changes. Axial alignment of the lens should be achieved by registering one optical surface against a machined mount surface. Centration should be established prior to insertion of the elastomer and maintained throughout the curing cycle with shims or an external fixture. The detail view of Figure 5.38 shows a simple ring-shaped tool, or fixture, that can be used to hold the lens in

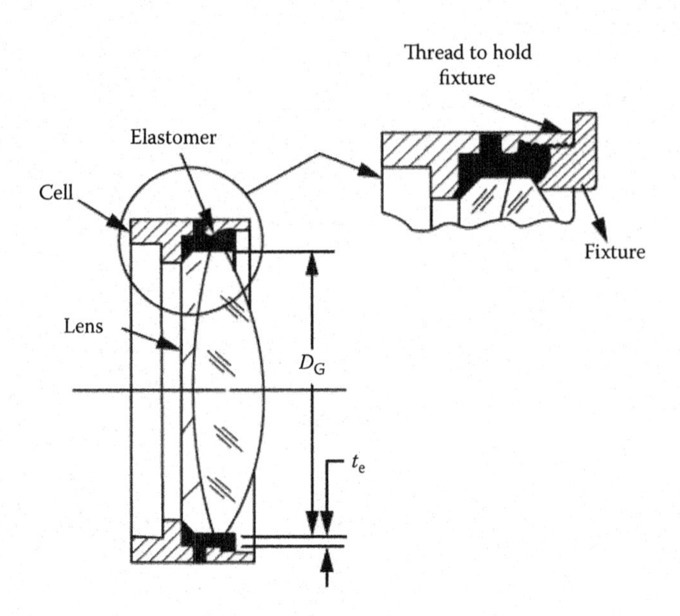

FIGURE 5.38 Technique for mounting a lens with an annular ring of cured-in-place elastomer. The detail view shows one means for holding the lens and constraining the elastomer during curing.

place while constraining the elastomer during curing. Made of Teflon or a similar plastic or metal coated with a release compound, it is removed after the elastomer has cured. The elastomer is typically injected through a hypodermic needle directly into the gap between the lens and the mount or through radially oriented holes in the mount until the space around the lens is filled. If shims are used to center the lens, they are removed after the curing cycle. The remaining small voids can be plugged with additional elastomer if the subassembly is to be sealed.

A circumferential adhesive or elastomeric bond around the lens is athermal if the radial stress in the optic is zero when the temperature changes. Without the elastomer, radial stress is induced in the lens when the temperature is lowered enough for the cell to come into contact. Since the CTE of the elastomer is higher than that of lens or cell, with the right radial thickness, the expansion or contraction of the elastomer with temperature just matches the change in radial clearance between cell and lens. For an athermal bond $\alpha_r > \alpha_M > \alpha_G$, where α_r is the CTE of the elastomer, α_M is the CTE of the cell material, and α_G is the CTE of the lens material. In deriving relationships for the athermal thickness, it is assumed that the CTEs of the materials are constant with temperature. While an athermal design is possible with material properties that vary with temperature, the design process becomes complex, requiring advanced numerical methods outside the scope of this discussion.

Considering only the radial change in clearance between the lens and cell width and assuming that the elastomer is in a zero stress condition, the optimum athermal radial elastomer thickness t_e is given by the Bayar equation:

$$t_{eBayar} = \frac{D_G}{2}\left(\frac{\alpha_M - \alpha_G}{\alpha_r - \alpha_G}\right), \tag{5.38}$$

where D_G is the lens diameter, α_M is the CTE of the cell material, α_G is the CTE of the lens material, and α_r is the CTE of the elastomer or adhesive.

The Bayar equation neglects the effects of Poisson's ratio and confinement of the elastomer. Both effects become increasingly important as the thickness-to-length ratio of the elastomer bond becomes larger. Equation 5.38 should be considered an approximation for preliminary analysis. It is also suitable for the design of discrete bond pads or a discontinuous series of bonds around the lens, where the elastomer is not confined.

If the elastomer is constrained in radial and axial directions, the required athermal radial thickness is given by the van Bezooijen equation. In the axial direction, the elastomer is constrained to the surfaces of lens and cell. This equation includes the effect of axial expansion or contraction of lens and cell. The optimum athermal elastomer thickness t_{eVB} is

$$t_{eVB} = \frac{D_G}{2} \frac{\alpha_M - \alpha_G}{\alpha_r - \alpha_M + \frac{2\nu_r}{1-\nu_r}\left(\alpha_r - \frac{\alpha_G + \alpha_M}{2}\right)}, \tag{5.39}$$

where ν_r is the Poisson's ratio of the elastomer.

If the elastomer is unconstrained in the axial direction, a modified form of Equation 5.39 gives the optimum radial thickness t_{eMVB} for the elastomer:

$$t_{eMVB} = \frac{D_G}{2} \frac{\alpha_M - \alpha_G}{\alpha_r - \alpha_M + \frac{\nu_r}{1-\nu_r}\left(\alpha_r - \frac{\alpha_G + \alpha_M}{2}\right)}, \tag{5.40}$$

Equations 5.39 and 5.40 define the upper and lower limits of the required elastomer thickness. An intermediate approximation which includes the bulging of the elastomer as the temperature changes is given by Monti's aspect ratio approximation. Although more complex than the other equations, the aspect ratio approximation is closest to results from finite-element analysis. The optimum athermal radial thickness is given by the aspect ratio approximation as

$$t_{eAR} = \frac{-b + \sqrt{b^2 - 4ac}}{2a}, \tag{5.41}$$

$$a = \frac{-\nu_r}{L}\left(\alpha_r - \frac{\alpha_G + \alpha_M}{2}\right), \tag{5.42}$$

$$b = (1-\nu_r)(\alpha_r - \alpha_M) + 2\nu_r\left(\alpha_r - \frac{\alpha_G + \alpha_M}{2}\right), \tag{5.43}$$

$$c = \frac{-D_G}{2}(1-\nu_r)(\alpha_M - \alpha_G), \tag{5.44}$$

where L is the length of elastomer (typically the edge thickness of the lens).

Lens mounts designed with the original Bayar equation (Equation 5.38) provide satisfactory results in many service environments. The inaccuracy of Equation 5.38 is offset by the low stiffness of the elastomers typically used to bond the lens into the cell. The radial thickness given by the Bayar equation is greater than that given by Equations 5.39 through 5.41, which provides additional radial compliance to reduce thermal stress. If the elastomer thickness is not optimum, the thermal stress σ_T in the optic is given approximately by

$$\sigma_T = -E_r \varepsilon \left(\frac{1 + \dfrac{\nu_r}{1 - 2\nu_r}}{1 + \nu_r} \right), \tag{5.45}$$

$$\varepsilon = \frac{\Delta t_e}{t_e + \Delta t_e} \left\{ \alpha_r - \frac{2\nu_r}{1 - \nu_r} \left[\frac{(\alpha_M - \alpha_G)}{2} - \alpha_r \right] \right\} \Delta T, \tag{5.46}$$

where E_r is the elastic modulus of the elastomer, Δt_e is the variation in elastomer thickness, and ΔT is the temperature change.

The accuracy of Equations 5.39 through 5.38 is limited by errors associated with the values of the elastomer properties. A variation of 10% in CTE in the elastomer changes the optimum radial thickness by about 15%. A change of 10% in the Poisson's ratio of the elastomer changes the optimum radial thickness by about 12%.

Elastomers used in these types of mountings typically have Poisson's ratios (ν_r) of 0.4300–0.4999. The maximum value of Poisson's ratio is 0.5000. Epoxies lie at the lower end of the range, while RTV elastomers lie at the top of the range.

Under radial acceleration, a lens surrounded by an elastomer displaces radially. The radial decenter δ for a radial acceleration a_G is given by

$$\delta = \frac{W_G a_G}{\dfrac{\pi}{2} D_G \dfrac{L}{t_e} \left(\dfrac{E_r}{1 - \nu_r^2} + G_r \right)}, \tag{5.47}$$

where W_G is the lens mass and G_r is the shear modulus of the elastomer.

The decentrations of small and modest-sized optics in a horizontal-axis orientation under normal gravity loading are usually quite small, but may grow significantly under shock and vibration loading. An elastomer is naturally somewhat elastic and so will tend to restore the lens to its unstressed location and orientation when the acceleration force dissipates.

Example 5.12

The mass of a 50 mm diameter lens (D_G = 50 mm) with an edge thickness of 17.4 mm and made of Schott N-SK16 glass is 124 g. The lens is mounted in a 6061 aluminum cell by using an athermal bond. The adhesive properties are E_r = 3.5 MPa, ν_r = 0.49, and $\alpha_r = 250 \times 10^{-6}$ K^{-1}. What is the athermal bond thickness using Bayar's method? What is the athermal bond thickness using Monti's method? Compare the lens stress for a temperature drop of 50 K between Bayar's and Monti's methods, assuming that the latter thickness is correct. What is radial decenter of the lens under self-weight for the bond line thickness given by Monti's method?

From Appendices 1 and 7, the material properties for aluminum and Schott N-SK16 are

$$\alpha_M = 23.6 \times 10^{-6} \mathrm{K}^{-1},$$

$$\alpha_G = 6.3\times10^{-6}\text{K}^{-1}.$$

The athermal bond thickness for Bayar's method is given by Equation 5.38:

$$t_{e\text{Bayer}} = \frac{50\times10^{-3}\,\text{m}}{2}\left(\frac{23.6\times10^{-6}\,\text{m}-6.3\times10^{-6}\,\text{m}}{250\,\times10^{-6}\,\text{m}-6.3\times10^{-6}\,\text{m}}\right)=1.78\times10^{-3}\,\text{m},$$

For Monti's method, the athermal bond thickness is given by Equations 5.41 through 5.44:

$$a=\frac{-0.49}{17.4\times10^{-3}\ \text{m}}\left(250\times10^{-6}\text{K}^{-1}-\frac{6.3\times10^{-6}\text{K}^{-1}+23.6\times10^{-6}\text{K}^{-1}}{2}\right)$$
$$=-6.6619\times10^{-3}\,\text{m}^{-1}\text{K}^{-1},$$

$$b=(1-0.49)(250\times10^{-6}\text{K}^{-1}-23.6\times10^{-6}\text{K}^{-1})$$
$$+2(0.49)\left(250\times10^{-6}\text{K}^{-1}-\frac{6.3\times10^{-6}\text{K}^{-1}+23.6\times10^{-6}\text{K}^{-1}}{2}\right)$$
$$=345.8\times10^{-6}\text{K}^{-1},$$

$$c=\frac{-\,50\times10^{-3}\,\text{m}}{2}(1-0.49)(23.6\times10^{-6}\ \text{K}^{-1}-6.3\times10^{-6}\text{K}^{-1})$$
$$=-220.6\times10^{-9}\,\text{m/K},$$

$$t_{e\text{AR}}=\frac{1}{2(-6.6619\times10^{-3}\,\text{m}^{-1}\text{K}^{-1})}\Big\{-345.8\times10^{-6}\text{K}^{-1}+\big[(345.8\times10^{-6}\text{K}^{-1})^2$$
$$-4(-6.6619\times10^{-3}\,\text{m}^{-1}\text{K}^{-1})(-220.6\times10^{-9}\,\text{m/K})\big]^{\frac{1}{2}}\Big\}=645.8\times10^{-6}\ \text{m}.$$

The stress for the bond line thickness variation is given by Equations 5.45 and 5.46:

$$\Delta t_e = 1.775\times10^{-3}\,\text{m}-645\times10^{-6}\,\text{m}=1.129\times10^{-3}\,\text{m},$$

$$\varepsilon=\frac{1.129\times10^{-3}\,\text{m}}{645\times10^{-6}\,\text{m}+1.129\times10^{-3}\,\text{m}}(50\text{K})\Bigg[\ \ 250\times10^{-6}\,\text{K}^{-1}$$
$$-\frac{2(0.49)}{1-0.49}\left(\frac{23.6\times10^{-6}\ \text{K}^{-1}-6.3\times10^{-6}\ \text{K}^{-1}}{2}-250\times10^{-6}\text{K}^{-1}\right)\Bigg]=0.023,$$

$$\sigma_T = -(3.5\times10^6\,\text{Pa})(0.23)\left[\frac{1+\dfrac{0.49}{1-2(0.49)}}{1+0.49}\right] = -1.36\times10^6\,\text{Pa}.$$

This low a tensile stress in the lens is unlikely to cause problems, although a check for stress optic effects is desirable.

The radial decenter is given by Equation 5.47:

$$G_r = \frac{3.5\times10^6\,\text{Pa}}{2(1+0.49)} = 1.12\times10^6\,\text{Pa},$$

$$\delta = (0.124\text{ kg})(9.81\text{ m/s}^2)\left\{\frac{\pi}{2}(50\times10^{-3}\,\text{m})\left(\frac{17.4\times10^{-3}\,\text{m}}{645\times10^{-6}\,\text{m}}\right)\left[\frac{3.5\times10^6\,\text{Pa}}{1-(0.49)^2}+1.174\times10^6\text{ Pa}\right]\right\}^{-1}$$

$$= 99.3\times10^{-9}\,\text{m}.$$

The radial decenter is about 0.1 μm, and this is well below the tolerance for even a high-performance lens. This example shows that athermal bonds using compliant adhesives are not terribly sensitive to variations in bond line thickness, while radial decenter is not an issue.

5.8 STRESS EFFECTS AT THE OPTOMECHANICAL INTERFACE

5.8.1 Compressive and Tensile Stresses in Optics

The contact stress at the interface between lens and mount is usually compressive. This compressive stress is accompanied by a tensile stress just outside the boundary of the region elastically compressed by the preload. Brittle optical materials are much weaker in tension than in compression, and the tensile stress arising from the compressive contact stress may be a concern. The tensile stress σ_{CT} arising from the compressive contact stress σ_C is given by

$$\sigma_{CT} = \frac{\sigma_C(1-2\nu_G)}{3}, \tag{5.48}$$

where ν_G is the Poisson's ratio of the lens material.

The Poisson's ratio ν_G of optical materials varies from 0.17 for fused silica to 0.343 for BaF_2. Using Equation 5.48, $\sigma_{CT} = 0.220\sigma_C$ to $0.0.105\sigma_C$. For most optical glasses in the visible, $\nu_G \approx 0.2$ and as a rule of thumb $\sigma_{CT} \approx 0.2\ \sigma_C$. Figure 5.39 is a plot of Equation 5.48 for Poisson's ratios of 0.15–0.35.

The tensile stress at the interface limits the amount of preload on a lens. A common assumption is that this tensile stress must be less than 7–10 MPa for long-duration loading. When the duration of the loading is short, such as is associated with a shock, the tensile strength of brittle materials increases. This increase arises from the strain rate sensitivity of brittle fracture; strength is greater at higher strain rates. If τ is the shock duration and f_n is the natural frequency of the lens in bending, the duration is "short" when $\tau f_n \le 0.17$. For short-duration loading, the lens may survive a tensile stress in the interface as high as 28 MPa.

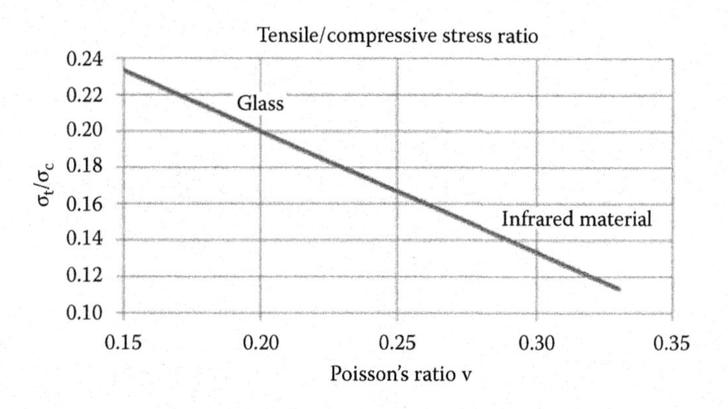

FIGURE 5.39 Ratio between tensile and compressive stress in a contact with lens as a function of Poisson's ratio of the lens material.

Example 5.13

Find the tensile stress in the lens mounted using a sharp corner contact in Example 5.10, where $\sigma_C = 330.7$ MPa and $\nu_G = 0.278$.

From Equation 5.47,

$$\sigma_{CT} = \frac{(330.7\times10^6\,\text{Pa})\left[1-2(0.278)\right]}{3} = 48.9\times10^6\,\text{Pa}.$$

The approximate fracture stress of germanium is 90 MPa, so the SF is about 90 MPa/48.9 MPa ≈ 1.8. With SF < 2 for a brittle material, a more detailed fracture analysis using Weibull statistics would be desirable.

5.8.2 Stress Birefringence

There is an OPD between two different perpendicularly polarized wave fronts passing through stressed glass. This OPD arises from the stress optic coefficients of the material. There are two stress optic coefficients for the two directions of the polarized light. Since only the total magnitude of the OPD is important, the two stress optic coefficients are summed and given in a single stress optic coefficient, K_S, with typical units of inverse megapascals (MPa^{-1+}). For optical glass in the visible, K_S is 0.02×10^{-6} MPa^{-1} for Schott SF57 to 3.90×10^{-6} MPa^{-1} for Schott N-KSF4. The OPD from the stress optic effect for a glass thickness h and stress σ is

$$\text{OPD} = K_S\sigma h. \tag{5.49}$$

Optical aberration is introduced into the wave front by the OPD from the stress optic effect. This optical aberration cannot be compensated for in the system. For some applications, the stress optic OPD limits the stress in the lens. Table 5.2 gives suggested tolerances for the stress optic OPD based on the application. Usually, the stress induced at the interface between the lens and mount is highly localized and does not extend far into the clear aperture. As a worst-case analysis, the stresses in the lens are root-sum-squared and applied at the thickest part of the lens within the clear aperture to determine the maximum stress optic OPD:

$$\text{OPD}_{\max} = K_S h_{\max}\left(\sum_{i=1}^{N}\sigma_i^2\right)^{\frac{1}{2}}, \tag{5.50}$$

TABLE 5.2
Suggested Tolerances for Stress Optic Effect in Mounted Lenses

Permissible OPD/cm Glass Path (nm/cm)	Application
<2	Polarization instruments
	Interferometers
5	Precision optics
	Astronomical optics
10	Photographic optics
	Microscope optics
20	Magnifying glasses
	Viewfinder optics
No requirement	Illumination optics

where OPD_{max} is the maximum stress optic OPD, h_{max} is the thickest part of the lens within the clear aperture, σ_i is the individual stress, and N is the number of stresses.

Example 5.14

The Schott N-SK16 lens in Example 5.12 is 22 mm thick, with a stress optic coefficient of 1.9×10^{-12} Pa^{-1}. In the example, the maximums stress for a bond line variation of 1.29 mm was found to be 1.174 MPa. What is the maximum OPD associated with this stress?

From Equation 5.49,

$$OPD = (1.9 \times 10^{-12} \text{ Pa}^{-1})(22 \times 10^{-3} \text{ m})(1.174 \times 10^{6} \text{ Pa}) = 23.2 \times 10^{-9} \text{ m}.$$

This OPD is about 23.2 nm/633 nm = 0.037 wave at a wavelength of 633 nm and is well below the normal quarter wavelength tolerance for diffraction-limited systems.

5.9 MOUNTING LOW-PRECISION LENSES

5.9.1 Low-Precision Lens Mounts

A low-precision lens is defined here (rather arbitrarily) as one requiring each surface to be centered to no better than, say, 30 arcmin deviation due to residual optical wedge or 250 mm (0.01 in.) maximum decentration relative to a common optical axis. The discussion of optomechanical designs for securing a single-element lens in a mechanical housing or cell in this section progresses from a low to a high degree of centration. The simplest designs involve springs that constrain the lens. All these designs can be classified as drop-in subassemblies in that the glass-to-metal interfaces in the mount are premachined to within some reasonable tolerances on ID and perpendicularity to the mechanical axis. The OD of the lens is dimensioned and toleranced to allow some radial clearance under all temperature conditions. The shape, or bending, of the lens is not of prime concern for low-precision mounts.

5.9.2 Spring-Loaded Mountings

Typical of optical mountings requiring a low level of precision are those for the condensing lens and heat-absorbing filter used in the illumination system of a projector. The design is driven to a large extent by the need for low cost, ease of maintenance, and free air circulation so that the optics can survive high temperatures from the nearby light source. Figure 5.40a shows such a simple mounting in which a lens made of heat-resistant glass (such as Pyrex) is held by three springs that engage the

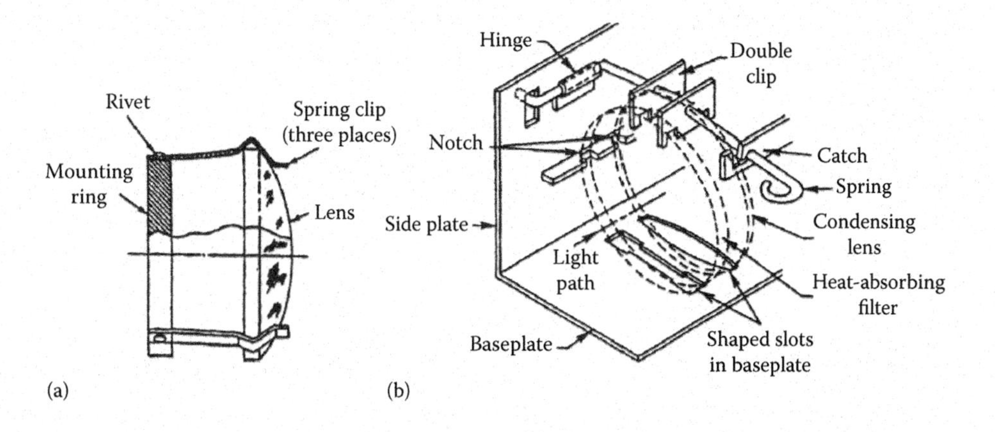

FIGURE 5.40 Spring mountings for lenses. (a) Concept using three leaf springs at 120° intervals around the lens rim and (b) mounting for a filter and lens used in a Kodak Ektagraphic slide projector. (From Yoder, P.R. Jr., *Mounting Optics in Optical Instruments*, 2nd ed., SPIE Press, Bellingham, WA, 2008.)

rim of the lens. The springs are located at 120° intervals. The symmetry of the mounting tends to keep the lens centered. The flexibility of the springs reduces the danger of breakage. Alignment of the optical elements depends on the axial locations and shapes of the detents on the springs, which can easily be controlled adequately during manufacture.

Figure 5.40b is a schematic of a low-cost, spring-loaded mounting for a heat-absorbing filter and a biconvex condensing lens as used in a Kodak Ektagraphic 35 mm slide projector. The rims of the optics fit into cutouts in the sheet-metal baseplate and into detents cut into a metal strip attached to the side of the subassembly. The cutout in the baseplate is shaped such that a lens with different first and second curvatures cannot be inserted backward. The spring holding the optics in place allows for ease of engagement and disengagement to facilitate assembly and servicing. Air from the cooling fan is directed across the optical surfaces to minimize temperature rise during use.

5.9.3 Burnished Cell Mountings

Burnishing of a lens into its cell is illustrated in Figure 5.41. This is accomplished by deforming the metal lip from the condition shown in the view in Figure 5.41a to that shown in the view in Figure 5.41) after the lens is inserted. The burnished mounting is most frequently used with small lenses such as those in microscope objectives, endoscopes, or short-focal-length camera

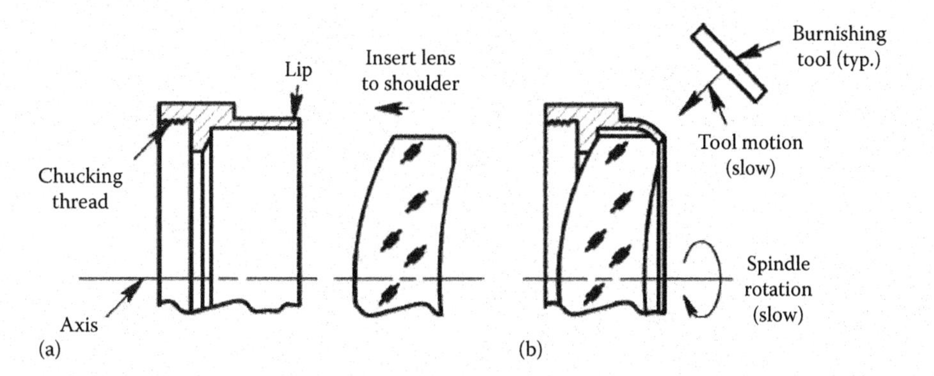

FIGURE 5.41 Lens burnished into a cell made of a malleable metal. (a) Typical cell and lens configurations. (b) Completed subassembly. (From Yoder, P.R. Jr., *Mounting Optics in Optical Instruments*, 2nd ed., SPIE Press, Bellingham, WA, 2008.)

objectives. It is inexpensive and compact and requires a minimum number of parts. Preload is uncertain. The mounting is permanent; removal of the lens without damage requires the cell to be destroyed.

The cell material must be malleable rather than brittle. Either brass or a soft aluminum alloy is appropriate for this reason. A radial clearance of 125 μm (0.005 in.) between the lens rim and the cell ID is common. One method for burnishing is performed by chucking the cell in a lathe, inserting the lens, slowly rotating the cell, and bringing three or more hardened cylindrical tools inclined at an angle against the projecting lip, bending it against the lens. Care must be exercised not to excessively deform or stress the lens during the burnishing operation. The lens must be held firmly against its seat while the oblique force is exerted on the metal so as not to decenter the lens within its radial clearance at the early stage of the process before the lens is held in place.

Alternately, a thin, narrow washer or a thin O-ring made of slightly resilient material such as Nylon or Neoprene is sometimes inserted against the exposed surface of the lens and the metal burnished over that resilient part rather than directly contacting the glass. This tends to seal the glass-to-metal interface and can offer some preloading action to hold the glass axially against the seat at higher temperatures when the metal lip tends to expand away from the glass.

Yet another version of the burnished cell mounting design is shown in Figure 5.42. Here a coil spring is inserted between the lens and the cell shoulder and slightly compressed as the lip is spun over. This method is suggested when the assembly is to be subjected to severe shock or when a low-cost means of avoiding stress in the lens is needed. A potential failure mode is rebound shock, since the spring is compressed by the initial shock. Yet another version has the spring between the lens and the burnished lip. This allows the lens to be registered directly against the shoulder, thereby somewhat enhancing location and alignment accuracy. A thin brass tube with staggered lateral slots can be used as the spring in either design. This avoids stress concentration at localized contacts between high points on the end of a coil spring and the lens that can occur even if the end of the spring is ground flat as shown in the figure. The spring must be stiffer than the lip, so the latter is bent rather than just compressing the spring while burnishing.

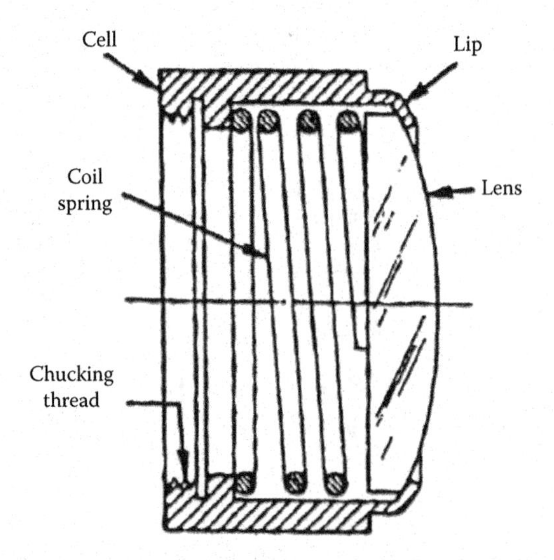

FIGURE 5.42 Variation of the burnished-in lens mounting with a spring to axially load the lens against a shoulder. (Adapted from Jacobs, D.H., *Fundamentals of Optical Engineering,* McGraw-Hill, New York, 1943.)

5.9.4 Snap Ring Mountings

A snap ring that drops into a groove machined into the inside surface of the cell is sometimes used in low-performance applications to hold lenses in a cell. Figure 5.43 shows such a design with a ring of circular cross section. The ring is cut through at some point on its periphery to allow it to spring into place. Cut edges must be smoothed to eliminate burrs that could damage the lens. Simplicity and low cost are the chief virtues. The preload, if any, exerted by the ring onto the lens is very unpredictable because variations in lens thickness, groove location and depth, and snap ring dimensions affect the interface between the lens and ring. Flat rings with rectangular cross sections are also used. Some of these have in-plane perforated ears as provisions for expanding the ring with a double-pronged tool for insertion and removal. They should be used only with flat beveled lenses.

A configuration for a ring-constrained lens mounting with a different form of groove is shown in Figure 5.44. Here, a continuous circular snap ring with a circular cross section is pressed into a tapered inside surface of the mount wall. The mount is made of plastic and is resilient. The groove is molded into the mount. The spring action of the wall holds the ring against both the lens surface and the ramp. This design is less sensitive to dimensional errors than those with metal mounts and conventional grooves. It is, however, hard to establish a particular preload with this technique.

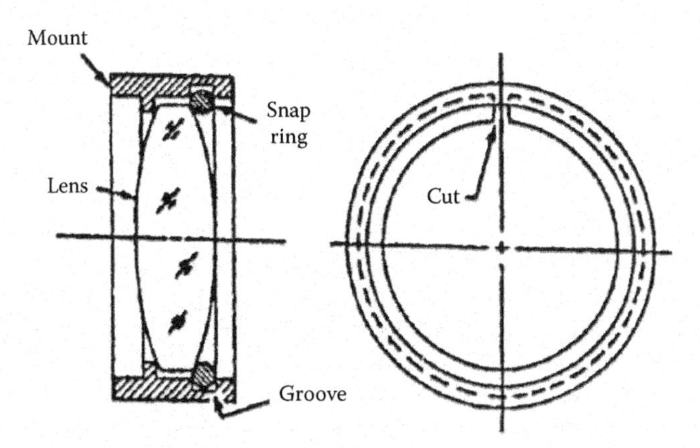

FIGURE 5.43 Technique for holding a lens in a cell with a circular cross-sectional snap ring located in a groove in the cell ID.

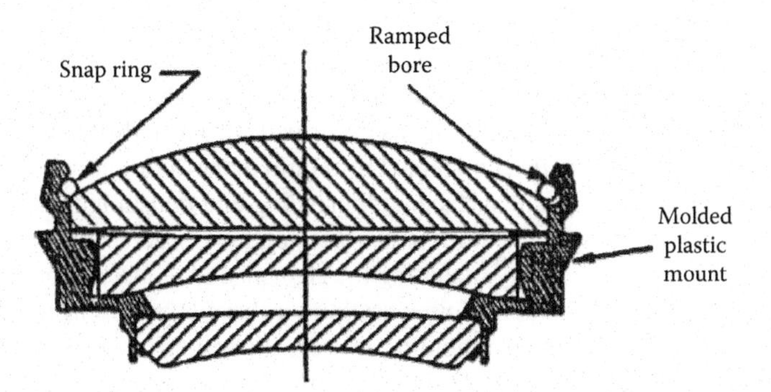

FIGURE 5.44 Lens mounting involving a circular cross-sectional continuous ring that snaps into a ramped groove in a molded plastic mount. (Adapted from Plummer, W.T., Precision: How to achieve a little more of it, even after assembly, *Proceedings of the First World Automation Congress (WAC '94)*, 1994, p. 193.)

FIGURE 5.45 Photograph of a four-element plastic objective with elements configured for conventional mechanical mounting. (From Lytle, J.D., Specifying glass and plastic optics–what's the difference? *Proc. SPIE*, 181, 93, 1979.)

5.10 MOUNTING PLASTIC LENSES

Most developments enabling plastics to be used in refractive optics have been accomplished by the private industry with the goal of fulfilling specific, high-volume product requirements. Polymethyl methacrylate (acrylic) (492574),* polysytrene (styrene) (590309), polycarbonate (Lexan) (585299), methyl methacrylate–styrene copolymer (NAS) (562335), copolymer styrene–acrylonitrile (567348), and allyldiglycol carbonate (CR-39) are some of the materials commonly used for optical applications.

Advantages of plastic optics are low density, ease of shaping both optical and mechanical surfaces and features, resistance to shock, and low costs in large quantities. A disadvantage is spotty and imprecise documentation of properties important to good engineering and design (such as refractive index, Young's modulus, Poisson's ratio, and CTE). This results largely from the fact that <1% of the total plastic market relates to optical applications, so plastic optics are generally ignored.

Three techniques are employed most often in fabricating plastic lenses: grinding and polishing, injection molding at high temperature and pressure, and single-point diamond turning (SPDT). The material type and quantity needed determine the best process to be used.

Injection/compression molding is the most common method for fabricating nonophthalmic plastic optics. These techniques can produce lenses configured as are most glass lenses, that is, with the equivalent of polished surfaces, cylindrical rims, and all types of bevels. Figure 5.45 shows a set of molded plastic lenses for a 28 mm (1.10 in.) focal length, *f*/2.8 photographic objective.[6] Each of these lenses is configured for conventional mounting, as described earlier in this chapter.

Molded plastic lenses are not usually constrained by requirements for rotational symmetry and ease of access to each optical and nonoptical surface with grinding and polishing tools. They can be provided with integral mounting flanges, locating pins, orienting tabs, spacers, and strategically located holes. These features can greatly simplify mounting the element and reduce the number and complexity of ancillary mechanical parts required. Compensation for thermoplastic shrinkage is an important consideration in developing an acceptable molding process. Figure 5.46 shows a sectional view of a typical injection machine for making multiple single-component biconvex lenses.[7] The cavity is shown with molded products in place awaiting ejection. Figure 5.47 shows a closer view of the closed mold.

* Refractive index Abbe number code as used in optical glass type designation.

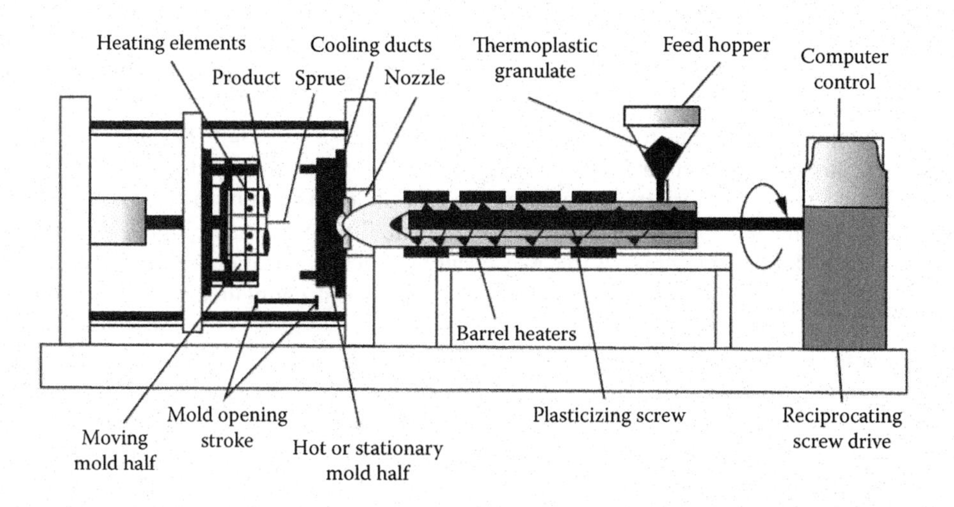

FIGURE 5.46 Section view of an injection molding machine. (Bauer, T., and Marschall, D., Tooling for injection molded optics, in Baumer, S., Ed., *Handbook of Plastic Optics*, pp. 35–63, 2005. Copyright Wiley-VCH Verlag GmbH & Co. KGaA. Reproduced with permission.)

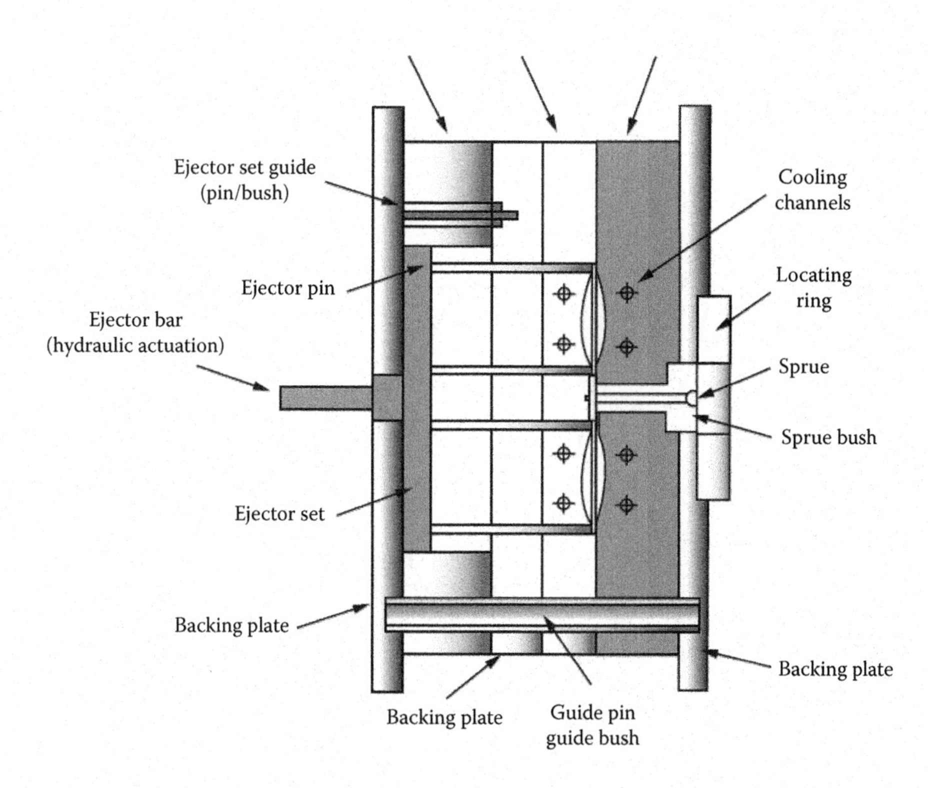

FIGURE 5.47 Detailed view of the molding cavity of the machine shown in Figure 5.46 (in closed position). (Bauer, T. and Marschall, D., Tooling for injection molded optics, in Baumer, S., Ed., *Handbook of Plastic Optics*, pp. 35–63, 2005. Copyright Wiley-VCH Verlag GmbH & Co. KGaA. Reproduced with permission.)

Assembly by solvent bonding, heat staking, or sonic welding techniques is feasible and economical. Plastic lenses can be nested into one another to form air-spaced doublets that align, center, and space automatically at assembly.

In designing optical systems to use molded plastic optics, it is especially important for the lens designer, mechanical engineer, and lens manufacturer to work closely together so that the advantages of the materials and fabrication processes can be fully realized. Proper design of the tooling and molding equipment is essential to economical producibility. For example, Figure 5.48 illustrates some good and bad lens configurations that may occur in typical applications. For reference, the element shown in the view in Figure 5.48a is a simple meniscus lens configured as it might be formed in glass. The equivalent plastic version of the same lens is shown in the view in Figure 5.48b. The axial thickness is here increased to allow the use of more liberal gate cross sections for introducing plastic material into the mold. This results in better surface contours than could otherwise be achieved in the plastic part. A positive element of the more strongly curved shape shown in the view in Figure 5.48c would mold better than weaker (longer-radius) types, since surface tension is then more effective.

The negative element shown in the view in Figure 5.48d would mold poorly, because the injected material would fill the thicker region at the edge more easily than the thin center region. Furthermore, gases would tend to be trapped at the center in this design. A preferred configuration would have greater axial thickness so that the edge-to-center thickness ratio does not exceed 3:1.

The configuration of the view in Figure 5.48e is reasonably easy to mold if care is taken to inject material into the lens region (outside the clear aperture) and not into the flange. Flanges of this type

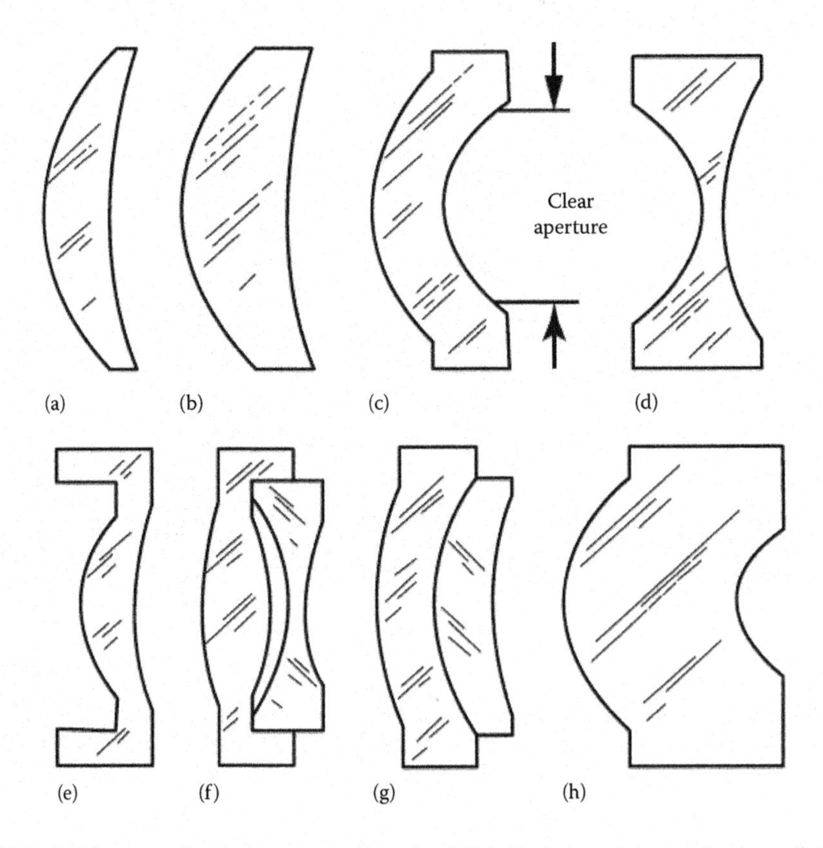

FIGURE 5.48 Different mechanical configurations of molded plastic lens elements that have distinct advantages and disadvantages as discussed in the text. (Adapted from Altman, R.M., and Lytle, J.D., Optical design techniques for polymer optics, *Proc. SPIE*, 237, 380, 1980.)

are useful in creating multiple-element subassemblies, as shown in the view in Figure 5.48f. This is essentially an edge-contact design. Cemented interfaces, as illustrated by the view in Figure 5.48g, should be avoided because the cemented interface can develop considerable stress owing to CTE differences. The view in Figure 5.48h shows a thick-meniscus lens that would tend to mold well if properly proportioned and carefully gated for material injection.

SPDT is feasible with some plastics (such as polymethylmethacrolate) and produces optical quality surfaces suitable for many visible light and near-infrared applications. Nonspherical surfaces can be made by this process on a one-off basis. They can be molded on plastic optics in quantity nearly as easily as spherical ones. Molds are simply made to the negative contour of the desired surface.

PROBLEMS

1. The dimensions of a convex-concave lens as shown in Figure 5.15d are $D = 60$ mm, $R_1 = 75$ mm, $R_2 = 50$ mm, $2y_2 = 40$ mm, and $t_A = 14$ mm. The lens is made of Schott F2 glass; $\rho = 3600$ kg/m^3. Find the mass of the lens and its CG location; the CG location is to be given from the vertex of the convex (75 mm radius) side. *Answer:* $W_{LENS} = 144 \times 10^{-3}$ kg, $CG_{LENS} = 10.1$ mm
2. Use the approximate methods to find the mass and CG location of the lens in problem 1. What are the relative errors in mass and CG location when using the approximate methods? *Answer:* $W_{LENS} = 144 \times 10^{-3}$ kg, error ≈ 0.26%; $CG_{LENS} = 9.99$ mm, error ≈ 1.1%
3. The dimensions of a double-convex lens as shown in Figure 5.15g are $R_1 = 52$ mm, $R_2 = -33$ mm, $D_1 = 50$ mm, $D_2 = 40$ mm, and $t_A = 22$ mm. The lens is made of zinc sulfide; $\rho = 4080$ kg/m^3. Find the mass of the lens and its CG location; the CG location is to be given from the vertex of the left (52 mm radius) side. *Answer:* $W_{LENS} = 102 \times 10^{-3}$ kg, $CG_{LENS} = 10.2$ mm
4. The left lens of a bonded doublet as shown in Figure 5.15h is made of Schott F2 glass ($\rho = 3600$ kg/m^3), and the right lens is made of Schott N-BK7 glass ($\rho = 2510$ kg/m^3). The dimensions of the F2 lens are $R_1 = 63$ mm and $D_1 = 50$ mm. The dimensions of the N-BK7 lens are $R_2 = 76$ mm and $D_2 = 40$ mm. For the bonded assembly, $t_A = 14$ mm; the center thickness of the right N-BK7 lens is 4 mm. The interface radius is plano (flat). Assume that the bond thickness is zero and that there are no bevels. Find the mass and CG location of the bonded doublet. *Answer:* $W_{LENS} = 88.4 \times 10^{-3}$ kg/m^3, $CG_{LENS} = 8.24$ mm
5. The diameter of a concave fused silica mirror is $D_1 = 300$ mm, with an optical radius of curvature of −1.200 m. The mirror edge thickness $L = 10$ mm, and the total thickness of the mirror from front edge to back is 50 mm (this is not t_A, but is measured from the front edge to back). The mirror back is tapered with $D_2 = 100$ mm. The density of fused silica glass is $\rho = 2205$ kg/m^3. Find the mass and CG of the mirror referenced to the front lip (radius side). *Answer:* $W_{MIRROR} = 3.83$ kg, $CG_{MIRROR} = 22.4$ mm
6. The mass of a 50 mm diameter biconvex lens is 98×10^{-3} kg. One radius is $R_1 = 75$ mm, with a contact at $y_1 = 24$ mm; the other radius is $R_2 = -100$ mm, with a contact at $y_2 = 23$ mm. If the friction coefficient between the lens and mount is $\mu = 0.165$, what preload force is needed to keep the lens in place under $a_G = 30g$ radial acceleration? *Answer:* $F_R = 115$ N
7. The mass of a 40 mm biconcave lens is 50.5×10^{-3} kg. One radius is $R_1 = -200$ mm, with a contact at $y_1 = 19$ mm; the other radius is $R_2 = 150$ mm, with a contact at $y_2 = 17$ mm. The friction coefficient between the glass and mount is $\mu = 0.375$. What preload force is necessary if the radial acceleration $a_G = 40g$? *Answer:* $F_R = 50.2$ N
8. A lens with a mass of 98×10^{-3} kg is subjected to an axial acceleration of $a_G = 30g$. A threaded aluminum retaining ring with a thread pitch diameter of $D_T = 55$ mm supplies the preload force, with SF = 1.5. If the friction coefficient between the lens and retaining ring

is $\mu_G = 0.165$ and that between the retainer and lens cell is $\mu_M = 0.19$, what is the torque necessary to provide the necessary preload force? Compare this torque with the approximate value given by Equation 5.20. *Answer:* For Equation 5.19, torque $M = 457 \times 10^{-3}$ N m; for Equation 5.20, torque $M = 476 \times 10^{-3}$ N m.

9. A lens with a mass of 225×10^{-3} kg is subjected to an axial acceleration of $a_G = 75g$. A threaded retaining ring made of 6061 aluminum with a thread pitch diameter of D_T = 25 mm supplies the preload force, with SF = 2.0. If the yield strength of 6061 aluminum is $\sigma_{YS} = 55$ MPa, what is the finest usable thread pitch? *Answer:* $p = 47.2 \times 10^{-6}$ m. Note that a thread this fine is impractical, so a coarser pitch would actually be used.
10. The mass of a 150 mm diameter lens is 1.6 kg; the lens is subjected to an axial acceleration of $a_G = 20g$ in service. An SF = 2 is required for the axial preload. The lens is held in place by a continuous flange retainer made of 17-4 PH stainless steel in condition H1025, with elastic modulus $E_M = 196$ GPa, Poisson's ratio $\nu_M = 0.272$, and yield strength σ_{YS} = 1.17 GPa. For the flange, $a = 77$ mm and $b = 73$ mm. Find the minimum thickness for the flange, assuming a safety against yield of SF = 2, and the flange deflection necessary for the required preload. *Answer:* The minimum thickness $t = 0.231$ mm, and the deflection $\Delta = 0.117$ mm.
11. A lens is held constrained axially by three equally spaced spring clips which provide a preload force of $F_A = 10$ N. The spring clips are made of beryllium copper, UNS C17200, in TD01 temper, with an elastic modulus $E_M = 125$ GPa, Poisson's ratio $\nu_M = 0.30$, and yield stress $\sigma_{ys} = 415$ MPa. The length of each clip is $L = 10$ mm, the width is $b = 4$ mm, and the thickness is $t = 0.5$ mm. Find the deflection required to produce the axial preload. What is the SF in the clips against yield? *Answer:* The required deflection $\Delta = 0.1$ mm, and the SF against yield is SF = 2.6.
12. A retaining ring with $y_2 = 24$ mm holds a biconvex lens in place against a seat with y_1 = 23 mm; the axial preload is $F_A = 18.044$ N. The edge thickness of the lens is $L = 5.46$ mm, and the center thickness is $t_A = 12.7$ mm. The Poisson's ratio of the lens material is ν_G = 0.206. Find the bending stress in the lens by using the average of the edge and center thickness. *Answer:* The bending stress $\sigma_B = 7.34$ kPa.
13. One element of an astronomical objective is made of F2 glass, with elastic modulus E_G = 57 GPa and density $\rho = 3600$ kg/m^3. The lens diameter is $d = 750$ mm, and it is used in the visible, where the wavelength $\lambda = 550$ nm and $n = 1.6237$. Gravity acts normal to the lens surface, or parallel to the optical axis. If the maximum OPD from the gravity deflection of the lens is $\lambda/8$, what is the minimum lens thickness assuming that the lens is approximated as a plane-parallel plate? *Answer:* The minimum thickness is $h = 4.35$ mm.
14. A biconvex lens made of Schott F2 glass is mounted in a cell with a sharp corner seat and retainer; the axial preload force is $F_A = 3.6$ N. The sharp corner radius is 50 µm ($D_2 = 100 \times 10^{-6}$ m). The optical radius of the first surface of the lens is 76.2 mm; the contact radius y_C for this surface is 24 mm, and the second surface is 114.3 mm, with a contact radius of 23 mm. The cell material is aluminum, with $E_M = 68.2$ GPa and $\nu_M = 0.332$; for the glass, E_G = 57 GPa and $\nu_G = 0.206$. Find the contact stresses for both surfaces; compare the contact stress with the approximation using Equation 5.37. *Answer:* Using Equation 5.35, for the 76.2 mm radius surface, $\sigma_C = 71.39$ MPa; for the 114.3 mm surface, $\sigma_C = 72.92$ MPa. Using Equation 5.37, for the 76.2 mm radius, $\sigma_C = 71.37$ MPa; for the 114.3 mm radius, $\sigma_C = 71.37$ MPa.
15. Using the properties from the previous problem, find the maximum preload for a compressive stress of 28 MPa. Hint: Use the approximate Equation 5.37. *Answer:* The maximum axial preload force is $F_A = 531 \times 10^{-3}$ N.
16. The lens in problem 14 is mounted using toroidal surfaces. Assume the same material properties, dimensions, and axial preload force of $F_A = 3.6$ N. What contact toroidal radius for each surface is necessary to limit the contact stress to 14 MPa? Hint: Use the approximate

Equation 5.37. *Answer:* For the first surface, with radius of 76.2 mm, the contact toroidal radius is 1.30 mm; for the second surface, with radius of 114.3 mm, the contact toroidal radius is 1.36 mm.

17. For the lens in problem 14; what are the surface tensile stresses from the compressive contact stresses? If the maximum thickness of the lens within the clear aperture is 12.7 mm, what is the total OPD from the contact compressive stresses? For Schott F2, the stress optic coefficient $K_S = 2.81 \times 10^{-12}$ Pa^{-1}. *Answer:* The surface tensile stress for the 76.2 mm radius is 14.0 MPa, and for the 114.3 mm radius, the surface tensile stress is 14.3 MPa; the OPD from the contact compressive stress is 3.64 μm.
18. A symmetric biconvex zinc sulfide lens is mounted using tangential contacts. The radius of both optical surfaces is R = 150 mm, and the lens central thickness is 15 mm. The mounting surfaces contact the lens at y_C = 28 mm. The maximum OPD tolerance from compressive stress is OPD = 100×10^{-9} m. The lens cell is made of 416 stainless steel, with E_M = 200 GPa and ν_M = 0.283. For zinc sulfide, E_G = 74.5 GPa and ν_G = 0.29; the stress optic coefficient is $K_S = 804 \times 10^{-15}$ Pa^{-1}. Using the worst-case OPD tolerance (hint: use lens center thickness), find the maximum compressive stress for each surface. What is the axial preload force for producing this compressive stress? What is the contact angle between the tangential interface and lens surfaces? *Answer:* The maximum compressive stress is σ_C = 5.86 MPa; note that since the lens is symmetric, the stress for each surface is 0.707 times the maximum compressive stress; the preload necessary to produce this stress is F_A = 48.1 N, and the contact angle $\psi = 79.2°$.
19. The diameter of an N-BK7 lens is D_G = 50.8 mm and it is mounted in a 416 stainless steel cell by using an athermal bond. For N-BK7, the thermal expansion is $\alpha_G = 7.1 \times 10^{-6}$ K^{-1}, and for 416 stainless steel, $\alpha_M = 9.9 \times 10^{-6}$ K^{-1}. The adhesive CTE $\alpha_r = 80 \times 10^{-6}$ K^{-1} and ν_r = 0.43. Find the upper and lower limits of the radial athermal bond thickness by using the van Bezooijen and modified van Bezooijen equations. *Answer:* The lower limit $t_{eVB} = 399.6 \times 10^{-6}$ m, and the upper limit $t_{eMVB} = 573.4 \times 10^{-6}$ m.
20. Using the properties from the previous problem, find the optimum athermal bond thickness using Monti's approximation. The lens edge thickness is L = 5.46 mm. Find the stress in the lens if the bond thickness varies by 10% for a temperature change of ΔT = 50 K; assume that the elastic modulus of the adhesive is E_r = 140 MPa. *Answer:* The optimum athermal bond thickness $t_{eAR} = 408.9 \times 10^{-6}$ m; for a 10% variation in thickness, the stress σ_T = −360 kPa.
21. A 250 mm diameter lens made of Schott N-BAK4 is mounted in a 6Al–4V titanium cell by using an athermal radial bond. The lens center thickness t_A = 25 mm, and the edge thickness L = 9.31 mm; the lens mass is 2.57 kg. For the titanium, $\alpha_M = 8.8 \times 10^{-6}$ K^{-1}. A compliant RTV-type adhesive is used for the athermal bond; $\alpha_r = 300 \times 10^{-6}$ K^{-1}, ν_r = 0.46, and E_r = 3.45 MPa. Find the optimum radial thickness using Bayar's and Monti's aspect ratio approximation. Assume that the aspect ratio approximation is correct; for a 75 K temperature drop, what is the stress in the lens if the thickness given by the Bayar equation is used? If the lens is used in the visible, where the wavelength is λ = 550 nm, is the OPD from the lens stress at 75 K less than $\lambda/8$? For N-BAK4, the stress optical coefficient is $K_S = 2.9 \times 10^{-12}$ Pa^{-1}. (Hint: Assume a worst-case thickness of t_A.) What is the self-weight deflection of the lens relative to the cell when gravity acts normal to the optical axis? *Answer:* The bond line thickness using Bayar's method is $t_{eBayar} = 772.2 \times 10^{-6}$ m, and the bond line thickness using Monti's aspect ratio approximation is $t_{eAR} = 289.6 \times 10^{-6}$ m. The stress in the lens for a temperature drop of 75 K is −714.5 kPa. The OPD associated with this stress is 51.8 nm, which is less than 550 nm/8 = 68.75 nm, so the lens stress is acceptable. The decenter under the lens self-weight is $\delta = 354 \times 10^{-9}$ m.

REFERENCES

1. Bayar, M., Lens barrel optomechanical design principles, *Opt. Eng.*, 20, 181, 1981.
2. Yoder, P.R. Jr., *Mounting Optics in Optical Instruments*, 2nd ed., SPIE Press, Bellingham, WA, 2008.
3. Yoder, P.R. Jr., Lens mounting techniques, *Proc. SPIE*, 389, 2, 1983.
4. Shigley, J.E., and Mischke, C.R., The design of screws, fasteners, and connections, Chapter 8 in *Mechanical Engineering Design*, 5th ed., pp. 409–458, McGraw-Hill, New York, 1989.
5. Yoder, P.R. Jr., Axial stresses with toroidal lens-to-mount interfaces, *Proc. SPIE*, 1533, 2, 1991.
6. Lytle, J.D., Specifying glass and plastic optics–what's the difference? *Proc. SPIE*, 181, 93, 1979.
7. Bauer, T., and Marshall, D., Tooling for injection molded optics, in Chap. 3, *Handbook of Plastic Optics*, Baumer, S., Ed., pp. 35–63, Wiley-VCH, Weinheim, 2005.

FURTHER READING

Baumer, S., *Handbook of Plastic Optics*, 2nd ed., Wiley-VCH, Weinheim, 2011.
Delgado, R.F., and Hallinan, M., Mounting of lens elements, *Opt. Eng.*, 14, S-11, 1975.
Karow, H.H., *Fabrication Methods for Precision Optics*, Wiley, Hoboken, NJ, 1993.
Monti, C.L., Athermal bonded mounts: Incorporating aspect ratio into a closed form solution, *Proc. SPIE*, 6665, 666503, 2007.
Schaub, M.P., *The Design of Plastic Optical Systems*, SPIE Press, Bellingham, WA, 2009.
Yoder, P.R. Jr., and Vukobratovich, D., *Opto-mechanical Systems Design*, 4th ed., CRC Press, Boca Raton, FL, 2015.

6 Mounting Multiple Lenses

6.1 INTRODUCTION

This chapter deals with various ways in which two or more lenses are mounted together to form optomechanical subassemblies or assemblies. Methods of establishing the proper spacing between elements and associated tolerances are discussed. Next are techniques for analyzing thermal effects on lens systems, including optical changes in focus, thermal stress in components, and preload variation with temperature. Lathe and subcell assemblies, as well as plastic lens assemblies, are covered.

6.2 LENS BARREL STRUCTURAL DESIGN

A lens barrel is the housing that holds multiple lenses in the correct position relative to each other. The mass and stiffness of the lens barrel are the most important performance parameters. Stiffness is determined by the mass of the optical elements and optical positioning tolerances, as well as the properties of the barrel material. Barrel radial wall thickness is set by stiffness. The mass of the barrel is a function of the radial wall thickness, overall length, and the density of the barrel material. System parameters influencing mass include the spacing between elements, element diameter, element mass, and allowable tolerance for decenter. Material parameters affecting mass include density and elastic or Young's modulus.

The tolerance for the position of the individual elements of an optical system includes the accuracy of location of the axis of the elements with respect to the overall optical axis. This is the centering tolerance for the element. The centering tolerance is allocated to errors associated with manufacture, location of the lens in the barrel, thermal distortion, and deflection from gravity. Deflection from gravity is usually a parameter that strongly influences the selection of the lens barrel material.

Deflection is at a maximum when gravity acts normal to the optical axis. Two loading conditions bound the design of a lens barrel: one where the lens is concentrated at the end of the barrel and another where the lens mass is uniformly distributed along the length of the barrel. For both loading conditions, the worst case is for a cantilever, with one end fixed and the other free.

If the lens mass is concentrated at the free end of the barrel, the deflection from gravity is

$$\delta = \frac{W_C L^3}{3EI}, \tag{6.1}$$

where δ is the deflection, W_C is the concentrated mass of the lens at the end of the barrel, L is the distance of the lens from the support, E is the elastic modulus of the barrel material, and I is the cross-sectional moment of the inertia of the barrel.

If the cell wall is thin with respect to the overall lens barrel diameter, $I \approx \pi r^3 t$, where r is the lens barrel radius and t is the radial wall thickness. Solving for the moment of inertia, and then substituting, the minimum value of t is given by

$$t = \frac{W_C L^3}{3\pi E r^3 \delta_C}. \tag{6.2}$$

The mass of the lens barrel W_{BC} is given by the volume times the lens cell material density ρ. The volume is the cross-sectional area A, which is $2\pi rt$ if the cell wall is thin with respect to the radius, times the length L. Then, the weight W_{BC} of the barrel is given by

$$W_{BC} = \frac{2}{3}\left(\frac{W_C}{\delta_C}\right)\left(\frac{\rho}{E}\right)\left(\frac{L}{r}\right)^2 L^2. \tag{6.3}$$

When the lens mass is uniformly distributed along the length of the barrel, a similar derivation gives the barrel mass W_{BC} for a total lens mass W_L and deflection tolerance δ_L:

$$W_{BC} = \frac{1}{4}\left(\frac{W_L}{\delta_L}\right)\left(\frac{\rho}{E}\right)\left(\frac{L}{r}\right)^2 L^2. \tag{6.4}$$

Equations 6.3 and 6.4 show that the lens barrel structural mass is linearly proportional to the mass of the lenses and inversely proportional to the deflection tolerance. Mass increases as the fourth power of the length and is inversely proportional to the square of the lens radius or diameter. Mass is also linearly dependent on the inverse specific stiffness, which is the density divided by the elastic modulus or ρ/E.

For common structural materials, including magnesium, aluminum, titanium, and steel, the ratio of ρ/E is nearly constant. Therefore, the weight of the cell is independent of the material used, which is a surprising result. There are exceptions; for example, the density of beryllium is comparable to that of magnesium, while its elastic modulus is 1.5 times higher than that of steel. Hence, ρ/E of beryllium is about 15% of that of a material such as aluminum, with a corresponding decrease in weight.

The equations for lens barrel mass also show that the mass is independent of the material strength. For common engineering materials, stiffness is independent of strength. For example, for low- to high-strength aluminum alloys such as A356, 6061, and 7075, the elastic modulus varies by less than 5%. Since stiffness is more important than strength in optomechanical design, and stiffness is independent of strength for most common structural materials, there is no reason to use high-strength-to-weight materials in lens barrels.

In some situations, the radial wall thickness is constrained by considerations of buckling or by the stress induced during fabrication. As an example, consider a small terrestrial telescope with an objective diameter of 50 mm and a focal length of 250 mm. The objective weighs about 100 g, and the maximum deflection tolerance is 10 μm. If the barrel is supported at its midpoint, the cantilever length is 125 mm. The required radial wall thickness of an aluminum barrel is about 19 μm. This is too thin for normal manufacturing processes; the minimum wall thickness for investment casting is about 750 μm, and a more reasonable value is 2 mm. Therefore, it may be necessary to significantly increase the radial thickness for ease of manufacture. Assuming that the minimum wall thickness provides sufficient stiffness, the lens barrel mass is determined by the material density.

Returning to the example of the 50 mm terrestrial telescope, if the aluminum barrel is replaced with a glass-reinforced polycarbonate plastic, the radial wall thickness increases to about 325 μm. Although still too thin for practical fabrication, the lower density of the plastic makes it possible to increase the wall thickness while still keeping the mass down. This is one reason for the use of glass-reinforced polycarbonate plastic as a lens barrel material in applications such as consumer camera lens barrels and binocular housings. The density of the glass-reinforced polycarbonate is about half that of aluminum, with corresponding savings in mass. Another reason for the use of polycarbonate plastics is a reduction in manufacturing costs since injection molding can be used to make the barrel.

Manufacturing cost may be more important than specific stiffness or density in selecting a material for a lens barrel. An example of cost taking precedence in selecting a barrel material is a microscope objective. Microscope objectives are small and relatively complex in design, and mass is not a concern. Since the microscope is normally employed in a static laboratory environment over a limited temperature range, thermal and dynamic material properties are not important. To minimize manufacturing cost, the barrel and subcells of a microscope objective are often made of brass. Brass is easy to machine with good friction characteristics. The low friction simplifies press fitting the individual subcells containing the lenses into the barrel. Brass does tarnish, and microscope objectives are usually plated with nickel or hard chrome to prevent this type of corrosion.

6.3 MULTIELEMENT SPACING CONSIDERATIONS

6.3.1 Lens Location Interfaces

Lens position is typically specified in the optical prescription or design by the vertex location of the optical surfaces of the system. The lens vertex is defined as the intersection of the optical axis with the optical surface of the lens. Axial position along the optical axis is given as part of the optical prescription, along with an associated tolerance. There is also a radial position tolerance called decenter, which is perpendicular to the optical axis.

The maximum accuracy of position is obtained when there is contact between the optical surface of the lens and the lens cell. Contact between the lens surface and cell is annular for an axisymmetric lens. This geometry is shown in Figure 6.1. The annular contact in Figure 6.1 is labeled d_{seat}. Also important is the diameter of the lens d_{lens}, and inside diameter (ID) of the lens cell d_{bore}. When the lens is centered in the cell, the radial clearance between the edge of the lens and the ID of the cell is $(d_{bore} - d_{lens})/2$. If the lens is not in the correct position, the lens will roll on its spherical optical surface, with a tilt and decenter.

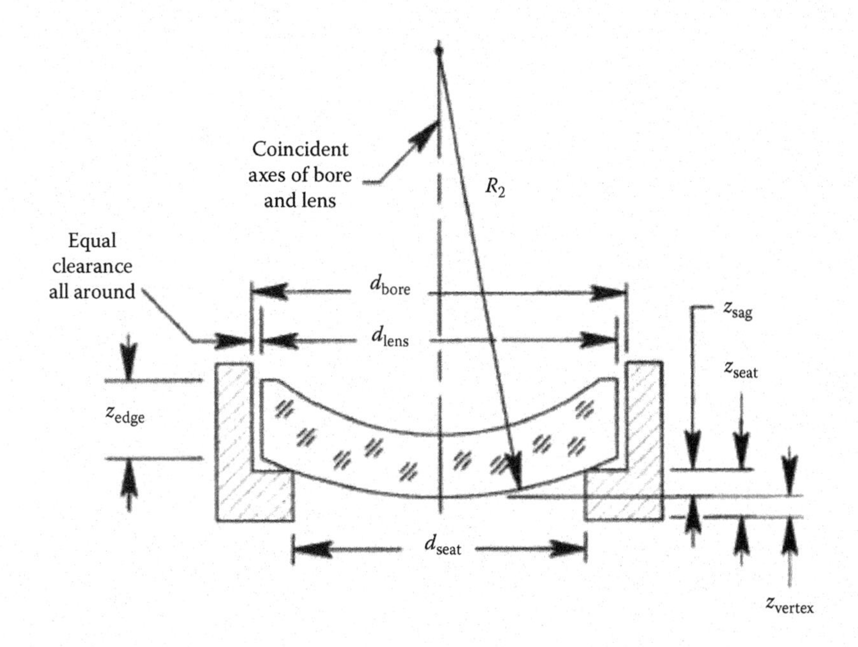

FIGURE 6.1 Meniscus lens in the correct position in its cell. (Adapted from Hopkins, C.L., and Burge, J.H., Application of geometric dimensioning and tolerancing for sharp corner and tangent lens seats, *Proc. SPIE*, 8131, 81310F, 2011.)

For a sharp corner interface, as shown in Figure 6.1, the distance between the lens vertex and reference surface on the lens cell z_{vertex} is given by

$$z_{vertex} = z_{seat} - R_{lens} + \sqrt{R_{lens}^2 - \left(\frac{d_{seat}}{2}\right)^2}, \tag{6.5}$$

where z_{seat} is the axial distance between the reference surface and lens seat and R_{lens} is the optical radius of the curvature of the lens.

When the seat is moved axially an amount Δz, the lens vertex moves the same distance. So the tolerance on seat location, assuming that the lens is perfect, is the same tolerance as for the axial vertex location of the lens.

The lens radial position varies by Δr, where the decenter is given by

$$\Delta r = R_{lens}\left[\sin^{-1}\left(\frac{d_{bore}}{2\ R_{lens}}\right) - \sin^{-1}\left(\frac{d_{lens}}{2\ R_{lens}}\right)\right]. \tag{6.6}$$

For most lenses, $d_{lens}/2R < 0.25$, and then $\sin^{-1}(d_{lens}/R_{lens}) \approx d_{lens}/2R_{lens}$ and $\Delta r \approx (d_{bore} - d_{lens})/2$, to an accuracy of about 1%. The radial error in positioning a lens, if relying on the fit of the lens to the cell, is therefore dependent on the minimum radial clearance between the lens edge and ID of the bore. For free assembly, a minimum radial clearance of about 25 μm is desirable. From experience, the absolute minimum radial clearance is about 5 μm. With this radial clearance, there is a significant risk of binding the edge of the lens during assembly, with possible damage by chipping.

Usually there is a small radius on the lens seat as a result of the machining process. As a rough rule of thumb for typical machining processes, this radius is 50–100 μm. Alternately, the lens may be mounted on a toroidal surface, as shown in Figure 6.2. If the radius of the toroidal seat is R_{toroid}, the shift in axial location Δz from the vertex location given by Equation 6.7 is

$$\Delta z \cong R_{toroid}\frac{d_{seat}}{2R_{lens}}. \tag{6.7}$$

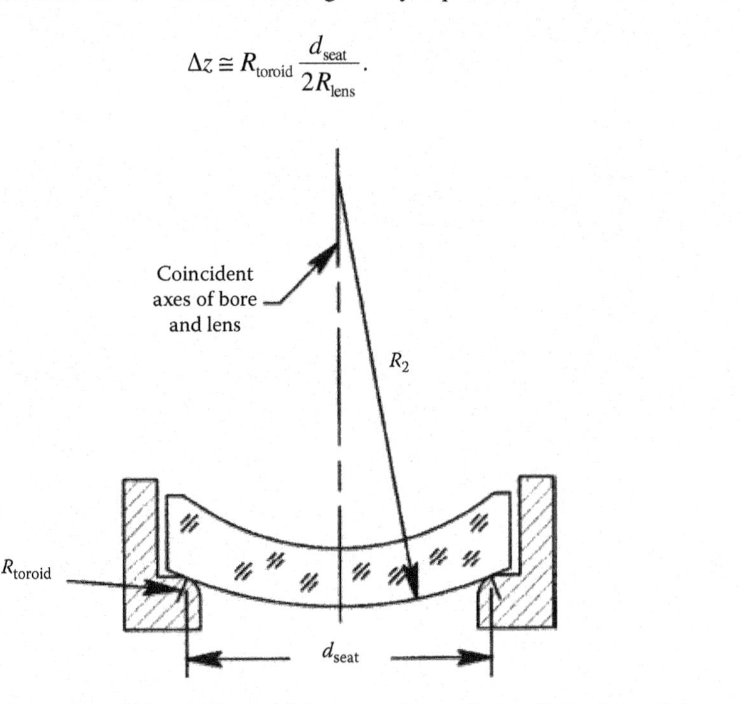

FIGURE 6.2 Modification of the cell from Figure 6.1 for a toroidal contact with the lens. (Adapted from Hopkins, C.L., and Burge, J.H., Application of geometric dimensioning and tolerancing for sharp corner and tangent lens seats, *Proc. SPIE*, 8131, 81310F, 2011.)

Example 6.1

A lens with a diameter of d_{lens} = 25 mm and a convex optical radius of curvature of R_{lens} = 40 mm rests on a sharp corner seat with d_{seat} = 23 mm. If the radial decenter tolerance is Δr = 25 μm, what is the maximum bore diameter of the cell? If the sharp corner contact radius is 75 μm, what is the displacement Δz of the lens vertex?

From Equation 6.6, the maximum bore diameter d_{bore} is given by

$$d_{bore} = 2(40\times10^{-3}\ \text{m})\sin\left\{\frac{25\times10^{-6}\ \text{m} + (40\times10^{-3}\ \text{m})\ \sin^{-1}\left[\dfrac{25\times10^{-3}\ \text{m}}{2(40\times10^{-3}\ \text{m})}\right]}{40\ \times10^{-3}\ m}\right\} = 25.0475\times10^{-3}\ \text{m}.$$

This is very close to the diameter of 25.05 × 10^{-3} m given by the approximation $d_{lens} + 2\Delta r$.

The displacement of the vertex is given approximately by Equation 6.7:

$$\Delta z = (75\times10^{-6}\ \text{m})\frac{23\times10^{-3}\ \text{m}}{2(40\times10^{-3}\ \text{m})} = 21.56\times10^{-6}\ \text{m}$$

For a tangential contact between lens and tangential contact diameter d_{tang}, the angle of contact θ is given by

$$\theta = \sin^{-1}\left(\frac{d_{tang}}{2R_{lens}}\right). \tag{6.8}$$

The axial distance z from the lens vertex to tangential contact is

$$z = R_{lens}\,(1 - \sin\theta). \tag{6.9}$$

If the angle of the tangential contact varies by $\Delta\theta$, the change in axial vertex position of the lens Δz is

$$\Delta z = R_{lens}\,\{[\sin(\theta+\Delta\theta) - \sin\theta]\tan(\theta+\Delta\theta) + \cos(\theta+\Delta\theta) - \cos\theta\}. \tag{6.10}$$

There are upper and lower limits on the tolerance associated with the tangential contact angle set by transition from tangential to corner contact. If the angle of the seat is too shallow, the lens will rest on the lower corner of the seat rather than making tangential contact; the minimum angle of the seat θ_{min} is given by

$$\theta_{min} = \sin^{-1}\left(\frac{d_A}{2\,R_{lens}}\right), \tag{6.11}$$

where d_A is the aperture diameter of the tangential seat.

There is also a maximum angle θ_{max} at which the lower corner of the lens makes contact with the tangential seat:

$$\theta_{max} = \sin^{-1}\left(\frac{d_{lens}}{2\,R_{lens}}\right). \tag{6.12}$$

Example 6.2

A lens with diameter d_{lens} = 32 mm is mounted on a tangential seat with d_{tang} = 30 mm and d_A = 28 mm; the lens optical radius of curvature in contact with the seat is convex with R_{lens} = 45 mm. What is the contact angle θ between the lens and tangential seat? If the tolerance on the contact angle is ±0.5°, what is the axial variation in the vertex position of the lens Δ*z*? What are the maximum and minimum contact angles?

From Equation 6.8, the contact angle θ is

$$\theta = \sin^{-1}\left[\frac{30\times10^{-3}\,\text{m}}{2(45\times10^{-3}\,\text{m})}\right] = 19.47°.$$

From Equation 6.10, the variation in the vertex position Δ*z* is

$$\Delta z = (45\times10^{-3}\,\text{m})\{[\sin(19.47°+0.5°)-\sin(19.47°)]\tan(19.47°+0.5°) + \cos(19.47°+0.5°)-\cos(19.47°)\} = 1.82\times10^{-6}\,\text{m}.$$

The minimum and maximum angles of the tangential seat angle are given by Equations 6.11 and 6.12:

$$\theta_{min} = \sin^{-1}\left[\frac{28\times10^{-3}\,\text{m}}{2(45\times10^{-3}\,\text{m})}\right] = 18.13°,$$

$$\theta_{max} = \sin^{-1}\left[\frac{32\times10^{-3}\,\text{m}}{2(45\times10^{-3}\,\text{m})}\right] = 20.83°.$$

6.3.2 Self-Centering

If the lens is clamped on its optical surfaces by seat and retainer, or seat and spacer, the lens will tend to self-center. The axial preload applied to the lens exerts a radial force when the lens is not centered. This radial force pushes the lens into alignment with the mechanical axis of the seat. Due to this self-centering, the accuracy of the lens location may be better than that predicted by Equation 6.6. The accuracy of self-centering is determined by the lens contact radii, lens optical radii, and the coefficient of friction between lens surface and seat.

If y_{c1} and y_{c2} are the respective surface contact radii between the lens and mount and R_1 and R_2 the respective optical surface radii, the condition for self-centering for a friction coefficient μ between lens and mount is

$$\frac{y_{c1}}{R_1} + \frac{y_{c2}}{R_2} \geq \mu. \tag{6.13}$$

Note that in Equation 6.13 the radii are positive for convex surfaces and negative for concave surfaces.

For uncoated glass in contact with metal, μ ≈ 0.15; then, one version of Equation 6.13 common in the literature is

$$\frac{y_{c1}}{2R_1} + \frac{y_{c2}}{2R_2} \geq 0.07. \tag{6.14}$$

Optical and mounting coatings may substantially increase the coefficient of friction. As an example, for optical glass with a MgF_2 antireflection coating on a black anodized aluminum seat, the friction coefficient is about $\mu \approx 0.38$. Although Equation 6.13 establishes the condition for self-centering, it does not provide the accuracy of centering. As a rough rule of thumb, for most lenses, where $d_{lens}/2R \le 0.25$, the self-centering error is 50–100 μm, with stronger optical curvatures (smaller radii) tending toward the lower value.

Example 6.3

A biconvex lens is mounted at y_{c1} = 14 mm with R_1 = 38 mm and y_{c2} = 13 mm with R_2 = 45 mm. If the friction coefficient between the lens and cell is μ = 0.30, will the lens self-center?

From Equation 6.14,

$$\frac{14\times10^{-3}\,\text{m}}{2(38\times10^{-3}\,\text{m})}+\frac{13\times10^{-3}\,\text{m}}{2(45\times10^{-3}\,\text{m})}=0.328>0.30;\text{ self-centering.}$$

So the lens will self-center. Note again that convex surfaces are positive and concave surfaces are negative for this analysis.

6.3.3 Calculating Spacing

In multielement lens assemblies, the elements frequently rest against shoulders or spacers whose axial lengths are carefully machined to obtain the required axial airspaces between surface vertices. A generic example is illustrated in Figure 6.3. The parameters indicated are those needed to find the length L_{JK} of the shoulder between the contact points P_J and P_K at heights y_J and y_K on the spherical surfaces with radii R_J and R_K to produce the vertex-to-vertex separation $t_{J,K}$ between the adjacent

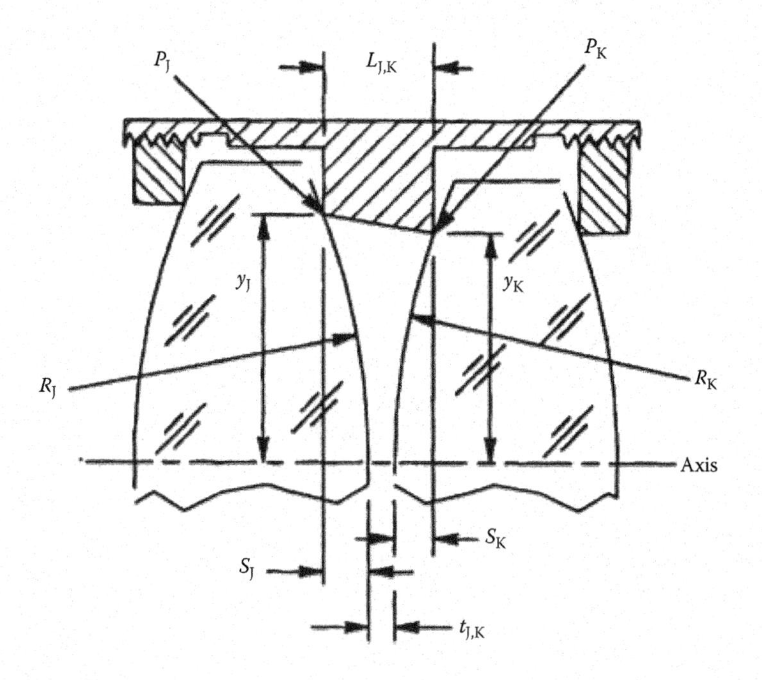

FIGURE 6.3 Parameters of importance when computing shoulder or spacer length to provide a given axial airspace between two lenses.

vertices. The sagittal depths of the surfaces measured from planes through P_J and P_K are S_J and S_K, respectively. They are assigned a positive sign if contact occurs to the right of the vertex and a negative sign if contact occurs to the left of the vertex. In the figure, S_J is negative, while S_K, $t_{J,K}$, and $L_{J,K}$ are positive. The surface sagittal depths and the shoulder length can be calculated by the following equations (note: absolute values are entered for the input dimensions):

$$S_J = R_J - \left(R_J^2 - y_J^2\right)^{1/2}, \tag{6.15}$$

$$S_K = R_K - \left(R_K^2 - y_K^2\right)^{1/2}, \tag{6.16}$$

$$L_{J,K} = t_{J,K} + S_J + S_K. \tag{6.17}$$

Example 6.4

The vertex-to-vertex separation of two lens elements is 4.978 mm. The radii of the adjacent convex surfaces are $R_J = -74.778$ mm and $R_K = +69.088$ mm. The contact heights are $y_J = 24.384$ mm and $y_K = 21.946$ mm. What is the spacer axial length $L_{J,K}$?

First, find the sagittal depths by using Equations 6.15 and 6.16:

$$S_J = -\left\{74.778\,\text{mm} - \left[(74.778\text{ mm})^2 - (24.384\text{ mm})^2\right]^{\frac{1}{2}}\right\} = -4.088\text{ mm},$$

$$S_K = \left\{69.088\text{ mm} - \left[(69.088\text{ mm})^2 - (21.946\text{ mm})^2\right]^{\frac{1}{2}}\right\} = 3.578\text{ mm}.$$

Then, using Equation 6.17, the spacer axial length is

$$L_{J,K} = 4.978\text{ mm} - (-4.088\text{ mm}) + (+3.578\text{ mm}) = 12.649\text{ mm}.$$

6.4 EFFECTS OF TEMPERATURE CHANGES

6.4.1 Radial Effects of Temperature Changes

Changes in temperature cause differential expansions or contractions for axisymmetric optics such as lenses with respect to their mountings in both the axial and radial directions. The following discussion assumes (1) rotational symmetry of the optics and mount; (2) that the clearances between the optic optical diameter (OD) and the ID of the mount are small; (3) that all components are at a uniform temperature before and after each temperature change; and (4) that the coefficients of thermal expansion (CTEs) of the optical materials and of the mount are α_G and α_M, respectively. In all cases, the temperature change is identified as ΔT.

The CTE of the mount usually exceeds that of the optic mounted in it. Usually, a drop in temperature will cause the mount to contract radially toward the rim of the optic. Any radial clearance between these components will decrease in size and, if the temperature falls far enough, the ID of the mount will contact the OD of the optic. Any further temperature decrease will then cause radial force to be exerted upon the rim of the optic. This force compresses (i.e., strains) the optic radially and creates radial stress. To the degree of approximation considered here, the strain and stress are symmetrical about the axis. If these are large enough, the performance of the optic will be adversely affected. Extremely large stresses may cause failure of the optic and/or plastic deformation of the mount.

If $\alpha_M > \alpha_G$, temperature increases will cause the mount to expand away from the optic, thereby increasing any existing radial clearance or creating such a clearance. Significant increases in radial clearance may allow the optic to shift under external forces such as shock or vibration. Alignment would then be affected, as would be the optical performance.

6.4.2 Radial Clearance Change with Temperature

The change in radial clearance $\Delta r(\Delta T)$ with temperature change ΔT between a lens of diameter D_G and its mount is given by

$$\Delta r(\Delta T) = (\alpha_M - \alpha_G)\frac{D_G}{2}\Delta T. \tag{6.18}$$

Equation 6.18 is used to determine the minimum radial clearance between the lens and lens mount for a specified lower temperature. If the radial clearance is smaller than that given by Equation 6.18, during the temperature change, the cell wall contacts the lens, inducing stress into the optical element. Analysis of the thermally induced radial stress in the lens is discussed in the next section.

6.4.3 Radial Thermal Stress in the Optic

If the temperature change is large enough for the cell to come into contact with the lens, stress may be induced in the optical element. The radial stress σ_R from the temperature change ΔT is given by

$$K_5 = 1 + \frac{2\Delta r}{D_G \Delta T (\alpha_M - \alpha_G)}, \tag{6.19}$$

$$\sigma_R = 0, \quad K_5 \leq 0, \tag{6.20}$$

$$\sigma_R = -K_4 K_5 \Delta T, \quad K_5 > 0 \tag{6.21}$$

$$K_4 = \frac{\alpha_M - \alpha_G}{\frac{1}{E_G} + \frac{D_G}{2E_M t_C}}, \tag{6.22}$$

where E_G is the elastic modulus of the lens material, E_M is the elastic modulus of the cell material, and t_C is the radial wall thickness of the cell.

If $K_5 > 0$, there is also a thermally induced hoop stress σ_M in the cell wall given by

$$\sigma_M = \frac{\sigma_R D_G}{2t_C}, \tag{6.23}$$

Example 6.5

A Schott BK7 lens with diameter $D_G = 40$ mm is mounted in a cell made of 6061 aluminum; the radial wall thickness of the cell $T_C = 6$ mm, and the initial assembly clearance $\Delta r = 25$ μm. The lens and cell assembly is subjected to a decrease in temperature of $\Delta T = -70$ K. The elastic moduli of the lens and cell are $E_G = 82$ GPa and $E_M = 68.2$ GPa, respectively. The CTEs of the lens and cell are $\alpha_G = 8.2 \times 10^{-6}$ and 23.6×10^{-6} K^{-1}, respectively. Does the cell come into contact with the lens during the specified decrease in temperature? If there is contact, what is the thermal stress in lens and cell?

First, calculate the parameter K_5 by using Equation 6.19:

$$K_5 = 1 + \frac{2(25\times10^{-6}\,\text{m})}{(40\times10^{-3}\,\text{m})(-70\text{K})[(23.6\times10^{-6}\,\text{K}^{-1})-(8.2\times10^{-6}\,\text{K}^{-1})]} = -0.16.$$

Since K_5 is negative, the cell contacts the lens during the 70 K drop in temperature, and there is stress in the lens and cell. The lens stress σ_R is given using Equations 6.21 and 6.22:

$$K_5 = \frac{(23.6\times10^{-6}\,\text{K}^{-1})-(8.2\times10^{-6}\,\text{K}^{-1})}{\dfrac{1}{82\times10^{9}\,\text{Pa}}+\dfrac{40\times10^{-3}\,\text{m}}{2(68.2\times10^{9}\,\text{Pa})(6\times10^{-3}\,\text{m})}} = 252.2\times10^{3}\,\text{Pa/K},$$

$$\sigma_R = -(-0.16)(252.2\times10^{3}\,\text{Pa/K})(-70\text{K}) = -2.816\times10^{6}\,\text{Pa}.$$

While there is very little risk of glass failure at this stress, it would be desirable to check for optical effects by using the stress optic coefficient for BK7.

Stress σ_M in the lens cell is given by Equation 6.23:

$$\sigma_M = \frac{(-2.816\times10^{6}\ \text{Pa})(40\times10^{-3}\,\text{m})}{2(6\times10^{-3}\,\text{m})} = -9.388\times10^{6}\,\text{Pa}.$$

This is a very low stress, and there is no risk of structural failure (the yield stress for 6061 aluminum is about 276 MPa).

6.4.4 Thermal Change in Spacing

After a thermal soak, with a uniform change in temperature ΔT, the sizes of the lens and mount change by an amount $\alpha\Delta T$, where α is the CTE of the material. The lens radius and size of the lens cell seat changes with temperature; these dimensional changes cause the sagittal depth of the lens to also change. There is therefore a shift in the vertex location of the lens with temperature change. If α_G is the CTE of the lens material and α_M is the CTE of the cell material, the sagittal depth after a temperature change ΔT is given by a modified form of Equation 6.15:

$$S(\Delta T) = R(1+\alpha_G\Delta T) - \sqrt{[R(1+\alpha_G\Delta T)]^2 - \left[\frac{d_{\text{seat}}(1+\alpha_M\Delta T)}{2}\right]^2}. \tag{6.24}$$

Similarly, the barrel length between two lenses will also change by an amount $(1 + \alpha_M\Delta T)$. In determining the change in lens vertex-to-vertex spacing, the length is the distance from the point of contact to the point of contact with the two lenses as given by Equation 6.17. For some geometries, it is also necessary to account for the axial expansion or contraction of the lens. The thermal expansion of the lens is given by $\alpha_G(t_A - S)\Delta T$, where t_A is the vertex-to-vertex axial thickness of the lens and S is the sagittal depth from the seat; S is positive if the lens is convex, and it is negative if the lens is concave. It is useful to make a simple sketch of the lens spacing geometry to ensure understanding of the spacing change with temperature before calculating the spacing change with temperature.

Example 6.6

The two lenses in Example 6.4 are made of Schott BK7, with $\alpha_{gBK7} = 8.2 \times 10^{-6}$ K^{-1} and Schott SF11 with $\alpha_{gSF11} = 6.1 \times 10^{-6}$ K^{-1}; the cell is made of 6061 aluminum, with $\alpha_M = 23.6 \times 10^{-6}$ K^{-1}. If the temperature change is −30 K, what is the change in vertex-to-vertex spacing?

From Equation 6.24, the sagittal depth after the temperature change for the two lenses is

$$\begin{aligned} S_j(\Delta T) &= (74.779\times10^{-3}\,\text{m})[1+(8.2\times10^{-6}\,\text{K}^{-1})(-30\,\text{K})] \\ &\quad -\langle\{(74.779\times10^{-3}\,\text{m})[1+(8.2\times10^{-6}\,\text{K}^{-1})(-30\,\text{K})]\}^2 \\ &\quad -\{(24.384\times10^{-3}\,\text{m})[1+(23.4\times10^{-6}\,\text{K}^{-1})(-30\,\text{K})]\}^2\rangle^{\frac{1}{2}} \\ &= 4.082\times10^{-3}\,\text{m}, \end{aligned}$$

$$\begin{aligned} S_K(\Delta T) &= (69.0088\times10^{-3}\,\text{m})[1+(6.1\times10^{-6}\,\text{K}^{-1})(-30\,\text{K})] \\ &\quad -\langle\{(69.088\times10^{-3}\,\text{m})[1+(6.1\times10^{-6}\,\text{K}^{-1})(-30\,\text{K})]\}^2 \\ &\quad -\{(21.946\times10^{-3}\,\text{m})[1+(23.4\times10^{-6}\,\text{K}^{-1})(-30\,\text{K})]\}^2\rangle^{\frac{1}{2}} \\ &= 3.574\times10^{-3}\,\text{m}. \end{aligned}$$

The spacer length after the temperature change is

$$L_{JK}(\Delta T) = (12.649\times10^{-3}\,\text{m})[1+(23.4\times10^{-3}\,\text{K}^{-1})(-30\text{K})] = 12.640\times10^{-3}\,\text{m}.$$

The vertex-to-vertex spacing after the temperature change is

$$t_A(\Delta T) = 12.640\times10^{-3}\,\text{m} - 4.082\times10^{-3}\,\text{m} - 3.574\times10^{-3}\,\text{m} = 4.984\times10^{-3}\,\text{m}.$$

The change from the original spacing is

$$4.978\times10^{-3}\ \text{m} - 4.984\times10^{-3}\ \text{m} = -5.84\times10^{-6}\ \text{m}.$$

This is a somewhat surprising result; the vertex-to-vertex spacing actually increased by about 5.8 μm after the drop in temperature. The decrease in the sagittal depth of the two lenses with temperature explains this change.

6.4.5 Change in Axial Preload with Temperature

Optical and mounting materials usually have dissimilar CTEs, with those of the mount greater than those of the lenses, so a temperature change of ΔT changes the axial preload P. The change in axial preload ΔP with temperature change ΔT is given by

$$\Delta P = K_3\Delta T. \tag{6.25}$$

In Equation 6.25, K_3 is the rate of the change in preload with temperature for the design, and it is sometimes called the temperature sensitivity factor of the design. Knowledge of K_3 of a given lens mounting design allows the estimation of actual preload P applied to the lenses at any temperature by combining ΔP with the assembly preload. In the absence of friction, this preload is the same at all surfaces of all lenses clamped by a single retaining ring or other noncompliant means.

If α_M exceeds α_G (as is usually the case), the metal in the mount expands more than the optic for a given temperature increase ΔT. Any axial preload P_A existing at assembly temperature T_A (typically 20°C [68°F]) will then decrease. If the temperature rises sufficiently, that preload will disappear. See Figure 6.4. If the lens is not otherwise constrained axially (as by an elastomeric sealant), it will then be free to move within the mount in response to externally applied forces. The temperature at which the axial preload goes to zero is defined as T_C. This temperature is

$$T_C = T_A - \frac{P_A}{K_3}. \tag{6.26}$$

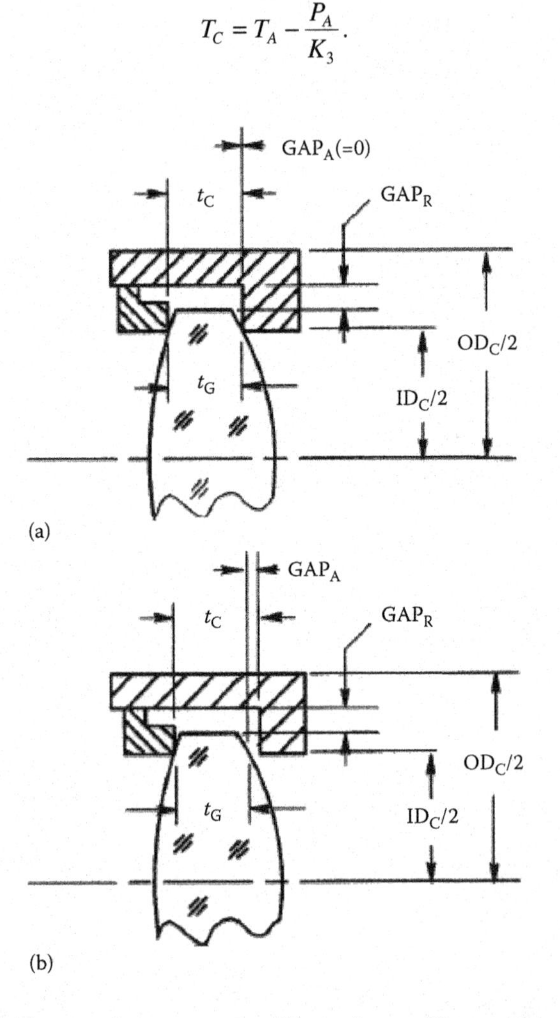

FIGURE 6.4 Effects of an increase in temperature within a subassembly comprising a glass lens and a metal cell where the CTE of the metal exceeds that of the lens: (a) at assembly temperature and (b) at maximum temperature. The resulting GAP$_A$ in (b) would allow the lens to move under vibration.

The mount maintains contact with the lens until the temperature rises to T_C. A further temperature increase introduces an axial gap between the mount and lens. This gap should not exceed the design tolerance for despace of the lens. The increase in axial gap ΔGAP_A for a system of n lenses and spacers is given by

$$\Delta GAP_A = \sum_{i=1}^{n} (\alpha_M - \alpha_i) t_i (T - T_C), \tag{6.27}$$

where t_i is the axial thickness of lens or spacer.

The increase in axial gap ΔGAP_A created in a single-element lens subassembly, a cemented doublet lens subassembly, an air-spaced doublet subassembly, and a general multilens subassembly as the temperature rises by ΔT above T_C can be approximated, respectively, as

$$\Delta GAP_A = (\alpha_M - \alpha_G) t_E (T - T_C), \tag{6.28}$$

$$\Delta GAP_A = [(\alpha_M - \alpha_{G1}) t_{E1} + (\alpha_M - \alpha_{G2}) t_{E2}](T - T_C), \tag{6.29}$$

$$\Delta GAP_A = [(\alpha_M - \alpha_{G1}) t_{E1} + (\alpha_M - \alpha_S) t_S + (\alpha_M - \alpha_{G2}) t_{E2}](T - T_C), \tag{6.30}$$

where the subscript "S" refers to the spacer and all other terms are as defined earlier. Figure 6.5 illustrates these cases. If the preload applied at assembly is large, the calculated value for T_C may exceed the maximum temperature for the system T_{max}. In this case, ΔGAP_A will be negative, indicating that glass-to-metal contact is not lost within the range $T_A \le T_C \le T_{max}$.

In nearly all applications, small changes in the position and orientation of a lens within axial and radial gaps created by differential expansion are tolerable. However, high accelerations (vibration or shock) applied to a lens assembly when clearance exists between the lens and its mounting surfaces may damage the lens from glass-to-metal impacts. This type of damage, associated with sustained vibration, is called "fretting" of the glass surfaces. To minimize this threat, the lens assembly should have sufficient residual preload at T_{max} to hold the lens firmly against the axial mechanical interface under the maximum expected acceleration. In other words, the direction of the preload should not change as a function of temperature. As indicated earlier in this book, the preload needed to constrain a lens of mass W under axial acceleration a_G is simply W times a_G. For preload to exist at T_{max}, the preload at assembly must be the sum of the acceleration-compensating minimum preload plus the preload decrease that is caused by the temperature increase from that at assembly (T_A) to T_{max}.

Such a case can be visualized as follows. Imagine a lens as assembled at T_A. No axial gaps exist between the lens surfaces at the retainer and the shoulder because preload is provided by the retainer. Now, the temperature rises above T_C. Because α_M is greater than α_G, the metal expands more than the glass, so the preload vanishes, and an axial gap develops in front of the lens, after the lens, or at both places. The lens is free to move somewhat under acceleration.

For $T \le T_C$, K_3 facilitates calculation of the effects of temperature changes on preload. But its value is difficult to quantify completely, even for a simple lens/mount configuration. For example, consider again the design shown schematically in Figure 6.5a. The contact stress developed within the lens by the preload is distributed approximately as indicated in Figure 6.6. The stressed

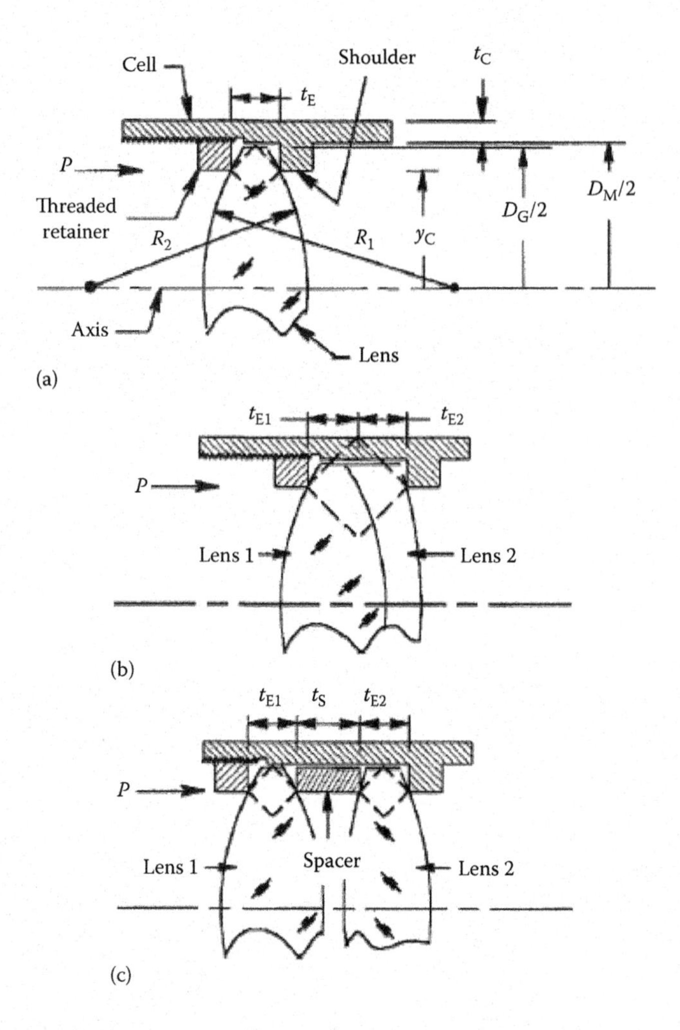

FIGURE 6.5 Schematics of lens mounting configurations for (a) a single lens element, (b) a cemented doublet lens, and (c) an air-spaced doublet. All show by dashed lines the propagation of contact stress within the lenses.

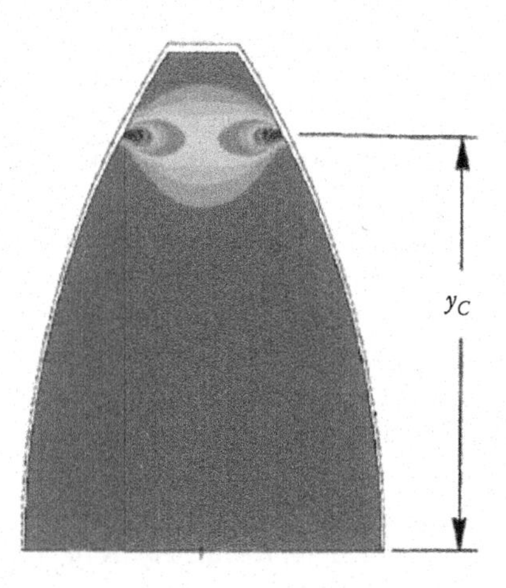

FIGURE 6.6 Finite-element analytical representation of stress distribution within a lens when preloaded as in Figure 6.5a. (Adapted from Genberg, V.L., Structural analysis of optics, Chapter 8 in *Handbook of Optomechanical Engineering*, Ahmad, A., Ed., pp. 255–328, CRC Press, Boca Raton, FL, 1997.)

(annular) zone is represented in section by the diamond-shaped region. K_3 for any mounted lens is given approximately as

$$K_3 = \frac{\sum_{i=1}^{n} (\alpha_M - \alpha_i) t_i}{\sum_{i=1}^{n} C_i}, \tag{6.31}$$

where C_i represents the compliance of one of the i elastic components in the subassembly. For each lens, compliance is approximated as $[2t_E/(E_G A_G)]_i$, that of a cell is $[t_E/(E_M A_M)]_i$, and that of a spacer is $[t_S/(E_S A_S)]_i$. The cross-sectional area A_i is the stress region in the glass and metal components. Figures 6.7 through 6.9 illustrate typical geometries for the cell and for the lens and two types of spacers, respectively.[1]

For any lens element, either of two cases can apply. If $(2y_C + t_E) \le D_G$, the stressed region shown as the diamond-shaped section in Figure 6.6 lies entirely within the lens rim and Equation 6.32 applies. The lens thickness t_E is measured at the height of contact y_C. This height is assumed the same on both sides of the lens. If $(2y_C + t_E) > D_G$, the stressed region is truncated by the rim and Equation 6.33 is used:

$$A_G = 2\pi y_C t_E, \tag{6.32}$$

$$A_G = \frac{\pi}{4}(D_G - t_E + 2y_C)(D_G + t_E - 2y_C), \tag{6.33}$$

where D_G is the outer diameter of the lens.

For the cell wall, A_M is given by

$$A_M = \pi t_C (D_M + t_C), \tag{6.34}$$

where D_M is the ID of the mount bore at the lens rim and t_C is the thickness of the cell wall adjacent to the lens rim.

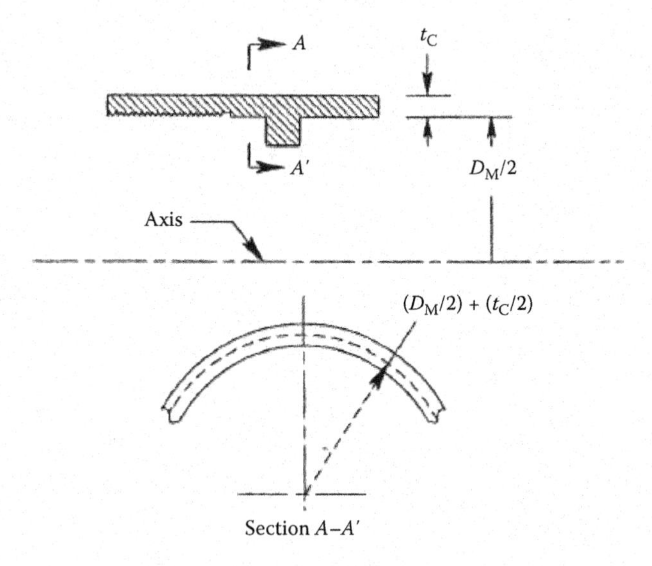

FIGURE 6.7 Geometric relationships used to estimate the cross-sectional area of the stressed region within a simple lens cell. (From Yoder, P.R. Jr., Advanced considerations of the lens-to-mount interface, *SPIE Crit. Rev.*, CR43, 305, 1992.)

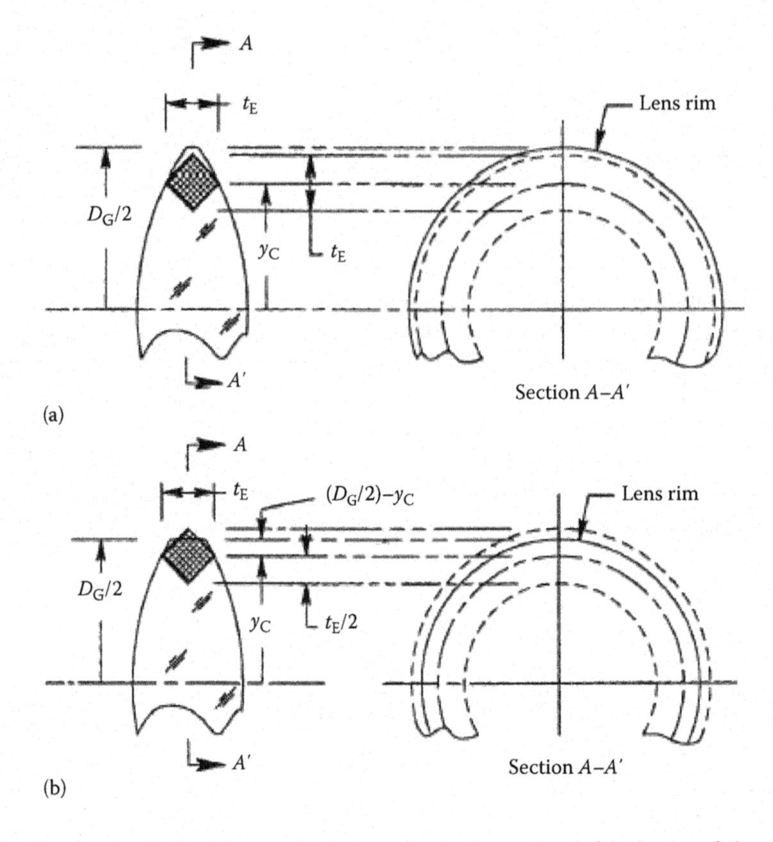

FIGURE 6.8 Geometric relationships used to approximate the cross-sectional area of the stressed region within a lens: (a) when complete within the lens rims and (b) when truncated by the rim. (From Yoder, P.R. Jr., Advanced considerations of the lens-to-mount interface, *SPIE Crit. Rev.*, CR43, 305, 1992.)

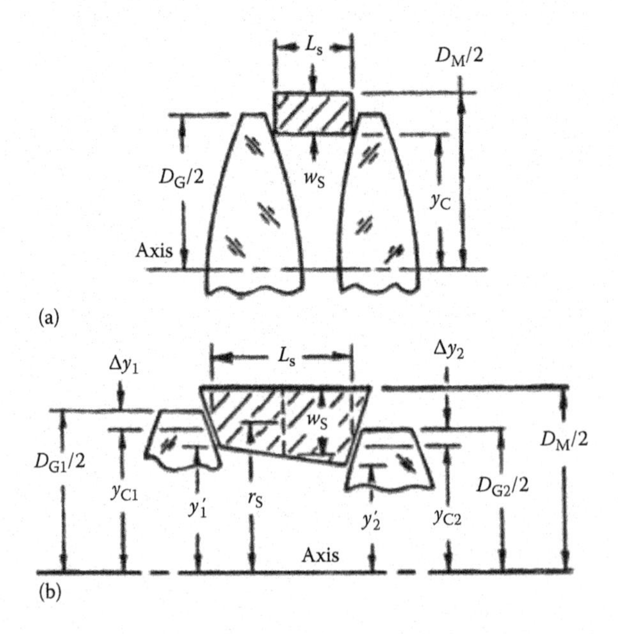

FIGURE 6.9 Schematics of two typical lens spacers: (a) cylindrical type with sharp corner interfaces and (b) tapered type with tangential interfaces. (From Yoder, P.R. Jr., Estimation of mounting-induced axial contact stresses in multi-element lens assemblies, *Proc. SPIE*, 2263, 332, 1994.)

For a spacer, the cross-sectional area A_S is given in general by

$$A_S = 2\pi r_S w_S, \tag{6.35}$$

$$r_S = \frac{D_M}{2} - \frac{w_S}{2}. \tag{6.36}$$

The cross-sectional parameter w_S for a spacer with a constant cross section as shown in Figure 6.9a is given by

$$w_S = \frac{D_M}{2} - y_C. \tag{6.37}$$

For the simple tapered spacer shown in Figure 6.9b, the cross-sectional parameter w_S is

$$\Delta y_i = \frac{(D_G)_i}{2} - (y_C)_i, \tag{6.38}$$

$$y_i' = (y_C)_i - (\Delta y)_i, \tag{6.39}$$

$$w_S = \frac{D_M}{2} - \frac{y_1' + y_2'}{2}. \tag{6.40}$$

Applying the preceding equations, as appropriate, the parameter K_3 is found using the following equations for a single lens element, for a cemented doublet, and for an air-spaced doublet:

$$K_3 = \frac{-(\alpha_M - \alpha_G)t_E}{\dfrac{2t_E}{E_G A_G} + \dfrac{t_E}{E_M A_M}}, \tag{6.41}$$

$$K_3 = \frac{-(\alpha_M - \alpha_{G1})t_{E1} - (\alpha_M - \alpha_{G2})t_{E2}}{\dfrac{2t_{E1}}{E_{G1} A_{G1}} + \dfrac{2t_{E2}}{E_{G2} A_{G2}} + \dfrac{t_{E1} + t_{E2}}{E_M A_M}}, \tag{6.42}$$

$$K_3 = \frac{-(\alpha_M - \alpha_{G1})t_{E1} - (\alpha_M - \alpha_S)t_S - (\alpha_M - \alpha_{G2})t_{E2}}{\dfrac{2t_{E1}}{E_{G1} A_{G1}} + \dfrac{t_S}{E_S A_S} + \dfrac{2t_{E2}}{E_{G2} A_{G2}} + \dfrac{t_{E1} + t_S + t_{E2}}{E_M A_M}}. \tag{6.43}$$

Example 6.7

The diameter of a Schott BK7 lens is D_G = 63.5 mm, and the distance from retainer to seat is t_E = 6.35 mm, and the contact radius between the retainer and lens is y_C = 29.3 mm. The lens is mounted in a cell made of 6061 aluminum with an ID of D_M = 63.55 mm and a radial wall thickness of $t_C = 2.5$ mm. The upper limit of service temperature is $T_{max} = +45°C$; the lower limit is $T_{min} = 0°C$, and assembly is done at a temperature of $T_A = 25°C$. The properties of Schott BK7 are $E_G = 82$ GPa, $\nu_G = 0.275$, and $\alpha_G = 8.2 \times 10^{-6}$ K^{-1}. The properties of 6061 aluminum are $E_M = 68.2$ GPa, $\nu_M = 0.33$, and $\alpha_M = 23.2 \times 10^{-6}$ K^{-1}. What is the minimum preload P_A to ensure that the preload does not go to zero at T_{max}? What is the change in preload ΔP at T_{min}?

First, find the stress areas by using Equations 6.32 through 6.34. Note that the following:

$$2y_C + t_E = 2(29.3\text{ mm}) + 6.35\text{ mm} = 64.95\text{ mm} > 63.5\text{ mm}.$$

Hence, the stress area in the lens A_G is found using Equation 6.33, while the A_M of the cell is found from Equation 6.34:

$$A_G = \frac{\pi}{4}[63.5\times10^{-3}\text{m} - 6.35\times10^{-3}\text{m} + 2(29.3\times10^{-3}\text{m})]$$
$$[63.5\times10^{-3}\text{m} + 6.35\times10^{-3}\text{m} - 2(29.3\times10^{-3}\text{m})] = 1.023\times10^{-3}\text{m}^2,$$

$$A_M = \pi(2.5\times10^{-3}\text{m})(63.55\times10^{-3}\text{m} + 2.5\times10^{-3}\text{m}) = 518.8\times10^{-6}\text{m}^2.$$

The change in preload with temperature K_3 is given using Equation 6.41:

$$K_3 = \frac{-(23.6\times10^{-6}\text{K}^{-1} - 8.2\times10^{-6}\text{K}^{-1})}{\frac{2(6.35\times10^{-3}\text{m})}{(82\times10^9\text{Pa})(1.023\times10^{-3}\text{m}^2)} + \frac{6.35\times10^{-3}\text{m}}{(68.2\times10^9\text{Pa})(518.8\times10^{-6}\text{m}^2)}} = -295.5\text{ N/K}.$$

From Equation 6.26, the minimum preload P_A at T_{max} is

$$P_A = -295.5\text{N/K}(298\text{K} - 318\text{K}) = 5.91\times10^3\text{ N}.$$

From Equation 6.25, the change in preload ΔP at T_{min} is

$$\Delta P = -295.5\text{N/K}(273\text{K} - 298\text{K}) = 7.388\times10^3\text{ N}.$$

Note that the total preload at T_{min} is $P_A + \Delta P$ or about 13.3 kN. This is a very high preload force, and it would be advisable to check for possible degradation of optical performance from stress optic effects as well as potential breakage. This example illustrates that thermal stresses can become very large even over relatively limited temperature ranges.

When the CTE of the lens cell material differs greatly from that of the lens, there is an associated large value of K_3, the change in preload with temperature. For large changes in temperature, this may lead to unacceptable values of the preload force P_A at low temperature, or excessive axial gaps between the lens and mount at high temperature. One remedy is selection of a lens cell material with a CTE closely matching that of the lens material.

The CTE of optical glasses in the visible is 4.5×10^{-6} to 12×10^{-6} K^{-1}. The CTEs of many of the commonly used optical glasses are limited to 6×10^{-6} to 9×10^{-6} K^{-1}. The CTE of titanium, 8.8×10^{-6} K^{-1}, is a good match with the CTEs of many glasses, and titanium is increasingly popular as a lens cell material. Offsetting the match of the CTE of titanium with that of optical glass is a relatively high fabrication cost. Substituting 6Al–4V titanium for 6061 aluminum in Example 6.7 reduces K_3 to −14.7 N/K and the preload change ΔP at T_{min} to more desirable value of 367.7 N. Another alternative lens cell material is type AISI 416 (UNS S416000) stainless steel, with a CTE

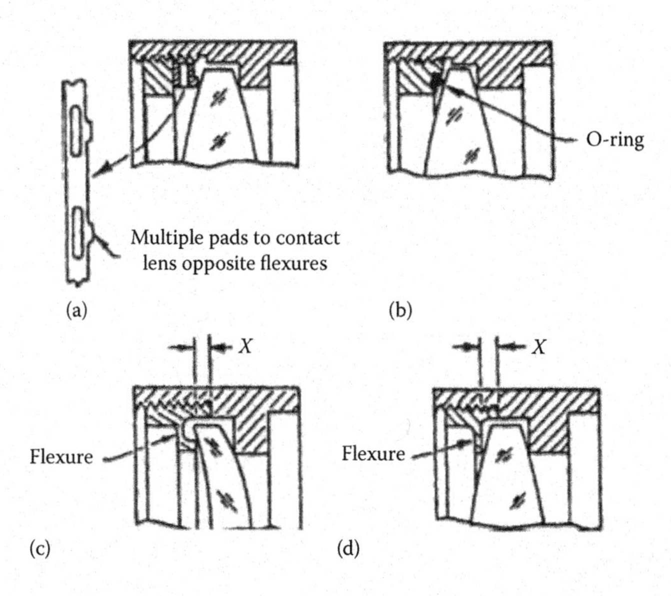

FIGURE 6.10 Four techniques for providing axial compliance in a lens mounting: (a) multiple flexures, (b) O-ring contact, (c) flexure retainer contacting a concave surface, and (d) flexure retainer contacting a convex surface. (From Yoder, P.R. Jr., *Mounting Optics in Optical Instruments*, 2nd ed., SPIE Press, Bellingham, WA, 2008.)

of 9.9×10^{-6} K^{-1}. Type 416 is considered to have the best machinability of any stainless steel, with a machining index of 50% (the machinability of type 1112 steel is 100%). A relatively high density of 7.75×10^3 kg/m^3 is the drawback of type 416 stainless.

Alternately, axial compliance is added to the retainer as a means of reducing the change in axial preload with temperature. Four possible axially compliant retainer configurations are shown in Figure 6.10.[2] In Figure 6.10b, an O-ring provides axial compliance; the O-ring acts as an elastic element. There is a gap between retainer and lens, with only the O-ring making contact with the lens. As a rough rule of thumb, the O-ring is compressed to about 50–70% of the nominal compression used for a seal (note that O-rings are normally compressed to less than 25% of their diameters for sealing purposes). Ideally, the O-ring material should be low in outgassing to minimize possible contamination of the lens surface. A common guide for selection based on outgassing is that the O-ring material should meet the National Aeronautics and Space Administration standard of less than 1.00% total mass loss (TML) and less than 0.10% collected volatile condensable material per ASTM E595. A drawback of the use of an O-ring is a high friction coefficient, typically around $\mu \approx 0.9$, which may make self-centering of the lens impractical.

In Figure 6.10c and d, the retainers incorporate diaphragm flexures as a means of reducing the axial preload change with temperature. Diaphragm flexure stiffness is calculated using either plate theory or spring design methods. Advantages of diaphragm flexures are low outgassing, since the retainer is made of metal, and a relatively low friction coefficient, which is desirable for self-centering of the lens. Drawbacks of the diaphragm flexure are a higher fabrication cost and normally a higher spring rate than that of an O-ring, with an associated larger value of K_3.

6.4.6 Thermal Stress in Bonded Lenses

Adhesive joints between lens elements using optical cements are subjected to stress buildup during curing and when the lens is subjected to temperature changes. Figure 6.11 shows an extreme case, and Figure 6.12 shows the failure of the mock-up. The change in temperature for the lens in Figure 6.12 was −82°C, and the difference in CTE between the two materials was $\Delta\alpha = 8.4 \times 10^{-6}$ K^{-1}.

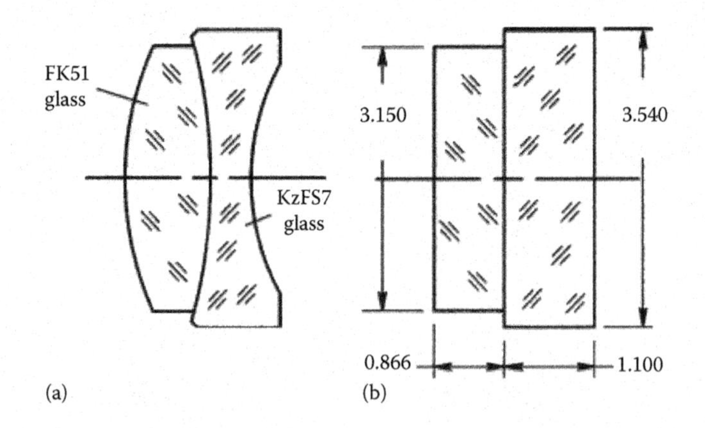

FIGURE 6.11 (a) Sectional view of a cemented doublet with thick lenses made of glasses with widely differing CTEs to be used in a military application. (b) Sectional view of a mock-up of the doublet made of plane-parallel plates and cemented with optical cement. This was intended for low-temperature testing to confirm the design. (From Yoder, P.R. Jr., *Mounting Optics in Optical Instruments*, 2nd ed., SPIE Press, Bellingham, WA, 2008.)

FIGURE 6.12 Photograph of the cemented doublet in Figure 6.11 after exposure to low temperature. Fractures occurred as anticipated due to differential thermal expansion effects. (From Yoder, P.R. Jr., *Mounting Optics in Optical Instruments*, 2nd ed., SPIE Press, Bellingham, WA, 2008.)

Two simplifying assumptions in estimating the thermal stress in a bonded doublet are that the adhesive stiffness is small compared to the stiffness of the lenses and that the lenses are constant-thickness, plane-parallel plates. The elastic modulus of even high-strength epoxy adhesives is less than about 1% of that of glass, so the assumption of low adhesive stiffness is reasonable. The assumption of constant thickness for the lenses is a concern, since there is a significant variation in lens thickness. Two approaches are to perform a worst-case stress analysis, with the thickness assumed to be the thickest part of each lens within the clear aperture, or alternately to pick a compromise thickness, normally at about one-half to two-thirds of the clear aperture. A lower, more realistic stress is usually found using the latter, compromise approach. The thermal stress is given by

$$\tau_S = \frac{2(\alpha_1 - \alpha_2)\Delta T G_e I_1(\beta R)}{t_e \beta (C_1 + C_2)}, \tag{6.44}$$

$$\beta = \left\{ \frac{G_e}{t_e} \left[\frac{(1-\nu_1^2)}{E_1 t_1} + \frac{(1-\nu_2^2)}{E_2 t_2} \right] \right\}^{\frac{1}{2}}, \tag{6.45}$$

$$C_1 = -\frac{2}{1+\nu_1} \left\{ \left[\frac{(1-\nu_1) I_1(\beta R)}{\beta R} \right] - I_0(\beta R) \right\}, \tag{6.46}$$

$$C_2 = -\frac{2}{1+\nu_2} \left\{ \left[\frac{(1-\nu_2) I_1(\beta R)}{\beta R} \right] - I_0(\beta R) \right\}, \tag{6.47}$$

where τ_S is the thermal shear stress in lenses and adhesive, α_1 and α_2 are the CTEs of the two lens materials, ΔT is the change in temperature, G_e is the shear modulus of the adhesive, t_e is the adhesive thickness, R is the radius of the lenses, E_1 and E_2 are the elastic moduli of the two lens materials, t_1 and t_2 are the thicknesses of the two lenses, and ν_1 and ν_2 are Poisson's ratios of the two lens materials.

The terms I_0 and I_1 refer to modified Bessel functions of the first kind. Bessel functions can be found in mathematical tables; some mathematical computer programs also evaluate Bessel functions. Polynomial approximations for the Bessel functions are given in the following. These are accurate to better than 1% up to $\beta R \approx 3.75$ and to 5% up to $\beta R \approx 5$; this accuracy is sufficient for initial stress estimates:

$$\begin{aligned} I_1(\beta R) &= 1 + 250\times10^{-3}(\beta R)^2 + 15.625\times10^{-3}(\beta R)^4 \\ &+ 433.94\times10^{-6}(\beta R)^6 + 6.80123\times10^{-6}(\beta R)^8 + 65.6017\times10^{-9}(\beta R)^{10}, \end{aligned} \tag{6.48}$$

$$\begin{aligned} I_0(\beta R) &= 500\times10^{-3}(\beta R) + 62.5\times10^{-3}(\beta R)^3 + 2.60419\times10^{-3}(\beta R)^5 \\ &+ 203.417\times10^{-6}(\beta R)^6 + 679.868\times10^{-9}(\beta R)^9 + 5.48303\times10^{-9}(\beta R)^{11}. \end{aligned} \tag{6.49}$$

Example 6.8

The doublet mock-up in Figure 6.11 is made of FK51 and KzFS57 glasses, with a diameter of $2R$ = 80 mm. For FK51, $\alpha_1 = 13.3 \times 10^{-6}$ K^{-1}, E_1 = 81 GPa, ν_1 = 0.274, and t_1 = 21.996 mm. For KzFS57, $\alpha_2 = 4.93 \times 10^{-6}$ K^{-1}, E_2 = 68 GPa, ν_2 = 0.293, and t_1 = 27.94 mm. For the adhesive, G_e = 386 MPa and t_e = 25.4 μm. Find the thermal stress for a temperature drop of −82°C.

First, find the parameter β by using Equation 6.45:

$$\beta = \left\{ \frac{386 \times 10^6 \,\text{Pa}}{25.4 \times 10^{-6} \,\text{m}} \left[\frac{1-(0.274)^2}{(81 \times 10^9 \,\text{Pa})(21.996 \times 10^{-3} \,\text{m})} + \frac{1-(0.293)^2}{(68 \times 10^9 \,\text{Pa})(27.94 \times 10^{-3} \,\text{m})} \right] \right\}^{\frac{1}{2}} = 123.293 \text{ m}^{-1}.$$

Then, βR = (123.293 m^{-1})(40 × 10^{-3} m) = 4.932. Finding the Bessel functions by using the approximations given in Equations 6.48 and 6.49,

$$I_0(4.932) = 1 + 250 \times 10^{-3}(4.932)^2 + 15.625 \times 10^{-3}\,(4.932)^4 + 433.940 \times 10^{-6}(4.932)^6 \\ + 6.80123 \times 10^{-6}(4.932)^8 + 65.6017 \times 10^{-9}(4.932)^{10} = 25.505,$$

$$I_1(4.932) = 500 \times 10^{-3}(4.932) + 62.5 \times 10^{-3}(4.932)^3 + 2.60419 \times 10^{-3}(4.932)^5 \\ + 203.417 \times 10^{-6}(4.932)^6 + 679.868 \times 10^{-9}(4.932)^9 + 5.48303 \times 10^{-9}(4.932)^{11} = 21.89.$$

Then, C_1 and C_2 are found using Equations 6.46 and 6.47:

$$C_1 = -\left\{ \frac{2}{1+0.274} \left[\frac{(1-0.274)(21.89)}{(123.293\text{m}^{-1})(40 \times 10^{-3}\,\text{m})} - 25.505 \right] \right\} = 34.981,$$

$$C_2 = -\left\{ \frac{2}{1+0.293} \left[\frac{(1-0.293)(21.89)}{(123.293\text{m}^{-1})(40 \times 10^{-3}\,\text{m})} - 25.505 \right] \right\} = 34.597.$$

Finally, the maximum shear stress τ_S is found using Equation 6.44:

$$\tau_S = \frac{2[(13.3 \times 10^{-6}\,\text{K}^{-1}) - (4.9 \times 10^{-6}\,\text{K}^{-1})](386 \times 10^6\,\text{Pa})(-\,82K)(21.89)}{(25.4 \times 10^{-6}\,\text{m})(123.293\text{m}^{-1})(34.981 + 34.597)} = 53.42 \times 10^6\,\text{Pa}.$$

A stress of 53.4 MPa is more than enough to cause fracture in the glass, which agrees with the test result.

Special optical adhesives are used for bonding lenses for visible wavelengths together; the important properties are index of refraction and elastic modulus. The index of refraction of an optical adhesive should be close to that of the glass to minimize transmission losses; for typical optical adhesives, n is 1.48–1.55. The elastic modulus of the optical adhesive should also be low relative to that of the glasses to minimize thermal stress; typically, for optical adhesives, E is between 1 and 4 GPa. The properties of typical optical adhesives are given in Table 6.1.

Canadian balsam, made from the sap of Canadian evergreen trees, is the oldest type of optical adhesive; it is not commonly used today. Canadian balsam is applied when heated; an advantage is that the lenses can be easily debonded by heating. Problems with quality and availability led to replacing Canadian balsam with a variety of synthetic adhesives.

Thermosetting cements were the first synthetic adhesives to replace Canadian balsam. These adhesives required heating to initiate curing. Typically, curing is initiated at about 70°C for 1–3 hours; this provides sufficient strength to tack the lenses in place, allowing limited handling. Final curing requires exposure to 70°C for about 16 hours. Faster curing is possible at higher temperatures, but

TABLE 6.1
Typical Properties of a Generic Optical Cement

Property	Typical Values
Refractive index n after cure	1.48–1.55
CTE (27–100°C)	63×10^{-6} K^{-1}
Elastic modulus E	4.3×10^{9} Pa
Shear strength	36×10^{6} Pa
Specific heat c_p	837 J/kg K
Water absorption (bulk material)	0.3% after 24 hours at 25°C
Shrinkage during cure	<6%
Viscosity	200–320 cP
Density	1.22 g/cm^3
Hardness (Shore D)	≈90
TML in vacuum	3–6%

higher curing rates also cause more shrinkage of the adhesive during curing, with higher residual stress.

The long cure times and high temperatures required by the thermosetting adhesives led to development of ultraviolet (UV) curing adhesives, which are now the most common in the optical industry. Curing is initiated in the UV curing adhesives by irradiation with UV light at wavelengths between 354 and 378 nm. A short intense UV illumination lasting 10–20 s tacks the lenses in place, permitting limited handling. Final curing is with a flood UV source, at a lower level, over a period of about an hour. As with the thermosetting adhesives, the shrinkage of UV curing adhesives is inversely proportional to the cure time. Short cure times induce considerable shrinkage and high levels of residual stress.

6.4.7 Focus Shift with Temperature

A uniform temperature change induces a focus shift in a lens. The thermal shift in focus arises from two material properties of the lens: the CTE α and change in the index of refraction with temperature dn/dT. Consider a simple thin lens with optical radii R_1 and R_2 and with an index of refraction n; the power K (reciprocal of the focal length, or $K = 1/f$) is given by

$$\frac{1}{f} = K = (n-1)\left(\frac{1}{R_1} - \frac{1}{R_2}\right). \tag{6.50}$$

Taking the derivative of Equation 6.50 with respect to temperature T,

$$\frac{dK}{dT} = \left(\frac{1}{R_1} - \frac{1}{R_2}\right)\frac{dn}{dT} - (n-1)\left(\frac{1}{R_1^2}\frac{dR_1}{dT} - \frac{1}{R_2^2}\frac{dR_2}{dT}\right). \tag{6.51}$$

Now $\alpha = (1/R)dR/dT$, so Equation 6.51 becomes

$$\frac{dK}{dT} = (n-1)\left(\frac{1}{R_1} - \frac{1}{R_2}\right)\left[\frac{1}{(n-1)}\frac{dn}{dT} - \alpha\right]. \tag{6.52}$$

Let

$$\frac{dK}{dT} = K\beta, \tag{6.53}$$

$$\beta = \frac{1}{(n-1)}\frac{dn}{dT} - \alpha, \tag{6.54}$$

$$df = -\beta f\, dT. \tag{6.55}$$

The term β is called the thermo-optical coefficient, and it is dependent only upon the lens material. Units of the thermo-optical coefficient are inverse temperature, typically inverse kelvins (K^{-1}). The size and shape of the lens do not influence the change in focus with temperature. For optical glass in the visible, the value of β is -32×10^{-6} to $+22 \times 10^{-6}$ K^{-1}. Note that there are several different relationships used for the thermo-optical coefficient; Schott, for example, used a version that predicts the change in optical path difference with temperature. Table 6.2 gives representative values of the thermo-optical coefficient for glasses in the visible, while Tables 6.3 and 6.4 give the thermo-optical coefficients for some infrared (IR) materials of interest used in the 3–5 and 8–12 μm wave bands.

Extending the preceding derivation to an achromatic doublet of focal length f_{DBLT} made of two different glass types with β_1 and β_2,

$$df_{DBLT} = -f_{DBLT}\, dT\left(\frac{\nu_1\beta_1 - \nu_2\beta_2}{\nu_1 - \nu_2}\right), \tag{6.56}$$

where ν_1 and ν_2 are the Abbe numbers of the two glass types in the achromatic doublet.

TABLE 6.2
Thermal Glass Constants of Selected Visible Glasses

Glass Type	Refractive Index n_d	Abbe Number ν_d	dn/dT_{rel} e-Line ($\times 10^{-6}$ K^{-1})	CTE α ($\times 10^{-6}$ K^{-1})	Thermal Glass Constant β ($\times 10^{-6}$ K^{-1})
N-PK51	1.52655	76.9	−6.7	12.35	−25
N-FK5	1.48794	76.41	−5.7	5.3	−17
N-FK5	1.48749	70.41	−1	9.2	−11
N-LAF2	1.74397	44.85	1	5.3	−4.0
N-SF15	1.69892	30.2	3.4	8.04	−3.2
N-SK16	1.62041	60.32	2.3	6.3	−2.6
N-SSK5	1.65844	50.88	3.2	6.8	−1.9
SF2	1.64769	33.85	4.6	8.4	−1.3
N-BK7	1.5168	64.17	3	7.1	−1.3
F2	1.62004	36.37	4.4	8.2	−1.1
N-BaSF52	1.66446	36	4.4	7.12	−5.0
N-BAK4	1.56883	55.98	3.9	6.99	−1.3
SF4	1.7552	27.58	9.2	8	4.2
LaFN7	1.7495	34.95	8.3	5.3	5.8
SF57	1.84666	23.83	12.5	8.3	6.5

Note: d*n*/d*T* is the relative value.

TABLE 6.3
Thermal Glass Constants for Selected Materials for the 3–5 μm IR Wavelength Region

Material	Refractive Index n	Abbe Number ν	dn/dT ($\times 10^{-6}$ K^{-1})	CTE α ($\times 10^{-6}$ K^{-1})	Thermal Glass Constant β ($\times 10^{-6}$ K^{-1})
Silicon	3.43	240	170	2.6	67.4
KRS5	2.38	230	−237	58	−230
AMTIR1	2.51	190	77	12	39.0
Zinc selenide	2.43	180	63	7.6	36.5
Arsenic trisulfide	2.41	160	9	2.4	3.98
Zinc sulfide	2.25	110	42	6.7	26.9
Germanium	4.02	100	410	6.1	130
Calcium fluoride	1.41	22	−8.1	18.9	−38.7
Magnesium oxide	1.67	12	19	10.5	17.9

TABLE 6.4
Thermal Glass Constants for Selected Materials for the 8–12 μm IR Wavelength Region

Material	Refractive Index n	Abbe Number ν	dn/dT ($\times 10^{-6}$ K^{-1})	CTE α ($\times 10^{-6}$ K^{-1})	Thermal Glass Constant β ($\times 10^{-6}$ K^{-1})
Germanium	4	860	404	6.1	129
Cesium iodide	1.74	230	−99.3	39.1	−173
Cadmium telluride	2.68	150	49	5.9	23.3
KRS5	2.37	170	−235	58	−230
AMTIR1	2.5	110	72	12	36.0
Gallium arsenide	3.28	110	150	5.7	60.1
Zinc selenide	2.41	58	60	7	35.6
Zinc sulfide	2.2	23	41	6.7	27.5
Sodium chloride	1.49	19	−25	40	−91.0

Assuming thin lenses for simplification, the change in the focal length df for a temperature change dt for a system of j lenses in a lens barrel made of a material with a CTE α_M is

$$df = -\left[f^2 \sum_{i=1}^{j}\left(\frac{\beta_i}{f_i}\right) + \alpha_M f\right]dt, \tag{6.57}$$

where f is the focal length of the optical system, β_i is the thermo-optical coefficient of the ith lens, and f_i is the focal length of the ith lens.

The change in the index of refraction with temperature dn/dT is measured with respect to either vacuum or air. When measured with respect to vacuum, dn/dT is called absolute and the value given a subscript "abs," or dn/dT_{abs}. When measured with respect to vacuum, dn/dT is called relative, with a subscript of "rel," or dn/dT_{rel}. Since most optical systems are used in air, the relative measurement is the most common. There is a significant difference in the absolute and relative values of dn/dT.

The index of refraction of air changes with wavelength, pressure, and temperature; like those of other optical materials do; there is a dn/dT for air. At a standard pressure of 1 atm (101.325 kPa) and a standard temperature of 15°C, the index of refraction of air with respect to wavelength is given by

$$n_{15} = 1 + \left[6432.8 + \frac{2949810\lambda^2}{146\ \lambda^2 - 1} + \frac{25540\lambda^2}{41\lambda^2 - 1} \right] \times 10^{-8}, \quad (6.58)$$

where λ is the wavelength in micrometers.

Then, the change in the refractive index of air with temperature $d/dt(n_{air})$ is given by

$$\frac{d}{dT}(n_{air}) = \frac{-0.003861(n_{15} - 1)}{(1 + 0.00366T)^2}, \quad (6.59)$$

where T is the air temperature in degrees Celsius.

In the visible, at $\lambda = 0.55$ μm, $n_{air} = 1.0002778$ at $T = 15$°C and $d/dT(n_{air}) = -964 \times 10^{-9}$ K^{-1}.

Maintenance of focus over temperature is called athermalization of the optical systems since the effect is to make the focus independent of temperature. For maintenance of focus over temperature, the structure must move the same amount as the focus shift, or $\beta = \alpha_S$, where α_S is the CTE of the structural material. This makes $df = 0$ in Equation 6.57. It is unlikely that there will be a structural material with a CTE match to the thermo-optical coefficient, especially for a complex multielement lens. If the structural material CTE differs from the optothermal coefficient, the system loses focus over temperature as shown in Figure 6.13. Two methods of athermalization are passive compensation, which uses a combination of structural materials, and active compensation, which involves motorized positioning of the focal plane or lens(es) over temperature.

Passive compensation uses the principle of a bimetallic compensator, with two materials of different CTEs used in tandem. The composite CTE α_C for the structure shown in Figure 6.14b[3] is

$$\alpha_C = \frac{\alpha_1 L_1 + \alpha_2 L_2}{L_1 + L_2}, \quad (6.60)$$

where α_1 and α_2 are the CTEs of the two materials and L_1 and L_2 are the lengths.

In Equation 6.60, a negative value of L_1 or L_2 implies a "folded-back" configuration as shown in Figure 6.14a. With the folded-back configuration, the structural effective α_C can match very low or negative values of the thermo-optical coefficient β. Generally, for athermalization of the location of the focal plane for a system of focal length f and thermo-optical coefficient β,

$$\beta f = \alpha_1 L_1 + \alpha_2 L_2, \quad (6.61)$$

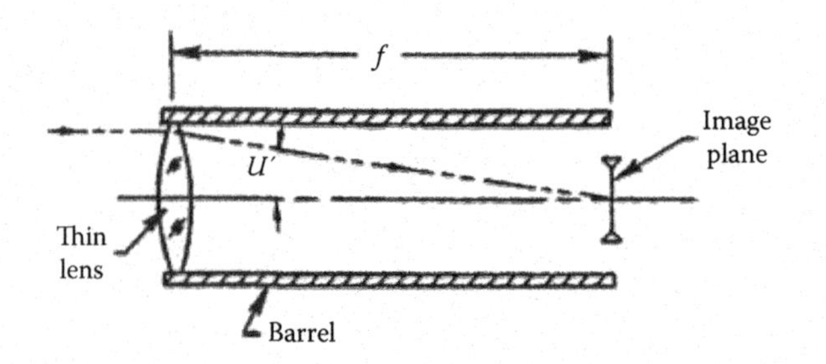

FIGURE 6.13 Schematic of a thermally uncompensated single thin lens and simple mount. (From Yoder, P.R. Jr., *Mounting Optics in Optical Instruments*, 2nd ed., SPIE Press, Bellingham, WA, 2008.)

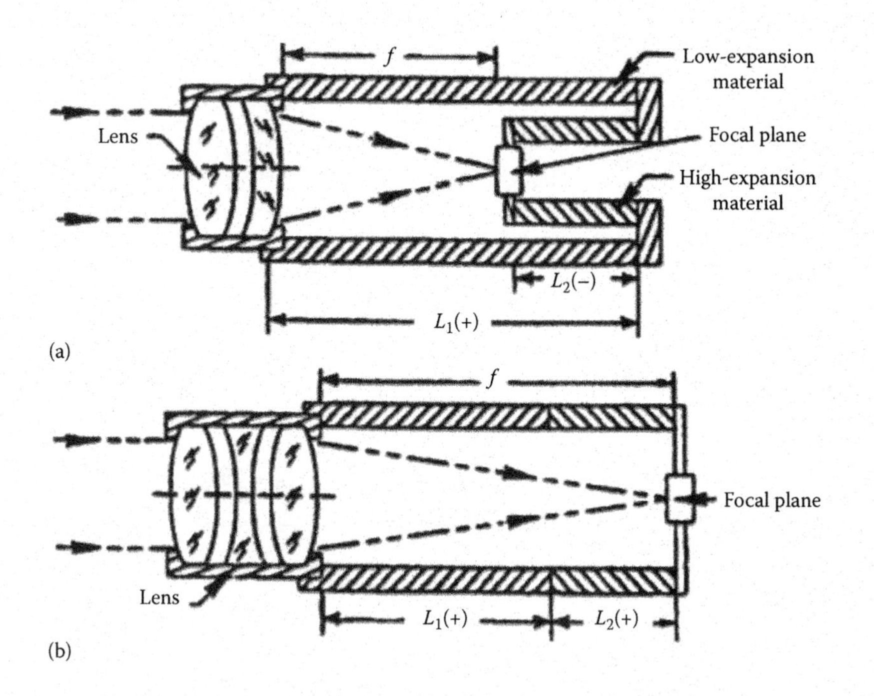

FIGURE 6.14 Schematic diagrams of simple dual-material (bimetallic compensated) mounts that athermalize the focus of lenses: (a) a reentrant dual-material compensator and (b) a mount where the two materials of the compensator are in series. (From Vukobratovich, D., Introduction to optomechanical design, SPIE short course SC014, 2014.)

$$L_1 = f - L_2, \tag{6.62}$$

$$L_2 = f\left(\frac{\alpha_1 - \beta}{\alpha_1 - \alpha_2}\right). \tag{6.63}$$

Although Figure 6.14 shows a concentric cylindrical structure or tandem structure, an alternative is a bimetallic compensator that moves a lens element to maintain focus. This type of design is called a metering structure. A concentric arrangement of high- and low-CTE rods provides the necessary motion over temperature.

An example of a bimetallic metering rod is shown in Figure 6.15. Shown is the DEIMOS (Deep Imaging Multi-object Spectrograph) camera lens. The largest lens is about 330 mm in diameter; three of the lenses are made of CaF_2.

The need for the athermalization mechanism resulted from the requirement that the scale of the image produced by the camera was to be constant over the operating temperature range of ±5°C. This requirement was in terms of the pixel diameter; the scale must be kept to 0.17 pixel diameter over the maximum image height of 4096 pixels, or $\delta p/p = 42 \times 10^{-6}$. The corresponding temperature sensitivity was $\delta p/p = 136 \times 10^{-6}$/K.

A passive bimetallic metering rod moved the fourth lens doublet axially, elements 7 and 8, with temperature. Elements 7 and 8 were mounted in a cell, and the cell was attached to the barrel by axially compliant flexures. The force for moving the cell was delivered by connecting it to an Invar rod with a concentric Delrin tube; both rod and tube were 440 mm long. The differential expansion of the rod and tube provided a length change of 36×10^{-6} m/K within ±3%, which was sufficient to compensate over the specified temperature range.

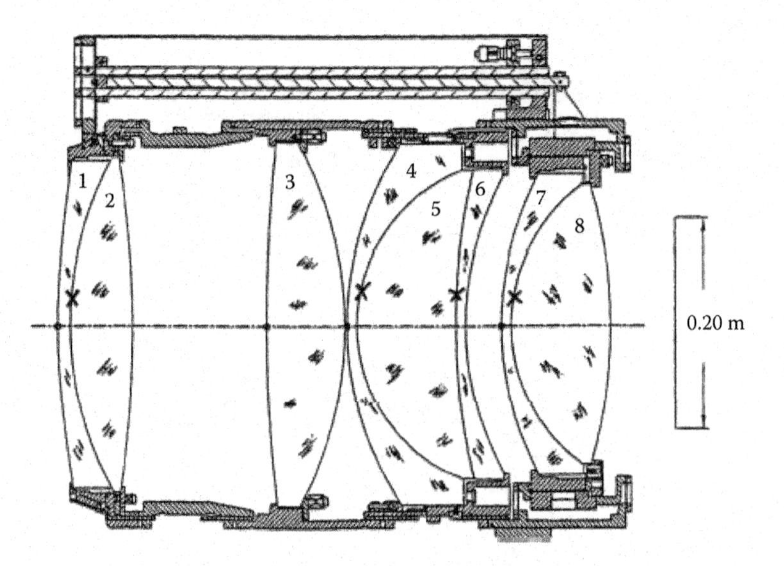

FIGURE 6.15 Optomechanical design layout for the DEIMOS camera assembly. (Adapted from Mast, T. et al., DEIMOS camera assembly, *Proc. SPIE*, 3786, 499, 1999.)

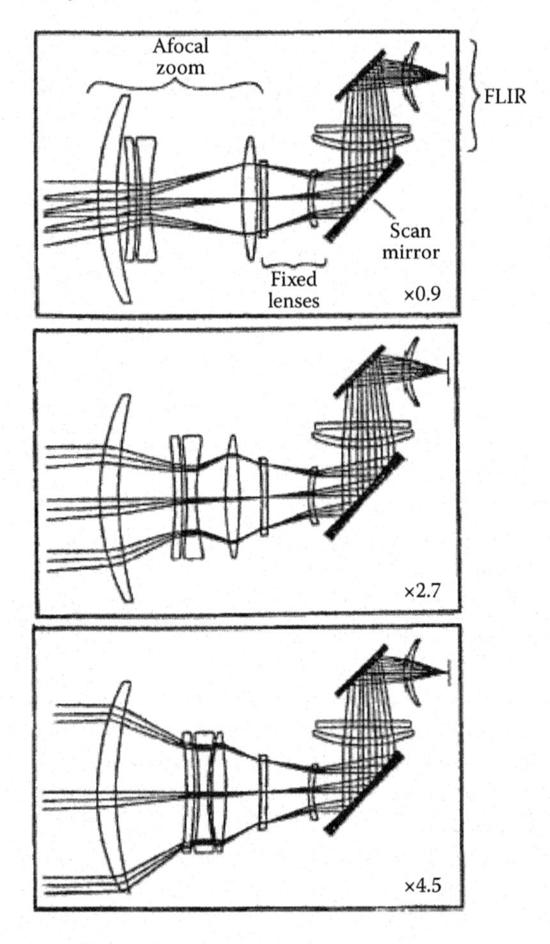

FIGURE 6.16 Optical system configurations of an afocal zoom lens at three different magnifications. (From Fisher, R.E., and Kampe, T.U., Actively controlled 5:1 afocal zoom attachment for common module FLIR, *Proc. SPIE*, 1690, 137, 1992.)

Active control of the focus involves moving one or more of the optical elements with a motorized actuator. Sensors determine the system temperature, and then either an algorithm or lookup table provides commands to move the actuator to the required position for temperature compensation. Monitoring of the image quality is a better system for sensing temperature correction, but requires a more complex sensor and control system.

Figure 6.16 shows a 5:1 afocal zoom for a military forward-looking infrared (FLIR) system operating in the spectral range of 8–12 μm.[4] Two components are moved by motorized actuators to athermalize the system: group 1, consisting of an air-spaced doublet, and lens 2, a singlet. All of the moving lenses are made of germanium.

Figure 6.17 shows that the movable elements are mounted in cells which are attached to linear bearings sliding axially on guide rods. Two stepper motors drive the two movable elements through spur

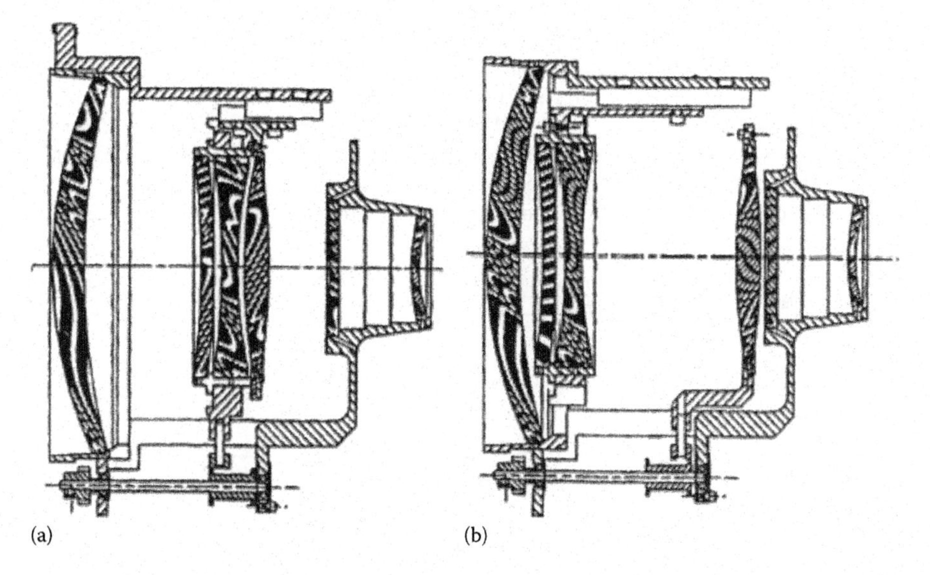

FIGURE 6.17 Section views through the zoom lens optomechanical system from Figure 6.16 at (a) a narrow field of view (×4.53) and (b) a wide field of view (×0.93). (From Fisher, R.E., and Kampe, T.U., Actively controlled 5:1 afocal zoom attachment for common module FLIR, *Proc. SPIE*, 1690, 137, 1992.)

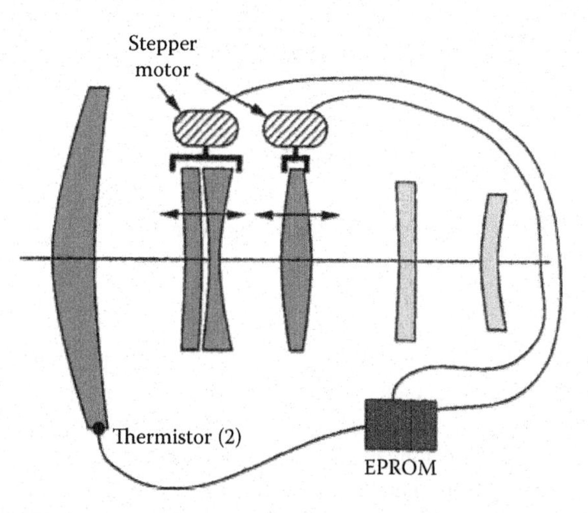

FIGURE 6.18 Schematic diagram of the temperature-sensing and motor drive system used to athermalize the zoom lens in Figure 6.16. (From Fisher, R.E., and Kampe, T.U., Actively controlled 5:1 afocal zoom attachment for common module FLIR, *Proc. SPIE*, 1690, 137, 1992.)

gear trains, as shown schematically in Figure 6.18. Two thermistors bonded to the lens housing sense the assembly temperature. Signals from the sensors go to the system electronics, where a lookup table (stored in the erasable programmable read-only memory [EPROM]) provides the information for the control system to move the motors to the desired position to minimize temperature effects.

Passive compensation is usually lower in cost, more reliable, and rugged than active control. Active control provides greater flexibility and is often better suited for large temperature variations. Nonlinear temperature effects are difficult to correct with passive compensation, but are readily accommodated with an active system. Zoom lenses, with multiple moving components, are also usually better candidates for active than passive athermalization.

Example 6.9

A doublet is made of Schott BK7 and F2 glass, with a focal length of 420 mm. A passive athermal compensator made of Invar 36 and 6061 aluminum maintains focus over temperature. For BK7, the thermo-optical coefficient $\beta_1 = -1.3 \times 10^{-6}$ K^{-1} and the Abbe number $\nu_1 = 64.7$. For F2, the thermo-optical coefficient $\beta_2 = -1.1 \times 10^{-6}$ K^{-1} and the Abbe number $\nu_2 = 36.4$. For Invar 36, the CTE $\alpha_{INVAR} = 1.26 \times 10^{-6}$ K^{-1}, and for 6061 aluminum, the CTE $\alpha_{Al} = 23.6 \times 10^{-6}$ K^{-1}. What lengths of Invar 36 and 6061 aluminum are required for the athermalized structure?

First, find the overall thermo-optical coefficient of the doublet by using Equation 6.56:

$$\beta_{DBLT} = \frac{(-1.3\times10^{-6}\,\text{K}^{-1})64.17-(-1.1\times10^{-6}\,K^{-1})36.4}{64.17-36.14} = -1.55\times10^{-6}\,\text{K}^{-1}.$$

The required lengths are found using Equations 6.62 and 6.63:

$$L_2 = 420\times10^{-3}\,\text{m}\left[\frac{(1.26\times10^{-6}\,\text{K}^{-1})-(-1.55\times10^{-6}\,\text{K}^{-1})}{(1.26\times10^{-6}\,\text{K}^{-1})-(23.6\times10^{-6}\,\text{K}^{-1})}\right] = -52.8\times10^{-3}\,\text{m},$$

$$L_1 = 420\times10^{-3}\,\text{m}-(-52.8\times10^{-3}\,\text{m}) = 472.8\times10^{-3}\,\text{m}.$$

The structure is a folded-back configuration with the outer Invar 36 tube length $L_2 = 472.8$ mm and the inner 6061 aluminum tube length $L_1 = -52.8$ mm. Equation 6.60 can be used as a check to see if the effective CTE α_C matches the thermo-optical coefficient of the doublet β_{DBLT}:

$$\alpha_C = \frac{(472.8\text{mm})(1.26\times10^{-6}\,\text{K}^{-1})+(-52.8\text{mm})(23.6\times10^{-6}\,\text{K}^{-1})}{472.8\text{mm}+(-52.8\text{mm})} = -1.55\times10^{-6}\,\text{K}^{-1}.$$

Since $\alpha_C = \beta_{DBLT} = -1.55 \times 10^{-6}$ K^{-1}, the system design checks.

6.5 LATHE ASSEMBLY

In the lathe assembly process, the lens elements are radially positioned by close fits to the IDs of the mating cell or housing. The OD of each element must be precision ground to a high degree of roundness and measured. The ID of the mating part is then finish-machined to fit that specific element. In multielement designs, the axial positions of the various elements are established by properly locating the machined seats while cutting the IDs. Since this machining process is traditionally done on a lathe or similar machine tool spindle, it has come to be known as "lathe assembly."

In a high-performance lens assembled in this manner, the diametric clearance between the OD of the element and the ID of the metal part may be as small as 5 μm (0.0002 in.). If the lens rim and the mounting seat are really round, this is just sufficient for the lens to slide in place. Since the radial clearance between the lens and cell is small, a drop in temperature may cause the cell wall to contact the lens, inducing a radial thermal stress. Methods from Section 6.4.3 can be used to analyze this stress.

As an example of the lathe assembly process, consider the ~57 mm (~2.25 in.) diameter air-spaced doublet subassembly in Figure 6.19.[5] Here, the conical seat D for the convex surface of lens 1 is to be cut so as to be tangent to the lens surface and to the proper depth to position the vertex of that surface at 57.150 ± 0.010 mm from the flange-mounting surface B. The axial airspace F is to be controlled to within ±0.025 mm of the nominal value from the lens design. Both lenses are to be constrained by a single threaded retaining ring acting through a pressure ring with a conical surface that is tangent to the second surface of the second lens. The purpose of this ring is to reduce the risk of the lens being rotated as the retainer is tightened. This is important if the elements have residual mechanical wedges that need to be rotationally positioned with respect to other wedged elements in the system during assembly, so these wedge effects tend to cancel—thereby improving optical performance.

The first step in the process is to measure the actual lenses to be assembled. The five boxes in Figure 6.19 indicate the five data items to be recorded. The required airspace (from the lens design)

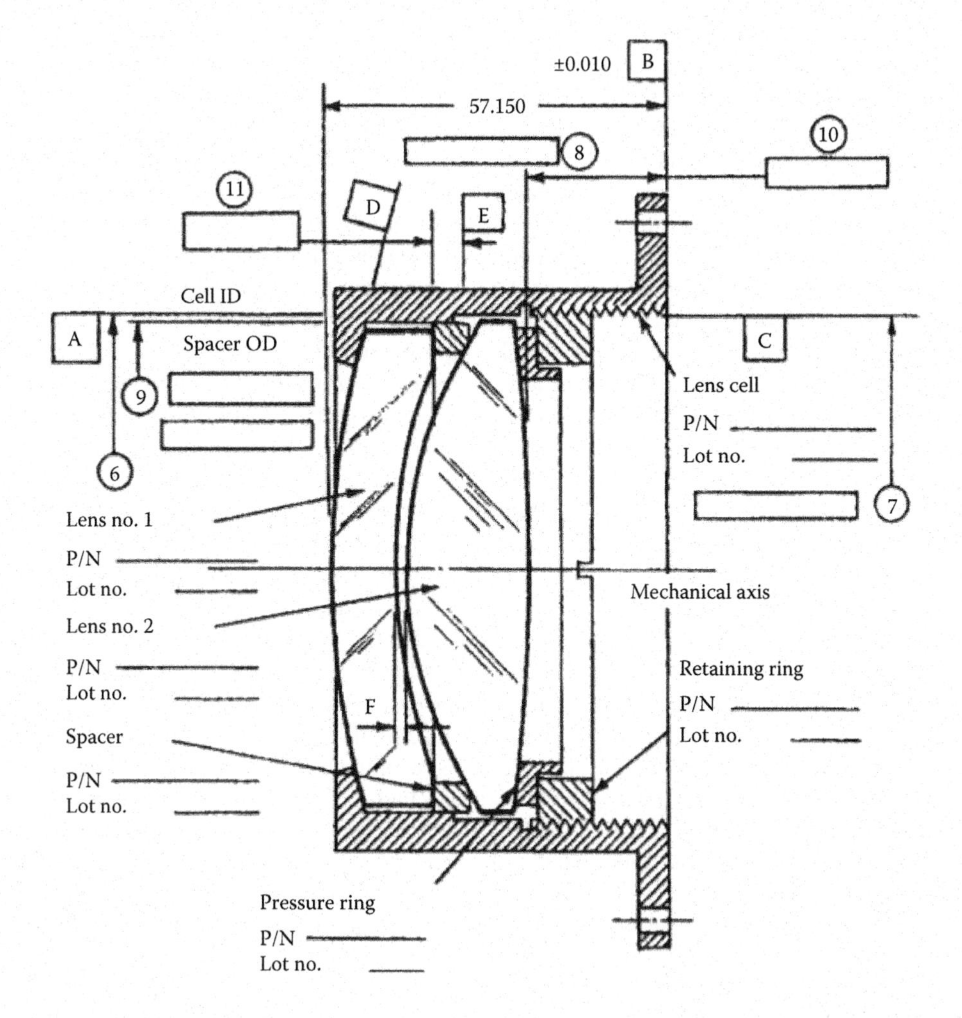

FIGURE 6.19 Optomechanical subassembly assembled by the lathe assembly process. The dimensions are in millimeters. (From Yoder, P.R. Jr., Lens mounting techniques, *Proc. SPIE*, 389, 2, 1983.)

is included. To protect the lens after these measurements are made, the lens surfaces could be protected temporarily with peelable lacquer, except for areas to be contacted by the mount or measured subsequently.

The lens cell flange is mounted via an adapter plate (not shown in Figure 6.19) to a precision lathe spindle. The cell ID (surface A) is machined perpendicular within 1.5 arcmin to flange surface B and to obtain 0.008–0.013 mm diametric clearance over the measured OD of lens 1. This cell ID is recorded as dimension 6. The second cell ID (surface C) is then machined concentric within 0.008 mm to surface A and to obtain 0.008–0.013 mm diametric clearance over the actual OD of lens 2. This ID is measured and recorded as dimension 7.

The angle for the conical surface D is calculated from Equation 5.36 and then machined. In this case, the angle is ~75° from the mechanical axis. A tolerance of ±0.5° applies to this angle. The axial location of surface D is determined by iterating the facing operation with trial installations of the lens and measuring the convex vertex location relative to the required axial location from the flange. When within tolerance, the actual value of dimension 8 is recorded.

The spacer is separately machined on a lathe-mounted mandrel so that its interface with lens 2 is a sharp corner, its OD is perpendicular within 15 arcmin to its surface E, and the diametric clearance with respect to the ID of the cell (dimension 6) is 0.008–0.013 mm. This OD is measured and recorded as dimension 7. The spacer and the pressure ring are inserted into the assembly so as to contact lens 1, and lens 2 is inserted. The actual axial location of the exposed surface of lens 2 relative to surface B is measured and recorded as temporary dimension 10. Subtracting the lens thicknesses (dimensions 3 and 4), the then existing airspace can be determined. This dimension will be slightly larger than needed. The spacer is removed, a small amount of material is removed from it, and the measurement is repeated. When the error in the airspace falls within the tolerance (from the lens design), the spacer dimension is recorded as dimension 11. The lenses and metal parts are then cleaned and reinstalled, and the pressure ring and retaining ring are installed to complete the subassembly. The wedges are then phased and the subassembly is ready for use.

A lathe-assembled unit is usually considered to be an inseparable subassembly because, if damaged, it cannot be repaired, but must be replaced. Because of their construction, such subassemblies are especially well suited for use in high-shock and high-vibration environments, such as those encountered in many military applications. Manufacturing cost is relatively high, while the iterations and associated precision machining operations are time consuming. For these reasons, lathe assembly is suited for production in limited numbers.

Another example of a lathe-assembled subassembly, designed for a severe environment, is the telescope objective shown in Figure 6.20. This lens is in the gunner's periscope on an armored vehicle. The three singlets are edged to the same nominal OD and fit into the cell with typically 0.005 mm radial clearances. All lenses are inserted from the right side. The first lens is Schott SF4 glass; it has a plano entrance face that registers against a flat shoulder on the cell. The first spacer is 0.066 ± 0.005 mm thick sheet material. Under preload, it conforms to the curves of the lenses to provide the proper axial spacing. The second lens is Schott SK16 glass, and the third is Schott SSK4 glass. The second spacer is shaped for tangential contact on the adjacent convex lens surfaces. The retainer is machined square to interface with a precision annular flat on the third lens. The threads on the retainer have a class 2 fit into the cell threads as recommended in Chapter 5. All metal parts are made of type 316 stainless steel and are black-passivated.

The mechanical wedge tolerances on the lenses and spacers are all 10 arcsec. The maximum edge thickness variation from the annular flat to the first surface on the third element is ±0.010 mm. During final assembly, the lens element wedges are iteratively phased by relative rotations about their common axis for maximum symmetry of the on-axis aerial image of a point light source located at infinity.

Another subassembly that can benefit from lathe assembly is the fixed focus relay lens illustrated in Figure 6.21. This is a fixed-aperture relay lens of the double-Gauss type comprising two cemented doublets and two singlets surrounding a fixed aperture stop. The dashed lines indicate the

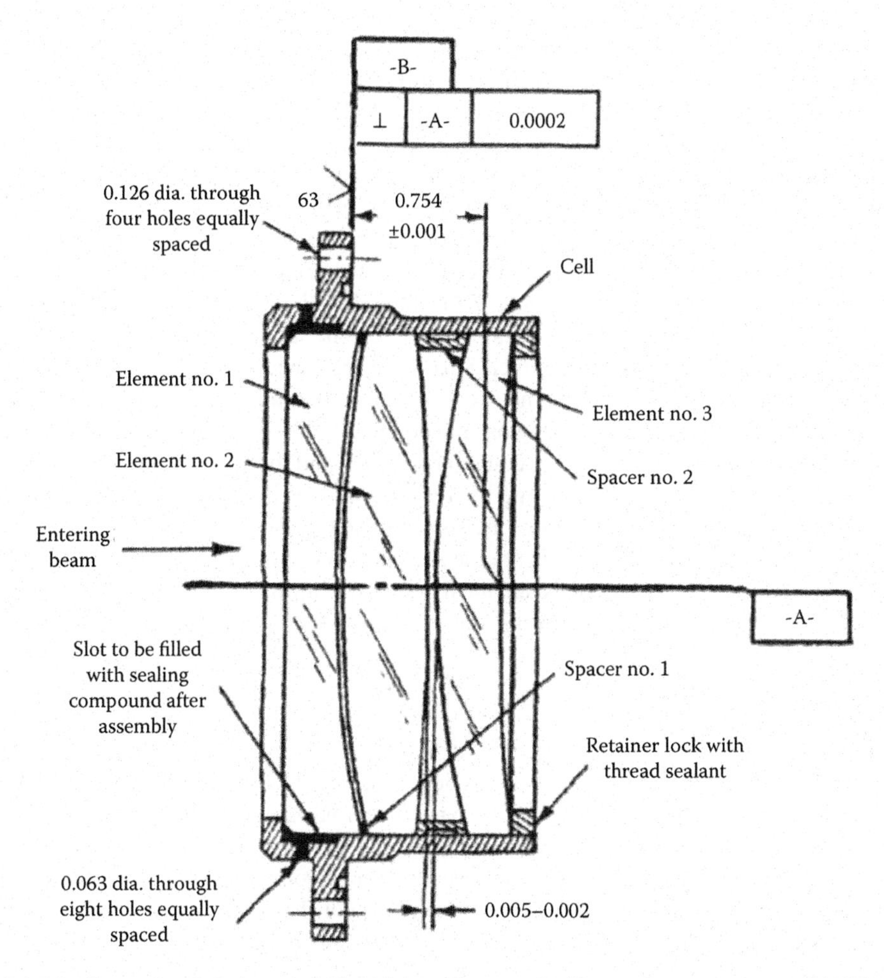

FIGURE 6.20 Example of a lathe-assembled high-performance military telescope subassembly. The dimensions are in inches. (From Yoder, P.R. Jr., Lens mounting techniques, *Proc. SPIE*, 389, 2, 1983.)

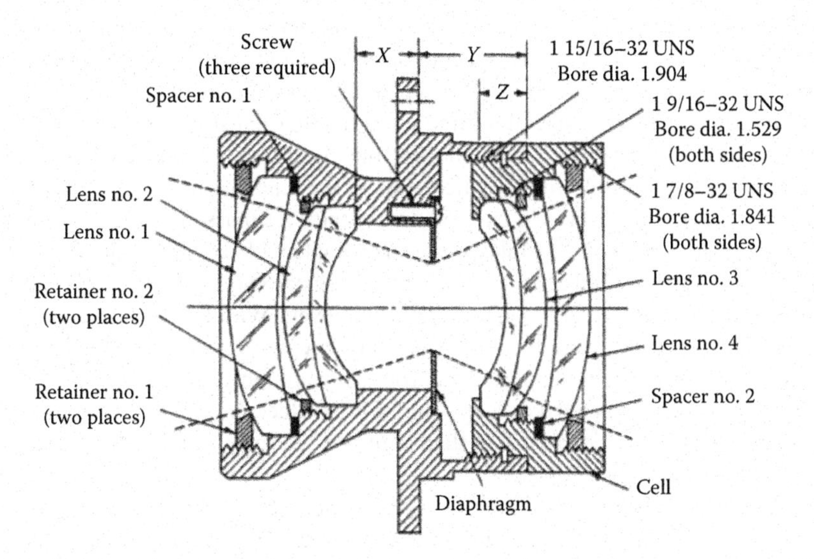

FIGURE 6.21 Configuration of a fixed-focus relay lens subassembly comprising two subassemblies threaded together. The dimensions are in inches. Dimensions *X*, *Y*, and *Z* are controlled to produce the required airspace.

apertures of the lenses to be used by the transmitted rays. The two lenses on the left are mounted directly into the aluminum lens barrel and held by threaded retaining rings. The two lenses shown at right are similarly mounted into a separate aluminum cell that screws into a mating thread on the barrel.

Each retaining ring is designed for sharp corner contact on the spherical surface of the lens approximately 0.25–0.50 mm outside the specified clear aperture of the element. The dimensions *X* and *Y* on the barrel and *Z* on the cell control the central airspace between surfaces 5 and 6, so their worst-case total error equals the allowable tolerance on the airspace. Alternatively, this airspace could be adjusted at assembly by machining the right end of the barrel to a computed value of *Y* that accounts for the measured actual values of *X* and *Z*. The latter approach adds an extra lathe machining operation, but allows three dimensions to be toleranced more loosely. Machining the spacers located between the singlets and their seats to suit actual dimensions measured at assembly allows the small airspaces between the singlets and doublets to be optimized. In production, a stock of spacers of slightly differing thicknesses could be provided and the appropriate spacers chosen for use in each individual assembly. By limiting the residual wedges in the lenses to small values, the need for pressure rings under the retaining rings is eliminated.

In this design, the ODs of the doublets are the same and those of the singlets are the same. The corresponding retaining rings can then be identical, thereby minimizing cost. These rings would not be interchangeable if designed for tangential contact because the lens surfaces have different radii.

6.6 SUBCELL ASSEMBLY

One of the most accurate lens-mounting methods is subcell mounting. Subcell mounting exploits the ability of modern fabrication methods to produce extremely round surface contours. The optical axis of a lens is accurately registered to the axis of a cylindrical subcell, and then the subcell is inserted into the barrel. The radial clearance between the subcell and barrel is nominally zero or even a light interference fit. Therefore, the only decentering error arises from the out-of-roundness condition of the outside diameter of the subcell and the ID of the barrel, as well as the location of the lens optical axis with respect to the subcell axis. With subcell mounting end to end, centering tolerances of ±5 μm or less are possible.

The use of the subcell mounting method requires locating or registering the optical axis of the lens with the mechanical axis of the subcell. Bringing the two axes into coincidence can be done in one of two ways: fixing the lens to an oversize subcell and then machining the subcell to final dimensions or by locating the lens with respect to the subcell and then fixing the lens in place. Both methods typically rely on optical metrology to locate the optical axis of the lens. A radial elastomeric bond between the lens and subcell minimizes radial thermal stress in most applications.

Mounting an individual lens into a cell and then fine machining the rim and faces of the cell so that the rim of the cell is centered to the optical axis of the lens and that one refracting surface is properly located axially with respect to one cell face creates a subassembly, or module, frequently called a "poker chip." Figure 6.22 illustrates a typical example with associated dimensions and tolerances summarized in Table 6.5. To achieve these demanding optomechanical specifications, a single-point diamond turning (SPDT) technique usually is employed. With this technique, extremely fine cuts are taken on the surface in work by using a specially prepared and oriented diamond crystal as the cutting tool. The workpiece is supported and rotated on a highly precise spindle (with an air or hydrostatic bearing). The tool is moved slowly across the surface on highly precise linear or rotary stages. Real-time interferometric control systems are used to ensure the location and orientation accuracy of the cutting tool at all times.

Key to the success of this process in making a poker chip module is the use of a centering chuck such as that shown in Figure 6.23. This device is usually made of brass because it can be easily SPDT-machined to precise dimensions. The surfaces to be made in this manner are indicated in the figure. They can be cut in the same setup, thus ensuring high accuracy relative to each other.

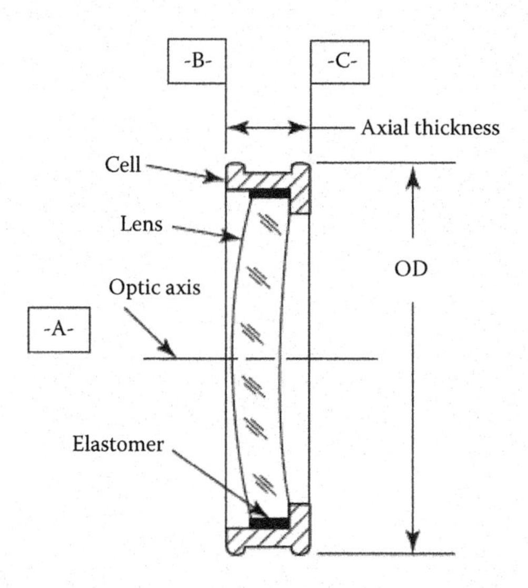

FIGURE 6.22 Section view of a high-precision lens-cell poker chip subassembly assembled, aligned, and final machined by SPDT as described in the text. (Adapted from Yoder, P.R. Jr., *Mounting Optics in Optical Instruments*, 2nd ed., SPIE Press, Bellingham, WA, 2008.)

TABLE 6.5
Specifications of the Poker Chip Lens Module Shown in Figure 6.22

Meniscus-shaped BK7 lens
Clear aperture: 3.000 ± 0.005 in. (76.200 ± 0.127 mm)
Axial thickness: 0.667 ± 0.004 in. (16.942 ± 0.102 mm)
Lens potted in place with General Electric RTV88 of radial thickness per Equation 5.38
Aluminum 6061-T6 cell
OD: 4.0000 ± 0.0002 in. (101.6000 ± 0.0051 mm)
OD concentric with lens axis within 0.0005 in. (0.013 mm)
OD parallel to axis within 10.0 arcsec
Axial thickness: 1.151 ± 0.0002 in. (29.2354 ± 0.0051 mm)
Surfaces -B- and -C- parallel within 5.0 arcsec

Other surfaces require only conventional machining. The conical interface is cut to the proper angle to interface properly with a convex surface of the lens to be mounted.

The centering chuck is made so as to fit snugly into a receptacle on a baseplate attached to the SPDT machine spindle. It seats against a surface with air path recesses for pulling a vacuum to secure the chuck in place. The finished lens is attached to the chuck with blocking wax (or a temporary UV curing adhesive) (see Figure 6.24). The lens is moved laterally to center its axis to the spindle axis before the wax solidifies (or the adhesive is cured). Initial alignment can be accomplished mechanically using precision indicators and then finalized using interferometric means. The measurement of the vertex distance indicated in the figure provides information useful in establishing the axial location of the machined fasces of the cell in a later step.

The cell into which the lens is to be mounted is placed over the rim of the lens as indicated in Figure 6.25. This cell has finished dimensions on all surfaces except those to be finished by SPDT to the tolerances indicated in Table 6.5. The cell is aligned mechanically with the spindle axis and waxed to the lens as shown.

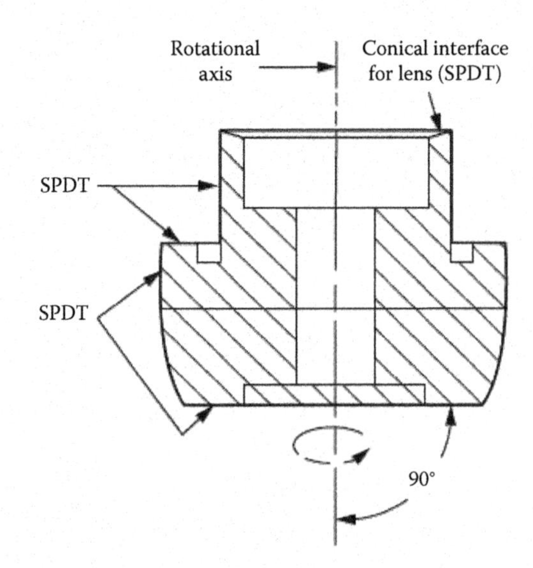

FIGURE 6.23 Removable centering chuck designed to interface a lens with the spindle of an SPDT machine. Surfaces marked SPDT are machined in one setup for maximum precision.

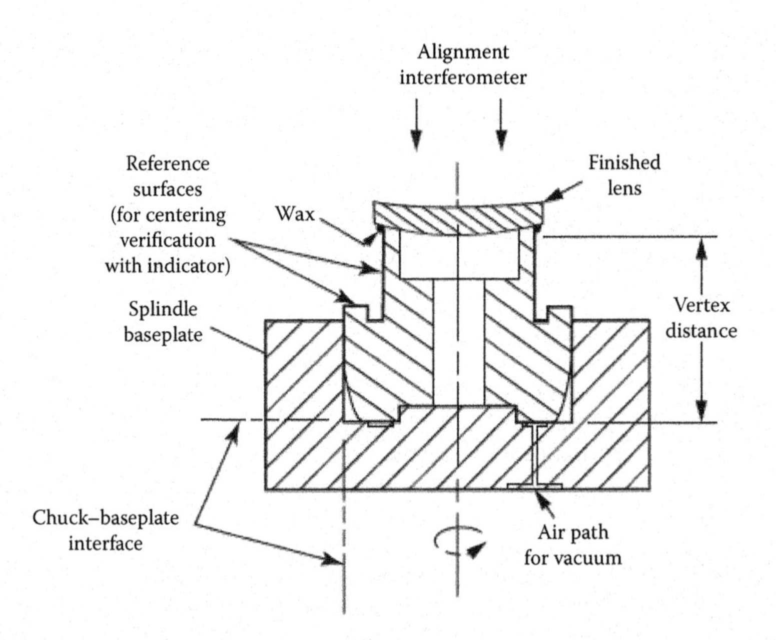

FIGURE 6.24 Centering chuck installed on the spindle baseplate of an SPDT machine with the lens waxed in place and centered interferometrically.

The next steps in the process are to remove the chuck and lens module from the spindle, invert it onto a horizontal surface as shown in Figure 6.26, and inject the elastomer into the annular cavity between the lens OD and the cell ID. Four radially directed holes are used for this purpose to ensure complete filling of the cavity. Note that this operation cannot be accomplished without inverting the subassembly because the elastomer must be constrained while curing. The use of the removable chuck allows the SPDT machine to be used for other purposes while the elastomer cures.

After the elastomer has completely cured, the subassembly is returned to the spindle baseplate and the exposed cell surfaces are turned to final dimensions (see Figure 6.27). As a precaution,

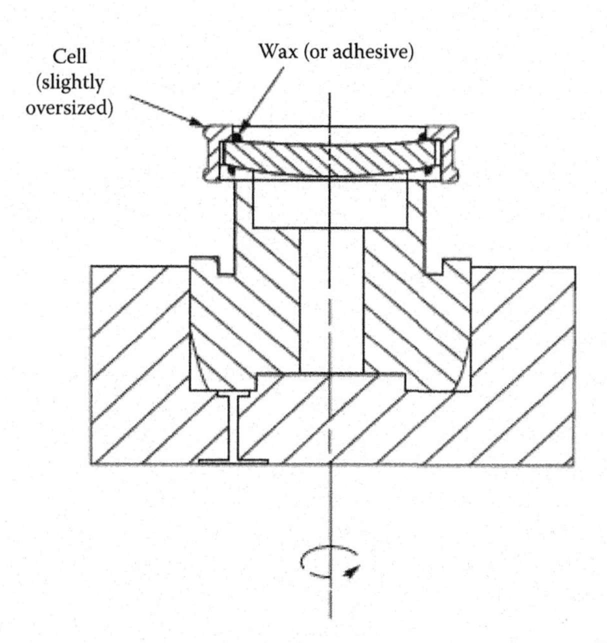

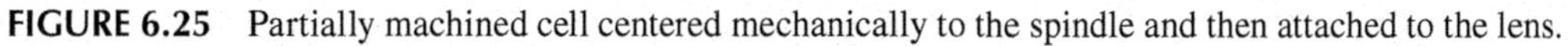
FIGURE 6.25 Partially machined cell centered mechanically to the spindle and then attached to the lens.

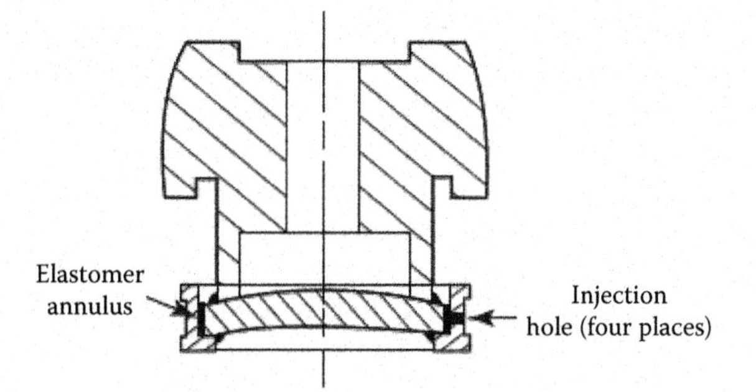

FIGURE 6.26 Centering chuck and lens removed from the spindle and inverted for injection of the elastomer.

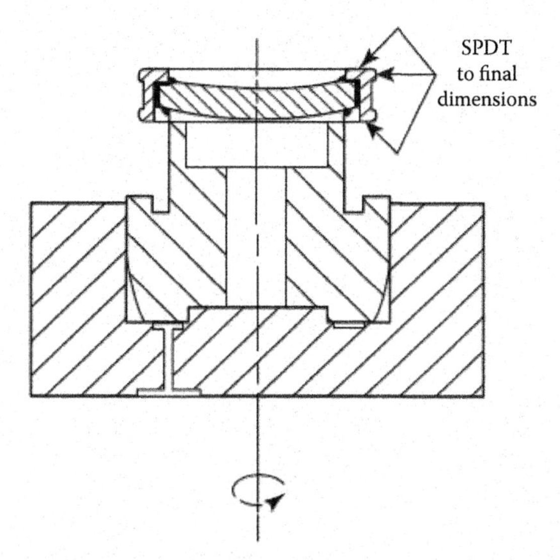

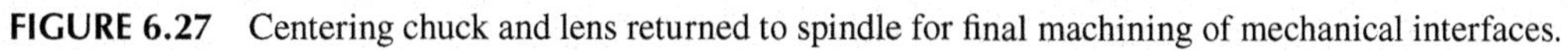
FIGURE 6.27 Centering chuck and lens returned to spindle for final machining of mechanical interfaces.

it would be advisable to verify the centration of the lens interferometrically before the subassembly is removed from the chuck. Removal is accomplished by heating gently to melt the wax or by applying a solvent to dissolve the temporary adhesive. Finally, the subassembly is cleaned, inspected, bagged, and identified for future use.

Alternately, the lens can be aligned with the subcell again using optical metrology and fixed in place. An example of this method is shown in Figure 6.28. Shown is a simplified sectional view of the assembly, which was to be used as a wide-field (110°) objective lens for bubble-chamber photography. A large amount of optical distortion of a particular form was designed into the lens and was expected to appear very precisely (i.e., within a few micrometers of the design values) in the completed lens over the entire image.

The technique was to mount each lens individually in a brass cell that had been machined true by SPDT on a precision air bearing. A rounded 0.25 mm radius "knife-edge" surface (actually a toroidal interface) was turned inside each cell to contact the spherical surface of the lens. All the machining and mounting of the lens were completed before removing the cell from the spindle. As shown in Figure 6.29, centering was monitored by observing Fresnel interference fringes between the lens surfaces and a spherical test plate held close to the exposed surface of the lens. Centration was judged to be correct when the fringe pattern appeared to remain stationary as the spindle was rotated slowly. This indicated that the rotating surface ran true to the spindle axis—and hence to the cell axis—within a fraction of a wavelength of the laser beam used (typically, $\lambda = 0.63$ μm). The lens was then bonded to the cell with a room-temperature-curing epoxy (not identified) that remained slightly flexible when cured. An epoxy layer thickness of about 0.1 mm was reported to be satisfactory for the intended application. The individual cemented subassemblies (poker chips) were assembled into a precisely bored barrel and secured with retaining rings. Upon evaluation, the authors reported that the centration errors of the system did not exceed 1 μm.

Another example of subcell or poke chip mounting is shown in Figure 6.30. This is a telecentric projection lens designed to provide near-diffraction-limited modulation transfer function (MTF) and distortion of less than 0.05% over its field of view. The lens is intended to be used at finite conjugates in the visible spectral range with a lateral magnification of 4.23 times. The high level of distortion correction is achieved in part by including an aspheric surface created by SPDT on the thin Cleartran™ zinc sulfide plate (lens 1 in the figure) located in front of the first doublet.

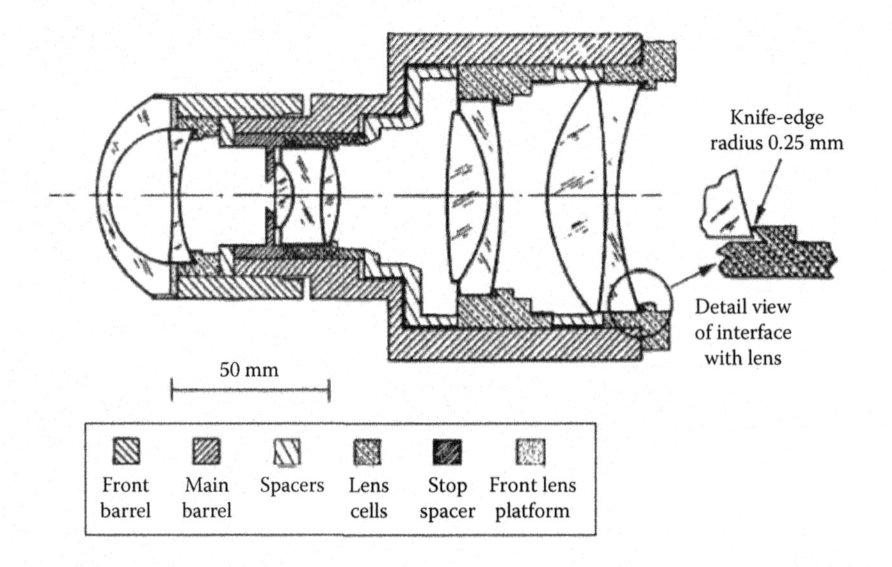

FIGURE 6.28 Optomechanical configuration (simplified) of a lens assembly requiring very precise centration of several components. (Adapted from Carnell, K.H. et al., *Opt. Acta*, 21, 615, 1974.)

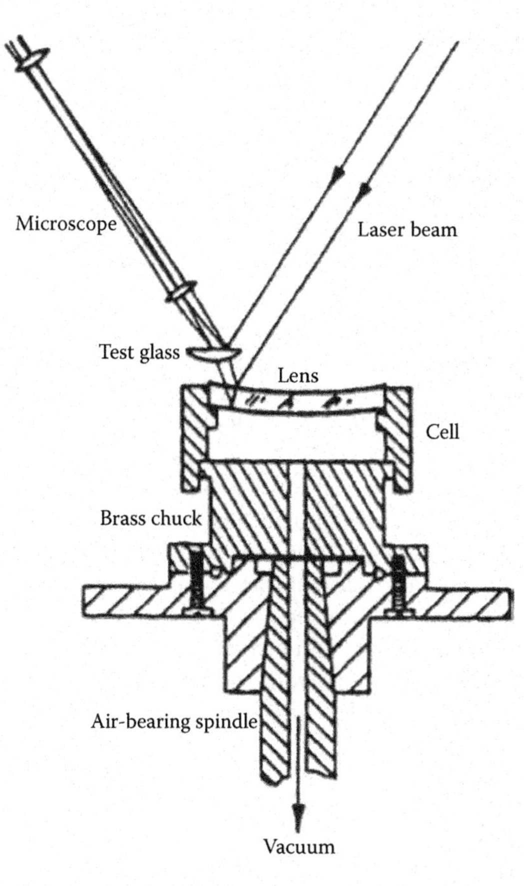

FIGURE 6.29 Fizeau interferometer instrumentation used to monitor the centration of a meniscus lens element on a precision spindle. (Adapted from Carnell, K.H. et al., *Opt. Acta*, 21, 615, 1974.)

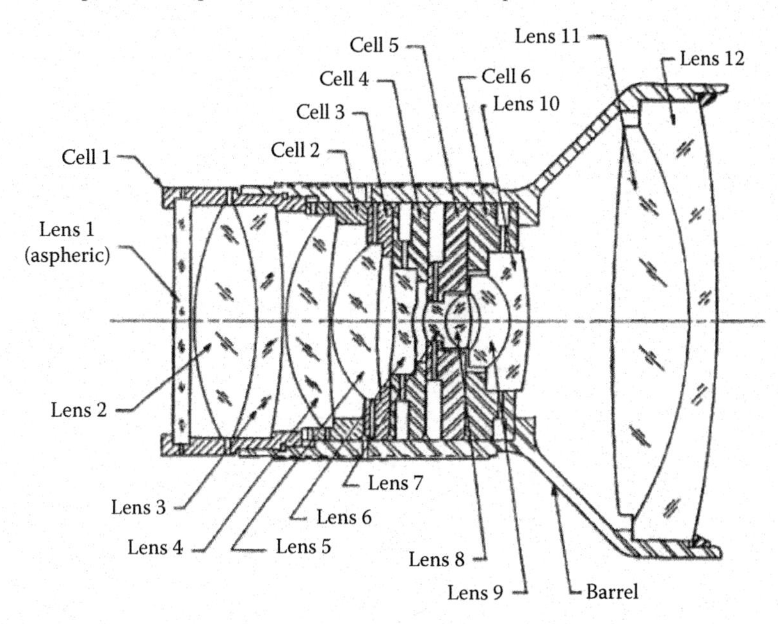

FIGURE 6.30 Optomechanical layout of a low-distortion telecentric projection lens assembly with poker chip lens subassemblies stacked together in the housing. (Adapted from Fischer, R.E., Case study of elastomeric lens mounts, *Proc. SPIE*, 1533, 27, 1991.)

The mountings for the lens elements to achieve the required alignment and preserve that alignment over an operational temperature range of 0–60°C posed a major design problem. Tolerances for some elements were as small as 0.013 mm decentration, 0.003 mm edge thickness runout due to wedge angle, and 0.008 mm surface edge runout due to tilt. The mechanical solution was to center each element or group of elements in precision stainless steel cells while rotating the cells on a rotary table and to inject a 0.381 mm thick annular ring of 3M 2216 epoxy to bond the lenses to their cells. After curing, the individual cell–lens subassemblies were inserted into the housing and held securely by retaining rings. Performance evaluation tests revealed that this lens assembly fully achieved the MTF and distortion requirements.

A large high-performance lens assembly that utilizes elastomeric lens mountings was developed at the Optical Sciences Center, University of Arizona, for the US Naval Observatory. The optical system, with 2059.99 mm focal length, *f*/10, and ±4.8° field-of-view lens and designed by R. R. Shannon, is shown schematically in Figure 6.31.[6] Its intended application was to be as an astrographic telescope objective.

The housing (see Figures 6.32 and 6.33a) was machined from a billet of 6Al–4V titanium. The technique used to mount the lens elements (diameters typically about 26.4 cm) was to bond them individually into 6Al–4V titanium cells (to approximate the CTEs of the glasses) with typically 6.35 mm wall thickness using an elastomeric ring approximately 5.08 mm thick. The elastomer used was Dow Corning 93-500. These cells then became poker chip subassemblies.

The cells were assembled into the titanium housing by pressing them in place with slight interference fits. They were further constrained with spacers and retaining rings. The photograph in

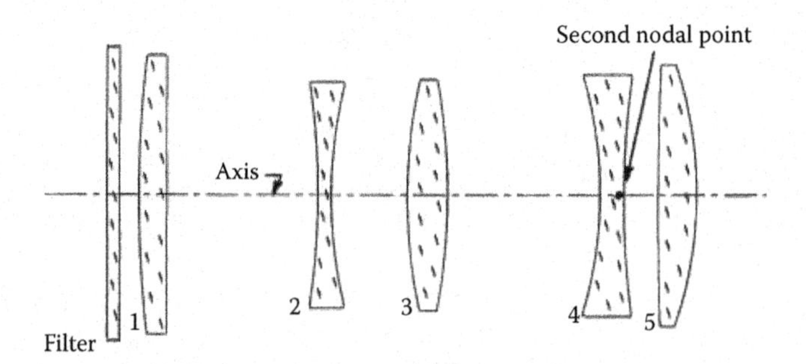

FIGURE 6.31 Optical schematic of a 2 m focal length, *f*/10 astrometric telescope objective. (From Vukobratovich, D., Design and construction of an astrometric astrograph, *Proc. SPIE*, 1752, 245, 1992.)

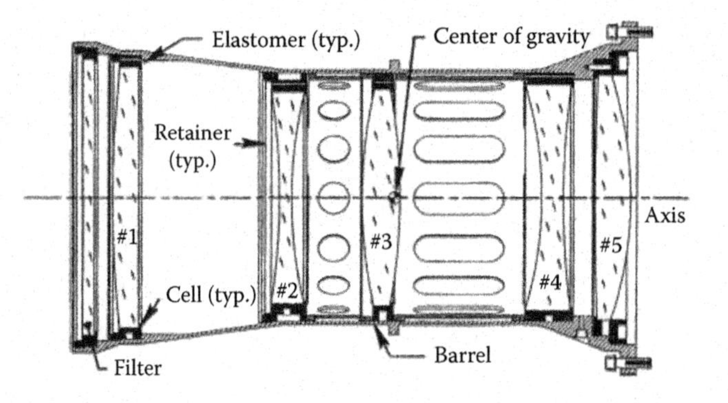

FIGURE 6.32 Sectional view of the astrographic telescope objective. The lenses are constrained in their cells, which are then pressed with interference fits into the barrel. (From Vukobratovich, D., Design and construction of an astrometric astrograph, *Proc. SPIE*, 1752, 245, 1992.)

FIGURE 6.33 Titanium components of the astrographic telescope objective: (a) main barrel; (b) barrel and cells for the six lenses (solid rings) and to spacers (slotted rings). (Courtesy of D. Vukobratovich, Optical Sciences Center, University of Arizona, Tucson, AZ.)

Figure 6.33b shows the mechanical components of the lens assembly. The overall weight of the assembly was about 44.6 kg, 21.9 kg of which was in the glass.

6.7 PLASTIC LENS MOUNTING

Optomechanical assemblies using plastic lenses, cells, retaining means, and structures have become important parts of a large variety of the consumer products in the world, as well as a smaller number of military and space applications. A few examples are camera viewfinders and objectives, magnifiers, television projection systems, compact disks and DVD readers, cell phone cameras, head-/helmet-mounted night-vision goggles, and telecommunication (fiber-optic) applications. In this section, some of the principles of design of such optomechanical assemblies are outlined and a few examples are discussed.

The 35 mm slide projector objective lens assembly shown in Figure 6.34 is designed to minimize cost. The housing is molded of plastic and has coarse integral threads for focusing. The lenses are separated by plastic spacers and held in place by a plastic retaining ring. Lens spacings and centering to a high accuracy are not essential to the function of this assembly. The illumination system, shown schematically on the left in the figure, comprises optics molded from heat-resistant glass (such as Pyrex). These optics are constrained with spring clips to allow for dimensional changes with temperature.

A hybrid lens system consisting of a glass element with acrylic lenses can, in combination with plastic housing, produce a lens with zero shift in back focal length with temperature. With the glass lens located in front, the internal plastic elements are protected from some adverse environmental conditions. This principle has been employed in many other more recent applications. A major contribution to the technology was the development of capabilities for creating aspheric surfaces (by molding or SPDT).

Mechanical features for mounting can be easily incorporated into plastic optics themselves. Cells and housings can also be configured to minimize the number of parts, as well as assembly labor, and to allow the use of mechanical fasteners, adhesives, or heat sealing. As an example, Figure 6.35

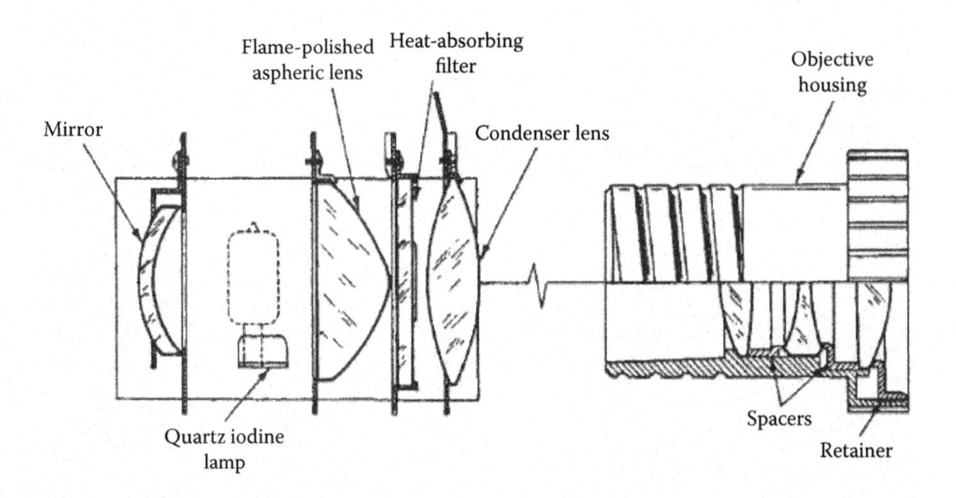

FIGURE 6.34 Schematic diagram of an optical system for projecting 2 in. × 2 in. (5.08 cm × 5.08 cm) slides onto a screen. It features an objective subassembly with lenses, housing, spacers, and retaining ring all made of plastic. The optics in the illuminating system are made of glass to resist heat. (Adapted from Horne, D.F., *Optical Production Technology*, Adam Hilger, Bristol, UK, 1972.)

FIGURE 6.35 Photograph of an all-plastic television projection lens assembly manufactured by U.S. Precision Lens, Inc. (now 3M Precision Optics, Inc., Cincinnati, Ohio). (From Yoder, P.R. Jr., *Mounting Optics in Optical Instruments*, 2nd ed., SPIE Press, Bellingham, WA, 2008.)

shows an air-spaced triplet objective of about 9 cm (3.5 in.) focal length designed and manufactured by U.S. Precision Lens, Inc. (Cincinnati, Ohio)* for use in a projection television system. It is labeled as a *Delta 20* design and is used to operate at a fixed relative aperture of *f*/1.2 with a nominal magnification of ×9.3. The dimensions of the assembly are 104.5 mm length, including the focus motion, and 117 mm diameter, not including the mounting flanges.

* Now 3M™ Precision Optics.

The three lenses are mounted in a longitudinally split (i.e., separately molded), two-part plastic mount that has integral axial locating tabs (acting as shoulders) and radial locating tabs. These tabs are sufficiently flexible to allow for slight variations in lens ODs and mount IDs occurring during the molding process. This type of mounting, sometimes called a "clamshell" mounting, is shown schematically in Figure 6.36. The mount material is thermally similar to the lens material. The two halves of the mount are symmetrical, allowing them to be cemented or taped together after the lenses are installed. Other mounts of this type are manufactured so that the two halves can be fastened together with self-tapping screws, by sliding an outer sleeve over the assembly with slight interference fit (see Figure 6.37[7]), or by heat sealing. Adhesives are not always suitable because their solvents can attack the plastic lens surfaces.

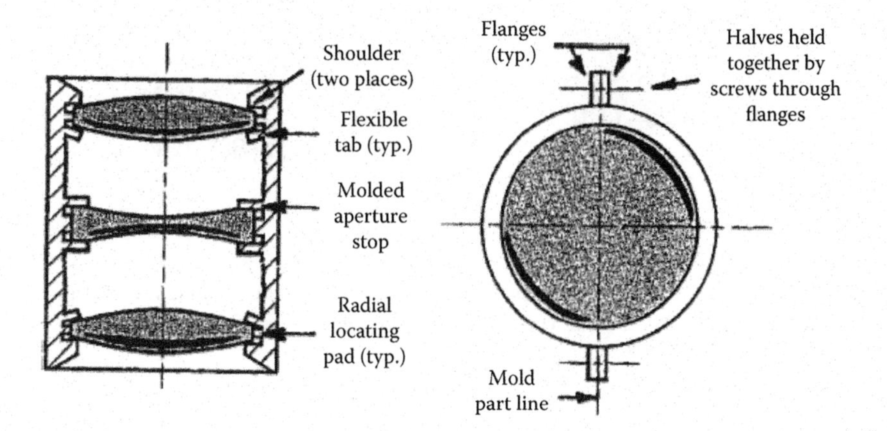

FIGURE 6.36 Schematic side and end views of an all-plastic clamshell lens mounting. (Adapted from U.S. Precision Lens, Inc., *The Handbook of Plastic Optics*, 2nd ed., U.S. Precision Lens, Inc., Cincinnati, OH, 1983.)

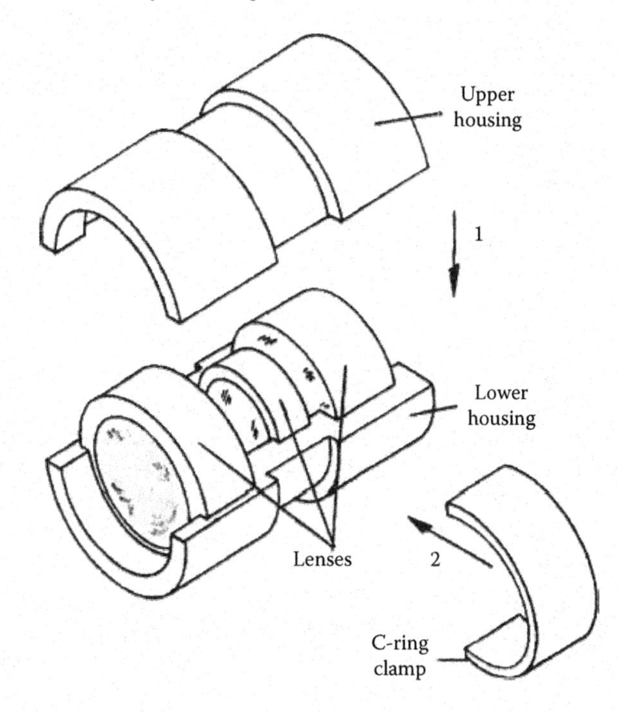

FIGURE 6.37 Technique for clamping housings of an all-plastic clamshell lens assembly using a slip-on C-ring. (From Lytle, J.D., Polymeric optics, Chapter 34 in *OSA Handbook of Optics*, Bass, M. et al., Eds., pp. 34.1–34.20, Vol. II, McGraw-Hill, New York, 1995.)

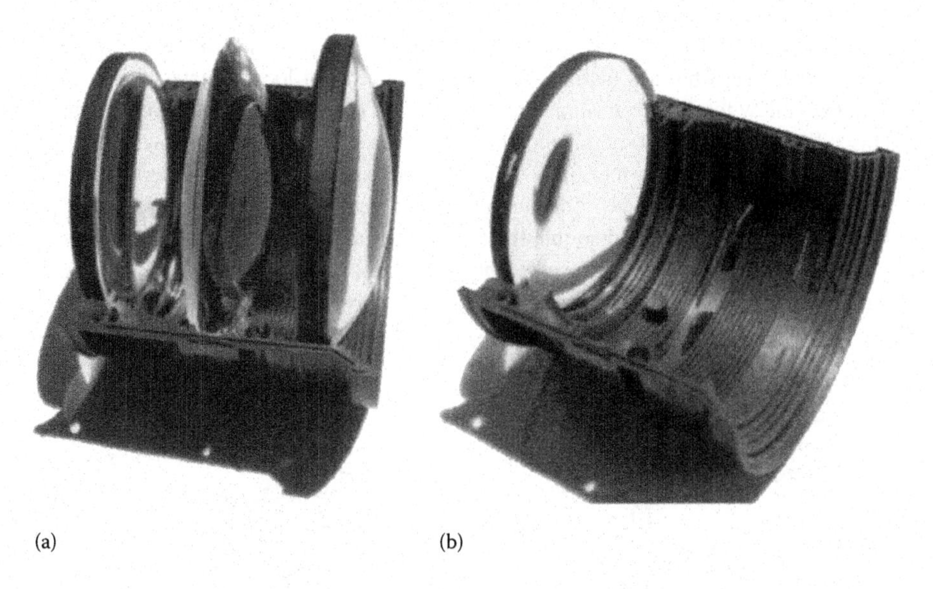

(a) (b)

FIGURE 6.38 Photograph of the interior of the lens assembly shown in Figure 6.35. View (a) has half the housing removed, while view (b) has the two lenses removed. Details of lens locating features and some stray-light-reducing grooves are shown. (From Yoder, P.R. Jr., *Mounting Optics in Optical Instruments*, 2nd ed., SPIE Press, Bellingham, WA, 2008.)

Figure 6.38 shows photographs of the interior of the assembly in Figure 6.35. The view in Figure 6.38a has one-half of the mount removed, while the view in Figure 6.38b has two of the lenses removed. Molded-in lens locating features and stray-light-reducing grooves can be seen in the figure. The mount halves are assembled with self-tapping screws; holes for two such screws are visible in the figure (in the shadows). The assembled lens cell fits snugly into the ID of an outer housing and is focused by sliding axially when rotated about the axis. Two screws passing through a helical cam slot molded into the housing wall at diametrically opposite locations translate the rotary motion into the axial motion. Wing nuts on one or both of these screws allow the adjustment to be clamped after focusing has been accomplished. The housing is designed to be attached to the structure of the television set by screws passing through three mounting ears, two of which are visible in Figure 6.35.

PROBLEMS

1. A camera lens is fixed at one end to the camera body. The 0.9 kg mass of the lenses is uniformly distributed along the length of the cell, which has a diameter of 60 mm and a length of 200 mm. The lens cell is made of a glass-reinforced polycarbonate plastic with a density of $\rho = 640$ kg/m^3 and an elastic modulus of 4.4 GPa. Assuming that the lens cell thickness is constant and that the decenter tolerance of the lens is 5 μm, what is the mass of the lens cell? What is the radial wall thickness of the lens cell? *Answer:* lens cell mass $W_{BC} = 0.114$ kg, radial wall thickness $t = 4.73$ mm
2. The mass of an astrometric astrograph lens is 45 kg, and it is located at the end of a cylindrical barrel, with the detector assembly at the other end. The lens focal length is 2 m, and its clear aperture is 200 mm. Assume that the lens is supported at a distance of half its focal length (the detector assembly weighs as much as the lens) and that the ID of the lens barrel is the same as the clear aperture. The deflection tolerance is 25 μm. If the lens barrel is made of stainless steel, with an elastic modulus of $E = 200$ GPa, what is the minimum radial wall thickness of the lens barrel? *Answer:* minimum radial wall thickness $t = 9.36$ mm

3. In problem 2, assume that the lens barrel material is 6Al–4V titanium, with an elastic modulus of E_{Ti} = 114 GPa and density $\rho_{Ti} = 4.43 \times 10^3$ kg/m^3. What is the ratio of the masses of the titanium and stainless steel barrels if the density of stainless steel is 7.8×10^3 kg/m^3? *Answer:* $W_{titanium}/W_{stainless} = 0.996$
4. A 50 mm diameter biconvex lens with 80 mm optical radius of curvature rests on a sharp corner seat with a diameter of 48 mm. The radius of the sharp corner of the seat is 50 μm, and there is a 25 μm radial wall clearance between the edge of the lens and ID of the cell. What is the radial accuracy of the position of the lens Δr, and what is the vertex displacement Δz? *Answer:* The radial accuracy of the location is Δr = 26.3 μm, and the vertex displacement Δz = 15 μm.
5. A 40 mm diameter biconvex lens with 60 mm radius of curvature rests on a tangential seat with a contact diameter of 38 mm; the seat ID is 36 mm. If the centration tolerance is Δr = ±20 μm, what is the maximum diameter of the bore? For an angular tolerance of ±0.5° for the tangential seat, what is the variation Δz in the vertex position of the lens? What are the upper and lower limits of the tangential surface angle? *Answer:* The maximum bore diameter d_{bore} = 40.038 mm; for a 0.5° tolerance, the variation in the vertex position Δz = 2.416 μm; and the minimum angle θ_{min} = 17.46° and the maximum angle θ_{max} = 19.47°.
6. The diameter of a biconvex lens is 50 mm. The lens is supported on a seat with a contact radius y_{c1} = 22 mm, and the optical radius in contact with the seat is R_1 = 80 mm; the retainer contact radius y_{c2} = 23 mm, and the optical radius in contact with the retainer is R_2 = 85 mm. If the friction coefficient μ between the lens and cell is 0.38, will the lens self-center? *Answer:* (22 mm)/(80 mm) + (23 mm)/(85 mm) = 0.546 ≥ 0.38; the lens will self-center
7. A concave–convex lens is mounted in a cell with the concave surface against the seat; the seat contact radius is y_{c1} = 9 mm, and its concave radius is R_1 = 40 mm. The convex surface is in contact with the retainer, with y_{c2} = 9 mm and with a radius of R_2 = 35 mm. Using the approximation from Section 6.3.2 ($\mu \approx 0.15$), will the lens self-center? *Answer:* (9 mm)/(–40 mm) + (9 mm)/(35 mm) = –0.119 < 0.15; the lens will not self-center
8. The diameter of a biconcave lens d_{lens} is 25 mm. The lens rests on a toroidal seat with R_{toroid} = 2 mm, and the radius of the optical surface in contact with the toroidal seat is R = –60 mm. The seat contact diameter is d_A = 24 mm. If the decenter tolerance is Δr = 20 μm, what is the maximum bore diameter? *Answer:* The maximum bore diameter is d_{bore} = 25.22 mm.
9. The vertex-to-vertex distance between two lenses is $t_{J,K}$ = 6.2 mm. Contact with the first lens is made against a concave surface, with y_J = 14 mm and R_J = 44 mm; contact with the second surface is made against a convex surface with y_K = 16 mm and R_K = 56 mm. What is the axial length of the spacer between the two lenses? *Answer:* The axial length of the spacer $L_{J,K}$ = 6.248 mm.
10. A spacer makes tangential contact with the convex surfaces of two lenses in a cell. For the first lens, y_J = 14 mm and R_J = 50 mm; for the second lens, y_K = 12 mm and R_K = 48 mm. The vertex-to-vertex axial spacing between the two lenses is $t_{J,K}$ 7.5 mm. If the angular tolerance of the tangential surfaces is ±0.5°, what is the variation in the spacer length $\Delta L_{J,K}$? *Answer:* The total variation in spacer length is $\Delta L_{J,K}$ = 3.77 μm.
11. A 50 mm diameter lens made of N-SK16 glass is mounted in a 416 stainless steel cell. The initial radial assembly clearance between the lens and cell is Δr = 25 μm. The CTE of N-SK16 is $\alpha_G = 6.3 \times 10^{-6}$ K^{-1}, and its elastic modulus is E_G = 89 GPa. The CTE of 416 stainless is α_M = 200 GPa, and its elastic modulus is E_M = 200 GPa. If the cell wall is 12 mm thick and the assembly is subjected to a –70°C temperature drop, what is the stress in the lens and cell? *Answer:* Since $K_5 = -2.968 \le 0$, $\sigma_R = 0$ and $\sigma_M = 0$.
12. A 40 mm diameter lens made of zinc sulfide is mounted in a 6061 aluminum cell. The CTE of zinc sulfide is $\alpha_G = 4.6 \times 10^{-6}$ K^{-1}, and the CTE of 6061 aluminum is $\alpha_M = 23.6 \times 10^{-6}$ K^{-1}.

For a –70°C temperature drop, what is the minimum radial clearance Δr to ensure that the cell wall does not come into contact with the lens? *Answer:* $\Delta r = 26.6$ μm

13. A 45 mm diameter germanium is mounted in a 6061 aluminum cell. The initial radial assembly clearance is $\Delta r = 25$ μm, and assembly takes place at a temperature of 25°C. In service, the lens is subjected to a lower temperature of –51°C. The CTE of germanium is $\alpha_G = 6 \times 10^{-6}$ K^{-1}, and its elastic modulus is $E_G = 103.7$ GPa. The CTE of 6061 aluminum is $\alpha_M = 23.6 \times 10^{-6}$ K^{-1}, and its elastic modulus is $E_M = 68.2$ GPa. If the cell wall thickness is $t_C = 10$ mm, what are the stresses in the lens and cell at the low temperature? *Answer:* The radial thermal stress in the glass $\sigma_R = 5.31$ MPa, and the stress in the aluminum cell is $\sigma_M = 11.95$ MPa.
14. Two germanium lenses are separated by a 6061 aluminum space; at 25°C, the vertex-to-vertex distance between the two lenses is $L_{JK} = 111$ mm. The first lens surface is concave with a radius of $R_J = 205$ mm and a contact radius of $y_J = 48$ mm; the second lens surface is convex with a radius of $R_K = 134$ mm and a contact radius of $y_K = 36$ mm. The CTE of germanium is $\alpha_G = 6 \times 10^{-6}$ K^{-1}, and the CTE of 6061 aluminum is $\alpha_M = 23.6 \times 10^{-6}$ K^{-1}. What is the vertex-to-vertex spacing of the lenses at –50°C and +70°C? *Answer:* At –50 K, the spacing $L_{JK} = 110.034$ mm, and at +75 K, the spacing $L_{JK} = 110.344$ mm.
15. The glass types N-SK16 and SF2 are used in a cemented doublet mounted in a 6061 aluminum cell. The edge thickness of the N-SK16 lens is $t_{E1} = 11.02$ mm; the CTE of this glass is $\alpha_{G1} = 6.3 \times 10^{-6}$ K^{-1}. The edge thickness of the SF2 lens is $t_{E2} = 6.48$ mm; its CTE is $\alpha_{G2} = 8.4 \times 10^{-6}$ K^{-1}. The CTE of 6016 aluminum is $\alpha_M = 23.6 \times 10^{-6}$ K^{-1}. What is the increase in axial gap ΔGAP_A when the temperature is 45 K above the temperature necessary for the preload to go to zero or $T - T_C$ is 45 K? *Answer:* The increase in the axial gap is $\Delta GAP_A = 13.01$ μm.
16. The diameter of an equiconvex lens made of Schott F5 glass is $D = 60$ mm; its center thickness is $t = 12$ mm, and the optical radii are $R = 120$ mm. The lens mass is 118×10^{-3} kg. Assembly of the lens into its cell is at a temperature of $T_A = 27$°C (300 K). The retaining ring preload supplies enough force to restrain the lens at an acceleration of $a_G = 75g$. The CTE of F5 is $\alpha_G = 8 \times 10^{-6}$ K^{-1}, and the elastic modulus is $E_G = 58$ GPa. The lens cell is made of 416 stainless steel with a radial wall thickness of $t_C = 6$ mm and contacts the glass at a diameter of $2y_C = 57$ mm; the cell ID is $D_M = 60.05$ mm (25 μm radial clearance between the lens and ID of cell) 68.2 mm. The CTE of 416 stainless steel is $\alpha M = 9.9 \times 10^{-6}$ K^{-1}, and the elastic modulus is $E_M = 200$ GPa. What is the preload force P_A on the lens? What is the temperature sensitivity factor K_3? What is the temperature T_C at which the preload goes to zero? *Answer:* $P_A = 86.74$ N, $K_3 = -45.75$ N/K, and $T_C = 28.9$°C (301.9 K).
17. A cemented doublet lens is made of Schott N-SK16 and SF2; it is mounted in a cell made of 6061 aluminum. The lens diameter is $D = 48$ mm, and the mounting diameter is $2y_C = 46$ mm. The thickness at the mounting diameter for the N-SK16 element is $t_{E1} = 7.31$ mm, and that for the SF2 element is $t_{E2} = 12.27$ mm. The properties of N-SK16 are $E_{G1} = 89$ GPa and $\alpha_{G1} = 6.3 \times 10^{-6}$ K^{-1}; those for SF2 are $E_{G2} = 55$ GPa and $\alpha_{G2} = 6.3 \times 10^{-6}$ K^{-1}. The ID of the aluminum cell is $D_M = 48.05$ mm, and the radial wall thickness is $t_C = 4$ mm. The properties of 6061 aluminum are $\alpha_M = 23.6 \times 10^{-6}$ K^{-1} and $E_M = 68.2$ GPa. Assembly of the lens into the cell is at a temperature of $T_A = 25$°C, and the preload at this temperature is $P_A = 101.4$ N. What is the temperature sensitivity factor K_3 for the lens? What is the preload at a temperature of –50°C? *Answer:* The temperature sensitivity factor $K_3 = -242.6$ N/K, and at –50°C, the preload is 18.29×10^3 N.
18. A Petzval lens is made of germanium and is housed in a 6Al–4V titanium cell with a tapered spacer made of the same type of titanium. The contact diameter of the first lens is $y_{C1} = 48$ mm, and the contact thickness is $t_{E1} = 6.11$ mm. For the second lens, $y_{C2} = 37$ mm and the contact thickness is $t_{E2} = 5.02$ mm; the spacing between the elements is $L_S = 111$ mm.

The cell ID is $D_M = 100.15$ mm, and the radial wall thickness of the cell is $t_S = 5$ mm. The properties of germanium are $E_G = 102.7$ GPa and $\alpha_G = 6 \times 10^{-6}$ K^{-1}. The properties of titanium are $E_M = 114$ GPa and $\alpha_M = 8.8 \times 10^{-6}$ K^{-1}. What is the temperature sensitivity factor K_3 for this assembly? *Answer:* The temperature sensitivity factor $K_3 = -59.66$ N/K.

19. The diameter of a cemented doublet is $D = 32$ mm. The first element is equiconvex with optical radii of 29.8 mm and a central thickness of 12 mm; it is made of N-BAK1 glass. The second element is concave flat and made of SF5; the front surface radius is –29.8 mm, and the central thickness is 2.8 mm. For N-BAK1, the elastic modulus is $E_{BAK1} = 73$ GPa, Poisson's ratio is $\nu_{BAK1} = 0.252$, and CTE is $\alpha_{BAK1} = 7.6 \times 10^{-6}$ K^{-1}. For SF5, the elastic modulus is $E_{SF5} = 56$ GPa, Poisson's ratio is $\nu_{SF5} = 0.233$, and CTE is $\alpha_{SF5} = 8.2 \times 10^{-6}$ K^{-1}. The elastic modulus of the adhesive used to cement the two lenses is $E_e = 1.034$ GPa, with a Poisson's ratio of $\nu_e = 0.43$ and thickness of $t_e = 25$ μm. The lens temperature change in service is 75 K. If the maximum thickness of the two lenses is used as a worst case, what is the thermal stress in the bond? Compare this with the thermal stress for the average thickness of the two lenses. The average thickness is computed by taking the average of the edge and center thicknesses. *Answer:* The thermal stress for the worst-case analysis (maximum thickness for both lenses) is 1.894 MPa; the thermal stress using the average thickness for both lenses is 1.547 MPa.
20. The diameter of a cemented doublet is $D = 86.6$ mm. The first element of the doublet is made of Schott N-BAK1, with a center thickness of 11.02 mm and an edge thickness of 4.02 mm. The second element is made of Schott N-SF8 with a center thickness of 8.15 mm and an edge thickness of 11.3 mm. The lenses are bonded together with an adhesive with a shear modulus of $G_e = 1.5$ GPa and a bond thickness of $t_e = 25$ μm. For N-BAK1, the elastic modulus is $E_{BAK1} = 73$ GPa, Poisson's ratio is $\nu_{BAK1} = 0.252$, and CTE is $\alpha_{BAK1} = 7.6 \times 10^{-6}$ K^{-1}. For N-SF8, the elastic modulus is $E_{SF8} = 88$ GPa, Poisson's ratio is $\nu_{SF8} = 0.245$, and CTE is $\alpha_{SF8} = 8.56 \times 10^{-6}$ K^{-1}. What is the temperature change required for a thermal stress of 5 MPa? Use the average thicknesses of the lenses, taking the average of edge and center thicknesses. *Answer:* For a thermal stress of $\tau_S = 5$ MPa, the temperature change is $\Delta T = 56.4$ K.
21. An achromatic doublet is made of Schott N-BAK1 and N-SF8 (this is the same optical design as problem 20) with a focal length of 508 mm. For N-BAK1, the CTE is $\alpha_{BAK1} = 7.6 \times 10^{-6}$ K^{-1}, the index of refraction is $n_{BAK1} = 1.5725$, the dispersion is $\nu_{dBAK1} = 57.55$, and the relative dn/dT is $dn/dT_{BAK1} = 2.50 \times 10^{-6}$ K^{-1}. For N-SF8, the CTE is $\alpha_{SF8} = 8.56 \times 10^{-6}$ K^{-1}, the index of refraction is $n_{SF8} = 1.68894$, the dispersion is $\nu_{dSF8} = 31.31$, and the relative dn/dT is $dn/dT_{SF8} = 2.6 \times 10^{-6}$ K^{-1}. What is the thermo-optical coefficient β for the doublet? A bimetallic compensator is used to passively maintain focus, made of 6061 aluminum ($\alpha_{Al} = 23.6 \times 10^{-6}$ K^{-1}) and 416 stainless steel ($\alpha_{CRES} = 9.9 \times 10^{-6}$ K^{-1}). What are the lengths of the bimetallic compensator? *Answer:* The thermo-optical coefficient of the doublet is $\beta = -1.38 \times 10^{-6}$ K^{-1}; the length of the 416 stainless is $L_1 = 926$ mm, and the length of the 6061 aluminum is $L_2 = -418$ mm (folded-back configuration).
22. An IR Petzval lens for a wavelength of 10.6 μm is made of germanium, with a focal length of100 mm. As a rule of thumb, the thermo-optical coefficient β of a Petzval lens is about 0.7 of that of a simple lens of the same focal length. For germanium at 10.6 μm, the index of refraction $n = 4.00$, $dn/dT = 424 \times 10^{-6}$ K^{-1}, and the CTE is $\alpha_{GE} = 6 \times 10^{-6}$K^{-1}. Using the rule of thumb given earlier, what is the thermo-optical coefficient of the Petzval lens with temperature? What are the lengths of a passive bimetallic compensator using Invar 36 with $\alpha_{INVAR} = 1.26 \times 10^{-6}$ K^{-1} and a polyethylene plastic with $\alpha_{PE} = 200 \times 10^{-6}$ K^{-1}? *Answer:* The thermo-optical coefficient of the Petzval lens is $\beta = 94.7 \times 10^{-6}$ K^{-1}; the length of the Invar 36 is $L_1 = 52.97$ mm, and the length of the polyethylene plastic is $L_2 = 47.03$ mm.

REFERENCES

1. Yoder, P.R. Jr., Advanced considerations of the lens-to-mount interface, *SPIE Crit. Rev.*, CR43, 305, 1992.
2. Yoder, P.R. Jr., *Mounting Optics in Optical Instruments*, 2nd ed., SPIE Press, Bellingham, WA, 2008.
3. Vukobratovich, D., Introduction to optomechanical design, SPIE short course SC014, 2014.
4. Fisher, R.E., and Kampe, T.U., Case study of elastomeric lens mounts, *Proc. SPIE*, 1690, 137, 1992.
5. Yoder, P.R. Jr., Lens mounting techniques, *Proc. SPIE*, 389, 2, 1983.
6. Vukobratovich, D., Design and construction of an astrometric astrograph, *Proc. SPIE*, 1752, 245, 1992.
7. Lytle, J.D., Polymeric optics, Chapter 34 in *OSA Handbook of Optics*, Bass, M. et al., Eds., pp. 34.1–34.20, Vol. II, McGraw-Hill, New York, 1995.

FURTHER READING

Bayar, M., Lens barrel optomechanical design principles, *Opt. Eng.*, 20, 181, 1981.

Chase, K.A., and Burge, J.H., Analysis of lens mount interfaces, *Proc. SPIE*, 8125, 81250F, 2011.

DeWitt, F. IV, and Nadorff, G., Rigid body movements of optical elements due to opto-mechanical factors, *Proc. SPIE*, 5667, 56670H, 2005.

Hopkins, C.L., and Burge, J.H., Application of geometric dimensioning & tolerancing for sharp corner and tangent lens seats, *Proc. SPIE*, 8131, 81310F, 2011.

Yoder, P.R. Jr., and Hatheway, A.F., Further considerations of axial preload variations with temperature and the resultant effects on contact stresses in simple lens mountings, *Proc. SPIE*, 5877, 38, 2005.

Yoder, P.R. Jr., and Vukobratovich, D., Shear stresses in cemented and bonded optics due to temperature changes, *Proc. SPIE*, 9573, 95730J, 2015.

Yoder, P.R. Jr., and Vukobratovich, D., *Opto-mechanical Systems Design*, 4th ed., CRC Press, Boca Raton, FL, 2015.

7 Techniques for Mounting Prisms

7.1 INTRODUCTION AND SUMMARY

The design of a mechanical mounting for a prism depends on a variety of factors such as the inherent rigidity of the optic; the tolerable movements and distortions of the surfaces (especially reflecting ones); the magnitudes, locations, and orientations of the steady-state forces holding the optic against its mounting reference surfaces during operation; the transient forces driving the optic against or away from those surfaces during exposure to extreme shock and vibration; thermal effects; the shape and quality of the mounting surfaces touching the optic; the sizes, shapes, orientations, and smoothness of those mounting surfaces (usually machined surfaces such as pads) on the mount; and the rigidity and long-term stability of the mount. In addition, the design must be compatible with assembly, adjustment, maintenance, package size, weight, and configuration constraints. Last, but not least, it must also be affordable in the context of the cost of the entire instrument.

In this chapter, a variety of techniques for mounting prisms are described. Included are kinematic, semikinematic, and nonkinematic interfaces. Generally, prisms are rigid solid polyhedrons with flat surfaces intersecting at various dihedral angles. Some prisms, such as Porro prisms, image-erecting prisms, axicons, and cube corner prisms, may have surfaces with curved apertures that conform to rotationally symmetric transmitted or reflected beam shapes for weight reduction and packaging reasons.

7.2 KINEMATIC, SEMIKINEMATIC, AND NONKINEMATIC PRINCIPLES

A rigid body in space has six degrees of freedom (DOFs), or ways in which it can move. These are translations along three mutually perpendicular coordinate axes and rotations about those axes. A body is constrained kinematically when each of these possible movements is singly prevented from occurring by point glass-to-metal preloaded contacts between the prism and the mount. If any one movement is constrained in more than one way as, for example, with two or more forces applied simultaneously to one surface to constrain a single DOF or if multiple sets of possible contacts control the location or orientation of a prism at a given temperature, then the body is overconstrained and may be overstressed or deformed. Both of these conditions are highly undesirable for optical components, so optomechanical engineers usually go to great lengths not to introduce overconstraints.

Transfer of a constraining force into the prism with point contacts may cause excessive localized internal contact stress and related optical surface deformation. To reduce these adverse effects, many successful mounts provide small area contacts to distribute the loads. Such mounts, designed to uniquely control all six DOFs without overconstraint, are semikinematic.

Application of kinematic or semikinematic principles is appropriate in many prism-mounting designs—especially ones intended for high-performance applications where precise relative locations of the optical components, minimal optical surface deformations, and stress-free optomechanical support under both friendly and adverse environmental conditions are needed. Many other optical devices meet their performance requirements with simpler, nonkinematic mountings for their optics.

7.3 PRISM MASS

Clamping forces used to locate prisms as well as adhesive bonds are determined by the prism mass and dynamic environment. The design load for a prism mount is equivalent to the prism mass multiplied by acceleration and applicable safety factors (SFs). Prism mass is the product of the prism material density and volume. The force required locating a prism, or force exerted on the adhesive bond, is given by

$$F_P = \rho V \, \mathrm{SF} \sum a_G. \tag{7.1}$$

where F_P is the force locating the prism, ρ is the prism material density, V is the prism volume, SF is the safety factor, and a_G are the accelerations acting on the prism.

Prism volumes are found from the equations in Table 7.1, referenced to the particular figures. As a rough rule of thumb for preliminary analysis, for glass in the visible, $\rho \approx 2.5$ g/cm^3. In US

TABLE 7.1
Prism Volume Equations

Type	Volume (Equation No.)		Figure No.
Right angle	$V = 0.5A^3$	(7.2)	Figure 7.1
Beam splitter	$V = A^3$	(7.3)	Figure 7.2
Rhomboid	$V = A^3 + A^2C$	(7.4)	Figure 7.3
Porro	$V = A^3 + A^2D$	(7.5)	Figure 7.4
Abbe–Porro	$V = A^3 + A^2D$	(7.6)	Figure 7.5
Amici	$V = 0.888\,A^3$	(7.7)	Figure 7.6
Pentaprism	$V = 1.500\,A^3$	(7.8)	Figure 7.7
Roof pentaprism	$V = 1.795\,A^3$	(7.9)	Figure 7.8
45° Bauernfeind	$V = 0.750\,A^3$	(7.10)	Figure 7.9
Porro erecting	$V = 2.200\,A^3$	(7.11)	Figure 7.10
Abbe–Koenig erecting	$V = 3.719\,A^3$	(7.12)	Figure 7.11
Dove	$V = AB(A+2a) - A(A+2a)^2 - Aa^2$	(7.13)	Figure 7.12
Roof Pechan	$V = 1.801\,A^3$	(7.14)	Figure 7.13

Note: Figure numbers refer to figures in this chapter.

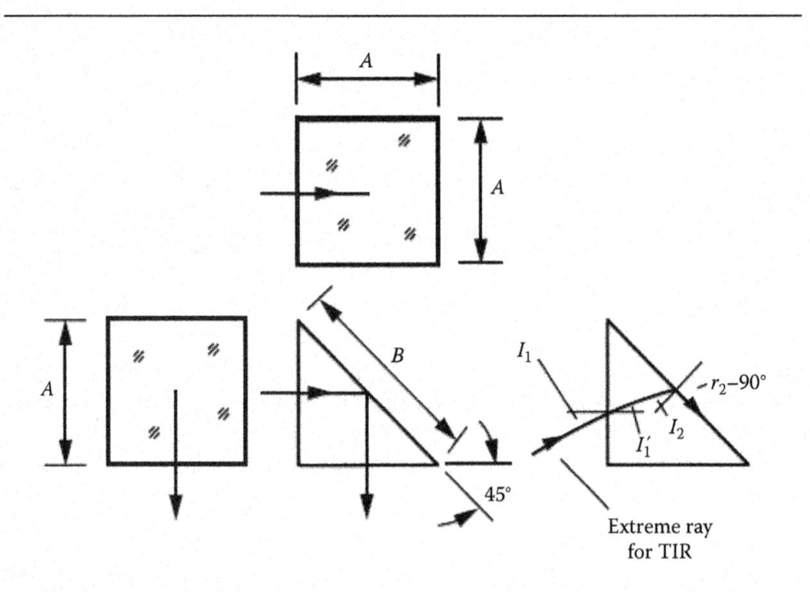

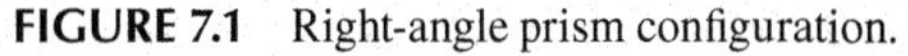

FIGURE 7.1 Right-angle prism configuration.

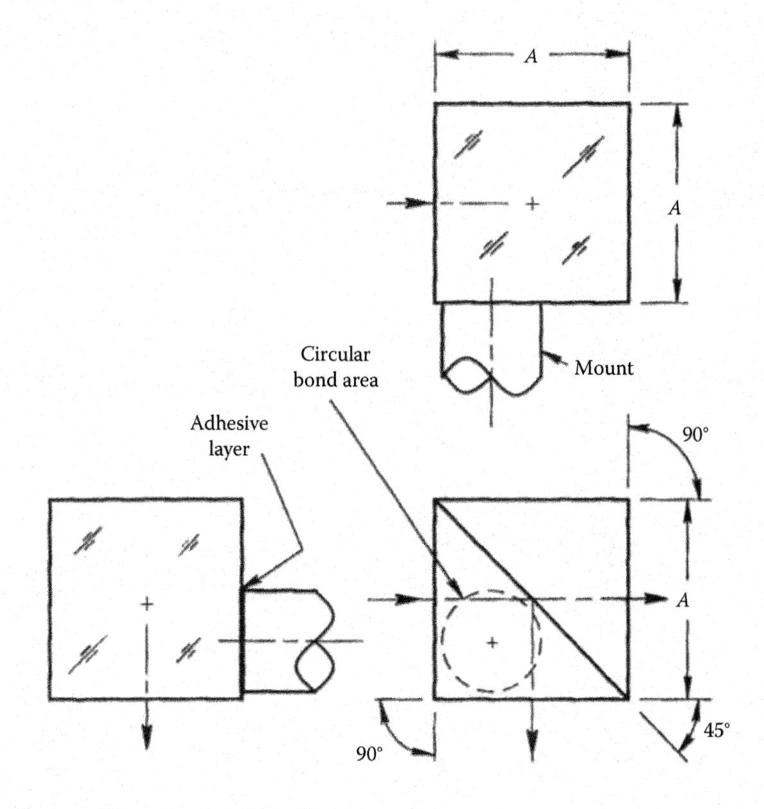

FIGURE 7.2 Beam splitter prism configuration.

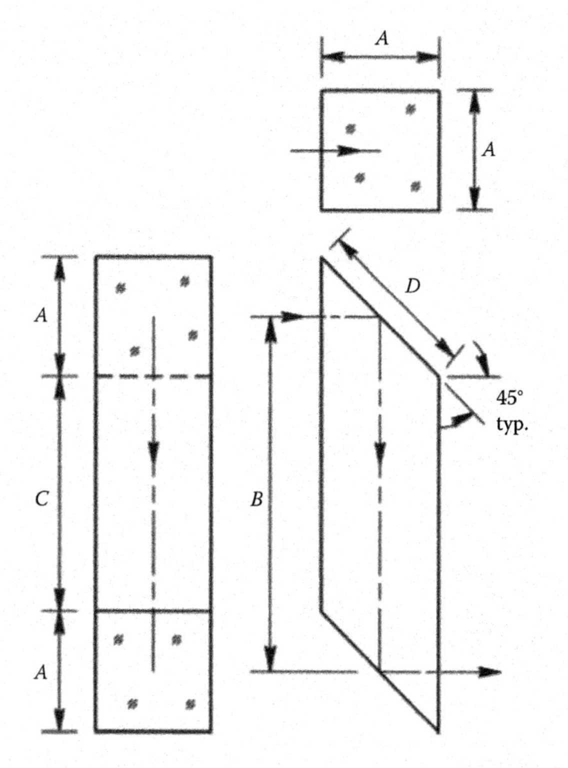

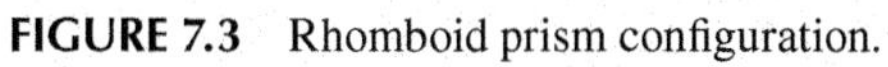

FIGURE 7.3 Rhomboid prism configuration.

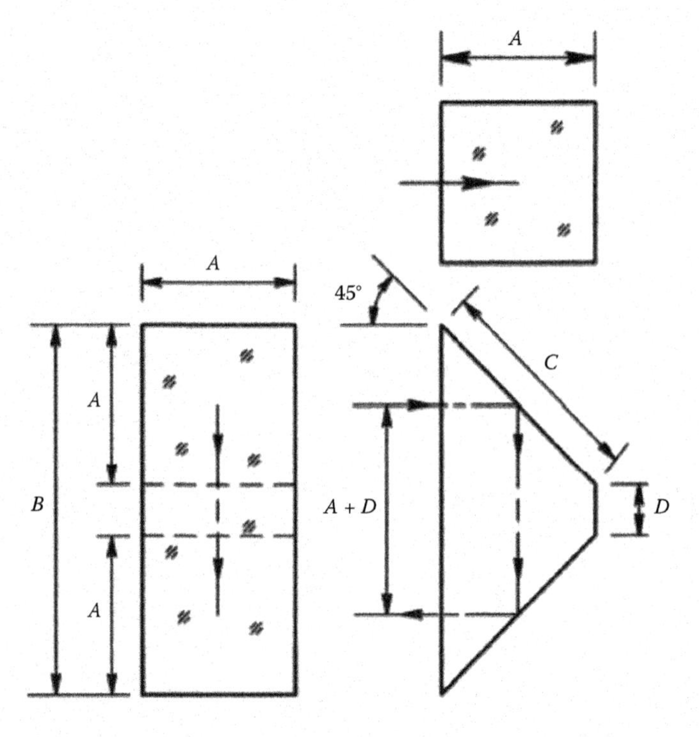

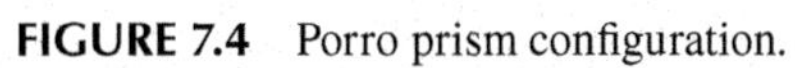

FIGURE 7.4 Porro prism configuration.

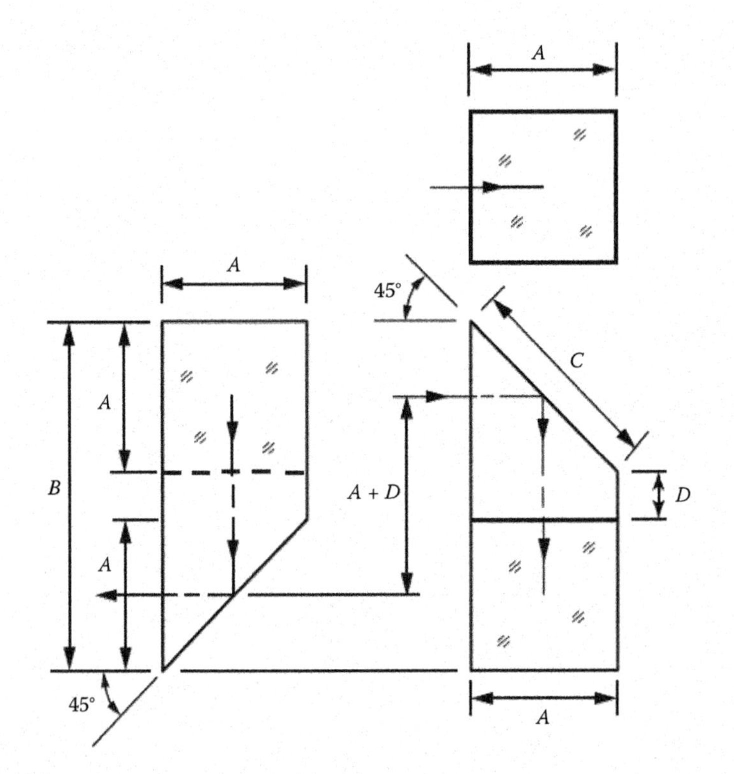

FIGURE 7.5 Abbe–Porro prism configuration.

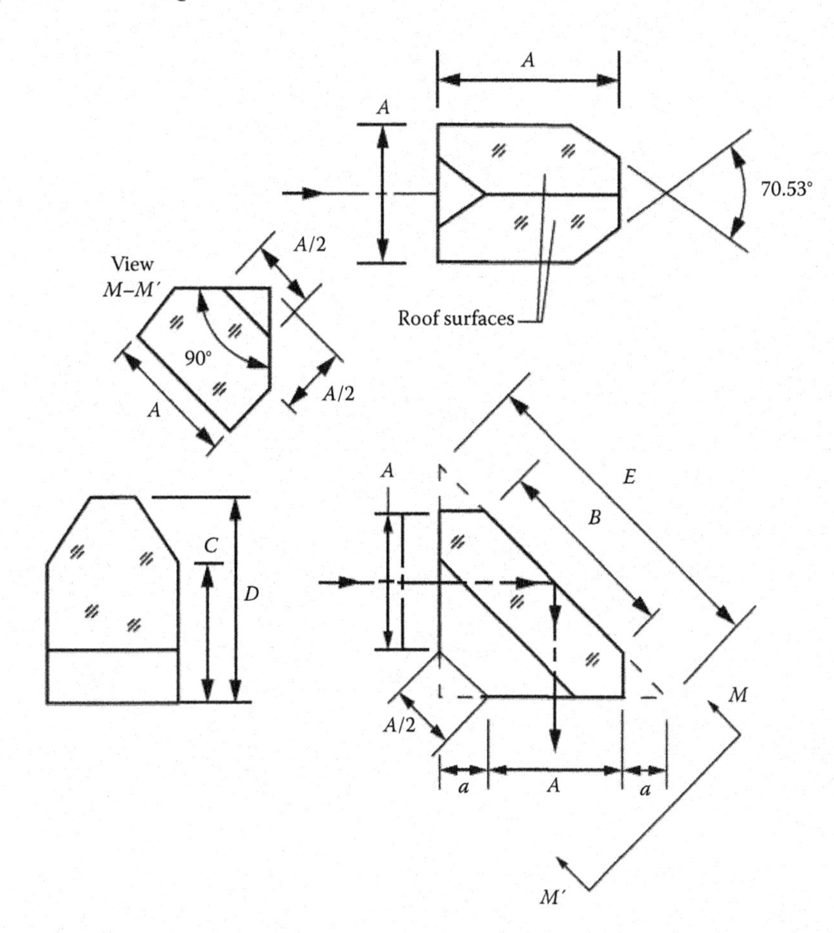

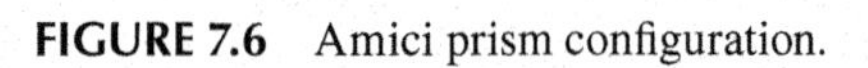
FIGURE 7.6 Amici prism configuration.

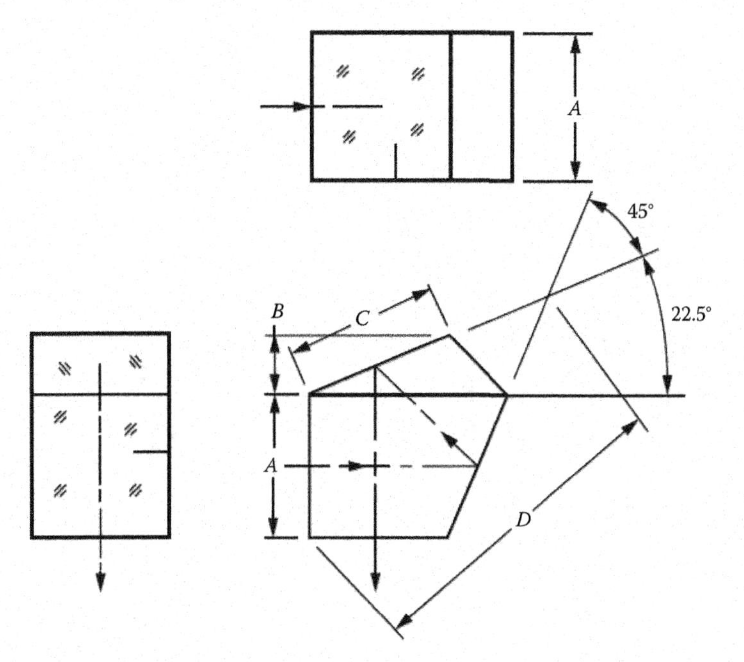

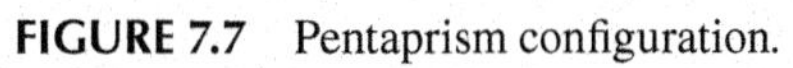
FIGURE 7.7 Pentaprism configuration.

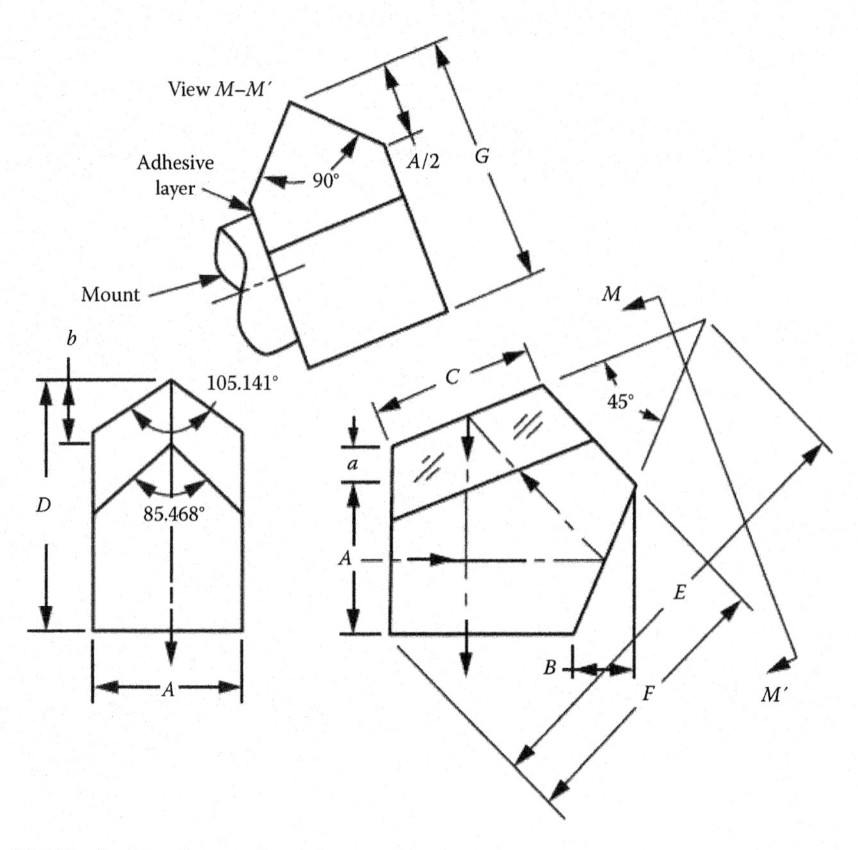

FIGURE 7.8 Roof pentaprism configuration.

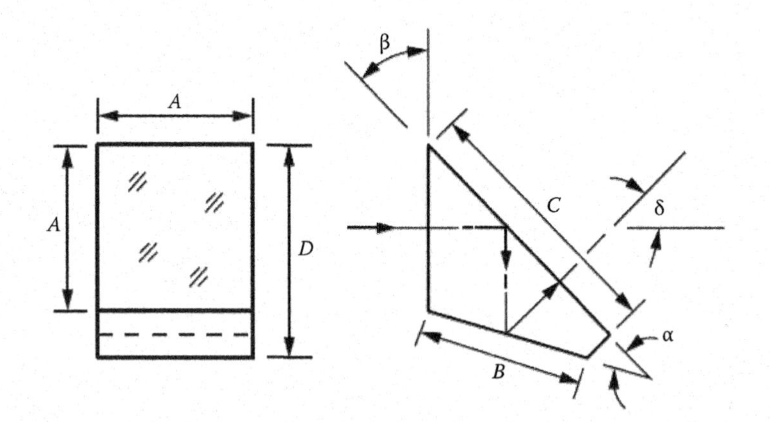

FIGURE 7.9 45° Bauernfeind prism configuration.

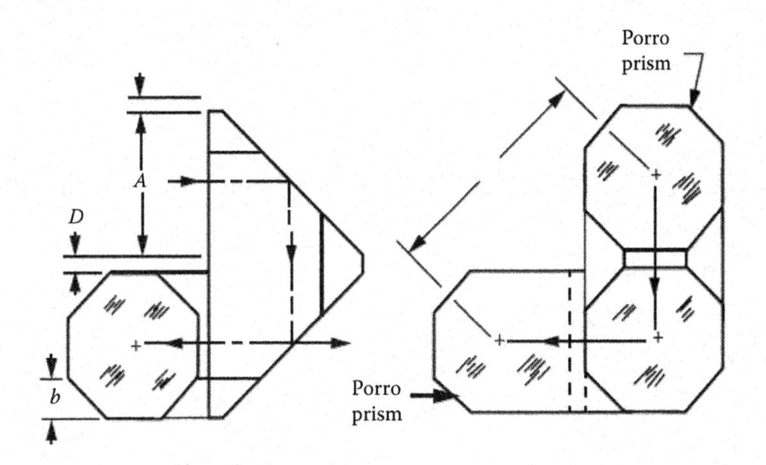

FIGURE 7.10 Porro erecting system configuration.

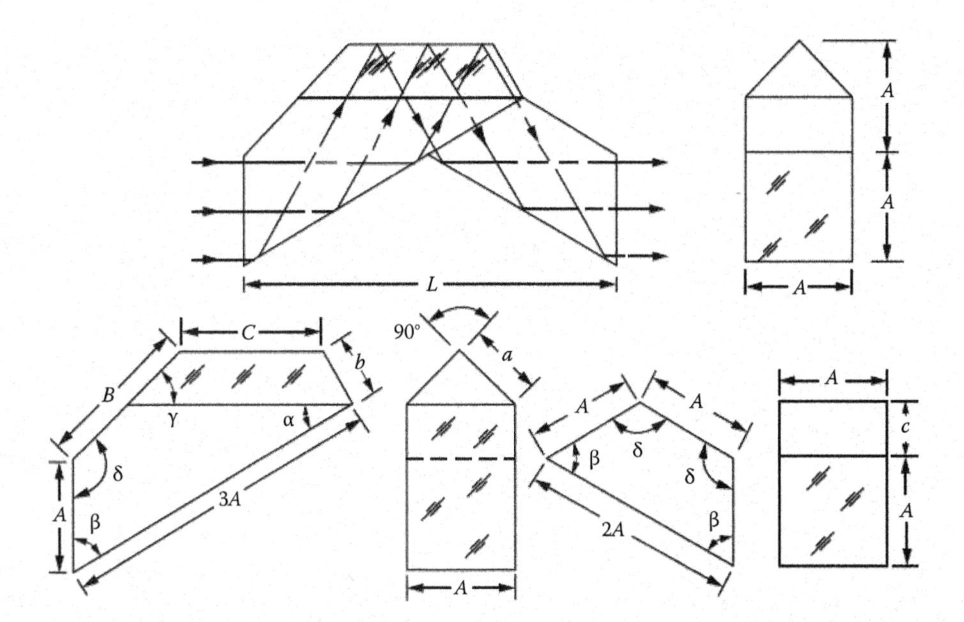

FIGURE 7.11 Abbe–Koenig erecting prism system configuration.

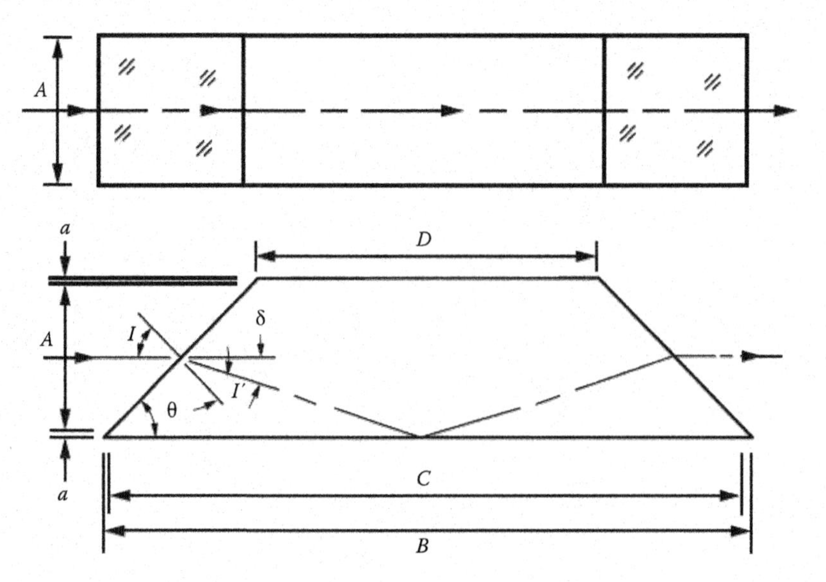

FIGURE 7.12 Dove prism configuration.

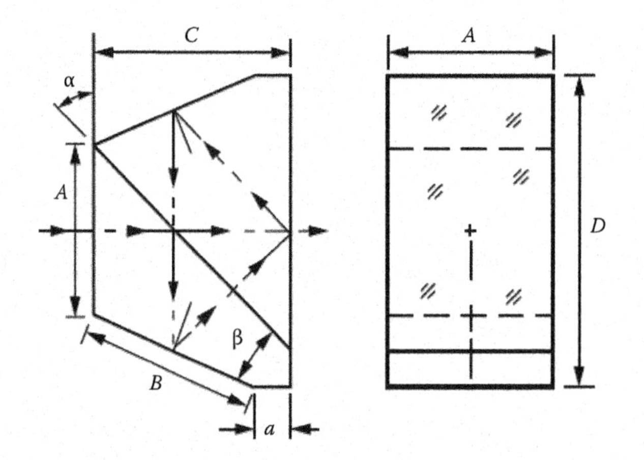

FIGURE 7.13 Pechan prism configuration.

customary units, a convenient approximation for glass density is 0.1 lb/in.^{3}. The densities of the most common prism glasses, Schott BK7 and BAK4, are 2.51 and 3.05 g/cm^3, respectively. Heavy flint glass such as Schott F2 is used in some prisms, with a density of 3.6 g/cm^3. High-index glass such as Schott N-LASF9HT, with a density of 4.41 g/cm^3, is used to minimize the length of Dove prisms.

7.4 MOUNTING PRISMS BY CLAMPING

7.4.1 Clamped Prism Mounts: Kinematic

Figure 7.14a indicates how the required six constraints are applied to a body with the shape of a rectangular parallelepiped such as a simple cube prism. A series of six balls are attached to three mutually orthogonal flat surfaces. If the prism is held in point contact with all six balls, it will be uniquely constrained. Three points on the X–Z plane define a plane on which the lower face of the prism rests; these points prevent translation in the Y direction and rotation about the X- and Z-axes. Two points on the Y–Z plane prevent translation along the X-axis and rotation about the Y-axis. The single point on the X–Y plane controls the last DOF (translation along the Z-axis). A single force exerted against a near point on the prism and directed toward the origin will hold the body against all six balls. This force should pass through the center of gravity of the body to minimize moments. If the force is light and gravity is ignored, the contact surfaces are not deformed elastically and the design is kinematic.

7.4.2 Clamped Prism Mounts: Semikinematic

Figure 7.14b illustrates one way of reducing the previously discussed mount concept to practice. The point contacts are replaced by small area contacts on pads. The pads are shown as rectangular, but their shapes are not of great importance. The multiple pads on the *X–Z* and *Y–Z* planes are machined carefully so that they are very accurately coplanar. The reference surfaces on all the pads and the prism surfaces contacted must have the proper angular relationships, i.e., orthogonal to each other, so that the area contacts with the cube prism do not degenerate into lines. This type of mounting is semikinematic because overconstraints and point contacts are avoided.

The minimum preload force F_{min} to be exerted against the prism to prevent its lifting off its mounting pads under maximum specified acceleration is calculated from Equation 7.1. In this equation, Σa_G is the vector sum, in the direction of F_{min}, of all the externally applied static and dynamic forces such as constant acceleration, random vibration, resonant vibration, shock, and gravity. The magnitude of each force is expressed as a multiple (a_G) of ambient gravity. The direction of F_{min} should pass through the center of gravity of the body. Because all types of external forces do not generally occur simultaneously, the magnitude of Σa_G can be expressed as the most probable value or the worst-case value. This value must be determined for each particular situation. Let the prism weight be $W = \rho V$. Assuming that a single value for a_G prevails and ignoring friction and moments acting at the small area contacts, Equation 7.1 then reduces to

$$F_{min} = W\,\mathrm{SF}\,a_G. \tag{7.15}$$

Note that if the prism weight is expressed in kilograms, this equation must include an additional multiplicative factor of 9.807 m/s^2 to convert the units. The force then is in newtons (N). The SF is typically 1.5–2.

Another important aspect of the semikinematic prism mounting is that surface distortions due to mounting forces must be carefully controlled. Unfortunately, no simple equations can be given here for relating surface deformations to applied forces. The best way to determine deformations for a given prism mounting being created or evaluated is to apply finite-element analysis (FEA) methods. The details of using FEA for this purpose are beyond the scope of this book.

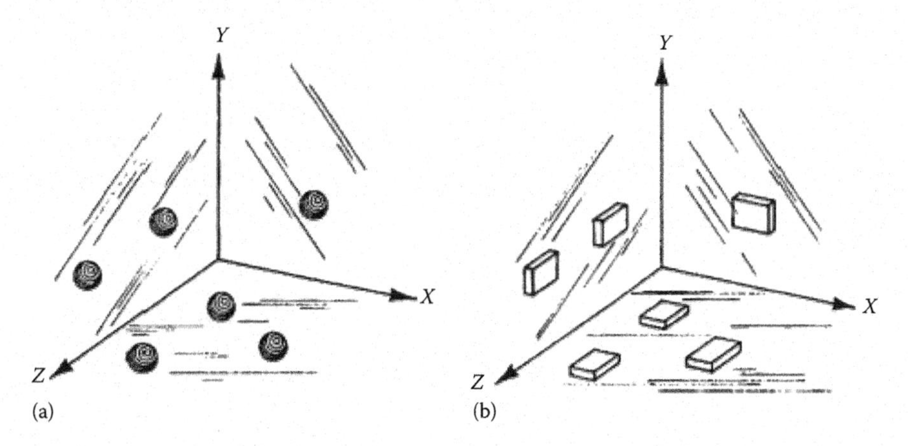

FIGURE 7.14 Position-defining reference surfaces for a rectangular parallelepiped optic such as a cube prism. (a) Kinematic mounting with point contacts on balls; (b) semikinematic mounting with small area contacts. (Adapted from Smith, W.J., *Modern Optical Engineering*, 4th ed., McGraw-Hill, New York, 2008.)

A mounting for a cube-shaped beam splitter prism in a semikinematic mount is shown in Figure 7.15.[1] The prism in Figure 7.15a is made of two similar right-angle prisms cemented together. Beam division occurs at a partially reflecting coating on one hypotenuse surface. The cemented prism may be treated as a rigid-body subject to the previously discussed six positional constraints. Surface distortions may result from improperly applied forces, and dimensional variations may occur when the temperature changes.

A prism of this type is frequently used to divide a beam converging toward an image plane into two parts. Each part forms an image, as indicated in Figure 7.15b. For these images to maintain a constant alignment relative to each other and to the structural references of the optical instrument, the prism must not translate on the plane of reflection (*X*–*Y*) nor rotate about any of the orthogonal axes. Pure translation along the *Z*-axis does not introduce any error, but should be kept small.

Once the optical system is aligned, the beam splitter prism must always be pressed against its five mounting pads, indicated by the symbol K_∞ in Figure 7.15a. These are raised areas (pads) on the instrument structure that are assumed to have spring rates approaching infinity. The pads constrain translations in two directions and rotations about the three axes. Five constraining forces (K_i in Figure 7.15a) have relatively low spring constants; they are illustrated as compression springs. These elements are located directly opposite the mounting pads to put the glass in pure compression, thereby minimizing surface distortion. The dashed lines in Figure 7.15b show how increased temperature will expand the cube. The light path to each image (at the *X* and *Y* detectors) is not deviated, so pointing errors are not introduced.

When the prism configuration is other than a cube, the mounting design may be more complex since it is more difficult to apply restraining forces directly opposite support pads. Figure 7.16a shows a right-angle prism semikinematically registered against its refracting faces and one edge. Three coplanar pads on the baseplate provide constraints in the *Y* direction, while three locating pins pressed into the baseplate add three more (*X*–*Y*) constraints. Note that a perforation is required in the baseplate for light passage. This is not shown in the figure. All pads and posts should contact the prism outside its optically active apertures (also not shown).

In the view in Figure 7.16b, the same prism is shown in side view. The preload forces F_1 and F_2 are oriented perpendicular to the hypotenuse face and touch the prism near the longer edges of the hypotenuse. F_1 is aimed symmetrically between the nearest pad (b) and the nearest pin (d), while F_2 is aimed symmetrically between pads a and c and pin e. Horizontal force F_1 holds the prism against

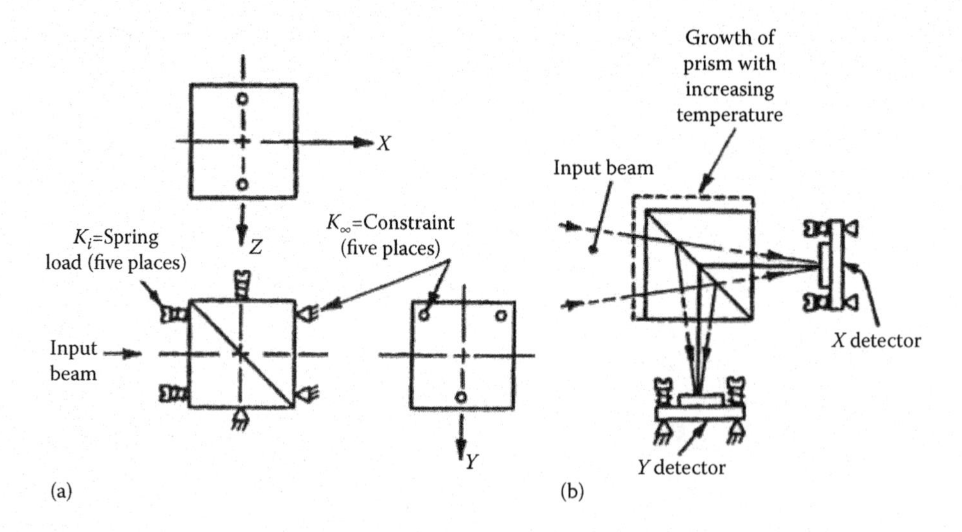

FIGURE 7.15 (a) Three views of a semikinematic mounting for a cube-shaped beam splitter prism. (b) Schematic of the typical optical function of the beam splitter showing the effect of a temperature rise. (Adapted from Lipshutz, M.L., *Appl. Opt.*, 7, 2326, 1968. With permission of Optical Society of America.)

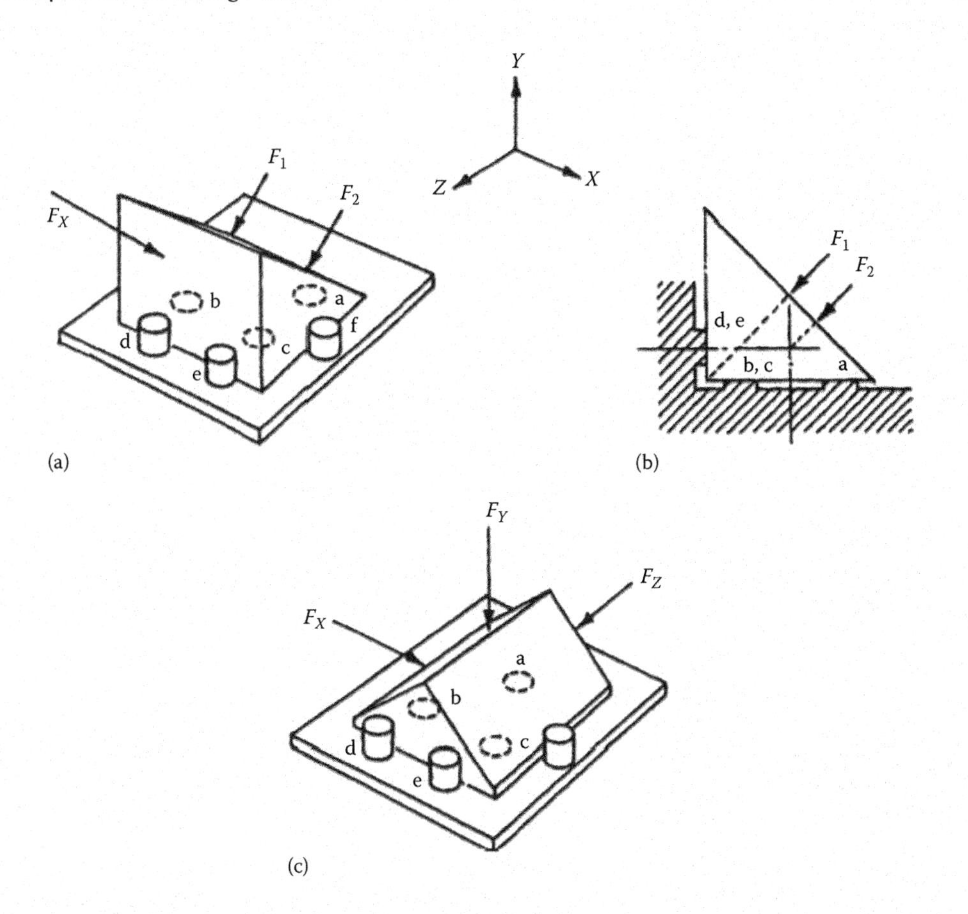

FIGURE 7.16 Schematics of semikinematic mountings for (a, b) a right-angle prism referenced to its refracting surfaces and one edge and (c) a Porro prism referenced to its hypotenuse surface, one triangular surface, and one edge. (Adapted from Durie, D.S.L., Stability of optical mounts, *Mach. Des.*, 40, 184, 1968.)

pin f, and the horizontal and vertical components of F_1 and F_2 hold the prism against the three pads and remaining two locating pins. Although not optimum in terms of freedom from bending moments, because the forces are not directed exactly toward the pads, this arrangement is adequate since the prism is relatively stiff.

In Figure 7.16c, the hypotenuse face of a Porro prism is positioned against three coplanar flat pads on a perforated plate (this perforation again not shown), while one triangular face touches two cylindrical pins and one beveled edge touches a third pin. The pin axes must be closely perpendicular to the plane of the pad surfaces. Optical clear apertures are not shown in the figure. A force F_Z directed parallel to and slightly above the baseplate holds the prism against two pins (marked d and e), while force F_X, also just above the plate, holds it against the third pin (marked f). A third force F_Y holds the prism against the three pads a, b, and c. This force acts against the dihedral edge of the prism at its center. Because the prism is stiff, surface distortion is minimal. Any distortion that does occur could be reduced by dividing F_Y into two parallel forces: one directed toward pad a and the other toward the midpoint of the line connecting the centers of pads b and c.

Figure 7.17 shows a right-angle prism with one triangular ground face pressed vertically against three coplanar raised pads and transversely against three locating pins on the baseplate. Three long screws threaded into the baseplate pull the clamping plate through a resilient (elastomeric) pad to clamp the prism against the three pads. A leaf spring anchored at both ends (called a *straddling* spring) presses the prism horizontally against the opposite locating pins. An attractive feature of this mounting is that it can be configured so that the circular clear apertures of optically active

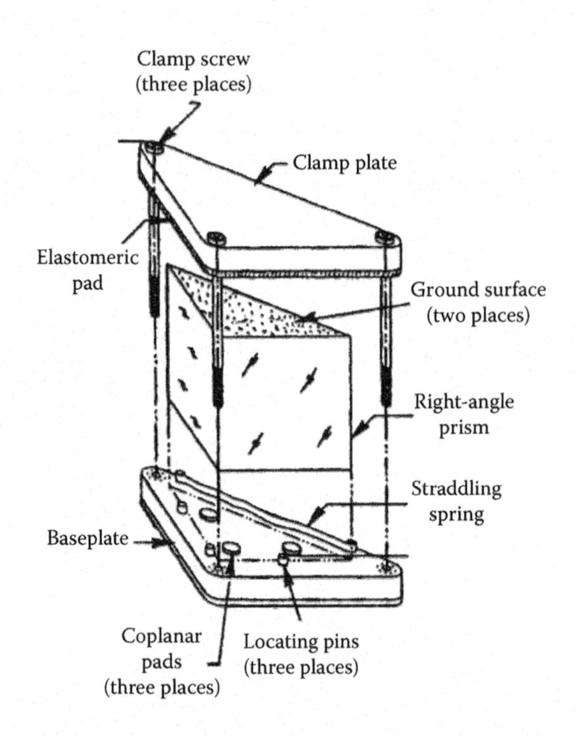

FIGURE 7.17 Semikinematic mounting for a right-angle prism preloaded by a compressed elastomeric pad. (Adapted from Vukobratovich, D., Optomechanical system design, Chapter 3 in *The Infrared and Electro-optical Systems Handbook*, Vol. 4, pp. 123–196, ERIM, Ann Arbor, MI; SPIE Press, Bellingham, WA, 1993.)

surfaces are not obscured and are not likely to be distorted by the imposed forces. The elastomeric pad (in compression) and the three screws (in tension) provide the vertical preload necessary to hold the prism in place under shock and vibration. The spring constant of the pad is considerably smaller than that of the screws acting together, so the latter can be ignored. Typical elastomeric materials have a limited elastic range, tend to creep with time, and may take a permanent set under sustained high compressive load, i.e., one greater than that for which the material ceases to act elastically. However, if a properly behaving elastomer is available, it can provide a convenient way to mount some prisms.

The spring rate k_E of an elastomeric pad is determined by the pad area A, pad thickness h, and effective compressive modulus $E_C(h)$ at thickness h:

$$k_E = \frac{E_C(h)A}{h}. \tag{7.16}$$

For most elastomeric materials, the elastic modulus $E_0 = 3G$, where G is the shear modulus of the elastomer. The effective compressive modulus E_C is related to the elastic modulus E_0 by

$$E_C = E_0(1 + 2\phi S^2), \tag{7.17}$$

$$\phi = 6.4044\, E_0^{-0.283}, \quad E_0 \text{ in kilopascals.} \tag{7.18}$$

Equations 7.17 and 7.18 are valid for 896 kPa $\le E_0 \le$ 9239 kPa and $0.1 \le S \le 100$. The shape factor S is the ratio of the area in compression or under to load to the area at the sides of the spring that is free to bulge during compression. If h is the pad thickness, the shape factors are given by

$$S = \frac{d}{4h}, \quad \text{circular pad,} \tag{7.19}$$

$$S = \frac{d_o - d_i}{4h}, \quad \text{annular pad,} \tag{7.20}$$

$$S = \frac{Lb}{2h(L+b)}, \quad \text{rectangular pad,} \tag{7.21}$$

where d is the pad diameter, d_o is the outer diameter of an annular pad, d_i is the inner diameter of an annular pad, L is the length of a rectangular pad, and b is the width of a rectangular pad.

The pad areas are

$$A = \frac{\pi}{4} d^2, \quad \text{circular pad,} \tag{7.22}$$

$$A = \frac{\pi}{4}\left(d_o^2 - d_i^2\right), \quad \text{annular pad,} \tag{7.23}$$

$$A = Lb, \quad \text{rectangular pad.} \tag{7.24}$$

The maximum deflection of an elastomer should be limited to less than 25%, ideally to around 15% or less. At larger deflections, elastomers become nonlinear with respect to stress and strain. In addition, larger deflections increase the compressive set. Elastomeric materials take a compressive set with time due to stress relaxation. Compressive set can reduce the force exerted by the elastomer as much as 50% in some cases. For example, at room temperature, a typical elastomer, such as Neoprene, may take a set anywhere from 17% to 45%, depending on the formulation and temperature. At higher temperatures, compressive set is greater and more rapid. Margin in the initial compressive force must be provided to allow for this compressive set.

Some elastomer suppliers provide information on compressive force deflection at some nominal deflection such as 25%. This information can be used as an alternative to Equations 7.17 through 7.24 to find the pad size. Compressive force deflection is assumed to be linear up to the nominal value. Normally, a limited range of elastomer thicknesses is available, so preliminary design begins with the selection of elastic modulus (or durometer), thickness, and acceptable deflection. Equations 7.17 through 7.24 are used to find the required pad area.

Another approach is to provide design curves of deflection versus load for different shape factors and durometers. Durometer is a measure of elastomer hardness and is related to the elastic modulus. Durometer is given in units of Shore hardness, either Shore A or Shore D. Shore hardness is determined from the process in ASTM 2240. Elastomer shore hardness is Shore 20–100. Empirical relationships between elastic modulus E_0 and Shore hardness (A or D) are given in the following:

$$E_0 = \exp[0.05423(\text{Shore A}) - 1.474], \quad 20 < \text{Shore A} < 80, \tag{7.25}$$

$$E_0 = \exp[0.05423(\text{Shore D} + 50) - 1.474], \quad 20 < \text{Shore D} < 85, \tag{7.26}$$

where E_0 is in megapascals.

Example 7.1

A pentaprism with a face edge length A = 38.1 mm, made of glass with a density of ρ = 2510 kg/m^3, is subjected to an acceleration of a_G = 6.7g. An elastomeric preload pad similar to the one in Figure 7.17 is used to provide preload with SF = 1.5. The elastomeric pad is 9.5 mm thick with a Shore A hardness of 30. If the pad does not compress more than 20%, what is the diameter of a circular pad?

The prism mass is

$$W = 1.5(2510\,\text{kg/m}^3)(38.1\times 10^{-3}\,\text{m})^3 = 0.208\ \text{kg}.$$

The preload force F_P from Equation 7.1 is

$$F_P = 1.5(0.208\text{kg})(6.7)(9.81\,\text{m/s}^2) = 19.7\,\text{N}.$$

The elastic modulus of the elastomer from Equation 7.25 is

$$E_0 = \exp[0.05423(30) - 1.474]\text{MPa} = 1.165\text{MPa}.$$

Then, from Equation 7.16,

$$\frac{F_P}{0.2h} = \frac{E_C A}{h},$$

$$A = 5\frac{F_P}{E_C}.$$

From Equations 7.17 through 7.19 and 7.22,

$$\frac{\pi}{4}d^2 = \frac{5(19.71\text{N})}{(1.165\text{MPa})1 + 2\left\{6.4044\left(\dfrac{1.165\text{MPa}}{1\text{kPa}}\right)^{-0.282}\left[\dfrac{d}{4(9.5\text{mm})}\right]^2\right\}}.$$

The preceding equation is solved numerically for d; d = 9.82 mm.

Another semikinematic mounting is illustrated schematically in Figure 7.18.[2] The pentaprism is pressed by three cantilevered springs against three coplanar circular pads on a baseplate. This constrains one translation and two tilts. The other translations and the third tilt are constrained by preloading the prism laterally against three locating pins with a straddling leaf spring.

Typically, the total preload force F_P exerted by all of the springs is equivalent to F_{min} as defined by Equation 7.15. If N springs provide the preload force F_P, the following equations give the deflection δ, stress σ, and angle of deflection φ (radians) for cantilever preload springs[3]:

$$\delta = 4F_P N\left(\frac{L^3}{bh^3}\right)\frac{(1-\nu^2)}{E}, \tag{7.27}$$

$$\sigma = 6F_P\frac{L}{bh^2}\frac{1}{N}, \tag{7.28}$$

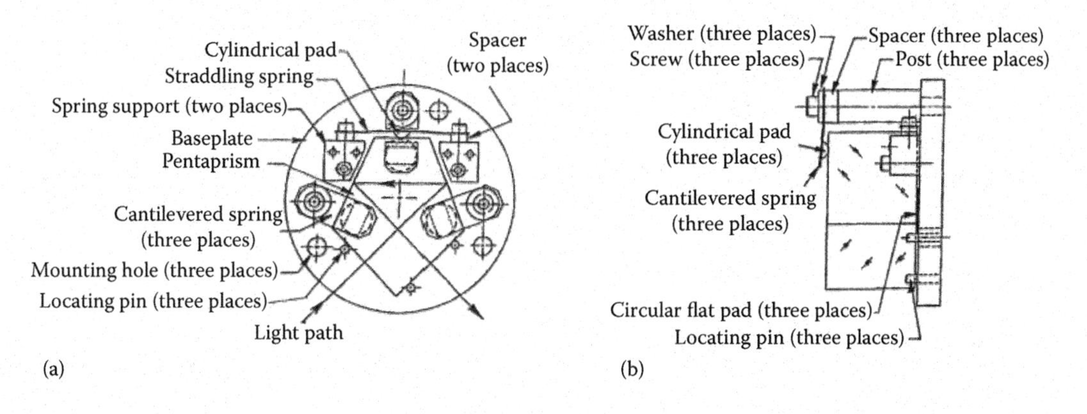

FIGURE 7.18 Semikinematic mounting for a pentaprism with cantilevered and straddling spring constraints: (a) plan view and (b) elevation view. (From Yoder, P.R. Jr., Mounting-induced contact stresses in prisms, *Proc. SPIE*, 3429, 7, 1998.)

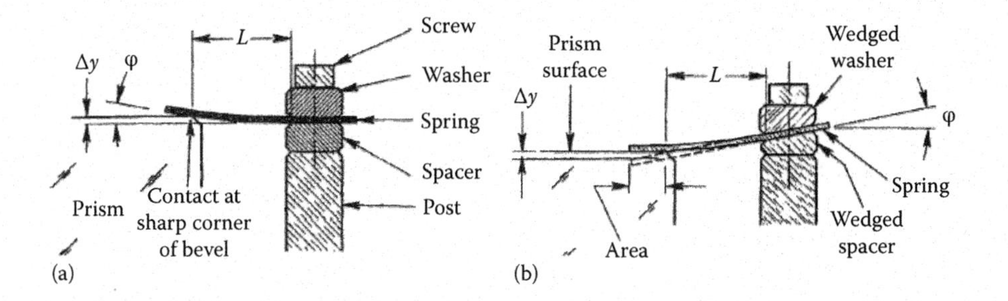

FIGURE 7.19 Two configurations for cantilevered spring interfaces with a prism: (a) spring touching prism bevel and (b) spring lying flat on prism surface. (From Yoder, P.R. Jr. *Mounting Optics in Optical Instruments*, 2nd ed., SPIE Press, Bellingham, WA, 2008.)

$$\varphi = 6F_P \frac{1}{N}\frac{L^2}{bh^3}\frac{(1-\nu^2)}{E}, \tag{7.29}$$

where L is the spring length, b is the spring width, h is the spring thickness, ν is the Poisson's ratio for the spring material, and E is the elastic modulus of the spring material.

Figure 7.19 shows the bending of cantilever preload springs in contact with a prism.[4] In Figure 7.19a, the bent spring is in contact with the edge bevel. High contact stresses can develop when the preload springs press against the edge bevel, with possible local fracture. An alternative is shown in Figure 7.19b, where the preload spring is deliberately tilted through an angle φ. When bent, the spring end lies flat against the prism surface, reducing contact stress.

Example 7.2

A pentaprism of mass W = 0.121 kg is subjected to an acceleration a_G = 8g. There is an SF = 1.5 in the mount design. The prism is held by three springs as shown in Figure 7.18. The spring dimensions are L = 9.525 mm, b = 6.35 mm, and h = 0.508 mm. BeCu is the spring material, with E = 128 GPa, ν = 0.35, and σ_{YS} = 1.069 GPa. What are the spring deflections? What is the bending stress in the spring? What is the ratio between the yield stress σ_{YS} and bending stress? How much is each spring bent?

From Equation 7.1, the required preload force F_P is

$$F_P = (0.121\text{kg}) \times 8 \times 1.5 \times 9.81\text{m/s}^2 = 14.24\text{N}.$$

From Equation 7.27, the deflection δ is

$$\delta = \frac{[1-(0.35)^2](4)(14.24\text{N})(9.525\times10^{-3}\text{m})^3}{3(128\times10^9\text{Pa})(6.35\times10^{-3}\text{m})(508\times10^{-6}\text{m})^3} = 136\times10^{-6}\text{m}.$$

From Equation 7.28, the bending stress σ is

$$\sigma = \frac{6(14.24\text{N})(9.525\times10^{-6}\text{m})}{3(6.35\times10^{-6}\text{m})(508\times10^{-6}\text{m})^2} = 165.5\text{MPa}.$$

The stress ratio is 1.069 GPa/(165.5 MPa) = 6.46.
From Equation 7.29, the bending angle of the spring φ is

$$\varphi = \frac{6[1-(0.35)^2](9.525\times10^{-6}\text{m})^2(14.24\text{N})}{3(128\times10^9\text{Pa})(6.35\times10^{-3}\text{m})(508\times10^{-6}\text{m})^3} = 21.35\times10^{-3}\text{rad}\,(1.223°).$$

Cylindrical pads between the springs and the prism surfaces in the mounting designs in Figure 7.20 ensure that line contact on the flat prism surface will occur in a reliable manner at each interface. In the designs shown in Figures 7.19 and 7.20, the angular orientation of the deflected spring relative to the top prism surface results from the wedge-shaped parts placed above and below the spring on the post. The angles of these wedges are determined by Equation 7.29. The axial lengths of the lower wedges determine the height of the cantilevered spring at its fixed end as well as the magnitude of the spring deflection.

Line contact at a rounded portion of the spring occurs if the end of the spring is bent to a convex cylindrical shape as indicated in Figure 7.20b. Because it may be difficult to form the spring into a smooth cylinder of a particular radius, a better interface results if a pad is machined integrally into the spring as indicated in Figure 7.20c. Note that a separately machined cylindrical pad also can be attached to the spring by screws, welding, or adhesive and achieve essentially the same function. Stress introduced at the interface between the convex cylinder and the flat prism surface can be estimated using Hertz contact mechanics, as discussed later.

The designs of the cylindrical pads for use on cantilevered or straddling springs should include consideration of the angular extent of the curved surfaces. Figure 7.21 shows the pertinent geometrical relationships. R_{CYL} is the radius of the pad, d_P is the width of the pad and α is one-half the angular extent of the cylindrical surface measured from its center of curvature. In Figure 7.21a, the center of the pad touches the prism surface with the spring straight instead of bent. There would then be no preload. To provide the required preload, the thickness of the spacer under the spring is reduced. In Figure 7.21b, the spring is bent to exert preload *P*, tilting the pad by the angle φ per Equation 7.29. If α is greater than φ, no sharp edge contact with the prism will occur. Once α is determined, the minimum value for d_P is ($2\,R_{CYL}\sin\alpha$).

Figure 7.22 shows how a straddling spring can be used instead of cantilevered spring clips to hold a prism in place. A convex pad is shown at the center of the spring as a means for distributing the force over a specific narrow region on the prism. A flat pad could be used if the spring action is symmetrical and the pad accurately lies flat on the prism. Dimensional errors or tolerance buildup could, however, tilt the pad to the wrong angle and cause stress concentration at a pad edge. Curving the pad

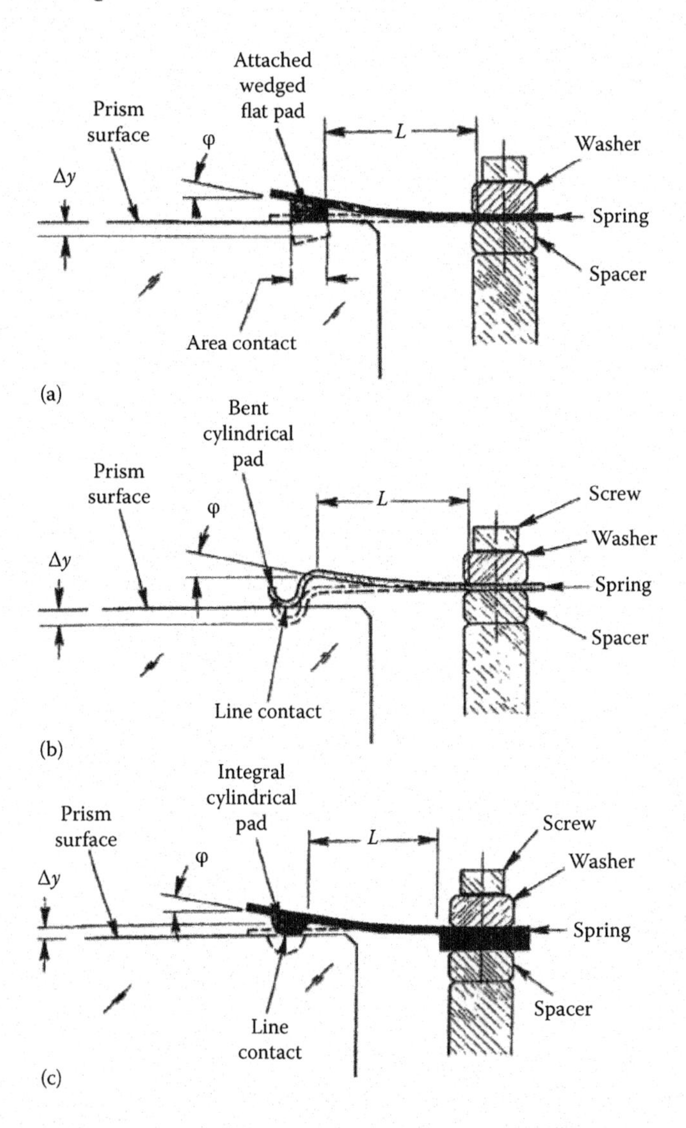

FIGURE 7.20 Configurations of a cantilevered spring interface with pads of various shapes pressing against a flat surface on the prism: (a) flat pad, (b) spring bent to convex cylindrical curve, and (c) integrally machined cylindrical or spherical pad. (From Yoder, P.R. Jr., *Mounting Optics in Optical Instruments*, 2nd ed., SPIE Press, Bellingham, WA, 2008.)

into a cylindrical or spherical shape eliminates this possibility. The parallel flexure blades incorporated into one post allow the post-to-post separation to change slightly as the spring bends.

The deflection Δy from the relaxed shape and the bending stress σ in the straddling spring are given by the following equations:

$$\Delta y = 0.0625\, F_P \left(\frac{L^3}{bh^3}\right)\frac{(1-\nu^2)}{E}, \tag{7.30}$$

$$\sigma = 0.750\, F_P \frac{L}{bh^2}. \tag{7.31}$$

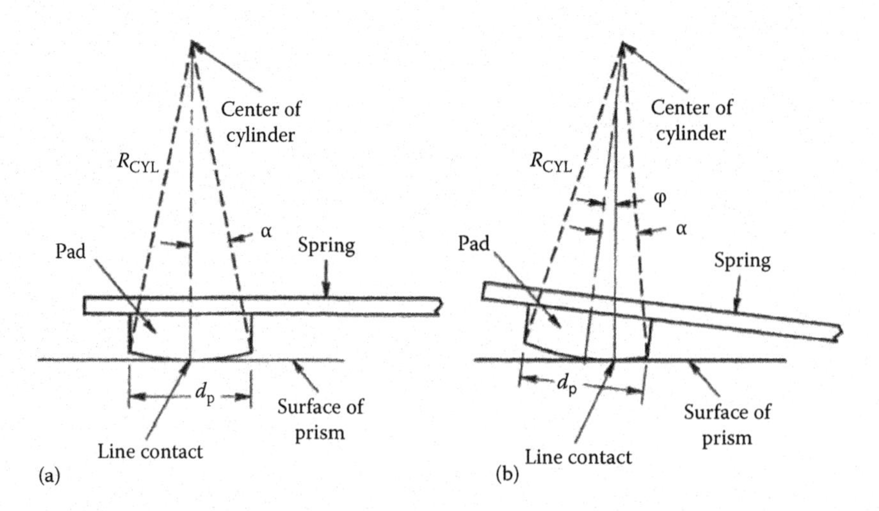

FIGURE 7.21 Geometrical relationships applied in the design of cylindrical pads for springs. The same geometry applies to spherical pads. (From Yoder, P.R. Jr., *Mounting Optics in Optical Instruments*, 2nd ed., SPIE Press, Bellingham, WA, 2008.)

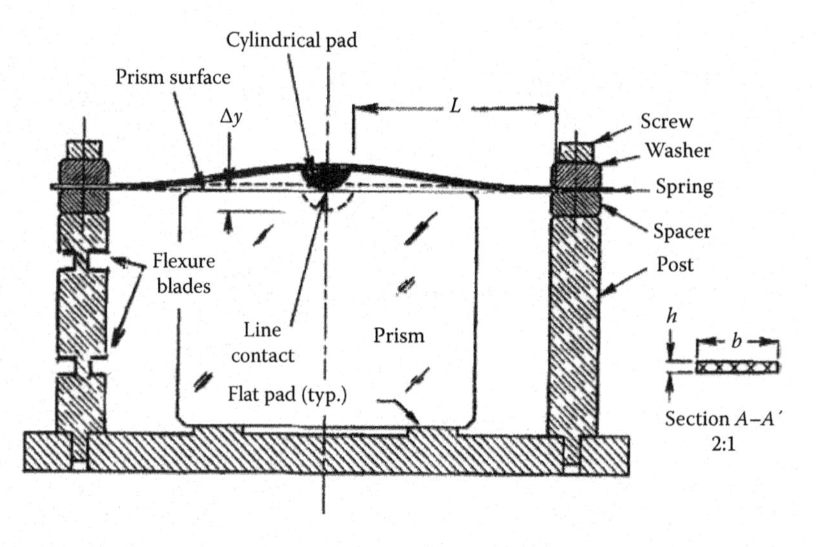

FIGURE 7.22 Straddling spring constraint for a prism shown with a cylindrical or spherical pad. The left post has two flexure blades that allow that post to bend into a thin S curve so as to accommodate the slight length change of the straddling spring as it bends to preload the prism. (From Yoder, P.R. Jr., *Mounting Optics in Optical Instruments*, 2nd ed., SPIE Press, Bellingham, WA, 2008.)

Example 7.3

A single BeCu straddling spring constrains a pentaprism with a mass of 0.121 kg (the same as in Example 7.2) with a mount shown in Figure 7.22. The prism is subjected to an acceleration of $a_G = 12g$, with SF = 1.5. The spring length is L = 52.83 mm, width b = 2.54 mm, and thickness h = 0.75 mm. For beryllium copper, $\nu = 0.35$, E = 128 GPa, and σ_{YS} = 1.07 GPa. What is the spring deflection Δy? What is the bending stress σ? What is the ratio between the yield stress and bending stress σ_{YS}/σ?

From Equation 7.1, the required preload force F_P is

$$F_P = 1.5(0.121\text{ kg})12(9.81\text{ m/s}^2) = 21.36\text{N}.$$

The spring deflection Δy is given by Equation 7.30:

$$\Delta y = \frac{0.0625[1-(0.35)^2](21.36\text{N})(52.83\times10^{-3}\text{ m})^3}{(128\times10^9\ Pa)(2.54\times10^{-3}\text{ m})(750\times10^{-6}\text{ m})^3} = 1.259\times10^{-3}\text{ m}.$$

The spring bending stress σ is given by Equation 7.31:

$$\sigma = \frac{0.75(21.36\text{N})(52.83\times10^{-3}\text{ m})}{(2.54\times10^{-3}\text{ m})(750\times10^{-6}\text{ m})^2} = 592.3\times10^6\text{ Pa}.$$

The stress ratio is 107 GPa/592.3 MPa = 1.806. This stress ratio is marginal, since a ratio of at least 2 is desirable. A redesign is probably necessary. A wider spring, with b = 5.08 mm, would reduce deflection and stress by a factor of 2.

An example of straddling springs is illustrated in Figure 7.23. This is a drawing of a portion of the housings for a commercial binocular offered by Swarovski Optik in Austria. The centers of the springs press against the apexes of the prisms and hold them against shoulders machined inside the housings. One end of each spring is attached with a screw to the housing, while the other end slips into a slot provided in the housing wall. This arrangement allows the springs to slide in the slots, as its chord length changes very slightly when the spring is bent to apply preload. It also removes the necessity for four threaded holes and four screws. Note that in this design, no pads are used on the springs. They merely press diagonally across the bevels on the prisms.

A cruder version of the straddling spring prism mounting is illustrated schematically in Figure 7.24.[5] Here, the refracting surfaces of an Amici prism are held against flat pads inside a military elbow telescope housing by a flat spring with bent ends. The screw pressing against the center of the spring forces the ends of the spring against ground surfaces on the prism. It is not apparent from the figure, but constraint perpendicular to the plane of the figure is provided by slightly resilient pads, typically Neoprene rubber or elastomer attached to the inside surfaces of triangular covers that are attached with screws to both sides of the cast housing. These covers are sealed to the housing with

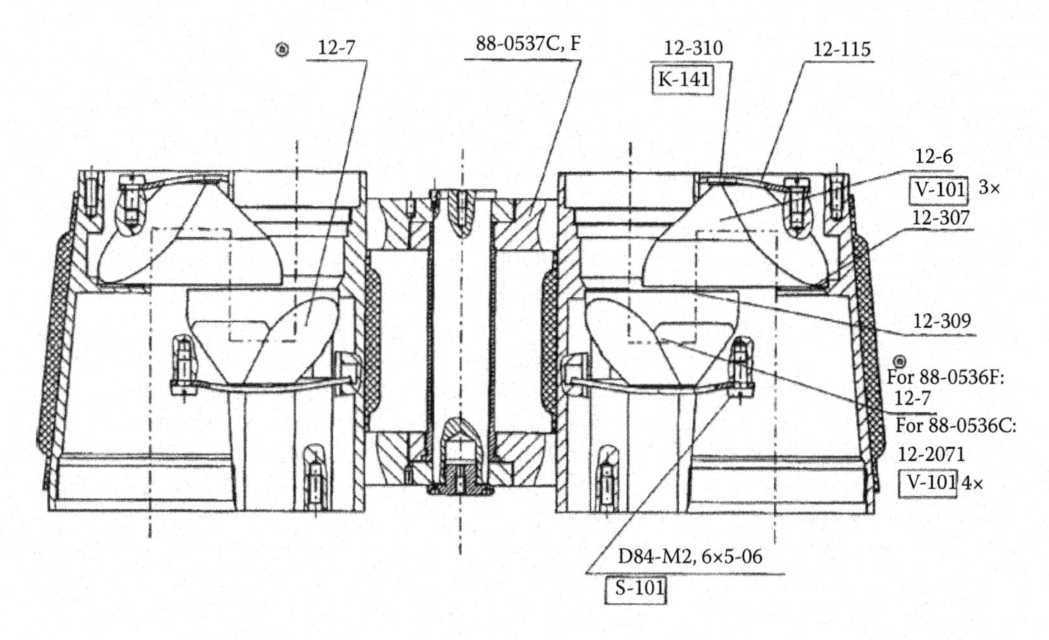

FIGURE 7.23 Porro prism constraints using straddling springs as found in a modern consumer binocular. (Courtesy of Swarovski Optik, Absam, Austria.)

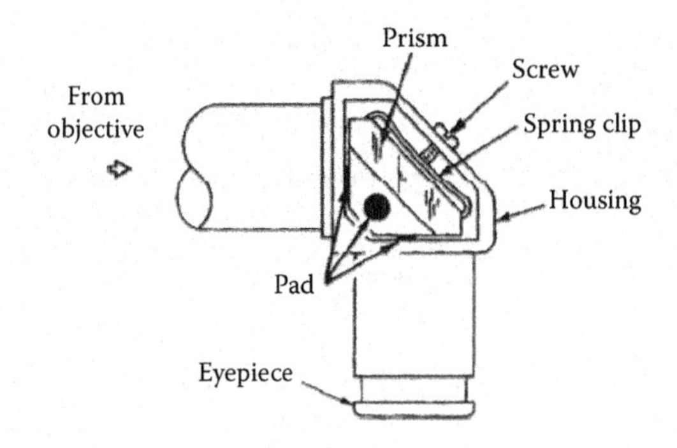

FIGURE 7.24 Schematic diagram of a small military telescope with an Amici prism spring loaded against positioning pads. (From Yoder, P.R. Jr., Optical mounts, lenses, windows, small mirrors, and prisms, Chapter 6 in *Handbook of Optomechanical Engineering*, Ahmad, A., ed., pp. 151–210, CRC Press, Boca Raton, FL, 1997.)

gaskets. The objective and eyepiece ends of the instrument, as well as the screw preloading the spring, are sealed with elastomeric sealants during assembly.

This design has a fundamental deficiency in that sharp and perhaps irregular edges on the spring ends might touch the glass. Failure of the glass might occur under extreme accelerations because of concentrated contact stress. It is also very important in this design that the screw not protrude far enough into the housing for the spring to touch the roof edge of the prism under vibration or shock, especially at extreme temperatures. The design should specify the correct screw length and appropriate tolerances to give sufficient preload to the prism without the possibility of spring contact with the prism. Explicit instructions to check the actual clearance after assembly should be included in the manufacturing procedures.

7.4.3 Clamped Prism Mounts: Nonkinematic

Another type of mount utilizes springs to clamp prisms against mounting surfaces in nonkinematic fashion. The first example is the Porro prism-erecting system used widely in military and commercial binoculars and telescopes. Figures 7.25 and 7.26 illustrate the design used in many 7 × 50 military binoculars of World War II (WWII) and later vintage. In this design, the prisms are made of light barium crown, type 573574 optical glass; the shelf is aluminum; the straps are spring-tempered phosphor bronze; the light shields are aluminum alloy painted matte black; and the pads are cork. Cork supports fungus growth and is no longer used for this purpose in military and high-quality consumer instruments. Neoprene rubber has been substituted in some models. The angled portions and the curved region of the spring in this design are all compliant and provide preload to hold the prism against the shelf. The shelf is configured so it can be registered against machined flat surfaces on three protrusions cast into the inside wall of the instrument housing and secured there with three screws and two pins.

The index of refraction of the prism glass is high enough for total internal reflection to take place, so the prisms are not silvered. The light shields reduce stray light entering the optical path through the reflecting surfaces. Tabs provided at the edges of the shields are bent inward, so the shields do not touch the glass within the reflecting apertures. The prisms fit into "flat-bottomed" recesses machined into both sides of the shelf. These recesses are contoured as racetrack shapes to fit the prism hypotenuse faces, thereby constraining rotation and translation of the prisms under the straps. Extra restraint is afforded in this particular design by injecting an elastomeric material with a syringe into the glass-to-metal gaps along the straight edges of the prisms. During and for a while after WWII, rubber cement was used. Today, a sealant such as a room-temperature-curing room-temperature-vulcanizing elastomer is used as an elastomeric restraint.

Another example of a nonkinematic clamped prism mounting is the periscope head prism arrangement shown in Figure 7.27. This subassembly is also from a military optical instrument and

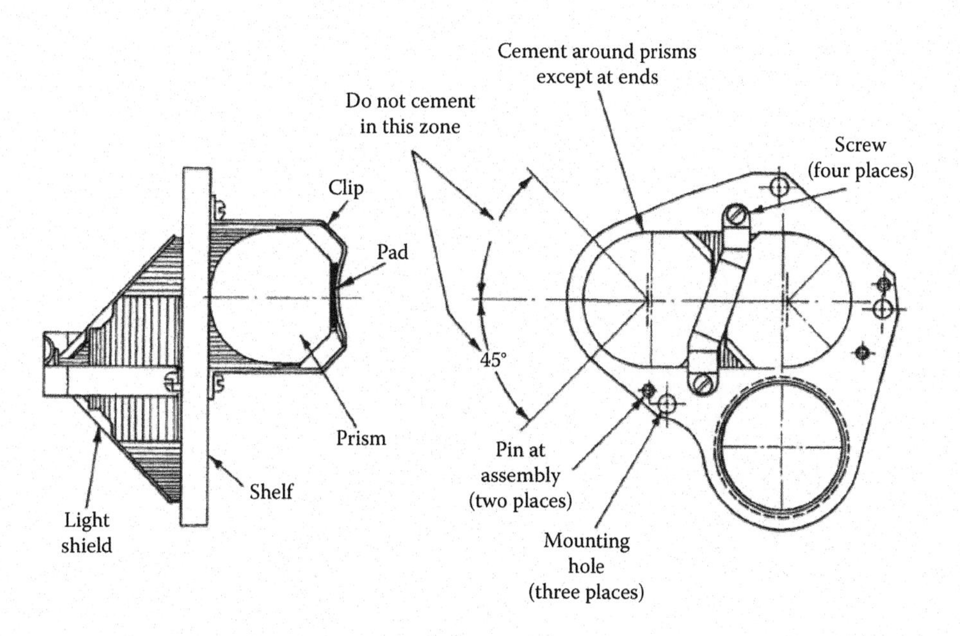

FIGURE 7.25 Schematic view of a typical strap mounting for a Porro prism-erecting subassembly used in telescopes and binoculars. (Adapted from Yoder, P.R. Jr., Non-image forming optical components, *Proc. SPIE*, 531, 206, 1985.)

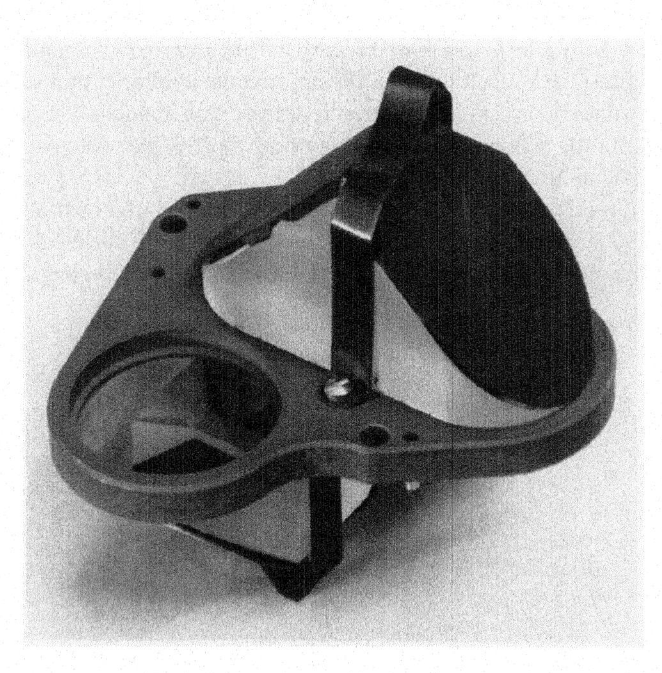

FIGURE 7.26 Photograph of a strap-mounted Porro prism-erecting subassembly of the type shown in Figure 7.25.

uses a modified single Dove prism with face angles of 35°, 35°, and 110° in lieu of the conventional 45°, 45°, and 90°. The prism can be tilted about a horizontal transverse axis to scan the line of sight in elevation from the zenith down to about 20° below the horizon.

The prism is held in its cast aluminum mount by four spring clips attached to the cast aluminum mount with screws adjacent to the entrance and exit prism faces. The edges of the reflecting surface

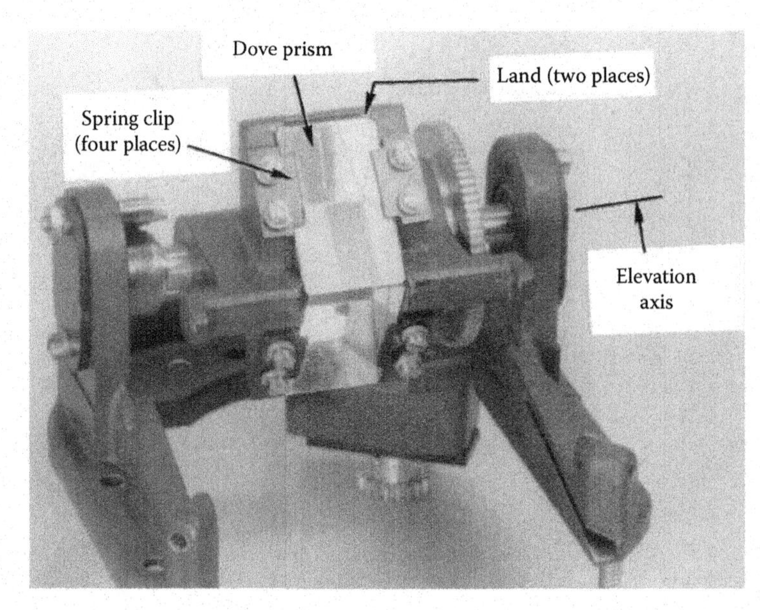

FIGURE 7.27 Clamped Dove prism of the type used in the elevation scanning head subassembly of a military periscope.

(hypotenuse) of the prism rest on narrow, nominally parallel lands machined into the casting. The prism faces protrude slightly a little less than 0.5 mm (0.02 in.) above the mount, and the undersides of the springs are stepped back locally so that the approximately proper preload is obtained when the springs bottom against the mount. Once centered, the prism cannot slide parallel to its hypotenuse surface because of the convergence of the clamping forces. The vector sum of these forces is nominally normal to the mounting surface.

Figure 7.28a illustrates the scanning function of the prism in the instrument shown in Figure 7.27. Rotation occurs about a horizontal axis located, as shown, a short distance below the prism center and slightly behind the reflecting surface. This motion is usually limited optically by vignetting of

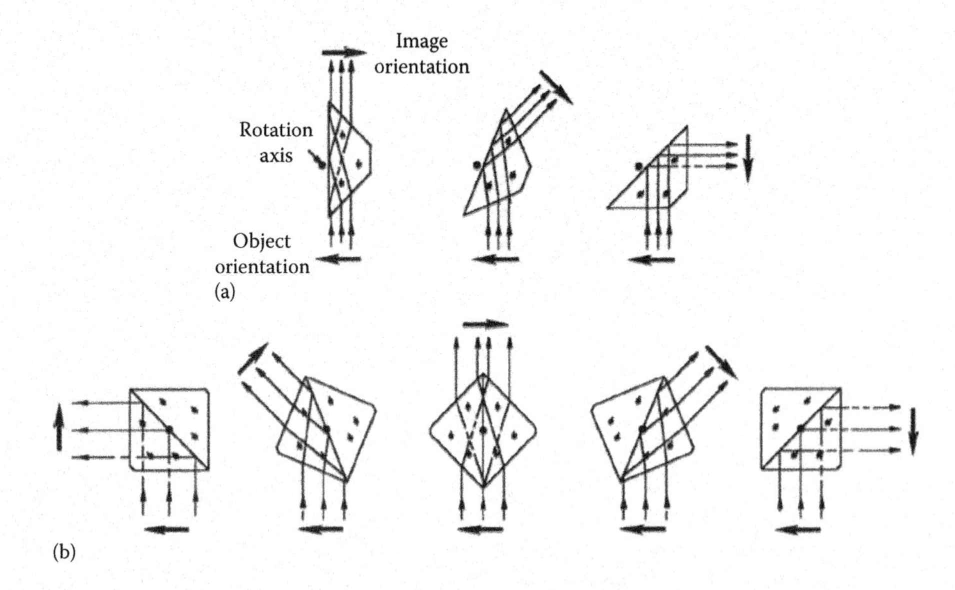

FIGURE 7.28 (a) Beam scanning with a Dove prism; (b) beam scanning with a double-Dove prism.

the refracted beam at the extreme angles. Mechanical stops are built into the instrument to limit physical motion, so that the vignetting at the end points is acceptable for the application.

Figure 7.28b shows beam scanning using a double-Dove prism. The total scan angle can exceed 180°. The transmitted beam can here be twice as large as that scanned by the Dove-type prism of the same aperture. The rotation axis is now located at the prism geometric center.

The most popular types of prisms used for image rotation or derotation about the axis are the Dove, double-Dove, Pechan, and delta. To function successfully, all these prisms must be mounted securely yet be capable of adjustment at the time of assembly to minimize lateral image motion during operation.

In Figure 7.29, we see a sectional view of a representative Pechan prism mounting.[6] If this type of prism is used in a collimated beam, it requires only angular adjustment of the optical axis relative to the rotation axis. In the application for this prism, the beam was converging, so both angular and lateral adjustments were needed. Bearing wobble would then cause angular errors. To minimize this, class 5 angular contact bearings, mounted back to back, were oriented with factory-identified high spots, matched, and then preloaded. Runout over 180° rotation was typically measured to be about 7.6 μm (0.0003 in.). The bearing axis was adjusted laterally by fine-thread screws (not shown) that permitted centration with respect to the optical system axis to better than 12.7 μm (0.0005 in.). The prism was adjusted laterally within the bearing housing on the plane of refraction by sliding it against a flat vertical reference surface with fine-thread screws pressing against the reflecting surfaces through pressure pads. A spherical seat with its center of rotation at the intersection of the hypotenuse face with the optic axis (to minimize axis cross coupling) was provided for angular adjustment. The tilt adjustment screws indicated in the figure controlled this movement.

In the mountings depicted in Figures 7.25 through 7.29, the prisms are pressed against machined surfaces on the mounts. Because it is virtually impossible for these surfaces to be as flat as the polished glass, contact will occur at the three highest points on the machined surfaces. Typically, these points are not directly in line with the clamping forces, so moments are applied to the glass, and surface distortions might occur. The designs work in their intended visual applications because the prisms are stiff enough not to bend significantly. If possible, it would be advisable to create

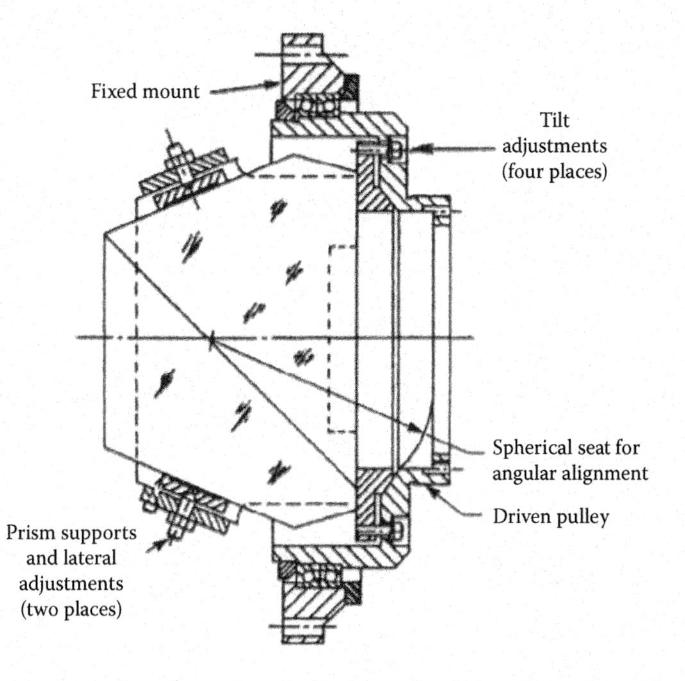

FIGURE 7.29 Optomechanical configuration of a Pechan prism image derotation subassembly. (Adapted from Delgado, R.F. The multidiscipline demands of a high performance, dual channel projector, *Proc. SPIE*, 389, 75, 1983.)

accurately machined coplanar lands on the metal surfaces to ensure that the clamping force vectors pass perpendicularly through these lands. Mountings should not depend on frictional forces between the glass and reference surfaces to constrain lateral motions. Independent positive lateral constraints should be provided, if possible.

7.5 MOUNTING PRISMS BY ADHESIVE BONDING

7.5.1 General Considerations

One popular technique for mounting prisms involves glass-to-metal bonds using adhesives. This design technique generally results in reduced interface complexity and compact packaging while providing mechanical strength adequate for withstanding the shock, vibration, and temperature changes characteristic of most military and aerospace applications. The technique is also frequently used in consumer applications because of its inherent simplicity and low cost.

Critical aspects of a glass-to-metal bond are the characteristics of the chosen adhesive; area and thickness of the adhesive layer; cleanliness of the surfaces to be bonded; dissimilarity of coefficients of thermal expansion (CTEs) for the adhesive and the materials bonded; environment that the bonded assembly will experience in storage, transportation, and use; and care with which the bonding operation is performed. The manufacturer's recommended procedures for applying and curing adhesives should be followed unless special requirements of the application dictate otherwise. The manufacturer should be consulted if there is any question about process or material suitability for any particular application. Experimental verification of the choice of adhesive and methods to be used in bonding is advisable in critical applications.

For maximum bond strength, the adhesive layer should have a specific thickness. In the case of 3M Epoxy EC-2216 B/A, a commonly used adhesive, experience has indicated a thickness of 0.075–0.125 mm (0.003–0.005 in.) to be appropriate. One method of ensuring achievement of the right layer thickness is to place spacers (wires, plastic fishing line, or flat shims) of the specified thickness at three places symmetrically located on one bonding surface (usually the metal one) before applying the adhesive. Care must be exercised to register the glass part against these spacers during assembly and curing. The adhesive should not extend between the spacers and either part to be bonded since this could affect the adhesive layer thickness. Another technique for obtaining a uniform thin layer of epoxy between the glass and metal surfaces is to mix a small quantity of small glass or plastic beads into the epoxy before applying it to the surfaces to be bonded. When the parts are clamped securely together, the largest beads contact both faces and hold those surfaces apart by their diameters. These beads can be obtained in a variety of closely controlled diameters. If the required size can be obtained, the achievement of specific thickness joints is relatively simple.

A thin adhesive bond is stiffer than a thick one. The Young's modulus of the adhesive can vary with the ratio of diameter to thickness of the bond by more than a factor of 100. Stiffness also depends on Poisson's ratio for the adhesive, which varies from ~0.430 (for some epoxies) to ~0.499 (for some elastomers). The thermal expansion of the elastomer also depends on diameter-to-thickness ratio. These parameters can affect the ability of the bond to give under shock and vibration as well as when the temperature changes.

The minimum area of the bond A_{min} for a prism of weight W is given by

$$A_{min} = \frac{W a_G \, SF}{\sigma_{eYS}}, \tag{7.32}$$

where σ_{eYS} is the yield strength of the adhesive.

Normally, the yield strength σ_{eYS} of the adhesive is the shear strength or tensile strength. For most adhesives, the shear and tensile strengths are equivalent. A minimum SF = 2 or preferably SF = 4 is recommended to allow for variations in adhesive properties.

Example 7.4

A roof pentaprism made of BK7 glass with a face width $A = 38.1$ mm is bonded to a metal bracket. The prism is subjected to an acceleration of $a_G = 250g$, and an SF = 4 is required. The density of Schott BK7 is 2510 kg/m³, and the yield strength of the adhesive is $\sigma_{eYS} = 14$ MPa. What is the diameter of a circular bond?

The prism mass W is

$$W = 1.795(38.1\times10^{-3}\,\text{m})^3(2.51\times10^{-3}\,\text{kg/m}^3) = 0.249\,\text{kg}.$$

The bond area is given by Equation 7.32:

$$A_{min} = \frac{4(0.249\,\text{kg})(250)(9.807\,\text{m/s}^2}{14\times10^6\,\text{Pa}} = 174.4\times10^{-6}\,\text{m}.$$

The bond diameter d is given by

$$d = \left(\frac{4}{\pi}174.4\times10^{-6}\,\text{m}^2\right)^{\frac{1}{2}} = 14.9\times10^{-3}\,\text{m}.$$

Because the dimensional changes of the adhesive bond during curing (shrinkage) and during temperature changes (expansion or shrinkage depending upon the algebraic sign of the temperature change) are proportional to the dimensions of the bond, it is advisable not to make the bond area too large. If a large area is necessary to hold a heavy optic, especially if high accelerations are expected, the bond should be divided into a group of smaller areas, such as a triangular or ring pattern of circles or spots. The aggregate area of the spots should equal that of the computed single bond.

If the bond area is small relative to the contact area, the "spot bond" equation gives the shear stress from the difference in CTE between prism and mount with temperature. This equation assumes that the stiffness of the adhesive is small compared to that of the prism or mount. For most adhesives and mounts, this assumption is valid, since the elastic modulus of a relative stiff epoxy adhesive is less than 1% of that of glass. For initial analysis, the tensile stress is considered equal in glass and metal. As a rough rule of thumb for most glasses and adhesives, the tensile stress τ_e from the temperature change ΔT should be less than 10 MPa and ideally less than 7 MPa, with an SF = 2. Tensile stress in a spot bond is given by

$$\tau = \frac{(\alpha_1 - \alpha_2)\,\Delta TG\,\tanh(\beta r)}{\beta h_e}, \tag{7.33}$$

$$\beta = \left\{\frac{G_e}{h_e}\left[\frac{(1-\nu_1^2)}{E_1 h_1} + \frac{(1-\nu_2^2)}{E_2 h_2}\right]\right\}^{\frac{1}{2}}, \tag{7.34}$$

where α_1 and α_2 are the CTEs of the prism and mount materials, E_1 and E_2 are the elastic moduli of the prism and mount materials, ν_1 and ν_2 are the Poisson's ratios of the prism and mount materials, h_1 and h_2 are the thicknesses of the prism and mount, ΔT is the change in temperature, G_e is the shear modulus of the adhesive, and h_e is the adhesive bond thickness.

An example of an unnecessarily large bond applied in a prism to baseplate mounting is illustrated in Figure 7.30. The prism was a fused silica beam splitter cube with $A = 35.00$ mm (1.375 in.).

FIGURE 7.30 Fused silica beam splitter cube attached with an excessively large bond area to a titanium base. The prism fractured during low temperature testing. The design was salvaged by substituting three smaller bonds plus one central bond in a triangular pattern. (From Yoder, P.R. Jr., *Mounting Optics in Optical Instruments*, 2nd ed., SPIE Press, Bellingham, WA, 2008.)

The bond was 3M EC-2216 adhesive, with a diameter of 30.1 mm (1.22 in.) and thickness of 010 mm (0.004 in.). The prism base had been ground after cementing so as to present a flat surface without any step to avoid adhesive layer thickness variations across the bond. The CTEs of the two materials were quite dissimilar ($\alpha_M = 8.82 \times 10^{-6}$ °C^{-1} [4.90×10^{-6} °F^{-1}]) and ($\alpha_G = 5.76 \times 10^{-7}$ °C^{-1} [0.32×10^{-6} °F^{-1}]) so the glass fractured when exposed to a temperature of approximately −42°C (approximately −44°F) or a temperature drop of about $\Delta T \approx 67$ K in design verification tests.

Analysis of the design shows (1) that the bond size was much larger than necessary to hold the components together under the expected worst-case acceleration, (2) that differential contraction at low temperature over this large bond introduced sufficient stress in the glass to cause the fracture, and (3) that reducing the bond area to four smaller circular bonds with three at the apexes of an equilateral triangle and the remaining one at the center of the triangle would eliminate the fractures at low temperatures while still providing sufficient strength to withstand the expected acceleration. These conclusions were substantiated by further testing of the new bond configuration. Example 7.5 provides an analysis of this design.

Example 7.5

A cubical fused silica prism with a face width $A = 35$ mm is bonded to a titanium base with 3M EC-2216 epoxy and is subjected to a 90 K decrease in temperature. The base is 26.7 mm thick. The bond thickness is 100 μm thick and is circular, with a 30 mm diameter. The elastic modulus of EC-2216 is $E_e = 690$ MPa, and Poisson's ratio is $\nu_e = 0.43$. For titanium, $\alpha_{Ti} = 8.8 \times 10^{-6}$ K^{-1}, $E_{Ti} = 114$ GPa, and $\nu_{Ti} = 0.31$. For fused silica, $\alpha_{FS} = 580 \times 10^{-9}$ K^{-1}, $E_{FS} = 73$ GPa, and $\nu_{FS} = 0.17$. What is the thermal stress in the original 30 mm diameter bond? If the bond is replaced with four smaller circular bonds, each with a 6.35 mm diameter, what is the stress in the smaller bonds?

The shear modulus G_e of the adhesive is given by

$$G_e = \frac{690 \times 10^6 \text{ Pa}}{2(1+0.43)} = 241 \times 10^6 \text{ Pa}.$$

Using Equation 7.34, find the parameter β

$$\beta = \left\{\frac{241\times10^6\,\text{Pa}}{100\times10^{-6}\,\text{m}}\left[\frac{1-(0.17)^2}{(73\times10^9\,\text{Pa})(35\times10^{-3}\,\text{m})}+\frac{1-(0.31)^2}{(114\times10^9\,\text{Pa})(26.7\times10^{-3}\,\text{m})}\right]\right\}^{\frac{1}{2}} = 40.42\text{m}^{-1}.$$

For the original 30 mm bond diameter, the shear stress τ_{max} is given by Equation 7.33:

$$\tau_{max} = \frac{[(8.8\times10^{-6}\,\text{K}^{-1})-(580\times10^{-9}\,\text{K}^{-1})](90\text{K})(241\times10^6\,\text{Pa})\tanh[(40.42\text{m}^{-1})(15\times10^{-3}\,\text{m})]}{(40.42\text{m}^{-1})(100\times10^{-6}\,\text{m})} = 23.9\times10^6\,\text{Pa}.$$

A stress of about 24 MPa is well above the nominal 14 MPa yield stress of EC-2216 and is also above the normal 10 MPa strength of fused silica. As discussed earlier, the prism mount did fail at low temperature. For the smaller bond, diameter d = 6.35 mm:

$$\tau_{max} = \frac{[(8.8\times10^{-6}\,\text{K}^{-1})-(580\times10^{-9}\,\text{K}^{-1})](90\text{K})(241\times10^6\,\text{Pa})\tanh[(40.42\text{m}^{-1})(6.35\times10^{-3}\,\text{m})]}{(40.42\text{m}^{-1})(100\times10^{-6}\,\text{m})} = 5.64\times10^6\,\text{Pa}.$$

A stress of 5.6 MPa is below the yield strength of both EC-2216 and fused silica. The prism mount was successful with the smaller bond diameter.

7.5.2 Typical Applications

Figure 7.31 shows a mounting for a large Porro prism used in a military telescope.[7] The prism is cantilevered from a nominally vertical mounting surface. The prism is made of Schott SKI6 glass and weighed 2.200 lb (1.000 kg). It is bonded to a type 416 stainless steel bracket with 3M EC-2216 B/A adhesive with σ_{eYS} ~ 17.2 MPa (2500 psi) over the maximum available surface area on one side of the prism. The CTEs of the glass, metal, and adhesive are α_G = 7.0 parts per million (ppm)/°C, α_M = 9.9 ppm/°C, and α_e = 103 ppm/°C. The final bonding surface finishes are fine ground (KH grit) on the glass and 63 on the metal.

Figure 7.32 shows a bond configuration appropriate to a two-part prism, in this case a Pechan prism. The central airspace was established temporarily with thin spacers and the two prisms attached to each other mechanically by a glass plate bonded to both prisms on the side opposite the glass-to-metal bond (not shown). The adhesive for the bond to the external structure was applied to

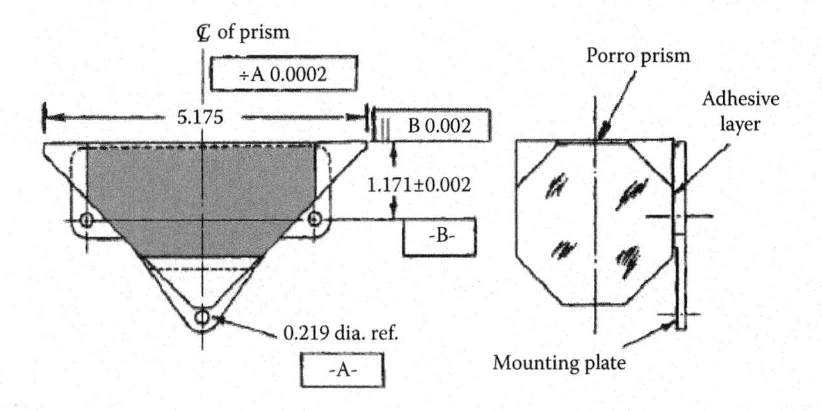

FIGURE 7.31 Design for bonding a large Porro prism in cantilevered fashion to a bracket so as to withstand large acceleration. (From Yoder, P.R. Jr., Non-image forming optical components, *Proc. SPIE*, 531, 206, 1985.)

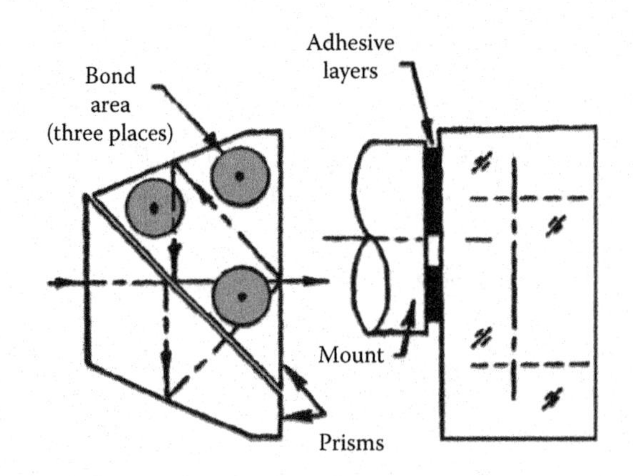

FIGURE 7.32 Triangular distribution of circular bonds on one prism of a two-component (Pechan) prism subassembly. (From Yoder, P.R. Jr., *Mounting Optics in Optical Instruments*, 2nd ed., SPIE Press, Bellingham, WA, 2008.)

only one prism because, as mentioned earlier, the ground surfaces on the adjacent components cannot be guaranteed to be coplanar. These bonds were arranged in triangular array.

7.5.3 Double-Sided Prism Supports

In some optomechanical designs, a prism is bonded to supports on both sides. A typical design involving a Schmidt prism, shown in Figure 7.33, illustrates the concept.[8] The prism was supported between nominally parallel inside surfaces of a U-shaped aluminum bracket with 3M EC-2216 B/A epoxy injected through centrally located holes (marked *P*) in the arms of the frame. It was thought that this arrangement would be stronger than a single-sided bond joint.

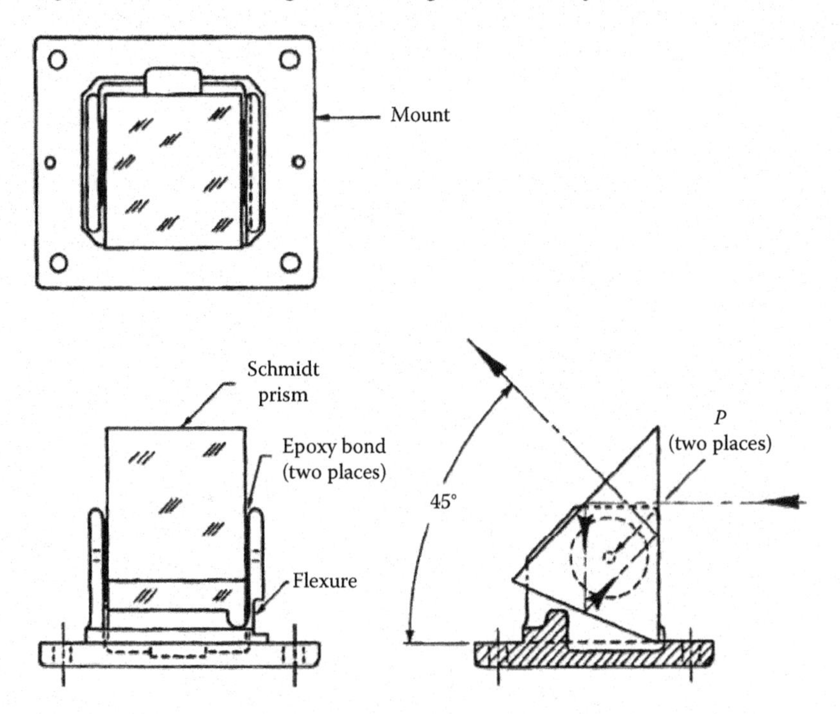

FIGURE 7.33 Schematic diagram of a Schmidt prism bonded on both sides to a U-shaped mount. (Courtesy of R. Willey, 1991.)

Initially, tests of several subassemblies at low temperature caused fractures to occur in the prisms at the tops of the bonded areas. Analysis indicated that differential contraction of the metal with respect to the glass caused the stiff arms to rotate about the hardened epoxy at the bottom of the bond area and to exert large tension forces at the top of each bond. The design was improved by locally reducing the thickness of one arm (indicated as "Flexure" in the figure) below the bond. This allowed the arm to flex with temperature changes sufficiently to reduce the stress within the bond, thereby solving the problem.

Another approach to double-sided mounting for prisms used successfully in the Netherlands is shown in Figure 7.34.[9] The prism was a crown glass beam splitter cube; it was to be bonded between the arms of a U-shaped frame. In this case, the frame was stainless steel, so thermal mismatch was minimal. An alternate technique to injecting epoxy through holes in the side arms directly into the interface region was sought after low-temperature tests produced fractures of the glass centrally under the injection holes. This was attributed to localized contraction of the plug of epoxy filling the injection holes. In a later approach, shown in Figure 7.34a and b, the prism was properly aligned (by fixturing) inside the frame and two short stainless steel plugs were inserted through slightly oversized holes in the arms. Each plug was centered temporarily to the holes with fixtures or spacers. The ends of the plugs were bonded in situ to the sides of the prism with layers of epoxy (labeled "First bond" in the figure). A small quantity of glass beads of appropriate diameter mixed into the epoxy assured proper bond thickness. After the plug-to-prism bonds were cured, the plugs were epoxied to the frame arms (labeled "Second bond"). This design reduces the requirement for close control of prism thickness between the bonding faces, precise parallelism of those faces, or precise parallelism between the frame arms.

Figure 7.34c and d shows a modification of the Beckmann design, in which only one plug is used, the prism being bonded directly to one arm of the mount during the initial bonding step. The end result is similar to that obtained with the approach shown in Figure 7.34a and b. Note that it is

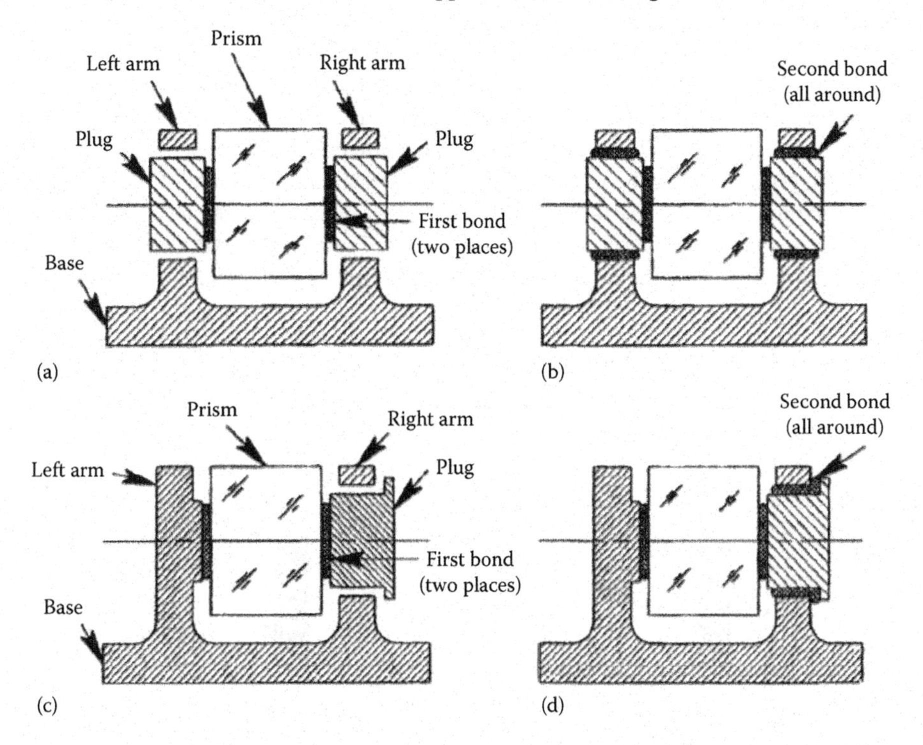

FIGURE 7.34 Two concepts for bonding a prism to parallel arms on a base. In (a) and (b), two plugs are bonded to the prism through holes in the arms. After alignment, the plugs are bonded into the arms. In (c) and (d), the prism is aligned and bonded to one arm. After curing that joint, a plug is bonded to the other side of the prism through a hole in the arm. After another cure cycle, the second plug is bonded to the second arm. (Courtesy of L. H. J. F. Beckmann, 1990.)

necessary that the bonding surface of the arm to which the prism is to be bonded directly is accurately positioned relative to the desired prism location and orientation because the only adjustments available in preparation for bonding are rotation about an axis perpendicular to that surface and two orthogonal translations parallel to that surface.

When using an epoxy bond between glass and a rigid material such as a metal, care should be exercised during application of the epoxy to ensure that fillets of excess adhesive are not formed around the glass-to-metal joint. Shrinkage during curing along the diagonal face of the fillet will tend to stress the glass. This shrinkage has been known to pull chunks of glass from the optic at a low temperature. The view in Figure 7.35a is a diagram of an undesirable fillet. The view in Figure 7.35b shows a preferred configuration of the adhesive joint. This configuration can be achieved by carefully controlling the volume of adhesive applied.

A different way to bond prisms is illustrated in Figure 7.36.[10] The individual prisms (items 466 and 467 in the figure) of a Schmidt–Pechan erector subassembly are inserted into close-fitting seats molded inside the filled-plastic housing of a commercial binocular. A plastic spacer (item 414) approximately 1.0 mm (0.04 in.) thick is positioned between the prisms to establish the airspace. An aperture in this spacer allows the light beam to pass through. The prisms are provisionally secured to the housing with a few dabs of ultraviolet curing adhesive applied through openings in the housing walls (see triangular areas on prism sides). After the temporary bonds have cured and alignment is confirmed, the prisms are firmly secured by adding several beads of urethane adhesive through the same wall openings. The slight resiliencies of both the housing and the adhesive accommodate the differential thermal expansion characteristics of the adjacent materials, thereby minimizing potential problems due to shrinkage in the adhesive beads. With precision-molded structural members and built-in reference surfaces, adjustments are not required.

Figure 7.37 shows two photographs of an assembly comprising a Porro prism-erecting system and a rhomboid prism mounted in a plastic housing by the same two-step bonding technique just described. In this design, one Porro prism is attached with temporary and (later) more permanent, but slightly flexible beads of adhesive to its plastic bracket. The view in Figure 7.37a shows that this bracket slides on two parallel metal rods attached to the main portion of the subassembly containing the second Porro prism. Sliding the movable prism provides axial movement of the first prism relative to the second for adjusting the focus of the associated optical instrument. The adhesive beads securing the prisms are most clearly shown in the view in Figure 7.37b, which shows the movable subassembly. Minimization of the number of components and ease of assembly are prime features

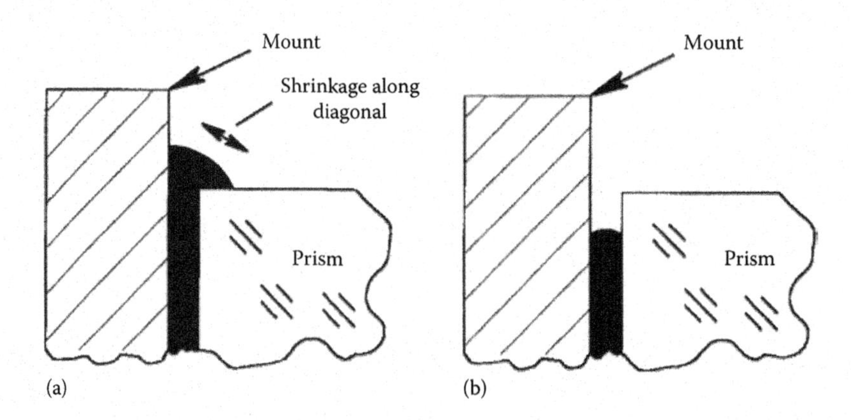

FIGURE 7.35 View (a) illustrates an undesirable adhesive joint configuration with a fillet of excess epoxy at the edge. Shrinkage along the diagonal of the fillet during cure or at low temperature can distort or fracture the prism. View (b) shows a more desirable configuration without a fillet.

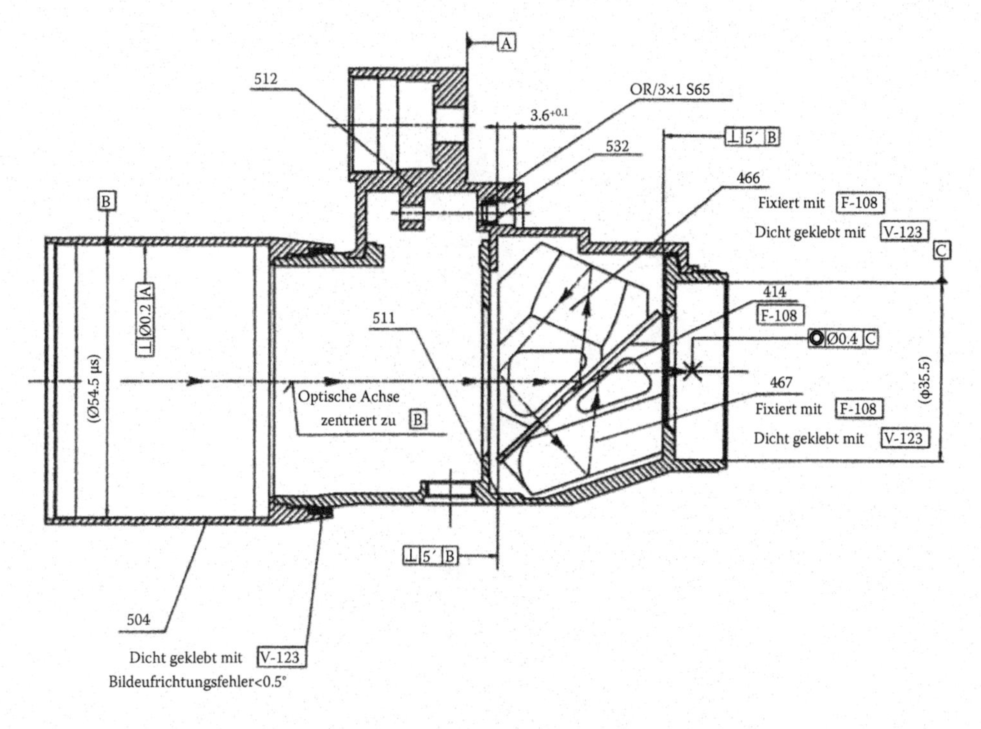

FIGURE 7.36 Schmidt–Pechan prism-erecting system fitted into a telescope housing and secured with beads of adhesive applied through access holes in the structure. (Courtesy of K. Seil, 1997.)

of this design. Customer acceptance of products assembled by this technique has demonstrated the durability and adequacy of the optomechanical performance achievable with this type of assembly.

PROBLEMS

1. A dove prism is made of Schott BK7 glass, with A = 38.1 mm, a = 1.905 mm, and B = 177.1 mm. The density of BK7 is ρ = 2510 kg/m^3. If the prism glass is changed to NSF1 with a higher index, the prism length becomes shorter, B = 152.9 mm, while A and a are the same. If the density of NSF1 is 3030 kg/m^3, is the prism mass reduced with the new glass? *Answer:* The mass of the BK7 prism is 0.542 kg, while that of the NSF1 is 0.537 kg. The NSF1 prism is slightly lighter.
2. What preload force F_P is needed for an Abbe–Koenig prism with A = 40 mm if the prism is made of N-BAK4 glass, with ρ = 3030 kg/m$_3$, when the prism is subjected to $a_G = 12g$, with SF = 2? *Answer:* 169.7 N
3. A right-angle prism is mounted with a circular bond area on one triangular side; the bond area just touches the edges of the prism. The prism is subjected to an acceleration of a_G = 1200g, with SF = 2. The adhesive yield strength is σ_{eYS} = 14 MPa, and the prism is made of glass with ρ = 2510 kg/m^3. What is the biggest prism that can be mounted (dimension of face A)? Hint: Find the inscribed circle diameter of the bond area to determine stress. *Answer:* A = 128 mm
4. A pentaprism with A = 30 mm is made of Schott F2 glass with ρ = 3600 kg/m^3. The prism is subjected to accelerations in three axes. In the X direction, $a_{GX} = 75g$; in the Y direction, $a_{GY} = 50g$; and in the Z direction, $a_{GZ} = 80g$. What preload force F_P is required if SF = 4? *Answer:* F_P = 689.3 N

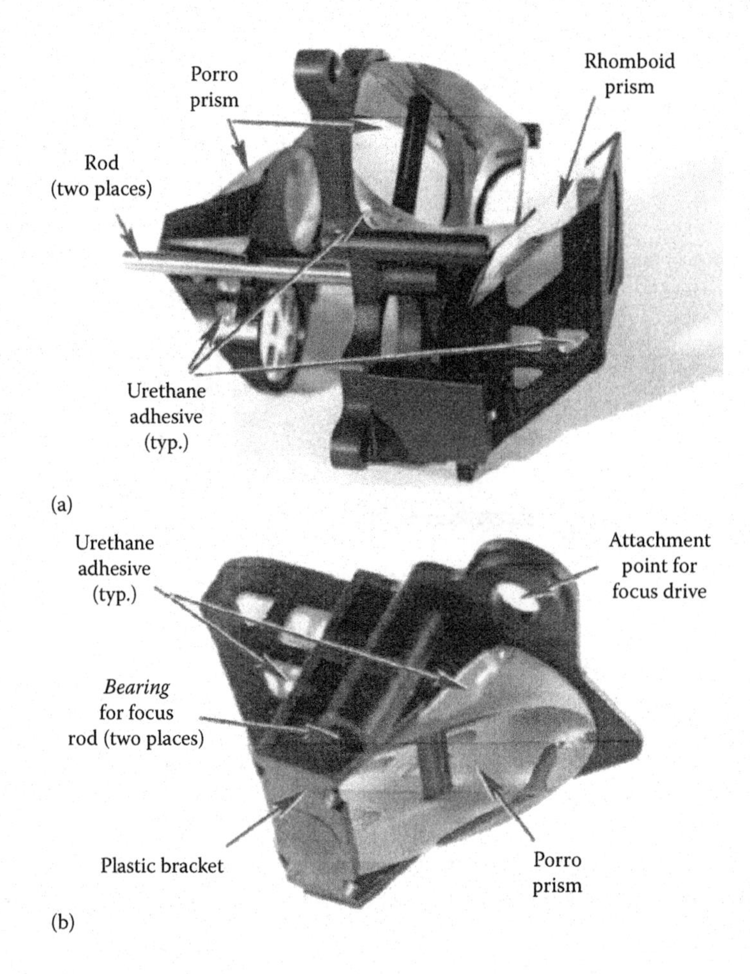

FIGURE 7.37 (a) Photograph of a Porro prism image-erecting system with variable separation and a rhomboid beam folding prism mounted in plastic housings by applying beads of adhesive to link the glass parts to the structure in the manner of Figure 7.36. (b) Close-up photograph of the movable Porro prism from the assembly shown in (a). The cured adhesive beads are clearly shown here. (Courtesy of K. Seil, 1997.)

5. A cubical prism with A = 25 mm is held down in its mount with an elastomeric pad that is in full contact with one face. The pad thickness is 8 mm, and the Shore A hardness of the elastomer is 40. If the pad is compressed 15%, what is the preload force on the prism? *Answer:* F_P = 358.7 N
6. A right-angle prism with A = 36 mm is held down with a 25 mm diameter circular elastomeric pad with Shore A hardness of 30. The prism mass is 0.53 kg, and it is subjected to an acceleration of $a_G = 15g$, with an SF = 4. What is the thickness of the pad necessary to provide the desired preload force F_P if the compression is 20%? *Answer:* pad thickness h = 6.27 mm
7. A right-angle prism with mass of 0.5 kg is held down with a circular elastomer pad that is 8 mm thick, with a diameter of 20 mm. The Shore A hardness of the pad is 35. What is the natural frequency f_n of the rigid-body motion of the prism in the direction normal to the plane of the elastomer pad? Hint: $f_n = (1/2\pi)(k/m)^{0.5}$. *Answer:* f_n = 92.6 Hz
8. A prism with a mass of 0.2 kg is subjected to acceleration of $a_G = 6g$, with an SF = 2. The prism is held in place by three cantilever springs made of 17-7 stainless steel that are 40 mm long and 3 mm wide. The yield strength of 17-7 stainless steel is σ_{YS} = 965 MPa,

and a factor of 2 in stress with respect to yield is necessary. What is the spring thickness? *Answer:* $h = 1.14$ mm

9. Using the information from problem 8, what is the natural frequency of the rigid-body motion of the prism in the direction normal to the plane of bending of the springs? For 17-7 stainless steel, $E = 204$ GPa and $\nu = 0.3$. Hint: See problem 7 and find the spring stiffness. *Answer:* $f_n = 12.8$ Hz
10. Three BeCu cantilever springs hold a 0.15 kg prism in place. The prism is subjected to an acceleration of $a_G = 6g$, with an SF of 2.5. The beryllium copper springs are 1 mm thick and 9 mm long. For beryllium copper, $E = 128$ GPa, $\nu = 0.35$, and $\sigma_{YS} = 1.07$ GPa. If the bending stress in the springs is half of yield, what is the spring width? What is the spring deflection? What is the spring angle in degrees? *Answer:* $b = 0.742$ mm, $\delta = 1.78$ mm, and $\varphi = 1.89°$
11. A straddling spring holds a right-angle prism with a face width of $A = 35$ mm against its mount. In service, the prism is subjected to acceleration of $a_G = 75g$ and an SF = 2 is required. The prism is made of Schott N-BAK4, with $\rho = 3050$ kg/m^3. The straddling spring is 50 mm long and 1 mm thick. The spring is made of AISI 301 stainless steel, A666, half-hard, with a yield strength $\sigma_{YS} = 758$ MPa, $E = 193$ GPa, and $\nu = 0.275$. If the maximum bending stress is half of yield, what is the spring width? How much does the spring deflect? *Answer:* $b = 9.52$ mm and $\Delta y = 0.378$ mm
12. Due to a mistake during assembly, a straddling spring is bent twice as much as the design deflection. The design deflection is 0.383 mm. The spring length is 40 mm, the width is 3 mm, and the thickness is 1 mm. The spring is made of 17-7 stainless steel, with $\sigma_{YS} =$ 965 MPa, $E = 204$ GPa, and $\nu = 0.3$. Is the yield strength of the spring exceeded? *Answer:* The yield strength is exceeded since $\sigma/\sigma_{YS} = 1.287$ GPa/965 MPa = 1.333.
13. A right-angle prism with $A = 32$ mm is made of BK7 glass, with $\rho = 2510$ kg/m^3. The prism is subjected to an acceleration of $a_G = 75g$, and an SF = 2 is required. The prism is to be bonded using 3M EC-2216 adhesive, with a yield strength of $\sigma_{eYS} = 14$ MPa. Three circular spot bonds are required. What is the diameter of each bond if the maximum stress in the adhesive is half of yield? *Answer:* $d = 1.92$ mm
14. A BK7 pentaprism is mounted to an aluminum plate by using three spot bonds of 3M EC-2216 adhesive. The spot bond diameter is 3 mm, and the thickness is 0.1 mm. The prism thickness is 25 mm, and the aluminum plate thickness is 15 mm. The prism is subjected to a temperature drop of 70 K after bonding. For BK7 glass, $E_{BK7} = 82$ GPa, $\nu_{BK7} =$ 0.205, and $\alpha_{BK7} = 8.2 \times 10^{-6}$ K^{-1}. For aluminum, $E_{Al} = 69$ GPa, $\nu_{Al} = 0.33$, and $\alpha_{Al} = 23 \times$ 10^{-6} K^{-1}. The shear modulus G_e of the adhesive is 241 MPa. What is the maximum shear stress? *Answer:* $\tau_{max} = 3.74$ MPa
15. A prism made of Schott N-BAK4 is mounted to a 416 stainless steel plate with 3M EC-2216 adhesive. The prism is 30 mm thick, and the stainless steel plate is 10 mm thick. There is a 75 K drop in temperature after assembly. The adhesive bond is 0.1 mm thick. For N-BAK4, $E_{BAK4} = 77$ GPa, $\nu_{BAK4} = 0.240$, and $\alpha_{BAK4} = 7 \times 10^{-6}$ K^{-1}. For 416 stainless steel, $E_{416} =$ 200 GPa, $\nu_{416} = 0.283$, and $\alpha_{416} = 9.9 \times 10^{-6}$ K^{-1}. The shear modulus G_e of the adhesive is 241 MPa. If the maximum shear stress in the adhesive is $\tau_{max} = 7$ MPa, what is the diameter of a circular spot bond? *Answer:* $d = 31.05$ mm

REFERENCES

1. Lipshutz, M.L., Optomechanical considerations for optical beam splitters, *Appl. Opt.*, 7, 2326, 1968.
2. Yoder, P.R. Jr., Mounting-induced contact stresses in prisms, *Proc. SPIE*, 3429, 7, 1998.
3. Young, W.C., *Roark's Formulas for Stress and Strain*, McGraw-Hill, New York, 1989.
4. Yoder, P.R. Jr. *Mounting Optics in Optical Instruments*, 2nd ed., SPIE Press, Bellingham, WA, 2008.

5. Yoder, P.R. Jr., Optical mounts, lenses, windows, small mirrors, and prisms, Chapter 6 in *Handbook of Optomechanical Engineering*, Ahmad, A., ed., pp. 151–210, CRC Press, Boca Raton, FL, 1997.
6. Delgado, R.F., The multidiscipline demands of a high performance dual channel projector, *Proc. SPIE*, 389, 75, 1983.
7. Yoder, P.R. Jr., Non-image forming optical components, *Proc. SPIE*, 531, 206, 1985.
8. Willey, R., private communication, 1991.
9. Beckmann, L.H.J.F., private communication, 1990.
10. Seil, K., private communication, 1997.

FURTHER READING

Department of Defense, Prism design, Chapter 13 in *Optical Design*, MIL-HDBK-141, Washington, DC, 1962.

Yoder, P.R. Jr., and Vukobratovich, D., *Opto-mechanical Systems Design*, 4th ed., CRC Press, Boca Raton, FL, 2015.

8 Factors Affecting Mirror Performance

8.1 INTRODUCTION AND SUMMARY

The most important performance specifications for a mirror are associated with the shape of and position of its optical surface. Secondary specifications include cost, weight, and technical risk. The design of a mirror and its mount requires balancing performance with other considerations. Simple approximate closed-form solutions are a useful means of rapidly evaluating performance during conceptual design. Following selection of conceptual designs, more detailed analysis methods, such as the finite-element method, can be used to more accurately determine performance.

8.2 TOLERANCES, ERROR BUDGETS, AND SUPERPOSITION

Mirror optical performance is specified by a pair of tolerances: optical figure and position. The optical figure tolerance determines the allowable deviation from the ideal form of the mirror. The position tolerance determines the location of the mirror surface with respect to the optical axis of the system. The optical figure of the mirror is influenced by the mirror material, its shape, size, and method of mounting, as well as environmental factors, including temperature. Mirror mounting controls the location of the mirror and is affected by dynamic effects such as mechanical shock and vibration, which are characteristics of the specified service environment.

Factors influencing the optical figure of the mirror are self-weight, excitation from vibration, temperature changes, temperature gradients, and dimensional instability. For optical applications, the deflection of a mirror from all causes is small, on the order of about 10^{-6} of the mirror diameter. Since the deflections are well below the nonlinear part of the stress–strain curve, the mirror material is within its linear range. It is therefore possible to deal with each source of deflection independently and add all deflections together using the principle of superposition. It is also assumed that the deflections are not correlated. For deflections that are not correlated, the overall deflection is found using the root-sum-square (RSS) of all deflections. Typically, the RSS is given in terms of the root-mean-square surface deflections. Root-mean-square is abbreviated as RMS for convenience.

The RMS of the optical surface deflections is defined as the square root of the sum of the deviations from the desired surface shape over the entire optical clear aperture of the mirror.[1] If the mirror optical surface deflections are measured at a series of points, as is standard practice in most metrology schemes, the overall RMS error of the optical surface is given by

$$\delta_{RMS} = \sqrt{\frac{1}{n}\sum_{i=1}^{n}\delta_i}, \tag{8.1}$$

where δ_{RMS} is the RMS deflection, n is the number of surface deflection measurements, and δ_i is the surface deflection measurement at some point i.

For mirrors the surface deflections are one-half the magnitude of the errors on the wave front. Wave front errors are often given in the form of an optical path difference, abbreviated to OPD. The OPD is the difference between the ideal and actual wave fronts. The OPD tolerances for the optical system must be converted to deflections for the mirror optical surface.

TABLE 8.1
Relationship between p-v and RMS

Aberration	p-v/RMS
Defocus	3.5
Third-order spherical at best focus	13.4
Fifth-order balanced spherical	57.1
Third-order coma	8.6
Third-order astigmatism, best focus	5.0
Smooth random errors	~5

The maximum surface deflection is given by the span of error from the height of the highest peak above the desired surface to the depth of the deepest valley below the desired surface. Maximum surface deflection is described in terms of peak to valley, which is abbreviated to p-v. The RMS surface deflections are related to the maximum surface deflection by using a constant:

$$\delta_{\text{p-v}} \approx C_{\text{RMS}}\delta_{\text{RMS}}, \tag{8.2}$$

where $\delta_{\text{p-v}}$ is the p-v deflection, C_{RMS} is a constant, and δ_{RMS} is the RMS deflection.

The value for the constant relating p-v to RMS is often given as being around 4.[2] Although it is a useful rule of thumb, this relationship is in error for many types of surface deflections or optical aberrations. This is shown in Table 8.1, which gives the relationship between p-v and RMS for several different types of optical aberrations.[3]

When designing a mirror, a top-level allocation of performance is provided in the form of an overall tolerance. This allocation is broken up to form a budget for the mirror performance. The budget values are determined using the RSS approach. It is possible to trade allocations, or shift some amount of tolerance from one part of the budget to another, again using the RSS method. The top-level budget items are the following:

1. Inertial mirror deflection, both static and dynamic
2. Mount-induced deflection from residual bending moments and forces
3. Temperature-induced deflection
4. Deflection from dimensional instability of the material
5. Optical figure errors from fabrication
6. Measurement errors associated with optical surface figure fabrication

The overall system performance budget includes allocations for individual elements. Each individual element is provided with its own budget. Alternately, the system budget is allocated to a variety of possible error sources, such as optical figure and thermal distortion. An example is the top-level wave front budget for the Keck telescope, shown in Table 8.2.[4]

Example 8.1

As given in Table 8.2, the overall angular error tolerance for the blur in the Keck telescope is 0.42 arcsec. The RSS of the primary mirror errors θ_P in Table 8.2 is

$$\theta_P = \sqrt{(0.24\,\text{arcsec})^2 + (0.15\,\text{arcsec})^2 + (0.10\,\text{arcsec})^2 + (0.12\,\text{arcsec})^2 + (0.10\,\text{arcsec})^2}$$

$$\theta_P = 0.34\,\text{arcsec}.$$

TABLE 8.2
Top-Level Angular Error Budget for Keck Telescope

Item	Angular Blur Diameter(arcsec)
Telescope total	0.42
Primary mirror total	0.34
Segment figure error	0.24
Segment thermal distortion	0.15
Segment support errors	0.10
Segment alignment, passive errors	0.12
Segment alignment, active errors	0.10
Secondary mirror total	0.16
Surface figure error	0.15
Support errors	0.05
Alignment errors	0.05
Misalignment	0.19
Tracking	0.05
Defocus	0.15
Other misalignment	0.10

The RSS of the secondary mirror errors θ_S in Table 8.2 is

$$\theta_S = \sqrt{(0.15\text{arcsec})^2 + (0.05\text{arcsec})^2 + (0.05\text{arcsec})^2}$$
$$\theta_S = 0.16\text{arcsec}.$$

The RSS of the misalignment errors θ_A in Table 8.2 is

$$\theta_A = \sqrt{(0.05\text{arcsec})^2 + (0.15\text{arcsec})^2 + (0.10\text{arcsec})^2}$$
$$\theta_A = 0.19\text{arcsec}.$$

Finally, the RSS of all of the errors for the telescope θ_T is

$$\theta_T = \sqrt{\theta_P^2 + \theta_S^2 + \theta_A^2}$$
$$\theta_T = \sqrt{(0.34\text{arcsec})^2 + (0.16\text{arcsec})^2 + (0.19\text{arcsec})^2}$$
$$\theta_T = 0.42\text{arcsec}.$$

Cost and schedule often prevent detailed design analysis of mirror performance, requiring the use of approximate methods. As a rough rule of thumb, errors of about 10% are acceptable in predicting mirror performance, and for preliminary design, errors as large as 20% may be allowable. Errors of this size are within normal budget margins. Caution is necessary since in a large error budget, the total increases as the square root of the number of errors. For example, in a budget with six items, each with an error of 10%, the total error would be $6^{½}(10\%) \approx 24\%$, which would exceed most design margins.

8.3 GRAVITY DEFLECTIONS

8.3.1 General Considerations

One of the most important sources of optical surface error is the self-weight deflection of a mirror, or deflection from the effect of gravity on the mass of the mirror material. This deflection must be considered even for satellite optics, which are made and tested on the surface of the Earth. In space, the mirror optical surface is different from that measured on the ground due to the phenomenon of gravity release. If the mirror performance is linear with respect to the magnitude of the gravity effect, the mirror optical surface figure after gravity release can be accurately predicted. In a terrestrial system, self-weight deflection is part of the performance budget if the direction of the gravity vector changes. An example of a system where the gravity vector direction changes in service is an astronomical telescope. Self-weight deflection is related to the fundamental frequency of the mirror, which determines response in a dynamic environment. An example of a system where frequency is important is a scanning mirror.

There are two components to mirror self-weight deflection: axial and radical. Axial deflections are associated with gravity acting perpendicular to the optical surface of the mirror, or parallel to the optical axis. Axial deflection is called "axis vertical" deflection since the worst case is for the mirror optical axis being in the vertical direction. Radial deflections are associated with gravity acting parallel to the optical surface of the mirror, or perpendicular to the optical axis. Radial deflection is called "axis horizontal" deflection since the worst case is for the mirror optical axis being horizontal with respect to the surface of the Earth.

If the mirror is subjected to a loading condition in which the direction of the gravity vector is at an angle to the optical axis, as shown in Figure 8.1, the resulting mirror surface deflections are given by

$$\delta_{\theta\text{-RMS}} = \sqrt{(\delta_{A\text{-RMS}} \cos\theta)^2 + (\delta_{R\text{-RMS}} \sin\theta)^2}, \tag{8.3}$$

where $\delta_{\theta\text{-RMS}}$ is the mirror RMS self-weight deflection at angle θ, $\delta_{A\text{-RMS}}$ is the mirror RMS self-weight deflection in the axis vertical position, $\delta_{R\text{-RMS}}$ is the mirror RMS self-weight deflection in

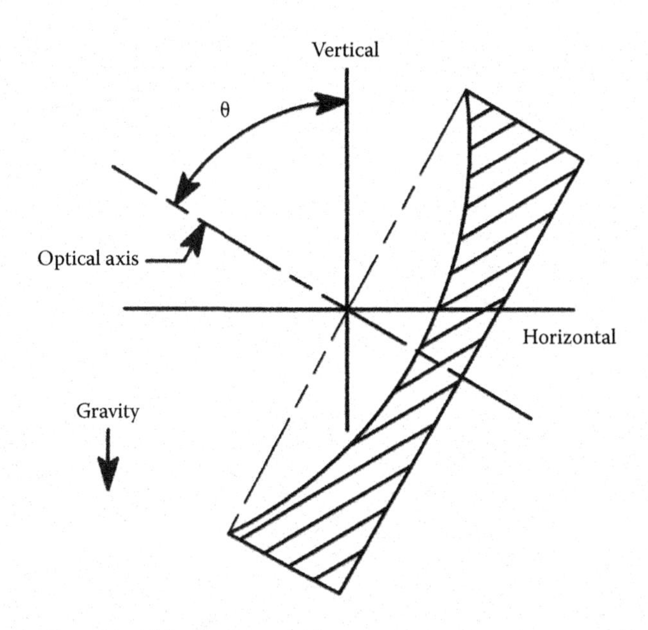

FIGURE 8.1 Mirror orientation with respect to gravity vector.

the axis horizontal position, and θ is the angle between the gravity vector and the mirror optical axis.

When the mirror is in the axis horizontal position, gravity produces a nonaxisymmetric surface deflection. For a concave mirror, this deflection is primarily astigmatism, with a change in the optical radius of curvature in the vertical direction, while the horizontal radius is relatively unchanged. A simple explanation is that gravity compresses the mirror in the vertical direction, changing the radius, while the horizontal direction is unaffected.

Schwesinger[5] showed that in the axis horizontal position, the approximate self-weight deflection of an axisymmetric concave mirror is given by

$$\delta_R = 2C_R \frac{\rho}{E} r^2, \tag{8.4}$$

where δ_R is the self-weight deflection of the mirror in the horizontal position, C_R is the support parameter, ρ is the mirror material density, E is the mirror material elastic modulus, and r is the mirror radius.

C_R varies with the diameter of the mirror, its thickness, and the optical radius of curvature. The mirror optical surface radius of curvature is proportional to the diameter and focal ratio (f_{no} or *f*/number) of the mirror, so the support parameter is also proportional to the focal ratio. The C_R parameter is determined by a series or polynomial approximation.

Self-weight deflection in the axial loading condition, or in axis vertical, is estimated using closed-form equations based on classic plate theory. The loading condition assumed for estimating axial plate deflection is a uniform or constant load over the plate surface, equivalent to the weight per unit area of the mirror. This loading condition is not equivalent to self-weight loading, where the mass is distributed through the thickness of the mirror.[6] Since the plate loading condition is not identical to the true loading condition, there is an error in calculating deflection, although this error is usually small, on the order of a few percent for most cases.

The general form of the equation for axial deflection of a mirror, based on classic plate theory, is

$$\delta_A = C \frac{qr^4}{D}, \tag{8.5}$$

where δ_A is the axial deflection, C is a parameter dependent on the support geometry, q is the weight per unit area of the mirror, r is the mirror radius, and D is the mirror flexural rigidity.

For a solid, constant-thickness axisymmetric mirror, the weight per unit area $q = \rho h$ and the flexural rigidity D is given by

$$D = \frac{Eh^3}{12(1-\nu^2)}, \tag{8.6}$$

where E is the elastic modulus of the mirror material, h is the mirror thickness, ν is the Poisson's ratio of the mirror material, and ρ is the density of the mirror material.

Substituting into the plate bending equation, the axial self-weight deflection δ_A for a solid mirror becomes:

$$\delta_A = C \frac{\rho(1-\nu^2)}{E} \frac{r^4}{h^2}. \tag{8.7}$$

TABLE 8.3
Inverse Specific Stiffness Values of Mirror Materials (Force Units)

Material	$\rho(1-\nu^2)/E$ ($10^{-9}\times$ m^{-1})
Beryllium I-220H	63.1
Silicon carbide	89.2
Silicon	144
Schott Zerodur	257
Molybdenum	282
Fused silica	290
Fused silica ULE	314
Pyrex 7740	333
Aluminum 6061	347
Titanium 6Al–4V	351
Stainless steel type 304	377
Invar 36	523
Copper OFC	661

In this equation, the material parameter associated with self-weight deflection is the ratio of the material density ρ to the elastic modulus E. This parameter is called the inverse specific stiffness. For common mirror materials, such as glass and aluminum, the inverse specific stiffness is nearly constant, with a range of values from 324×10^{-9} to 390×10^{-9} m^{-1}. This neglects variation in Poisson's ratio, which changes the inverse specific stiffness by less than 7% for most common materials. Materials with lower values of the inverse specific stiffness parameter include beryllium and silicon carbide. Table 8.3 gives the inverse specific stiffness parameter $\rho(1-\nu^2)/E$ for a variety of commonly used mirror materials.

Force per unit area is the loading condition for self-weight deflection. Force per unit area is the mirror mass per unit area times the acceleration from gravity. In calculating mirror deflections, there is a difference between mass and weight, the latter being mass times acceleration. For Système International (SI), the correct units for the density are newtons per cubic meter (N/m^3), not kilograms per cubic meter (kg/m^3), when calculating self-weight deflection.

Example 8.2

The mass W of a mirror is found from $W=\rho hA$, where A is the mirror area. For a circular mirror, show how the mass varies with material at constant deflection and with identical support.

The thickness h of the mirror is given from

$$h=r^2\sqrt{C\frac{1}{\delta_A}\frac{\rho(1-\nu^2)}{E}}.$$

Since $A=\pi r^2$ for a circular mirror, W is

$$W=\rho(\pi r^2)r^2\sqrt{C\frac{1}{\delta_A}\frac{\rho(1-\nu^2)}{E}}=\pi r^4\sqrt{C\frac{1}{\delta_A}\frac{\rho^3(1-\nu^2)}{E}}.$$

At constant deflection, mirror mass varies as the fourth power of diameter. Since $\rho^3(1-\nu^2)/E$ is nearly constant for most common materials, mirror mass is usually independent of material. This rule of thumb is not correct for materials with unusual values of inverse specific stiffness, such as beryllium, composites, and silicon carbide.

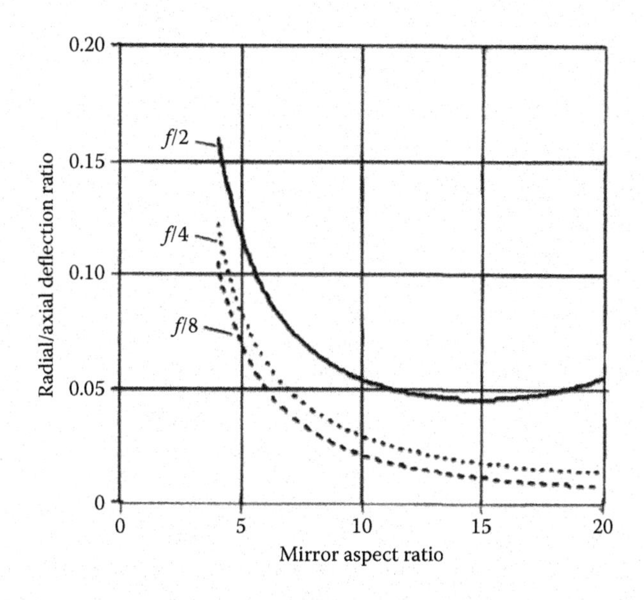

FIGURE 8.2 Ratio of radial to axial deflection for aspect ratios d/h between 4 and 20, and focal ratios f_{no} = $f/2$, $f/4$, and $f/8$ for solid circular mirrors made of Corning Pyrex 7740.

The axis horizontal component of gravity deflection is often neglected during conceptual design analysis of mirror performance, particularly for smaller mirrors. The ratio of the axial and radial self-weight deflections scales as the square of the mirror aspect ratio. This is shown by the following equation:

$$\frac{\delta_A}{\delta_R} = \frac{C_A \frac{\rho(1-\nu^2)}{E}\left(\frac{r}{h}\right)^2 r^2}{2C_R \frac{\rho}{E} r^2} = \frac{C_A}{C_R}\frac{(1-\nu^2)}{2}\left(\frac{r}{h}\right)^2. \tag{8.8}$$

Figure 8.2 gives the ratio between radial and axial deflections for concave axisymmetric mirrors made of Pyrex 7740, for aspect ratios varying from 4 to 20 and focal ratios varying from 2 to 8. A two-point radial support, with the support points 90° apart, and a three-point axial support, with three points equally spaced on a 64.5% diameter, is assumed in this analysis. The assumption of a two-point horizontal support is a worst case in that other horizontal supports produce less self-weight deflection. This figure shows that for small axisymmetric mirrors and aspect ratios above 6:1, the radial deflection is normally below 10%.

If the radial or axis horizontal deflection is neglected, the mirror self-weight deflection δ_θ at any angle θ between mirror axis and local vertical is given approximately by

$$\delta_\theta \approx \delta_A \cos(\theta). \tag{8.9}$$

8.3.2 Axial Gravity Deflection

If shear is neglected, the approximate p-v self-weight bending deflection of a constant-thickness circular mirror at any angle is given by

$$\delta_{\theta\text{p-v}} = C_A \frac{\rho(1-\nu^2)}{E}\frac{d^4}{h^2}\cos(\theta). \tag{8.10}$$

TABLE 8.4
Support Parameters for Different Mirror Mount Geometries

Support Geometry	C_A
Ring at 67.9% of diameter ($d_S/d = 0.679$)	2.38×10^{-3}
Six points equally spaced on 68.1% diameter ($d_S/d = 0.681$)	2.59×10^{-3}
Three points equally spaced at 64.5% diameter ($d_S/d = 0.645$)	19.9×10^{-3}
Edge simply supported	49.2×10^{-3}
Continuous support along diameter (scanning mirror)	59.2×10^{-3}
"Central support" (stalk or mushroom mount, $d_S/d = 0$)	77.5×10^{-3}
Three points equally spaced at edge ($d_S/d = 1.00$)	85.3×10^{-3}

Note: d_s is the support diameter, d is the mirror diameter, and h is the mirror thickness.

Table 8.4 gives the support parameter C_A for a variety of mounting geometries.

For the special case of the three-point axial support, the deflection coefficient C_A for any ratio of support diameter d_S/d is given approximately by the following empirical polynomial[7] (the average accuracy is 3% and the maximum error is about 9%):

$$C_A \cong .3830\left(\frac{d_S}{d}\right)^3 - 0.3626\left(\frac{d_S}{d}\right)^2 - 0.00795\left(\frac{d_S}{d}\right) + 0.0718, \tag{8.11}$$

where d_S is the support diameter and d is the mirror diameter.

Example 8.3

A small 200 mm diameter axisymmetric mirror with a 6-to-1 aspect ratio, made of 7740 Pyrex, is supported on three points equally spaced at 64.5% of the mirror diameter, with a 45° angle between optical axis and the local gravity vector. What is the p-v self-weight deflection in units of waves of a HeNe laser, with $\lambda = 633$ nm?

From Table 8.3 for 7740 Pyrex, $\rho(1 - \nu^2)/E = 333 \times 10^{-9}\ \mathrm{m}^{-1}$. The mirror thickness is $h = 200$ mm/6 = 33.3 mm. From Table 8.4, $C_A = 19.9 \times 10^{-3}$ and $\theta = 45°$:

$$\delta_A = C_A \frac{\rho(1-\nu^2)}{E}\frac{d^4}{h^2}\cos\theta$$

$$\delta_A = (19.9\times10^{-3})(333\times10^{-9}\mathrm{m}^{-1})\frac{(200\times10^{-3}\,\mathrm{m})^4}{(33.3\times10^{-3}\,\mathrm{m})^2}\cos(45°)$$

$$\delta_A = 6.8\times10^{-9}\,\mathrm{m}$$

$$\frac{\delta_A}{633\times10^{-9}\,\mathrm{m}} = 0.011.$$

The p-v self-weight deflection is 0.011 λ, $\lambda = 633$ nm.

8.3.3 Optimum Axial Support Locations

Minimizing self-weight deflection requires an optimum location for the support. For a continuous ring support at the optimum location, the deflection at the center of the mirror equals the deflection at the edge. Considering bending only, the optimum location d_s for a continuous ring support is about 0.679 of the outside diameter of the mirror, or $d_s/d \approx 0.68$.[8]

The location of optimum point supports is based on semi-infinite plate theory. For a semi-infinite plate, minimum self-weight deflection is with an array of point supports arranged as a pattern of equilateral triangles.[9] Self-weight deflection $\delta_{\text{p-v}}$ for a mirror of area A on this type of support geometry (p-v) is given by

$$\begin{aligned}\delta_{\text{p-v}} &= 4.95\times10^{-3}\frac{q}{D}\left(\frac{A}{N}\right)^2\\ &= 4.95\times10^{-3}\frac{\rho h}{\dfrac{Eh^3}{12(1-\nu^2)}}\left(\frac{\pi r^2}{N}\right)^2\\ &= 0.586\frac{\rho(1-\nu^2)}{E}\frac{r^4}{h^2}\frac{1}{N^2}.\end{aligned} \tag{8.12}$$

Real mirrors are not semi-infinite plates, and lack of support at the mirror edge increases the self-weight deflection substantially above the theoretical value. For an optimum three-point support, the actual self-weight deflection is about five times greater than the theoretical value. Self-weight deflection does not approach the theoretical value until the number of point support equals or exceeds 18. A modified form of the self-weight deflection relationship for optimum point supports where $3 \le N \le 36$, with empirical corrections for edge effects, is

$$\begin{aligned}\delta_{\theta\text{p-v}} &= \frac{3.9701}{N^{2.387}}\frac{\rho(1-\nu^2)}{E}\left(\frac{r}{h}\right)^2 r^2\\ \delta_{\theta\text{p-v}} &\cong \frac{1}{4N^{2.4}}\frac{\rho(1-\nu^2)}{E}\left(\frac{d}{h}\right)^2 d^2.\end{aligned} \tag{8.13}$$

Semi-infinite plate theory also does not indicate the location of the optimum point supports. Plate bending theory can be used to find optimum locations for the point supports. For a single ring of point supports, an optimum ring diameter can be found. The difference in deflection for a ring of six supports ($N = 6$) and a continuous ring is small. There is no reason therefore to use more than six supports on a single ring, although influencing the total axial self-weight deflection shear does not significantly affect the location of the optimum supports. Figure 8.3 shows the relative deflection of supports with 3, 4, 6, and 12 equally spaced point supports on a common diameter. For $N = 3$, the optimum ring location is $d_S/d = 0.645$, and for $N = 6$, the optimum ring location is $d_s/d = 0.681$.

8.3.4 Factors Influencing Axial Self-Weight Deflection

Factors influencing the accuracy of the plate bending equation are a central hole (for circular mirrors), variations in thickness, mounting pad size, and shear. Central holes are found in primary mirrors for Cassegrain systems and other folded axisymmetric optical systems. The change in axial self-weight deflection for a central hole is small, on the order of 10% when the central hole diameter is less than 50% of the mirror diameter. This effect is shown in Figure 8.4, which gives the deflection ratio for circular mirrors with and without central holes for different materials and for central hole sizes up to 60% of the diameter, for an edge ring support.

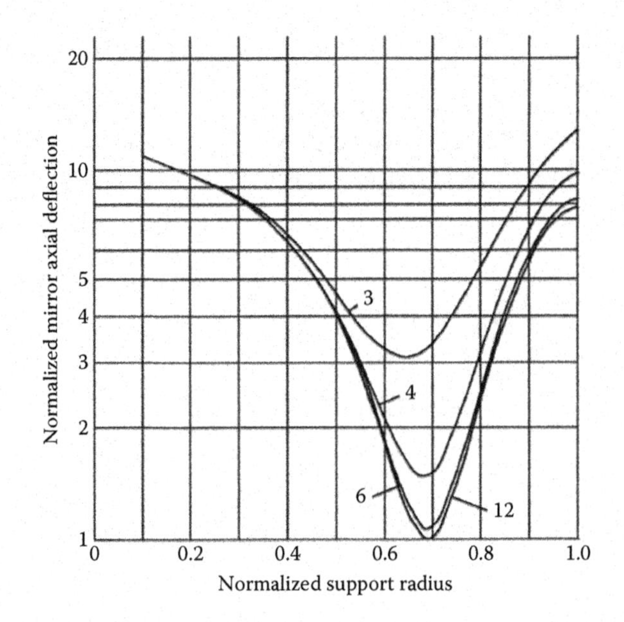

FIGURE 8.3 Normalized axial self-weight deflection for rings of equally spaced multiple point supports. The ring radius is normalized to the outer radius of the circular mirror, and deflection is normalized to minimum deflection for $N = 12$. Each curve is for a single ring support; shear is neglected.

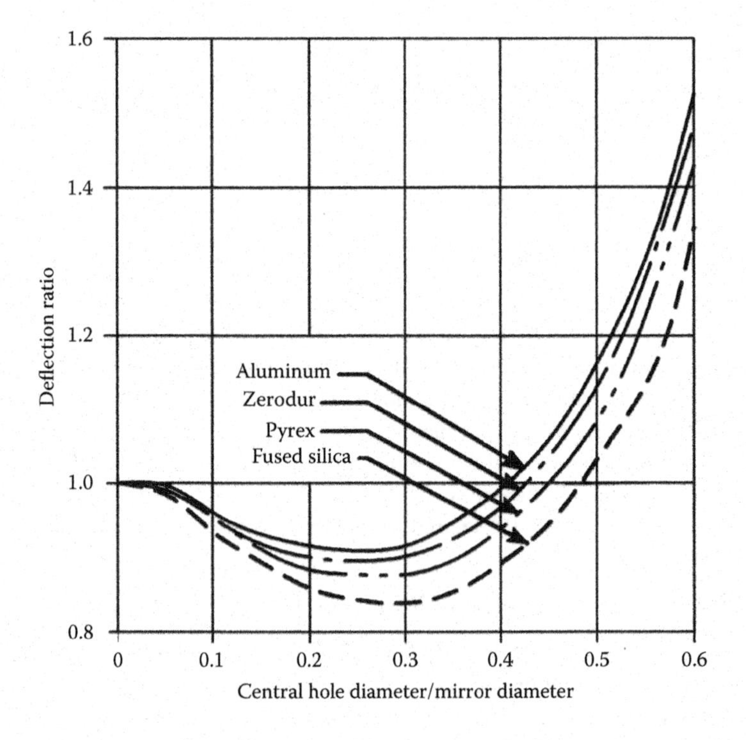

FIGURE 8.4 Effect of central hole in mirror deflection compared to the deflection of mirror without central hole, for central hole diameters from 0.0 to $0.5d$, where d is the mirror diameter. Materials are fused silica, $\nu = 0.17$; Corning 7740 Pyrex, $\nu = 0.2$; Schott Zerodur, $\nu = 0.24$; and 6061 aluminum, $\nu = 0.3$. The mirror is simply supported on an edge ring. Shear is neglected.

A curving optical surface changes the thickness of a circular mirror with a flat back. When the thickness variation is less than 10%, the error associated in calculating axial self-weight deflection assuming constant thickness is also less than 10%.[10] Most curved optical surfaces can be approximated by a spherical sector. The maximum sagittal depth δ_S of a spherical sector is given approximately by $\delta_S \approx r^2/(2R)$, where R is the optical radius of curvature. Now $R = 2df_{no}$, where d is the mirror diameter ($r = d/2$) and f_{no} is the focal ratio. Substituting, $\delta_S \approx d/(16f_{no})$. The ratio between the sagittal depth and mirror thickness h is δ_S/h. The aspect ratio of the mirror is d/h, so $h = d/(d/h)$ using the aspect ratio and solving for a thickness variation of 0.1:

$$0.1 = \frac{\delta_S}{h} = \frac{\dfrac{d}{16f_{no}}}{\dfrac{d}{\left(\dfrac{d}{h}\right)}} \tag{8.14}$$

$$f_{no} = 0.625\left(\frac{d}{h}\right).$$

To keep the thickness variation below 10%, the focal ratio must be greater than 0.625 times the mirror aspect ratio. For a 6-to-1 aspect ratio, the minimum focal ratio is $f_{no} = 0.625(6) = 3.75$, or about 4.

Plate theory assumes that the size of the supports is very small, hence the use of the term *point supports*. In reality, the size of mirror supports is finite. The finite size of the supports reduces the amount of axial self-weight deflection. For constant-thickness circular mirrors on arrays of point supports, where the point supports have threefold symmetry, Lemaitre[11] showed that the reduction in axial bending deflection is

$$\delta_A = \delta_{AP}\left\{1 - \frac{4}{3}\left(\frac{a^2}{b^2}\right)\left[1 - \ln\left(\frac{a^2}{b^2}\right)\right]\right\}, \quad 0 \le \frac{a}{b} \le 0.549, \tag{8.15}$$

where δ_{AP} is the axial bending deflection assuming point supports, a is the diameter of the support pad, and b is the pad spacing or distance between supports.

Lemaitre's relationship is shown in Figure 8.5. When a/b is 0.12, the reduction in deflection is about 10%. For conservatism, the effect of pad size is usually ignored when calculating axial self-weight deflection.

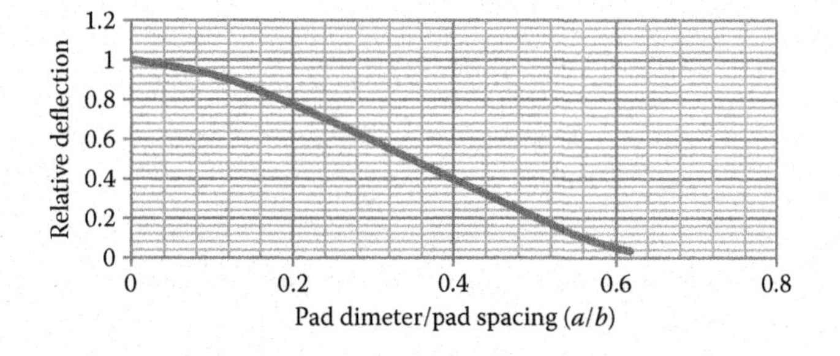

FIGURE 8.5 Relative deflection as a function of the ratio of individual support pad diameter a to support spacing a/b, $0 \le a/b \le 0.6184$, normalized to deflection for $a/b = 0$.

Shear deformations of mirrors depend on the aspect ratio (diameter-to-thickness) and the Poisson's ratio of the mirror material. As a rough rule of thumb, shear deflection can no longer be neglected when the aspect ratio is less than about 10.[12] Shear deflections can become as large as bending deflections in some circumstances. Shear deflection correlates with bending, so the axial deflection δ_A is found by adding bending δ_B and shear deflection δ_S contributions:

$$\delta_A = \delta_B + \delta_S = \delta_B\left(1 + \frac{\delta_S}{\delta_B}\right). \tag{8.16}$$

Two common mirror support geometries are the continuous edge ring and multipoint. Simple support for an edge ring means that the edge is free to rotate. Including shear effects, the approximate p-v self-weight deflection $\delta_{\theta\text{p-v}}$ of a constant-thickness circular mirror simply supported around its edge is[13]

$$\delta_{\theta\text{p-v}} = \frac{3}{256}\frac{\rho}{E}\frac{d^4}{h^2}\left[(5+\nu)(1-\nu) + \frac{16}{3}(3+\nu)\left(\frac{h}{d}\right)^2\right]\cos(\theta). \tag{8.17}$$

Shear slightly changes the location of the optimum support diameter to about $d_s/h = 0.674$ when $d/h \approx 2$. For most applications, the approximation of $d_s/d \approx 0.68$ is sufficiently accurate. When supported at an optimum ring location of $d_s/d \approx 0.68$, the p-v self-weight deflection $\delta_{\theta\text{p-v}}$ including the shear effects of a circular mirror is given as approximately

$$\delta_{\theta\text{p-v}} \cong 0.00156\frac{\rho(1-\nu)}{E}\left(\frac{d}{h}\right)^2 d^2\left[\frac{44.8}{1-\nu}\left(\frac{h}{d}\right)^2 + 2.09\nu + 1\right]\cos(\theta). \tag{8.18}$$

Cowper[14] provides correction factors for shear in Timoshenko beams; these factors can also be used to correct for shear effects in plate bending. Using Cowper's correction factors, Seibert[15] gave an approximate relationship for the self-weight axial deflection of constant-thickness mirrors on arrays of optimum point supports:

$$\begin{aligned} \delta_A &= 0.025q\frac{r^4}{D}\left(\frac{3}{N}\right)^2 + 0.65q\frac{r^2}{S_C G h}\left(\frac{3}{N}\right), \\ D &= \frac{Eh^2}{12(1-\nu^2)}, \\ S_C &= \frac{10(1+\nu)}{12+11\nu}, \\ G &= \frac{E}{2(1+\nu)}, \end{aligned} \tag{8.19}$$

where q is the weight per unit area of the mirror, r is the mirror radius, D is the mirror flexural rigidity, E is the elastic modulus of the mirror material, h is the mirror thickness, ν is the Poisson's ratio of the mirror material, N is the number of support points, G is the shear modulus of the mirror material, and S_C is the shear correction (from Cowper).

In Equation 8.19, the first term is the deflection from the mirror bending, or δ_B, and the second term is the shear deflection δ_S. Then, the ratio δ_S/δ_B becomes

$$\frac{\delta_S}{\delta_B} = \frac{0.025q\dfrac{r^4}{\dfrac{Eh^3}{12(1-\nu^2)}}\left(\dfrac{3}{N}\right)^2}{0.65q\dfrac{r^2}{\left[\dfrac{10(1+\nu)}{12+11\nu}\right]\left[\dfrac{E}{2(1+\nu)}\right]h}} = \frac{13}{90}\left(\frac{h}{r}\right)^2 N\frac{12+11\nu}{1-\nu^2}. \tag{8.20}$$

Since $r = d/2$,

$$\frac{\delta_S}{\delta_B} = \frac{52}{90}\left(\frac{h}{d}\right)^2 N\frac{12+11\nu}{1-\nu^2}. \tag{8.21}$$

The Poisson's ratios of most metallic and glassy mirror materials are between 0.17 and 0.3. For this range of Poisson's ratios, δ_S/δ_B lies between $2.06(h/r)^2N$ and $2.43(h/r)^2N$. In comparison, Nelson et al.[16] suggested an empirical correction factor of $2(h/r)^2N$ for shear. Figure 8.6 is a plot of the ratio δ_S/δ_B for a Poisson's ratio of 0.25, for aspect ratios (d/h) varying between 5 and 20 and for supports where N = 3, 6, 18, and 36. Figure 8.6 shows that shear corrections exceed 10% when the mirror aspect ratio is less than about d/h = 15 for most multipoint supports.

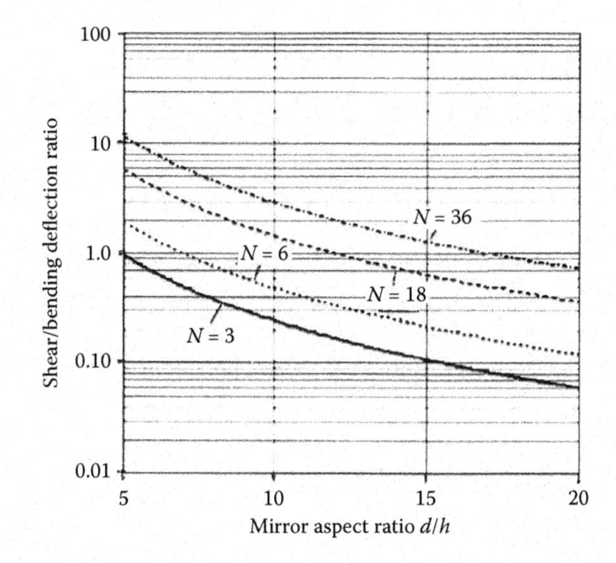

FIGURE 8.6 Shear-to-bending deflection ratio for solid mirrors where Poisson's ratio = 0.25 (Corning Pyrex 7740), with aspect ratios (d/h) varying from 5 to 20 and optimum multipoint supports (N from 3 to 36).

The accuracy of Seibert's relationship degrades rapidly when the number of point supports increases, especially for $N > 6$. For more complex support geometries, a better estimate is to find the bending deflection coefficient and then correct for shear as shown in the following:

$$\delta_{\theta\text{p-v}} = C_{\text{A}} \frac{\rho(1-\nu^2)}{E} \frac{d^4}{h^2} \left[1 + \frac{52}{90} N \left(\frac{h}{d} \right)^2 \frac{12+11\nu}{1-\nu^2} \right] \cos(\theta). \tag{8.22}$$

For a quick estimate, assuming optimum multipoint supports,

$$\delta_{\theta\text{p-v}} \cong \frac{1}{4N^{2.4}} \frac{\rho(1-\nu^2)}{E} \left(\frac{d}{h} \right)^2 d^2 \left[1 + 8N \left(\frac{h}{d} \right)^2 \right] \cos(\theta). \tag{8.23}$$

8.3.5 Small-Mirror Approximation

If the mirror is "small," it may be possible to neglect axial deflection. Mirror performance is limited only by diffraction and hence is independent of surface error from deflection when the wave front error satisfies the Rayleigh criterion, or the p-v wave front error $\delta_{\text{Wp-v}} \le 0.25\lambda$, where λ is the wavelength.[17] For demanding applications, the entire system wave front error must meet the Rayleigh criterion, and the tolerable surface error becomes much smaller. A reasonable mirror surface deflection tolerance in this situation is about 0.05λ, for a wave front error of 0.1λ. The axial deflection equations can be solved to find the maximum mirror size before self-weight deflection exceeds these deflection tolerances. A mirror below this maximum size is considered small and does not need an elaborate support; in addition, gravity deflection can be ignored in most applications.

Solving for the maximum diameter d_{max} as a function of the deflection tolerance δ_{TOL},

$$d_{\text{max}} = \frac{h}{d} \left[\frac{\delta_{\text{TOL}}}{C_{\text{A}}} \frac{E}{\rho(1-\nu^2)} \right]^{\frac{1}{2}}. \tag{8.24}$$

Figure 8.7 gives the maximum diameter of mirrors supported on three points for an optimum support diameter $d_S/d = 0.645$ as a function of the mirror aspect ratio when the tolerance is $\delta_{\text{TOL}} = 0.25\lambda$ and $\delta_{\text{TOL}} = 0.05\lambda$ ($\lambda = 633$ nm). Corning 7740 Pyrex is the mirror material in this analysis; $\rho(1 - \nu^2)/E = 333 \times 10^{-9}\ \text{m}^{-1}$ (from Table 8.3). From this graph, for $\delta_{\text{TOL}} = 0.25\lambda$ and an aspect ratio of 6:1 ($d/h = 6$), the maximum diameter for a small mirror is about 0.58 m.

Another criterion was derived by Danjon and Couder.[18] It defines a "rigid" mirror as one small enough to maintain its shape when supported by its hard points. When using this criterion, a mirror is rigid if $D^4/h^2 < 5\ \text{m}^2$, where D and h are in meters. This applies to mirrors made of conventional glass. Assuming a 6:1 aspect ratio, a mirror that meets this criterion must be smaller than about 0.37 m in diameter to be rigid.

8.3.6 Mirror Deformation from Mounting

Forces and moments induced by mounting can create excessive stresses and distort mirrors. For mirrors, OPD errors on the reflected wave front are produced from surface distortions due to forces and moments.

Stress and distortion in mirrors arise from improper mounting. For minimum stress and optical surface distortion, mirror mounts must be statically determinate. The position of a mirror in

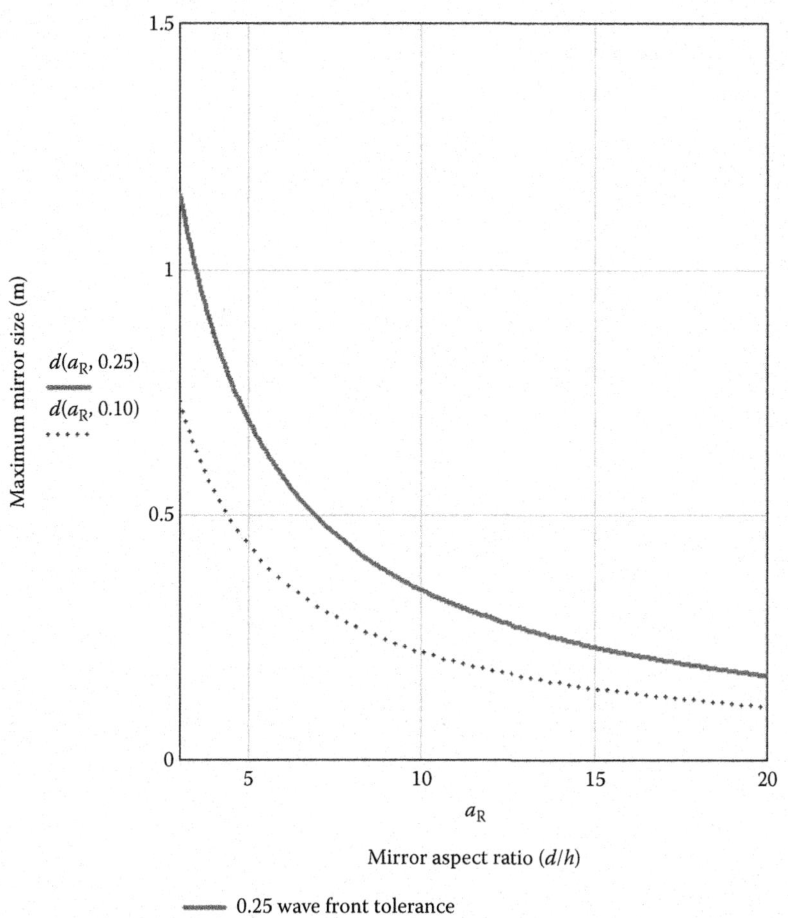

FIGURE 8.7 Maximum diameter for mirror self-weight deflection to meet wave front tolerances of δ_{TOL} = 0.25λ and δ_{TOL} = 0.05λ (λ = 633 nm), as function of mirror aspect ratio $3 \le d/h \le 20$, and Pyrex 7740 mirror material ($\rho(1 - \nu^2)/E = 333 \times 10^{-9}\ \text{m}^{-1}$).

a statically indeterminate mount is unstable. A statically indeterminate mount is overconstrained and may induce forces and moments in the mirror large enough to produce unacceptable surface deformations.

Semikinematic mounting principles are often applied to mirror mounts, with small, finite areas, and typically, flat pads are employed to reduce contact stress at each interface to an acceptable level. If the mounting pads are not accurately flat, or if they are not parallel to the mating surfaces on the mirror, the mirror is deformed when placed into the mount. Forcing the mounting pads into intimate contact with the mirror causes distortion of both the mirror and mount. This induces bending moments into the mirror and adversely influences the overall optical figure of the mirror.

This problem is called "the mounting pad coplanarity problem." Reducing the mirror deformation to an acceptable level requires additional degrees of rotational freedom between the mirror and mounting interfaces. Since the moment cannot be reduced to zero, it is necessary to estimate how much moment can be safely exerted on the mirror without producing an unacceptable change in optical surface figure.

Allowable bending moments in a mirror are estimated using the principle of superposition. Influence functions relate the bending moments to the surface deflection of the mirror.[19] For an array of N contact points supporting a mirror, where all of the contact points are on the same plane, the maximum surface deflection δ_{max} is given by[20]

$$\delta_{max} = \left\langle \sum_{i=1}^{N} \left\{ \left[M_{ri} Z_{ri}(x_i, y_i) \right]^2 + \left[M_{hi} Z_{hi}(x_i, y_i) \right]^2 \right\} \right\rangle^{\frac{1}{2}}, \tag{8.25}$$

where M_{ri} is the radial bending moment at the ith point, Z_{ri} is the radial influence function of the ith point, M_{hi} is the hoop radial bending moment at the ith point, Z_{hi} is the hoop influence function at the ith point, and x_i and y_i are surface locations.

For the special case of a right circular cylinder mirror supported by N_s points equally spaced on a common diameter, the preceding equation becomes

$$\delta_{max} = \left\{ N_s \left[(M_{rs} Z_{rs})^2 + (M_{hs} Z_{hs})^2 \right] \right\}^{\frac{1}{2}}, \tag{8.26}$$

where M_{rs} is the maximum moment in the radial direction at any support point, Z_{rs} is the influence function in the radial direction for any support point, M_{hs} is the maximum moment in the hoop direction at any support point, and Z_{hs} is the influence function in the hoop direction at any support point.

Influence functions are derived from finite-element analysis of the mirror. When a right circular cylinder mirror is mounted by three equally spaced mounts on a common diameter attached to its back, the deflection associated with the out-of-plane condition of the mounts can be found by an approximate method. This method scales influence functions from reference mirror designs. The scaling relationship is

$$\delta = \left(\frac{D_{REF}}{D} \right) \left(\frac{d}{d_{REF}} \right)^2 Mf\left(\frac{r}{r_0} \right), \tag{8.27}$$

where REF is a subscript referring to the reference mirror, D is the mirror flexural rigidity, d is the diameter of the mirror, M is the moment applied at the mounting point, $f(r/r_0)$ is an empirical influence function for the reference mirror, and r/r_0 is the ratio of the support radius/mirror radius.

Then, the mirror surface deflection is given by

$$\delta\left(M, \frac{r}{r_0} \right) = \sqrt{3} \left(\frac{D_{REF}}{D} \right) \left(\frac{d}{d_{REF}} \right)^2 \left\{ \left[M_r f_r\left(\frac{r}{r_0} \right) \right]^2 + \left[M_h f_h\left(\frac{r}{r_0} \right) \right]^2 \right\}^{\frac{1}{2}}, \tag{8.28}$$

where M_r and M_h are the moments in the radial and hoop directions, respectively; $f_r(r/r_0)$ is an empirical function for the moment effect in the radial direction, and $f_h(r/r_0)$ is an empirical function for the moment effect in the hoop direction.

The following are the empirical influence functions for a reference mirror where $d_{REF} = 600$ mm, $h_{REF} = 63.5$ mm, $E_{REF} = 69$ GPa, and $\nu_{REF} = 0.3$. The reference flexural rigidity is $D_{REF} = 1.617 \times$

10^6 N m. The influence functions are given in units of RMS waves per newton-meter, where 1 wave = 633 nm:

$$f_r\left(\frac{r}{r_0}\right)=\left[-0.0250\left(\frac{r}{r_0}\right)^2+0.0444\left(\frac{r}{r_0}\right)-0.0013\right]\text{RMS waves/N m},$$
$$0.2\le\frac{r}{r_0}\le 0.9, \tag{8.29}$$

$$f_h\left(\frac{r}{r_0}\right)=\left[0.004392\left(\frac{r}{r_0}\right)^2+0.008603\left(\frac{r}{r_0}\right)-0.000690\right]\text{RMS waves/N m},$$
$$0.2\le\frac{r}{r_0}\le 0.98. \tag{8.30}$$

Example 8.4

A right circular cylinder mirror with diameter of d = 300 mm and thickness of h = 40 mm is mounted on three equally spaced points against its back; the support points are located at 0.65 of the mirror diameter (r/r_0 = 0.65). The mirror is made of Schott Zerodur, where E = 90.6 GPa and ν = 0.24. The maximum surface deflection from the moments induced by the mounting is 0.1 RMS wave (1 wave = 633 nm). It is assumed that the moment effect is stronger in the hoop than radial direction, and, therefore, the radial moment is allowed to be three times bigger than the hoop direction moment, or $M_r = 3 \times M_h$. To determine the maximum allowable moments from the mounts, the scaling relationship is solved for the moments:

$$\frac{\delta_{TOL}}{\sqrt{3}\left(\frac{D_{REF}}{D}\right)\left(\frac{d}{d_{REF}}\right)^2}=\left\{\left[M_r f_r\left(\frac{r}{r_{REF}}\right)\right]^2+\left[\frac{M_r}{3}f_h\left(\frac{r}{r_{REF}}\right)\right]^2\right\}^{\frac{1}{2}}$$

$$M_r=\left\{\frac{1}{\left[f_r\left(\frac{r}{r_{REF}}\right)\right]^2+\left[\frac{1}{3}f_h\left(\frac{r}{r_{REF}}\right)\right]^2}\right\}^{\frac{1}{2}}\frac{\delta_{TOL}}{\sqrt{3}\left(\frac{D_{REF}}{D}\right)\left(\frac{d}{d_{REF}}\right)^2}.$$

where δ_{TOL} is the mirror surface deflection tolerance.

From the empirical reference equations, $f_r(0.65) = 17 \times 10^{-3}$ RMS wave/N m and $f_h(0.65) = 6.76 \times 10^{-3}$ RMS wave/N m. The flexural rigidity D of the mirror is given by

$$D=\frac{Eh^3}{12(1-\nu^2)}=\frac{90.6\,\text{GPa}(40\,\text{mm})^3}{12(1-0.24^2)}=5.127\times10^5\,\text{N m}.$$

The maximum radial moment at each support point is then M_h = 4.40 N m, and M_r = 13.2 N m.

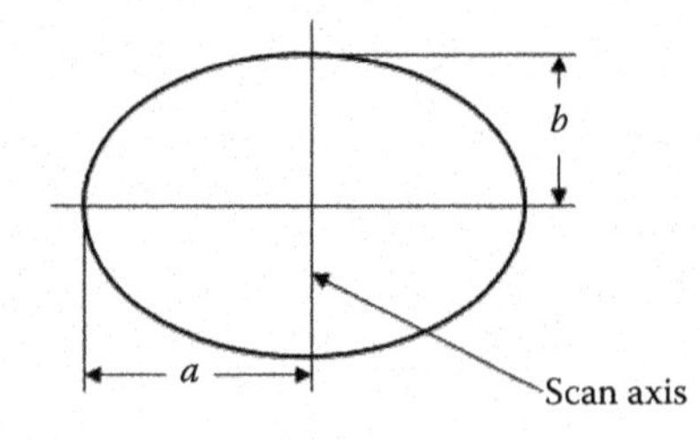

FIGURE 8.8 Elliptical scan mirror geometry. For circular mirrors, $a = b = r$.

8.3.7 Scanning Mirror Deformation

In scanning applications, the mirror oscillates through some angle about a diameter to move the scanning beam. As the mirror oscillates, it is subjected to varying angular acceleration normal to the mirror surface. The angular acceleration varies with the distance from the axis of rotation, with maximum acceleration at the greatest distance from the axis, or the edge of the mirror. Maximum deflection is at the edge, while the mirror surface near the center of rotation is unaffected. A scan mirror becomes astigmatic, with a change in curvature in only one direction, normal to the scan axis.

For a circular scanning or stabilized mirror revolved at an angular acceleration ω (radians/s^2) about its diameter as shown in Figure 8.8, the maximum deflection is at the greatest distance from the axis of rotation and is given by[21]

$$\delta = 0.628\frac{\rho(1-\nu^2)}{E}\frac{r^5}{h^2}\frac{\omega}{g}, \text{ circular,}$$
$$\delta = 0.628\frac{\rho(1-\nu^2)}{E}\frac{ba^4}{h^2}\frac{\omega}{g}, \text{ elliptical,} \tag{8.31}$$

where g is the acceleration of the gravity field of the Earth, a is the semimajor axis of the elliptical mirror, and b is the semiminor axis of the elliptical mirror.

For a rectangular scan mirror rotated about its middle where s is the width and h is the thickness, the maximum deflection is given by[22]

$$\delta = 0.00542\frac{\rho(1-\nu^2)}{E}\frac{s^5}{h^2}\frac{\omega}{g}. \tag{8.32}$$

8.4 DYNAMIC EFFECTS ON MIRRORS

8.4.1 General Considerations

The response of a mirror to dynamic excitation is governed by its fundamental frequency and damping. Types of dynamic excitation include vibration, angular acceleration (for a scanning system), and mechanical shock. Dynamic excitation induces deflection in the optical surface of the mirror and rigid-body displacement of the mirror relative to its mount. Under severe dynamic loading conditions (characteristic of mechanical shock), the surface deflection may temporarily exceed the design tolerance. After the mechanical shock has subsided, the critical parameter is the settling time or time necessary to recover from the subsiding vibration induced by the shock.

There are two motions, each with its own frequency, associated with a mirror: the diaphragm bending frequency of the mirror and the rigid-body frequency of the mirror on its mount. The diaphragm bending frequency of the mirror is associated with the self-weight deflection of the mirror. Diaphragm bending changes the optical surface figure of the mirror. The rigid-body frequency is associated with physical displacement of the mirror relative to its mount. The rigid-body motion of the mirror changes the alignment of the mirror optical surface relative to the optical axis and other components of the system.

8.4.2 Diaphragm Frequency of a Mirror

The fundamental diaphragm bending or drumhead frequency f_n in the direction perpendicular to the plane of the optical surface of an axis vertical mirror is related to the self-weight deflection δ of the mirror by the following approximate equation[23]:

$$f_n = \frac{1}{2\pi}\sqrt{\frac{g}{\delta}}, \tag{8.33}$$

where f_n is the fundamental or drumhead frequency (in hertz), g is the acceleration of the gravity of Earth (9.807 m/s^2), and δ is the self-weight deflection (maximum or p-v).

The preceding relationship is plotted in Figure 8.9. A convenient rule of thumb is that a self-weight deflection of 1 μm occurs at ≈500 Hz. With a typical tolerance for self-weight deflection of mirrors 0.125 λ (λ ≈ 633 nm), the diaphragm frequency is about 1800 Hz.

Example 8.5

A 200 mm diameter axisymmetric mirror is 33 mm thick and is made of Pyrex 7740. It is mounted at three equally spaced points at its edge. The axial p-v deflection is about 171 nm. The diaphragm bending frequency is given by $(2\pi)^{-1}(g/171\text{ nm})^{-1/2} \approx 1.205$ kHz.

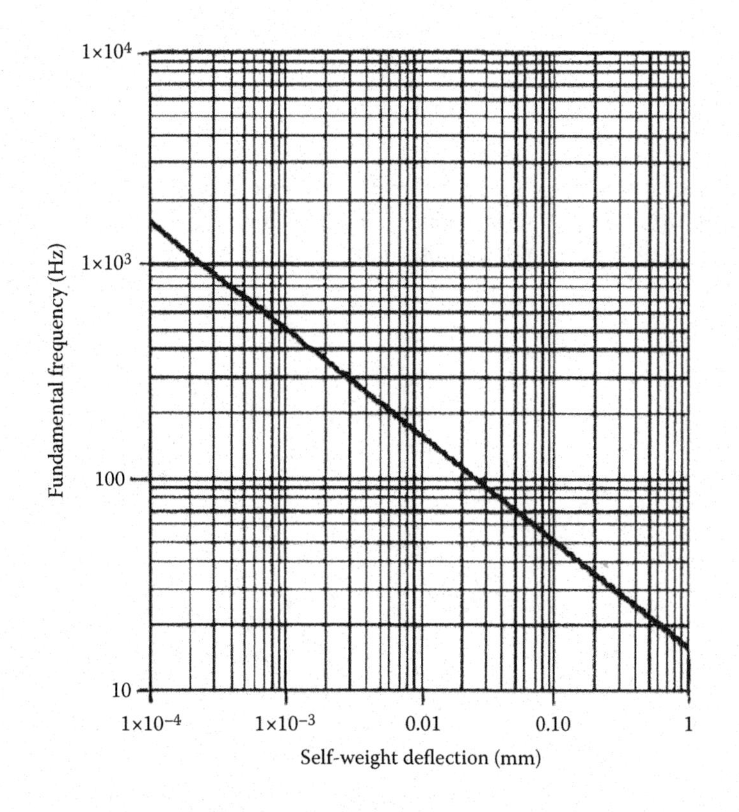

FIGURE 8.9 Frequency (Hz) as a function of self-weight deflection in millimeters.

8.4.3 Rigid-Body Frequencies of an Axisymmetric Mirror

The rigid-body frequencies of a mirror in translation are determined by the mirror mass and spring stiffness of the mount. Since there are normally a number of support points in a mirror mount, finding the overall spring stiffness of the mount requires summing the spring stiffnesses of the individual mounting points. The rigid-body frequencies in rotation are determined by the mirror mass moment of inertia, the spring stiffnesses of each individual mounting point, and the locations of the mounting points. For axisymmetric mirrors with equally spaced mounting points, the rigid-body frequencies in translation are equal in both axes on the plane of the mirror. Similarly, the rigid-body frequencies in rotation about both axes on the plane of the mirror are equal.

The spring stiffnesses in translation of an axisymmetric mirror shown in Figure 8.10 supported at N equally spaced points are given by

$$k_X = k_Y = \frac{N}{2}(k_R + k_T), \tag{8.34}$$

$$k_Z = Nk_A, \tag{8.35}$$

where k_X is the spring stiffness in the X direction, k_Y is the spring stiffness in the Y direction, k_A is the spring stiffness in the Z direction (axial), k_R is the radial spring stiffness, and k_T is the tangential spring stiffness.

Then the mirror fundamental frequencies in translation are[24]

$$\begin{aligned} f_{nX} = f_{nY} &= \frac{1}{2\pi}\sqrt{\frac{3}{2}\frac{k_R + k_T}{m}}, \\ f_{nZ} &= \frac{1}{2\pi}\sqrt{\frac{3k_A}{m}}, \end{aligned} \tag{8.36}$$

where f_{nX} is the natural frequency in the X direction (mirror surface plane), f_{nY} is the natural frequency in the Y direction (mirror surface plane), and f_{nZ} is the natural frequency in the Z direction (normal to surface; in the direction of the optical axis).

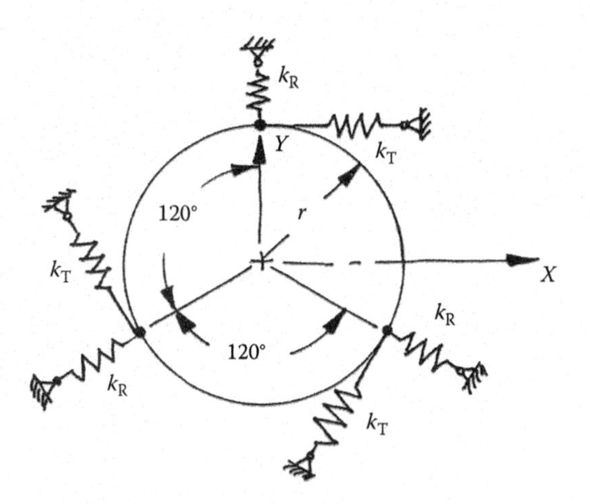

FIGURE 8.10 Three-point mount for a mirror, where k_R is the radial stiffness of each mounting point and k_T is the tangential stiffness; the center of the coordinates coincides with the mirror center or optical axis.

Often $k_T \gg k_R$; then,

$$f_{nX} = f_{nY} \approx \frac{1}{2\pi}\sqrt{\frac{3}{2}\frac{k_T}{m}}. \tag{8.37}$$

One common type of athermal mount for small mirrors is the three-point flexure mount. A rectangular cross-sectional flexure blade is used at each mounting point. If the end of each flexure is attached to the mirror by a ball joint (hence is free to rotate) as a way of reducing induced moments in the mirror, the stiffnesses of the individual flexures are given by

$$\begin{aligned} k_R &= \frac{Ebh^3}{4L^3}, \text{ radial,} \\ k_T &= \frac{Ebh}{L}, \text{ tangential,} \\ k_A &= \frac{Eb^3h}{4L^3}, \text{ axial,} \end{aligned} \tag{8.38}$$

where E is the flexure elastic modulus, b is the flexure width in the radial direction, h is the flexure depth (in the direction of the Z-axis or optical axis), and L is the flexure length.

If the flexure ends are fixed and cannot rotate,

$$\begin{aligned} k_R &= \frac{Ebh^3}{L^3}, \text{ radial,} \\ k_T &= \frac{Ebh}{L}, \text{ tangential,} \\ k_A &= \frac{Eb^3h}{L^3}, \text{ axial,} \end{aligned} \tag{8.39}$$

Example 8.6

A 100 mm diameter right circular cylinder aluminum mirror weighing 424 g is mounted at three equally spaced points at its circumference, using single blade titanium flexures. The flexure thickness $h = 0.5$ mm, width $b = 10$ mm, and length $L = 20$ mm. The flexures are compliant in the radial direction, and the long axis of the flexures is parallel to the optical axis. For titanium, $E = 110$ GPa. The spring stiffness of the flexures is given by

$$\begin{aligned} k_R &= \frac{Ebh^3}{4L^3} = \frac{110\,\text{GPa}(10\,\text{mm})(0.5\,\text{mm})^3}{4(20\,\text{mm})^3} = 4.3\times10^3\ \text{N/m}, \\ k_T &= \frac{Eb^3h}{4L^3} = \frac{110\,\text{GPa}(10\,\text{mm})^3(0.5\,\text{mm})}{4(20\,\text{mm})^3} = 1.7\times10^6\ \text{N/m}, \\ k_A &= \frac{Ebh}{L} = \frac{110\,\text{GPa}(10\,\text{mm})(0.5\,\text{mm})}{(20\,\text{mm})} = 2.8\times10^7\ \text{N/m}. \end{aligned}$$

The fundamental frequencies in translation are

$$f_{nX} = f_{nY} = \frac{1}{2\pi}\sqrt{\frac{N}{2}\frac{k_R + k_T}{m}} = \frac{1}{2\pi}\sqrt{\frac{3}{2}\frac{4.3\text{kN/m} + 1.7\text{mN/m}}{424\text{g}}} = 393\text{Hz},$$

$$f_{nZ} = \frac{1}{2\pi}\sqrt{N\frac{k_A}{m}} = \frac{1}{2\pi}\sqrt{3\frac{28\text{mN/m}}{424\text{g}}} = 2220\text{Hz}.$$

8.4.4 Damping

Damping is another system property determining response to dynamic excitation. There are a variety of ways to describe it. Most common are the critical damping factor, loss coefficient, and quality factor ("*Q*"). Three types of damping in optomechanical systems are viscous, structural, and coulomb friction damping. Viscous damping is proportional to velocity, structural damping is proportional to displacement, and coulomb friction damping is proportional to friction. In optomechanical systems, the damping is usually small and all three types of damping are essentially equivalent.

Damping is the energy loss from the system during oscillation. A system is critically damped if it does not oscillate when released after being perturbed from equilibrium. The amount of damping C_C required for critical damping is given by $2(km)^{½}$, where k is the system spring stiffness and m is the system mass. Alternatively, $C_C = 2m\omega_N = 4\pi m f_n$, where ω_n is the angular natural frequency (in radians per second) and f_n is the natural frequency (in hertz). Optomechanical systems are generally poorly damped and their damping is given in terms of the critical damping ratio ζ, which is the ratio of the actual system damping C to C_C required for critical damping, i.e., $\zeta = C/C_C$.

In aerospace engineering, the quality factor or Q is used to describe system damping. For lightly damped systems, the quality factor is the amplification of the system at resonance. Alternatively, the quality factor is the ratio of the "half-power width" at resonance $\Delta\omega$ to the system fundamental frequency ω_n. The quality factor and critical damping ratio are related by

$$\frac{\Delta\omega}{\omega_n} = \frac{1}{Q} = 2\zeta, \tag{8.40}$$

where $\Delta\omega$ is the half-power width of the response at resonance, ω_N is the fundamental frequency, Q is the quality factor, and ζ is the critical damping ratio.

Damping in an optomechanical system is difficult to estimate using analytical methods. Testing, or extrapolation from previous designs, is needed to find the damping in most structures. Rules of thumb are commonly used to estimate damping. For example, the Q of a printed circuit board is approximately equal to the square root of its fundamental frequency.

Material damping is proportional to stress.[25] Stiff optomechanical structures are lightly stressed. Therefore, material damping is also small. Damping in some metals is high; magnesium is an example with $Q \approx 13$. However, for most structural metals, damping is small.[26] For the various steel alloys, Q ranges from about 125 for low-carbon steel to about 630 for austenitic stainless steel. For high-strength materials used in aerospace (such as 2024 aluminum and titanium), $Q \approx 3000$.

Lower values of Q are associated with complex, built-up structures. Such structures embody multiple joints, with losses occurring at each joint. In aerospace, a common rule of thumb is that the Q for a complex structure will be between 20 and 40. An average value for spacecraft structures is $Q \approx 42$ up to a frequency of 195 Hz.[27] Even higher values of Q may occur in stiff satellite structures.

For example, the frequency of the first bending mode of the Hubble Space Telescope truss is about 17 Hz and its $Q \approx 125$.[28]

Damping changes the natural frequency. For the small amounts of damping in mirror mounts, the effect on frequency can be ignored. This is shown by considering the change in frequency from damping in a single-degree-of-freedom (SDOF) system. Let f_n be the undamped natural frequency and f_{nd} be the damped natural frequency, while Q is the damping; then,

$$f_{nd} \approx f_n\sqrt{1-\zeta^2} = f_n\sqrt{1-\frac{1}{4Q^2}}. \tag{8.41}$$

Assume that damping is relatively high, $\zeta \approx 0.10$, or $Q \approx 5$. Then, from the preceding equation,

$$f_{nd} = f_n\sqrt{1-(0.10)^2} = 0.995 f_n. \tag{8.42}$$

A change of 0.5% is within the errors associated with ordinary engineering analysis and can be neglected.

8.4.5 Rigid-Body Response to Dynamic Excitation

The rigid-body response of a mounted mirror to dynamic excitation is determined by the fundamental frequency of the mirror and its damping as well as the type of excitation. A simplified approach to system response, using the concept of an SDOF system, is often used in conceptual design. The SDOF approximation provides considerable insight into system response and is a primary tool in design.

Three important types of dynamic excitation are periodic vibration, random vibration, and mechanical shock. Periodic vibration is characterized by a frequency f, which is the inverse of the period, and maximum amplitude x. Random vibration is given as a power spectral density or PSD, which is the area, and hence power, under the vibration curve over some frequency interval, with units of acceleration due to Earth's gravity squared per Hertz (g^2/Hz). Finally, a mechanical shock is a sudden transition of state or displacement, where the transition time τ is much less than the system period or $\tau < 1/f_n$. System response to all of these dynamic excitations is determined by fundamental frequency f_n and damping Q.

The maximum displacement response x_0 of an SDOF system with fundamental frequency f_n to a periodic excitation of maximum amplitude x_1 and frequency f is given by[29]

$$x_0 = x_1\left[\frac{1+\left(\frac{f}{f_n}\frac{1}{Q}\right)^2}{\left(1-\frac{f^2}{f_n^2}\right)^2+\left(\frac{f}{f_n}\frac{1}{Q}\right)^2}\right]^{\frac{1}{2}}. \tag{8.43}$$

If the damping is small, Q is large, and the preceding equation simplifies to

$$x_0 \approx \text{abs}\left(\frac{x_1}{1-\frac{f^2}{f_n^2}}\right). \tag{8.44}$$

The RMS acceleration a_{RMS} and displacement response x_{RMS} of an SDOF system to random vibration are given approximately by[30]

$$a_{RMS} = \sqrt{\frac{\pi}{2} f_{nx} Q \, PSD},$$

$$x_{RMS} = \frac{a_{RMS}}{4\pi^2 f_n^2} = \left(\frac{1}{32\pi^3} \frac{Q \, PSD}{f_n^3} \right)^{\frac{1}{2}}, \tag{8.45}$$

where PSD is the power spectral density at the fundamental frequency f_n.

For excitation by random vibration, the worst case is assumed to be the 3-sigma value, which corresponds to a 99.7% probability of occurrence. For maximum 3-sigma response, the RMS values should be multiplied by a factor of 3. Then, $x_{max} = 3x_{RMS}$.

Example 8.7

In a previous example, the radial and axial frequencies of a flexural mounted 100 mm diameter aluminum mirror were found to be 393 and 2220 Hz, respectively. The damping of this assembly is $Q = 35$, and it is subjected to a random vibration of PSD = 0.02 g²/Hz. The RMS acceleration a_{rRMS} and displacement x_{rRMS} responses in the radial direction are

$$a_{rRMS} = \sqrt{\frac{\pi}{2} f_{nX} Q \, PSD} = \sqrt{\frac{\pi}{2}(393\,Hz)35(0.02g^2/Hz)} = 21g,$$

$$x_{rRMS} = \frac{a_{rRMS}}{4\pi^2 f_n^2} = \frac{21g}{4\pi^2 (393\,Hz)^2} = 33 \times 10^{-6}\,m.$$

Similarly, the RMS acceleration a_{aRMS} and displacement x_{aRMS} responses in the axial or Z direction are

$$a_{aRMS} = \sqrt{\frac{\pi}{2} f_{nZ} Q \, PSD} = \sqrt{\frac{\pi}{2}(2220\,Hz)35(0.02g^2/Hz)} = 49g,$$

$$x_{aRMS} = \frac{a_{aRMS}}{4\pi^2 f_n^2} = \frac{49g}{4\pi^2 (2220\,Hz)^2} = 2.5 \times 10^{-6}\,m.$$

The three-sigma values would be three times greater than the RMS values. In this example, the worst-case deflection would be in the radial direction and is about 100 μm, while the worst-case acceleration would be in the axial direction and is about 148g.

When the shock duration is short, defined as $f_n\tau \le 0.5$, the maximum displacement response x_{max} of an SDOF system to an acceleration shock a_s is independent of the shape of the shock and is given by[31]

$$x_{max} = \frac{1}{\sqrt{1+\frac{1}{Q^2}}} \frac{a_s}{4\pi^2 f_n^2}; \; f_n\tau \le 0.5 \tag{8.46}$$

If the damping is small, Q is large and the maximum displacement response is given approximately by

$$x_{max} \approx \frac{a_s}{4\pi^2 f_n^2}, \; f_n\tau \leq 0.5. \tag{8.47}$$

A mechanical shock causes the system to vibrate or ring at its fundamental frequency, with the vibration decaying exponentially. The time $t(1\%)$ for the response amplitude to decay to 1% of its initial maximum is called the settling time of the system and is given by[32]

$$t(1\%) \approx 1.466\frac{Q}{f_n}. \tag{8.48}$$

Example 8.8

The radial fundamental frequency of a 100 mm diameter mirror mounted on three titanium flexures is 393 Hz, with a quality factor of 35. The settling time of this mirror in the radial direction would be 1.466(35/393 Hz) = 131 ms.

8.5 THERMAL EFFECTS

8.5.1 General Considerations

Changes in temperature can adversely affect the performance of a mounted mirror. Four thermal effects must be considered in a mounted mirror: a uniform change in temperature; temperature gradients, anisotropy of the coefficient of thermal expansion (CTE) of the mirror material, and bimetallic bending effects from plating of the mirror substrate. A secondary concern is the rate of change in the mirror temperature, or thermal time constant.

8.5.2 Uniform Change in Temperature

A uniform change in temperature or "soak" condition causes the mirror to change size with respect to its mount because the mirror and mount are likely to have different CTEs. This change in size may cause forces and moments to act on the mirror. Methods for mounting mirrors to minimize these effects, i.e., make the mirror/mount system act more athermally, are discussed in Chapters 9 and 10. Unit moment analysis is a useful analytical tool for designing an athermal mounting.

The change in mirror focal length Δf with a uniform change in temperature ΔT is proportional to the product of the optical radius R, temperature change, and mirror substrate CTE α. Then, $\Delta R = R\Delta T\alpha$. Since the mirror focal length $f = R/2$, the change in focal length with a uniform change in temperature is $\Delta f = f\Delta T\alpha$. To minimize focal shift with temperature, the CTE must be reduced; this is one reason for the use of low-CTE materials in mirrors.

8.5.3 Temperature Gradients

A temperature gradient causes different parts of the mirror to change size by different amounts due to thermal expansion. This distorts the surface of the mirror. Temperature gradient effects are generally independent of the mount and arise from the specific mirror geometry, the properties of the mirror material, the distribution of the gradient, and the magnitude of the gradient.

The simplest thermal gradient is through the thickness of the mirror or perpendicular to the optical surface. With an axial temperature gradient, there is a variation in expansion from front to back. One side of the mirror increases in surface area, while the other side of the mirror decreases in surface area. If the mirror is unconstrained by its mount, it will spring out of shape, with the "hot" side becoming more convex (thereby increasing its surface area), while the "cold" side becomes more concave (thereby decreasing its surface area). The optical radius of curvature of the mirror can be changed significantly.

An axial gradient is characteristic of a mirror used in a high-energy laser system, such as ones used for cutting and welding materials. The laser beam heats the reflective (front) side of the mirror, while the back of the mirror remains relatively cool. Even with a coating designed to be highly reflective at the wavelength of the laser, some energy is absorbed by the mirror. Similarly, the primary mirror of an astronomical telescope develops an axial thermal gradient because the optical surface loses heat by radiation to the night sky and by convection currents within the telescope. At the same time, the back and sides of the mirror are usually more insulated from temperature changes by the mount.

A linear axial gradient in a mirror produces a change in radius proportional to $h/(\alpha \Delta T_G)$, where h is the mirror thickness, α is the mirror substrate, and ΔT_G is the temperature difference through the thickness of the mirror. At steady state, the heat flow from the gradient through the mirror is $P = k\Delta T_G/h$, where P is the power (watts per square meter [W/m^2] in SI) and k is the mirror substrate thermal conductivity. The relationship between focal length f and optical radius of curvature is $2f = R$. Then, the change in focal length f due to a linear axial temperature gradient through the mirror is[33]

$$\Delta f = \frac{2f^2 \alpha \Delta T_G}{h} = 2f^2 P \frac{\alpha}{k}. \tag{8.49}$$

The ratio α/k is a figure of merit for the thermal distortion of the mirror. This is called the thermal distortion index of the mirror material. To keep the distortion from a gradient small, the thermal distortion index must also be small. The units of the thermal distortion index are length/power or, in SI units, meters per watt.

The thermal conductivities of "glasses" used in mirrors vary from about 1.13 W/m K for Pyrex 7740 to about 1.46 W/m K for Zerodur. Because the thermal conductivities of different glasses do not vary by much, reducing the magnitude of distortion from thermal gradients requires selecting a glass with a low CTE. The situation is different with metals, where high thermal conductivities more than offset the higher CTEs as compared to those of glasses. For example, the thermal distortion index of 6061 aluminum is 141×10^{-9} m/W compared with 2.92×10^{-6} m/W for Pyrex 7740. Some exotic materials with exceptional values of the thermal distortion index are silicon (with $\alpha/k = 16 \times 10^{-9}$ m/W) and reaction-bonded silicon carbide (with $\alpha/k = 12.3 \times 10^{-9}$ m/W). The very low thermal distortion indices of silicon and silicon carbide make these materials good candidates for high-energy laser mirrors. Oxygen-free high-thermal conductivity (OFHC) copper mirrors are also used in high-energy laser cutting and machining applications because this material can be diamond turned to reduce fabrication cost and its thermal distortion index is about 43×10^{-9} m/W. Table 8.5 gives the thermal distortion indices of some mirror materials.

The surface of a flat mirror subjected to an axial temperature gradient develops a spherical radius of curvature. If the axis of an incident optical beam is not normal to that surface, the radius of curvature on the plane of reflection is different from the radius in the direction perpendicular to that plane. This introduces astigmatism into the beam and creates an OPD in the reflected

TABLE 8.5
Thermal Distortion Indices of Mirror Materials

Material	Thermal Distortion Index α/k ($\times 10^{-9}$ m/W)
Silicon carbide, cast reaction bonded	12.3
Silicon	16.0
Corning fused silica ULE 7972	22.9
Copper, OFHC	43.4
Beryllium, S-200 FH	52.8
Schott Zerodur	68.5
Aluminum alloy, 6061-T6	141
Corning fused silica 7980	400
Corning Pyrex 7740	2920

wave front when measured on the plane of reflection and perpendicular to that plane. This OPD is given by

$$\text{OPD} = \left(\frac{P}{\pi}\right)\left(\frac{\alpha}{k}\right)(\sin^2\theta)(\varepsilon), \tag{8.50}$$

where P is the laser power incident on the mirror, α is the mirror substrate CTE, k is the mirror substrate thermal conductivity, θ is the laser beam angle of incidence on the mirror surface (relative to the surface normal), and ε is the mirror surface coating absorption.

Example 8.9

A 200 mm diameter circular mirror made of Pyrex 7740 in a solar telescope is exposed to solar radiation flux of 1050 W/m^2. The mirror coating is 95% reflective. There is a 45° angle between the optical beam and surface normal of the mirror. What is the OPD on the reflected wave front induced by thermal distortion of the mirror surface?

From Table 8.5, for Pyrex 7740, $\alpha/k = 2920 \times 10^{-9}$ m/W:

$$\text{OPD} = \left(\frac{P}{\pi}\right)\left(\frac{\alpha}{k}\right)(\sin^2\theta)(\varepsilon)$$

$$\text{OPD} = \left[\frac{(1050\,\text{W/m}^2)\frac{\pi}{4}(0.2\,\text{m})^2}{\pi}\right](2920\times10^{-9}\,\text{m/W})(\sin^2 45°)(1-0.95)$$

$$\text{OPD} = 767\times10^{-9}\,\text{m}.$$

Temperature gradients in an axisymmetric concave or convex mirror can be estimated using a method developed by Pearson and Stepp.[34] This method is applicable to linear gradients in the mirror. Kremer[35] modified the method for paraboloid mirrors. The mirror distortions are given in terms of optical aberrations referenced to a Cartesian coordinate system coincident with the vertex of the optical surface. The Z-axis is coaxial with the optical axis of the mirror. The gradient in the mirror is defined by

$$\begin{aligned} T_{(X,Y,Z)} &= C_0 + C_{1(X)} + C_{2(Y)} + C_{3(Z)} \\ C_1 &= \frac{\Delta T_X}{d}, \\ C_2 &= \frac{\Delta T_Y}{d}, \\ C_3 &= \frac{\Delta T_Z}{h}, \end{aligned} \tag{8.51}$$

where C_0 is the soak or uniform temperature difference, ΔT_X is the temperature gradient in the X-axis (mirror plane), ΔT_Y is the temperature gradient in the Y-axis (mirror plane), ΔT_Z is the temperature gradient in the Z direction (through mirror thickness), d is the mirror diameter, and h is the mirror thickness.

The thermal deformations $\delta(r, \theta)_i$ of the mirror resulting from these gradients are expressed as aberrations, and other defects in sequence—spherical, coma, focus, tilt, and piston—are as follows:

$$\begin{aligned} \sum \delta(r,\theta)_i &= \delta_{\text{spherical}} + \delta_{\text{coma}} + \delta_{\text{focus}} + \delta_{\text{tilt}} + \delta_{\text{piston}} = \left(\frac{\alpha C_3}{8R^2} r^4\right) \\ &+ \left(\frac{\alpha C_1}{2R} r^3 \cos\theta + \frac{\alpha C_2}{2R} r^3 \sin\theta\right) + \left(\frac{\alpha C_3 h_0}{2R} - \frac{\alpha C_3}{2} + \frac{\alpha C_0}{2R}\right) r^2 \\ &+ \left[(\alpha C_1 h_0 r \cos\theta) + (\alpha C_2 h_0 r \sin\theta)\right] + \left[\left(\frac{\alpha C_3 h_0^2}{2}\right) + (\alpha C_0 h_0)\right], \end{aligned} \tag{8.52}$$

where $\delta(r, \theta)$ is the position on the mirror surface, $r = (x^2 + y^2)^{1/2}$ is the radial position on the mirror surface, $\theta = \arctan(y/x)$ is the angular position on the mirror surface, α is the CTE of the mirror material, and R is the mirror optical radius of curvature.

The Z-axis coincides with the optical axis.

Example 8.10

An axisymmetric, concave aluminum ($\alpha = 23 \times 10^{-6}\ \text{K}^{-1}$) mirror is used in a cryogenic environment at a temperature of 233 K. The mirror diameter D = 150 mm, its thickness h_0 = 15 mm, and the optical radius of curvature is 1200 mm. The initial temperature of the mirror is 300 K. There is a 3 K axial temperature gradient in the mirror, and the mirror initial temperature is 300 K. What is the spherical aberration and focus shift induced at 233 K?

Since the initial temperature is 300 K, C_0 = 300 K − 233 K = 67 K. The gradient in the Z direction is C_1 = (3 K)/15 mm = 200 K/m. The spherical aberration induced by the axial gradient is

$$\delta_{\text{spherical}} = \frac{\alpha C_3}{8R^2} r^4 = \frac{(23\times10^{-6}\text{K}^{-1})(200\text{K/m})}{8(1.2\text{m})^2}(0.075\text{m})^4 = 12.6\times10^{-9}\,\text{m}.$$

The focus shift in going from 300 to 233 K is (ignoring the piston displacement)

$$\begin{aligned} \delta_{\text{focus}} &= \frac{\alpha r^2}{2}\left(\frac{C_3 h_0}{R} - C_3 + \frac{C_0}{R}\right) = \frac{(23\times10^{-6}\text{K}^{-1})(0.075\text{m})^2}{2}\left[\frac{(200\text{K/m})0.075\text{m}}{1.2\text{m}} - 200\text{K/m} + \frac{67\text{K}}{1.2\text{m}}\right] \\ &= -8.5\times10^{-6}\,\text{m}. \end{aligned}$$

8.5.4 Anisotropy of Coefficient of Thermal Expansion

There is a spatial variation in CTEs of most materials.[36] This represents anisotropy, or change in CTE from point to point within the material. It induces deformation in the mirror surface figure when the temperature changes. Anisotropy is independent of temperature gradients and occurs even when the temperature change is uniform. It arises from alterations in the chemical or physical composition of the material. For example, in aluminum alloys, a variation in grain structure from processing may create a change in optical surface figure with temperature.

Accurate knowledge of the anisotropy of the CTE requires testing of coupons taken from different parts of the bulk material. The spatial variation of low-thermal expansion mirror materials such as Zerodur and ULE were characterized by Jedamzik et al.[37] and by Arnold,[38] respectively. Testing of materials to be used in a specific optical component is frequently not possible, so rules of thumb based on experience are used to estimate the anisotropy. One such rule, from Pellerin,[39] is that the spatial variation in the CTE of most materials is about 5% of the overall CTE. This is only an approximation because spatial variations depend on the sizes of the components. Larger mirrors are likely to have a greater anisotropy than smaller ones. Table 8.6 gives approximate values of the thermal expansion inhomogeneity of various mirror materials.

If the spatial variation in CTE is on the plane of the mirror surface, astigmatism is induced by a change in temperature. The difference in CTE in two axes causes a change in radius in the two axes when the mirror temperature is changed. A rough-order-of-magnitude estimate of the surface deflection for an axisymmetric mirror in this situation is

$$\delta \cong \frac{r^2}{h}(\alpha_X - \alpha_Y)\Delta T, \tag{8.53}$$

where δ is the mirror surface deflection, r is the mirror radius, h is the mirror axial thickness, α_X is the CTE along the X-axis, α_Y is the CTE along the Y-axis, and ΔT is the temperature change.

If the CTE varies through the mirror thickness, a rough estimate of the surface deflection for an axisymmetric mirror when the temperature changes is

$$\delta \cong \frac{r^2}{h}\Delta\alpha_Z \Delta T, \tag{8.54}$$

TABLE 8.6
Anisotropy of CTEs of Mirror Materials

Material	CTE α ($\times 10^{-6}$ K^{-1})	Inhomogeneity of CTE $\Delta\alpha$ ($\times 10^{-9}$ K^{-1})
Aluminum alloy, 6061-T6	23.0	(60)
Beryllium VHP I-70A	11.5	130
Beryllium HIP I-70A	11.5	(30)
Borosilicate Schott Duran	3.2	30
Borosilicate Ohara E-6	3.0	50
Fused quartz Heraeus Amersil T0iE	0.50	1
Fused silica Corning 7940	0.56	2.0
Fused silica Corning ULE	0.03	40
Glass ceramic Schott Zerodur	0.05	10
Glass ceramic Cervit C-101	0.03	15

where δ is the mirror surface deflection, r is the mirror radius, h is the mirror axial thickness, $\Delta\alpha_Z$ is the variation in thermal expansion through the thickness (Z-axis), and ΔT is the temperature change.

Example 8.11

A 150 mm diameter aluminum mirror with a 25 mm axial thickness is used at 233 K, after being fabricated at 300 K. There is a difference in the CTE on the plane of the mirror of 30×10^{-9} K^{-1}. What is the predicted surface deflection at 233 K?

$$\delta \cong \frac{r^2}{2h_0}\Delta\alpha_Z \Delta T = \frac{(75\text{mm})^2}{2(25\text{mm})}(30\times10^{-9}\text{K}^{-1})(300\text{K}-233\text{K}) = 226\times10^{-9}\,\text{m}.$$

This is 226 nm/633 nm = 0.36 wave on the surface, or about 0.72 wave on the wave front. A surface deflection of this magnitude would seriously degrade the performance of most optical systems. This example shows that anisotropy of CTE is an important issue for cryogenic systems subjected to a large change in temperature.

8.5.5 Thermal Inertia

Thermal inertia causes a temperature differential between the temperature of the interior and surface of a mirror after a change in the temperature of the exterior environment. This temperature differential from interior to exterior of a mirror takes the form of a gradient, with resulting change in optical surface figure. In many instances, such a gradient is severe enough to degrade mirror performance.

One of the most famous historical examples of gradient induced degradation of optical performance was reported by Noyes.[40] It occurred during the commissioning of the 2.5 m aperture Hooker Telescope at the Mount Wilson observatory in 1917. A temperature gradient in the glass primary mirror of the telescope was inadvertently produced after some workmen left the dome open during the day and allowed sunlight to strike the mirror. Much to the concern of Hale, the director of the observatory, the telescope image quality right after sunset was terrible. By early morning, performance improved dramatically because the mirror had begun to reach thermal equilibrium. Thereafter, precautions were taken to ensure that the telescope was kept during the day as nearly as possible at the average nighttime temperature to minimize gradient effects. This effect has driven design of observatory domes and procedures throughout the world.

Gradient effects are important in many optomechanical applications, including optical fabrication. Friction experienced during the optical fabrication process changes the temperature of the optical surface of a mirror relative to the interior and back of the mirror. If the mirror is tested immediately after polishing or diamond turning, the temperature gradient will distort the optical surface and lead to erroneous test results. Knowledge of the time required to reach thermal equilibrium or a uniform temperature condition after a machining or polishing run is essential.

Accurately predicting the time required for a mirror to reach thermal equilibrium requires evaluating heat transfer from the outer surfaces by both radiation and convection, as well as conduction from the interior. Since heat transfer rates from the mirror surface are usually poorly understood, the time to reach thermal equilibrium usually is estimated using an approximate approach. A method described by Strong[41] assumed that the temperature of the surface of the mirror changes instantaneously and that the primary mechanism for heat transfer is by conduction from the interior to surface.

A figure of merit for the time required to reach thermal equilibrium is called the thermal time constant of the mirror. This is the time required for the mirror to change its interior temperature by a factor of $e \approx 2.72$ with respect to the surface. The thermal time constant τ is proportional to

TABLE 8.7
Thermal Diffusivities of Mirror Materials

Material	Thermal Diffusivity β (m²/s × 10⁻⁶)
Corning Pyrex 7740	0.483
Schott Zerodur	0.721
Corning fused silica ULE 7972	0.776
Corning fused silica 7980	0.788
Beryllium, S-200 FH	60.5
Aluminum alloy, 6061-T6	69.0
Silicon	92.9
Copper, OFHC	114
Silicon carbide, cast reaction bonded	411

the square of the distance over which the heat is conducted. This distance is normally taken to be the mirror thickness. It depends upon the thermal diffusivity of the material. The relationships used are

$$\tau = \frac{h^2}{\pi^2 \beta}, \quad \beta = \frac{k}{\rho c_\rho}, \tag{8.55}$$

where h is the mirror thickness, β is the mirror material diffusivity, k is the thermal conductivity of the mirror material, ρ is the mirror material density, and c_ρ is the specific heat of the mirror material.

The thermal diffusivities of metals are greater than those of glasses, so metals have shorter thermal time constants. For example, the thermal diffusivity of 6061 aluminum is 69×10^{-6} m²/s, versus 483×10^{-9} m²/s for Pyrex 7740. The thermal time constant for an aluminum mirror would be $(483 \times 10^{-9})/(69 \times 10^{-6}) = 0.007$, compared to that for a Pyrex mirror. Hence, the time required for the metal mirror to reach equilibrium would be about 0.7% of that for the glass mirror, assuming identical thickness. Table 8.7 includes the thermal diffusivities of selected mirror materials.

The interior of the mirror approaches the surface temperature exponentially. It is commonly assumed that thermal equilibrium is reached after the temperature difference decays to a differential of 1%. This occurs at 4.605 time constants. A rule of thumb states that a mirror is in thermal equilibrium after "five thermal time constants." The temperature of the interior at a time T_i after the change in surface temperature T_S is given by

$$T_i = T_S \left[1 - \exp\left(\frac{-t}{\tau}\right)\right], \tag{8.56}$$

where T_i is the interior temperature of the mirror after time t.

This equation is plotted in Figure 8.11.

Example 8.12

In Noyes' classic account of the 1917 commissioning of the 2.5 m Hooker Telescope, it is stated that the image quality of the telescope improved after about 3 hours. The exact composition of

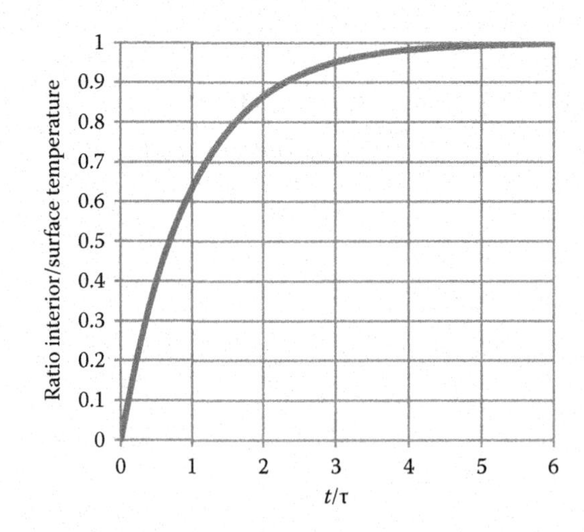

FIGURE 8.11 Ratio of mirror interior temperature T_i to surface temperature T_s after an instantaneous change in surface temperature as a function of the ratio of time after change t to mirror thermal time constant τ (t/τ). Approximate thermal equilibrium is reached ($T_i/T_s \approx 1.0$) after 4.605 thermal time constants ($t/\tau \approx 4.605$).

the glass used in the primary mirror of the 2.5 m Hooker Telescope is not known, although it is a "crown" glass. The thermal properties of most crown glasses are nearly the same, so assume that the properties of the glass in the primary mirror are those of Schott BK-7. For BK-7, ρ = 2510 kg/m^3, k = 1.11 W/m K, c_p = 858 J/kg K, and $\alpha = 7.1 \times 10^{-6}$ K^{-1}. Then, the thermal diffusivity is $D = 515 \times 10^{-9}$ m^2/s. The mirror thickness is about 300 mm, so the thermal time constant is τ = (300 mm)^2/(π^2 × 858 J/kg K) = 4.9 hours. Then, equilibrium is reached in 4.650 × 4.9 hours = 22.6 hours. If the rule of thumb of five thermal constants is used, the time required to reach equilibrium is 24.5 hours, which is close to the usually quoted figure of 24 hours. Atop Mount Wilson, the air temperature changes about 10 K in the first hour after sunset, so this is used as an approximation of the temperature gradient from surface to interior. The gradient is likely to be from front to back in the mirror since the upper surface of the mirror can lose heat via radiation and convection, while the back and sides of the mirror are insulated by the mirror cell. According to Pearson and Stepp,[34] spherical aberration is induced from an axial temperature gradient. The primary mirror focal ratio is *f*/5, so the optical radius is 2 × 5 × 2.5 m = 25 m. Using a 10 K differential, the spherical aberration is given by the following (converting the aberration into waves in the visible [546 nm]):

$$\delta(t) = \frac{23\times10^{-6}\,\mathrm{K}^{-1}}{8}\,\frac{10\,\mathrm{K}}{300\,\mathrm{mm}}\,\frac{(1250\,\mathrm{mm})^4}{(25\,\mathrm{m})^2}\,\frac{1}{0.546\,\mu\mathrm{m}} = 0.23\ \text{wave}.$$

This is a surface aberration, so the aberration on the wave front is twice as big or 0.43 wave, which is well above the Rayleigh criterion and will dramatically degrade image quality. After 3 hours, the temperature differential decreases by a factor of exp(−3 hours/4.9 hours) = 0.542, or to 5.4 K. This drops the aberration to about 0.23 wave on the wave front, which satisfies the Rayleigh "quarter wave" requirement. This analysis supports Noyes' account.

PROBLEMS

1. The maximum surface error tolerance of a mirror is 0.125 wave. Contributions to this tolerance are self-weight deflection, fabrication error, thermal distortion, and alignment. If the tolerance allocation for fabrication error is twice the size of any of the other errors, what is the maximum size of the surface figure error?

   Answer: There are four sources of error, and one is twice as big as any of the others; thus,

$$\varepsilon_{\text{self-weight}} = \varepsilon_{\text{thermal}} = \varepsilon_{\text{align}} = \varepsilon,$$

$$\varepsilon_{\text{fab}} = 2\varepsilon,$$

$$0.125 \text{ wave} = \sqrt{(2\varepsilon)^2 + 3\varepsilon^2}$$

$$\varepsilon = \frac{0.125 \text{ wave}}{7^{\frac{1}{2}}} = 0.047 \text{ wave}$$

$$2\varepsilon = \varepsilon_{\text{fab}} = 0.094 \text{ wave}.$$

   The maximum fabrication error is about 0.094 wave.

2. A limiting factor in optical error budgets is metrology. If the individual errors in a system are at or below the metrology error of 0.02 wave, how many items can there be in the budget where the top-level error is 0.05 wave?

   Answer: Since all of the errors are at or below 0.02 wave, assume that there are N errors which are all equal to 0.02 wave. Then,

$$0.05 \text{ wave} = \sqrt{N(0.02 \text{ wave})^2}$$

$$N = \frac{(0.05 \text{ wave})^2}{(0.02 \text{ wave})^2} = 6.25.$$

   Rounding, there can be six items in the budget, each with a value of 0.02 wave.

3. A telescope mirror is used in a range of elevation from 0° to 90°. In the axis horizontal position, the self-weight deflection is 0.05 RMS wave (1 wave = 633 nm), and in the axis vertical position, the self-weight deflection is 0.1 RMS wave. At what angle is the maximum self-weight deflection, and what is the value of the self-weight deflection at that angle?

   Answer: The self-weight deflection at any angle θ is given by

$$\delta_{\theta\text{-RMS}} = \sqrt{(\delta_{\text{A-RMS}} \cos\theta)^2 + (\delta_{\text{R-RMS}} \sin\theta)^2}$$

$$\delta_{\theta\text{-RMS}} = \sqrt{\left[(0.1 \text{ wave})\cos\theta\right]^2 + \left[(0.05 \text{ wave})\sin\theta\right]^2}.$$

There are several ways to find the maximum value of this relationship, but one that provides insight is to graph it, as shown in the following:

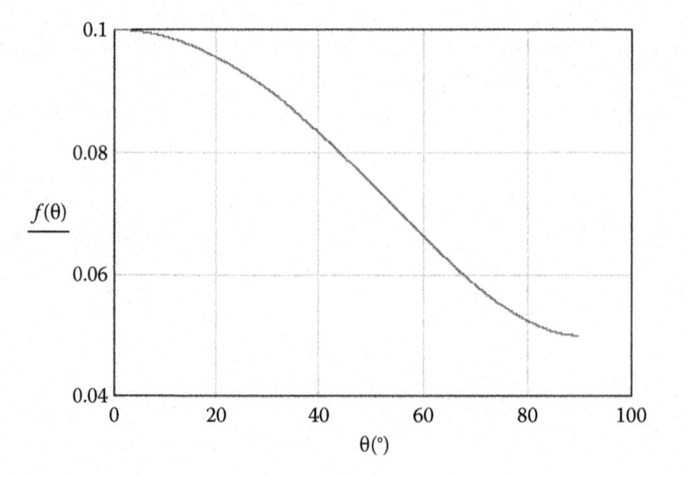

From the graph, it is seen that the maximum value is at $\theta = 0°$, corresponding to the vertical position, and is 0.1 RMS wave.

4. A telescope mirror is used in a range of elevation from 0° to 90°. In the axis horizontal position, the self-weight deflection is 0.10 RMS wave (1 wave = 633 nm), and in the axis vertical position, the self-weight deflection is 0.1 RMS wave. At what angle is the maximum self-weight deflection, and what is the value of the self-weight deflection at that angle?

   Answer: The self-weight deflection at any angle θ is given by

$$\delta_{\theta\text{-RMS}} = \sqrt{(\delta_{\text{A-RMS}} \cos\theta)^2 + (\delta_{\text{R-RMS}} \sin\theta)^2}$$

$$\delta_{\theta\text{-RMS}} = \sqrt{\left[(0.1\text{ wave})\cos\theta\right]^2 + \left[(0.1\text{ wave})\sin\theta\right]^2}.$$

There are several ways to find the maximum value of this relationship, but one that provides insight is to graph it, as shown in the following:

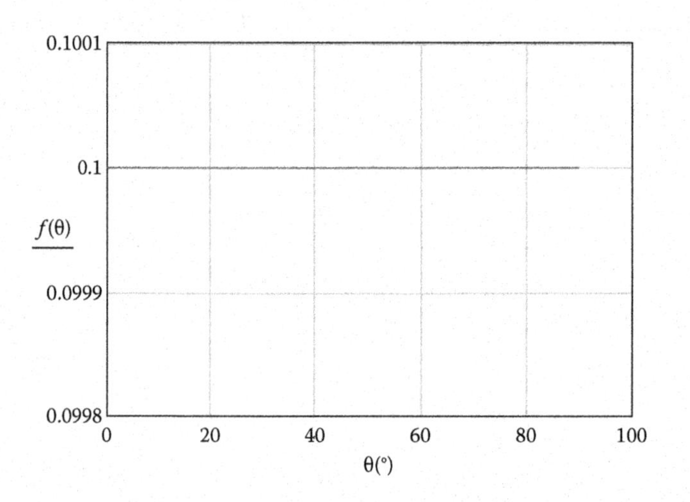

There is no change in the mirror self-weight deflection with elevation angle, and it remains constant at 0.1 RMS wave.

5. Assuming a 0.125 wave surface deflection (1 wave = 633 nm), what is the largest aluminum mirror with a 6:1 aspect ratio (diameter/thickness) that can be mounted on three axial support points at the edge? Neglect shear effects.

   Answer: From Table 8.3, for aluminum, $\rho(1 - \nu^2)/E = 347 \times 10^{-9}\ \text{m}^{-1}$. For a three-point edge mount, $C_A = 85.3 \times 10^{-3}$ from Table 8.4. Solving the equation for diameter by using the bending deflection equation,

$$d_{max} = \frac{h}{d}\left[\frac{\delta_{TOL}}{C_A}\frac{E}{\rho(1-\nu^2)}\right]^{\frac{1}{2}}$$

$$d_{max} = \left(\frac{1}{6}\right)\left[\frac{(0.125)(633\times10^{-9}\ \text{m})}{85.3\times10^{-3}}\frac{1}{347\times10^{-9}\ \text{m}^{-1}}\right]^{\frac{1}{2}}$$

$$d_{max} = 273\,\text{mm}.$$

6. At constant frequency, how does the weight of a circular mirror vary with diameter?

   Answer: The relationship between weight of a circular mirror and self-weight deflection is

$$W = \pi r^4\sqrt{C\frac{1}{\delta_A}\frac{\rho^3(1-\nu^2)}{E}}.$$

Now, frequency is related to self-weight deflection by

$$f_n = \frac{1}{2\pi}\sqrt{\frac{g}{\delta}}.$$

Substituting,

$$W = \pi\frac{d^4}{16}\sqrt{C\frac{4\pi^2 f_n^2}{g}\frac{\rho^3(1-\nu^2)}{E}}$$

$$W = \frac{\pi^2}{8}f_n d^4\sqrt{C\frac{1}{g}\frac{\rho^3(1-\nu^2)}{E}}.$$

At constant frequency, the weight of a mirror varies as the fourth power of its diameter.

7. A 150 mm diameter mirror made of Schott Zerodur is supported at its center by using a stalk mount. How thick must the mirror be if the maximum axial self-weight deflection is 0.05 wave (1 wave = 633 nm) p-v, neglecting shear?

Answer: From Table 8.4, $C_A = 77.5 \times 10^{-3}$ for a center support, while from Table 8.3, for Schott Zerodur, $\rho(1-\nu^2)/E = 257 \times 10^{-9}\ \text{m}^{-1}$. Then,

$$h = \sqrt{C_A \frac{\rho(1-\nu^2)}{E} \frac{d^4}{\delta_{\theta p\text{-}v}} \cos(\theta)}$$

$$h = \sqrt{(77.5\times10^{-3})(257\times10^{-9}\ \text{m}^{-1}) \frac{(0.15\,\text{m})^4}{(0.05)(633\times10^{-9}\ \text{m})} \cos(0°)}$$

$$h = 17.8\,\text{mm}.$$

The mirror must be at least 17.8 mm thick.

8. The diameter of the primary mirror for the Hale Telescope, at Mount Palomar Observatory in California, is 5 m and is made of Pyrex. Although the mirror is an open-back ribbed structure, testing showed that the bending stiffness is equivalent to that of a solid mirror about 370 mm thick. The mirror is supported axially at 36 optimum points. If the Poisson's ratio for Pyrex is 0.2, what is the p-v axial self-weight deflection including the effects of shear?
   Answer: From Table 8.3, for Pyrex 7740, $\rho(1-\nu^2)/E = 333 \times 10^{-9}\ \text{m}^{-1}$. Then,

$$\delta_{\theta p\text{-}v} = \frac{1}{4N^{2.4}} \frac{\rho(1-\nu^2)}{E} \left(\frac{d}{h}\right)^2 d^2 \left[1 + 8N\left(\frac{h}{d}\right)^2\right] \cos(\theta)$$

$$\delta_{\theta p\text{-}v} = \frac{1}{4(36)^{2.4}} (333\times10^{-9}\ \text{m}^{-1}) \left(\frac{5\,\text{m}}{0.370\,\text{m}}\right)^2 (5\,\text{m})^2 \left[1 + 8(36)\left(\frac{0.370\,\text{m}}{5\,\text{m}}\right)^2\right] \cos(0°)$$

$$\delta_{\theta p\text{-}v} = 180\ \text{nm}$$

$$\delta_{\theta p\text{-}v} = \frac{180\,\text{nm}}{633\,\text{nm}} = 0.284\,\text{wave}.$$

9. If the support pad diameter-to-spacing ratio a/b is 0.33 for the Palomar primary mirror, how much reduction in axial deflection can be expected?
   Answer: The correction for finite pad size is

$$\delta_A = \delta_{AP} \left\{1 - \frac{4}{3}\left(\frac{a^2}{b^2}\right)\left[1 - \ln\left(\frac{a^2}{b^2}\right)\right]\right\}$$

$$\delta_A = \delta_{AP} \left\{1 - \frac{4}{3}(0.33)^2\left[1 - \ln(0.33)^2\right]\right\}$$

$$\delta_A = 0.533\delta_{AP}.$$

This would reduce the deflection in the previous problem to 0.533 (180 nm) = 96 nm or 0.152 wave p-v.

10. Using semi-infinite plate theory, what is the self-weight deflection of the Hale Telescope primary mirror? Assume Pyrex, $d = 5$ m, $h = 0.37$ m, and 36 optimum support points.

    Answer: The self-weight deflection from semi-infinite plate theory is

$$\delta_{\text{p-v}} = 0.586 \frac{\rho(1-\nu^2)}{E} \frac{r^4}{h^2} \frac{1}{N^2}$$

$$\delta_{\text{p-v}} = 0.586(333 \times 10^{-9}\ \text{m}^{-1}) \frac{\left(\frac{5\,\text{m}}{2}\right)^4}{(0.37\,\text{m})^2} \frac{1}{(36)^2}$$

$$\delta_{\text{p-v}} = 43\ \text{nm}.$$

This is about four times smaller than the self-weight deflection predicted when including shear effects, showing that deflection from shear is an important factor.

11. An aluminum mirror is mounted with three back supports at $d_s/d = 0.8$; the mirror diameter is 150 mm, and its thickness is 25 mm. The tolerance for mounting deflection is 50 RMS nm. If the hoop and radial moments are equal, what are the tolerances for moments in hoop and radial directions? Use $E = 69$ GPa and $\nu = 0.3$ for aluminum.

    Answer: First, find the flexural rigidity and then calculate the influence functions. Next, allocate tolerances to the hoop and radial directions. Finally, use the influence functions to find the tolerances.

    The flexural rigidity D is

$$D = \frac{Eh^3}{12(1-\nu^2)} = \frac{69\,\text{GPa}\,(0.025\,\text{m})^3}{12\left[1-(0.3)^2\right]} = 98.7 \times 10^3\ \text{N m}.$$

Since $d_s/d = 0.8$, the influence functions are

$$f_r\left(\frac{r}{r_0}\right) = \left[-0.0250\left(\frac{r}{r_0}\right)^2 + 0.0444\left(\frac{r}{r_0}\right) - 0.0013\right]\text{RMS wave/N m}$$

$$f_r(0.8) = \left[-0.0250(0.8)^2 + 0.0444(0.8) - 0.0013\right]\text{RMS wave/N m}$$

$$f_r(0.8) = 18.2 \times 10^{-3}\ \text{RMS wave/N m},$$

$$f_h\left(\frac{r}{r_0}\right) = \left[0.004392\left(\frac{r}{r_0}\right)^2 + 0.008603\left(\frac{r}{r_0}\right) - 0.000690\right]\text{RMS wave/N m}$$

$$f_h(0.8) = \left[0.004392(0.8)^2 + 0.008603(0.8) - 0.000690\right]\text{RMS wave/N m}$$

$$f_h(0.8) = 9.00 \times 10^{-3}\ \text{RMS wave/N m}.$$

Now the radial moments M_r and M_h are equal, so $M_r = M_h$ and

$$\delta_{TOL}(M,0.8) = \sqrt{3}\left(\frac{D_{REF}}{D}\right)\left(\frac{d}{d_{REF}}\right)^2\left\{\left[M_r f_r(0.8)\right]^2 + \left[M_h f_h(0.8)\right]^2\right\}^{\frac{1}{2}}$$

$$\delta_{TOL}(M,0.8) = \sqrt{3}\left(\frac{D_{REF}}{D}\right)\left(\frac{d}{d_{REF}}\right)^2\left\{M_r^2\left[f_r(0.8)^2 + f_h(0.8)^2\right]\right\}^{\frac{1}{2}}$$

$$M_r = \frac{\delta_{TOL}(M,0.8)}{\sqrt{3}\left(\frac{D_{REF}}{D}\right)\left(\frac{d}{d_{REF}}\right)^2\left[f_r(0.8)^2 + f_h(0.8)^2\right]^{\frac{1}{2}}}$$

$$M_r = \frac{\left(\frac{50\,\text{nm}}{633\,\text{nm}}\right)\text{RMS wave}}{\sqrt{3}\left(\frac{1.617\times10^6\,\text{N m}}{98.7\times10^3\,\text{N m}}\right)\left(\frac{0.150\,\text{m}}{0.600\,\text{m}}\right)^2\left[(18.2\times10^{-3}\,\text{RMS wave/N m})^2 + (9.00\times10^{-3}\,\text{RMS wave/N m})^2\right]^{\frac{1}{2}}}$$

$$M_r = M_h = 2.19\,\text{N m}.$$

So the tolerance for the radial and hoop moments when mounting the mirror is about 2.2 N m.

12. A scanning mirror is elliptical with a semiminor axis a = 10 mm and semimajor axis = 14 mm. The mirror is 2 mm thick and is made of aluminum. If the surface deflection tolerance when scanning is 0.1 wave (1 wave = 633 nm), what is the limiting angular acceleration?

    Answer: From Table 8.3, for aluminum, $\rho(1 - \nu^2)/E = 347 \times 10^{-9}\ \text{m}^{-1}$. Then,

$$\delta = 0.628\frac{\rho(1-\nu^2)}{E}\frac{ba^4}{h^2}\frac{\omega}{g},$$

$$\omega = \frac{\delta g}{0.628\frac{\rho(1-\nu^2)}{E}\frac{ba^4}{h^2}}$$

$$\omega = \frac{(0.1)(633\,\text{nm})(9.81\,\text{m/s}^2)}{0.628(347\times10^{-9}\,\text{m}^{-1})\frac{(0.01\,\text{m})(0.014\,\text{m})^4}{(0.002\,\text{m})^2}}$$

$$\omega = 29.7\times10^3\,\text{rad/s}^2.$$

13. A mirror is mounted with three tangential flexures attached to its edge by ball joints. The mirror mass is 3 kg. Each flexure is 50 mm long, 25 mm deep (axial direction), and 1 mm wide (radial direction). The flexures are made of stainless steel, where E = 193 GPa. If the mirror is excited by a sinusoid with amplitude of 0.1 mm and frequency of 60 Hz, what is the maximum displacement of the mirror in the axial direction, assuming that damping is negligible?

Answer: Since the flexures are free to rotate at the end (ball joint), the axial stiffness k_A is given by

$$k_A = \frac{Eb^3h}{4L^3}$$

$$k_A = \frac{(193\,\text{GPa})(25\,\text{mm})^3 1\,\text{mm}}{4(50\,\text{mm})^3}$$

$$k_A = 6.031 \times 10^6\,\text{N m}.$$

Then, the fundamental frequency f_{nZ} is given by

$$f_{nZ} = \frac{1}{2\pi}\sqrt{\frac{3k_A}{m}} = \frac{1}{2\pi}\sqrt{\frac{3(6.031 \times 10^6\,\text{N m})}{3\,\text{kg}}} = 391\,\text{Hz}.$$

If the damping is small, the maximum displacement is

$$x_0 \approx \text{abs}\left(\frac{x_1}{1 - \frac{f^2}{f_n^2}}\right)$$

$$x_0 = \frac{100 \times 10^{-6}\,\text{m}}{1 - \frac{(60\,\text{Hz})^2}{(391\,\text{Hz})^2}}$$

$$x_0 = 102 \times 10^{-6}\,\text{m}.$$

So the axial displacement of the mirror is 102 μm, which is almost the same as the amplitude of the excitation.

14. How long will it take for the Hubble Space Telescope to "ring down" after an impulse excitation? Assume that the amplitude after ring down is 1% (ring down is the same as the settling time defined in the text) of the initial impulse, and use the values for the Hubble frequency and damping given in the text.
Answer: From the text, f_n = 17 Hz and Q = 125. Then, the ring-down time (1%) is

$$t(1\%) \approx 1.466\frac{Q}{f_n} = 1.466\frac{125}{17\,\text{Hz}} = 10.8\,\text{s}.$$

After a sudden impulse, it takes about 11 s for the Hubble to recover.

15. A small mirror has an axial frequency (rigid body) of 500 Hz, and a Q = 40. The focus tolerance for the mirror is ±50 μm (100 μm total). Will the mirror stay within its focus tolerance if subjected to random vibration in the axial direction where PSD = $0.02g^2$/Hz?

Answer: The RMS response of the mirror is given by

$$x_{\text{RMS}} = \left(\frac{1}{32\pi^3} \frac{Q\,\text{PSD}}{f_{\text{n}}^3} \right)^{\frac{1}{2}}$$

$$x_{\text{RMS}} = \left[\frac{1}{32\pi^3} \frac{(40)(0.02\,g^2/\text{Hz})}{(500\,\text{Hz})^3} \right]^{\frac{1}{2}} = \left\{ \frac{1}{32\pi^3} \frac{(40)(0.02)(9.81\,\text{m/s})^2(1\,\text{s}^{-1})}{(500\,\text{s}^{-1})^3} \right\}^{\frac{1}{2}}$$

$$x_{\text{RMS}} = 24.9 \times 10^{-6}\,\text{m}.$$

Maximum displacement $x_{max} = 3x_{RMS}$, or 3(25 μm) = 75 μm. Since x_{max} = 75 μm < 100 μm tolerance, the mirror stays in focus.

16. Free convection from the front surface of a telescope mirror induces a heat flow of 2 W/m^2. For a focal length of 1 m, what focus shift can be anticipated from this heat flow for a Pyrex 7740 mirror?

Answer: From Table 8.5, for Pyrex 7740, $\alpha/k = 2920 \times 10^{-9}$ m/W. Then, the focus shift Δf is given by

$$\Delta f = 2f^2 P \frac{\alpha}{k} = 2(1\,\text{m})^2(2\,\text{W/m}^2)(2920 \times 10^{-9}\,\text{m/W}) = 11.7 \times 10^{-6}\,\text{m}.$$

The focus shift is about 12 μm.

17. Amateur astronomers often claim that their telescope mirrors are subjected to excessive cooldown times, requiring an hour or more after sunset to reach thermal equilibrium. Assume that a large amateur telescope mirror is 300 mm in diameter, with a 6:1 aspect ratio, it is made of Pyrex, and the focal length f = 1.5 m. If the instantaneous air temperature change after sunset is 10 K and the mirror loses heat by conduction along its axis, estimate the time required for the OPD from spherical aberration to be less than 0.125 wave (1 wave = 550 nm). Use 3.3×10^{-6} K^{-1} for the thermal expansion of Pyrex 7740.

Answer: The OPD for spherical aberration due to an axial temperature difference is given by

$$\delta_{\text{spherical}} = \frac{\alpha C_3 r^4}{8R^2} = \frac{\alpha \Delta T r^4}{8hR^2}$$

$$\Delta T = \frac{\delta_{\text{spherical}} 8hR^2}{\alpha r^4} = \frac{(0.125)(550\,\text{nm})8\left(\frac{0.3\,\text{m}}{6}\right)\left[2(1.5\,\text{m})\right]^2}{(3.3 \times 10^{-6}\,\text{K}^{-1})\left(\frac{0.3\,\text{m}}{2}\right)^4}$$

$$\Delta T = 148\,\text{K}.$$

A temperature difference from front to back in the mirror of 148 K is necessary to cause a 0.125 wave spherical aberration error. This temperature difference is greater than the 10 K air temperature change, so the mirror can be used immediately, without waiting for cooldown. Hence, it is not necessary to calculate the thermal equilibrium time. It is possible that there is some other thermal effect on the mirror, but axial gradients are not likely to be a problem.

18. A copper folding flat is used at a 30° angle to the beam in a laser machining system, with a 1 μm wavelength laser. The optical coating on the mirror reflects 99% of the incident power. If the OPD tolerance from thermal distortion of the mirror is 0.05 wave (1 wave = 1 μm), how much power can the mirror handle without exceeding this tolerance?

Answer: From Table 8.5, $\alpha/k = 43.4 \times 10^{-9}$ m/W for OFC copper. The OPD from thermal distortion is

$$\text{OPD} = \left(\frac{P}{\pi}\right)\left(\frac{\alpha}{k}\right)(\sin^2\theta)(\varepsilon).$$

Solving for the power P,

$$P = \frac{\pi\,\text{OPD}}{\left(\frac{\alpha}{k}\right)(\sin^2\theta)(\varepsilon)}$$

$$P = \frac{\pi(0.1)(1\times10^{-6}\,\text{m})}{(43.4\times10^{-9}\,\text{m/W})\left[\sin(30°)\right]^2(1-0.99)}$$

$$P = 2895\,\text{W}.$$

The mirror can handle about 2.9 kW of laser power without exceeding the OPD tolerance of 0.1 μm.

19. The Russian BTA-6 telescope primary mirror diameter is 6 m, with an aspect ratio of 10:1. There are reports that thermal inertia is a problem with this mirror. Estimate the time to reach thermal equilibrium of this mirror, assuming that the properties of the mirror material are similar to those of Pyrex 7740.

Answer: From Table 8.7, the thermal diffusivity of Pyrex 7740 is $\beta = 0.483 \times 10^{-6}$ m²/s. The thermal time constant τ is given by

$$\tau = \frac{h^2}{\pi^2\beta} = \frac{\left(\frac{6\,\text{m}}{10}\right)^2}{\pi^2(0.483\times10^{-6}\,\text{m/s}^2)} = 75.5\times10^3\,\text{s}.$$

Thermal equilibrium is reached (1% temperature difference from interior to surface) after 4.65 thermal time constants, so the equilibrium time t is

$$t = \frac{4.65(75.5\times10^3\,\text{s})}{(24\,\text{hours/day})(3600\,\text{s/hour})} = 4.06\,\text{days}.$$

With a thermal equilibrium time of about 4 days (!), the mirror interior temperature will always lag behind the diurnal cycle, hence the thermal inertia issue.

20. A giant 8.1 m diameter telescope with a thickness of 200 mm mirror is made of Corning ULE. The measured anisotropy of CTE is about 5×10^{-9} K⁻¹. The tolerance on surface distortion of the mirror is about 15 nm. If the anisotropy of CTE is across the diameter of the mirror (varies in the mirror surface plane), can the tolerance be held for a 10 K change in temperature?

Answer: The surface deflection for and in-plane variation is given by

$$\delta \cong \frac{r^2}{h}(\alpha_X - \alpha_Y)\Delta T.$$

Let $\alpha_X - \alpha_Y = 5 \times 10^{-9}\ \text{K}^{-1}$. Then,

$$\delta = \frac{\left(\frac{8.1\,\text{m}}{2}\right)^2}{0.2\,\text{m}}(5\times10^{-9}\,\text{K}^{-1})(10\,\text{K})$$
$$\delta = 4.1\times10^{-6}\,\text{m}.$$

Since $\delta = 4.1\ \mu\text{m} \gg 15$ nm, the tolerance for deflection from anisotropy cannot be held. In a real telescope mirror, much of this distortion would be a focus shift, which could be minimized by adjusting focus. Even so, this problem shows that anisotropy of CTE is a major issue for very large telescope mirrors.

REFERENCES

1. Fischer, R.E., and Tadic-Galeb, B., *Optical System Design*, McGraw-Hill, New York, 2000.
2. Miller, J.L., and Friedman, E., *Photonics Rules of Thumb*, 2nd ed., McGraw-Hill, New York, 2003.
3. Shannon, R.R., Image quality, in *Optical Engineer's Desk Reference*, Wolfe, W.L., Ed., pp. 423–432, Optical Society of America, Washington, DC, 2003.
4. Nelson, J.E., and Mast, T.S., Very large telescopes and their instrumentation, *Proc. ESO Conf.*, ESO, Garching, Germany, Proc. No 30, 7, 1988.
5. Schwesinger, G., Optical effect of flexure in vertically mounted precision mirrors, *J. Opt. Soc. Am*, 44, 417, 1954.
6. Schwesinger, G., and Knohl, E.D., Comments on a series of articles by L.A. Selke, *Appl. Opt.*, 11, 200, 1972.
7. Craig, L.D., and Boulet, J.A.M., *Deflections of a Uniformly Loaded Circular Plate with Multiple Point Supports*, NASA/TN-1999-209631, Marshall Space Flight Center, Huntsville, AL, 1999.
8. Selke, L.A., Theoretical elastic deflections of a thick horizontal circular mirror on a ring support, *Appl. Opt.*, 9, 149, 1970.
9. Arnold, L., Optimized axial support topologies for thin telescope mirrors, *Opt. Eng.*, 34, 567, 1995.
10. Wan, D-S., Angel J.R.P., and Parks, R.E., Mirror deflection on multiple point supports, *Appl. Opt.*, 28, 2, 354, 1989.
11. Lemaitre, G.R., *Astronomical Optics and Elasticity Theory*, Springer-Verlag, Berlin, 2009.
12. Selke, L.A., Theoretical elastic deformations of a thick horizontal circular plate having interrupted peripheral arc supports, *Int. J. Solids Struct.*, 7, 241, 1971.
13. Timoshenko, S., and Woinokowsky-Krieger, S., *Theory of Plates and Shells*, McGraw-Hill, New York, 1959.
14. Cowper, G.R., The shear coefficient in Timoshenko's beam theory, *J. Appl. Mech.*, 33, 2, 335, 1966.
15. Seibert, G.E., Design of lightweight mirrors, SPIE short course SC18, 1990.
16. Nelson, J.E. et al., Telescope mirror supports: Plate deflections on point supports, *Proc. SPIE*, 332, 212, 1982.
17. Kingslake, R., *Optical System Design*, Academic Press, Cambridge, MA, 1983.
18. Danjon, A., and Couder, A., *Lunettes et Telescopes*, Blanchard, Paris, France 1935.
19. Tyson, R.K., and Ulrich, P.B., Adaptive optics, in *Emerging Systems and Technologies, Vol. 8, The Infrared & Electro-optical Systems Handbook*, Robinson, S.R., Ed., pp. 167–240, SPIE Press, Bellingham, WA, 1993.
20. Vukobratovich, D., and Coronato, P., Unit moment analysis as a guide to mirror mount design, *Proc. SPIE*, 9573, 95730R, 2015.

21. Montagu, J.I., Galvanometric and resonant low-inertia scanners, *Optical Scanning*, Marshall, G.F. ed., Marcel Dekker, New York, 1991.
22. Brosens, P.J., Dynamic mirror distortions in optical scanning, *Appl. Opt.*, 11, 2987, 1972.
23. Blevins, R.D., *Formulas for Natural Frequency and Mode Shape*, Van Nostrand Reinhold, New York, 1979.
24. Vukobratovich, D., and Richard, R.M., Flexure mounts for high-resolution optical elements, *Proc. SPIE*, 959, 18, 1989.
25. Lazan, B.J., *Damping of Materials and Members in Structural Mechanics*, Oxford, NY: Pergamon, 1968.
26. James, D.W., High damping metals for engineering application, *Mater. Sci. Engr.*, 4, 1, 1969.
27. Simonian, S.S., Survey of spacecraft damping measurements: Applications to electro-optical jitter problems, in *The Role of Damping in Vibration and Noise Control, Proceedings of the ASME, Symposium of the Role of Damping in Vibration and Shock Control*, New York, 287, 1987.
28. Wada, B.K., and DesForges, D.T., Spacecraft damping considerations in structural design, in *Proceedings of the 48th Meeting of the AGARD Structures and Materials Panel (AGARD Proc. 277)*, Williamsburg, VA, April 2–3, NATO Advisory Group for Aerospace Research and Development, Neuilly-Sur-Seine, France, 6, 1979.
29. Snowdon, J.C., *Vibration and Shock in Mechanical Systems*, John Wiley & Sons, Hoboken, NJ, 1968.
30. Lalanne, C. *Random Vibration, Mechanical Shock and Vibration, Vol. III*, Taylor & Francis, London, 2002.
31. Ayre, R.S., Transient response to step and pulse functions, in *Shock and Vibration Handbook*, 2nd ed., Harris, C.M., and Crede, C.E., Eds., pp. 8.1–8.51, McGraw-Hill, New York, 1976.
32. Smith, S.T., and Chetwynd, D.G., *Foundations of Ultraprecision Mechanism Design*, Gordon & Breach, Philadelphia, PA, 1992.
33. Wilson, R.N., *Reflecting Telescope Optics II*, Springer, New York, 1999.
34. Pearson, E., and Stepp, L., Response of large optical mirrors to thermal distributions, *Proc. SPIE*, 748, 215, 1987.
35. Kremer, R.M., Response of parabolodial surfaces to linear thermal gradients, *Proc. SPIE*, 5176, 9, 2003.
36. Paquin, R.A., Dimensional instability: An overview, *Proc. SPIE*, 1335, 2, 1990.
37. Jedamzik, R., Muller, R., and Hartman, P., Homogeneity of the linear thermal expansion coefficient of Zerodur measured with improved accuracy, *Proc. SPIE*, 6273, 06, 2006.
38. Arnold, W.R., Study of small variations of coefficient of thermal expansion in Corning ULE glass, *Proc. SPIE*, 5179, 28, 2003.
39. Pellerin, C.J. et al., New opportunities from materials selection trade-offs for high precision space mirrors, *Proc. SPIE*, 542, 5, 1985.
40. Noyes, A., *Watchers of the Sky*, F.A. Stokes, New York, 1922.
41. Strong, J., *Procedures in Applied Optics*, Marcel Dekker, New York, 1989.

9 Design and Mounting of Small Mirrors

9.1 INTRODUCTION AND SUMMARY

In this chapter, some general aspects of designing smaller mirrors are discussed. Common applications are listed to illustrate the use of small mirrors. Next, the design of mirrors is considered from the geometric, image orientation, and system function viewpoints. Methods for determining the required apertures of flat reflecting surfaces for diverging, converging, and collimated beams of given sizes are summarized. The formation of ghost images by second-surface mirrors and a method for estimating their intensities relative to those of the main images are explained. The influences of selected types of reflecting coatings on main image and ghost image intensities are summarized.

Flat as well as curved mirrors must, of course, be mounted to incorporate them in an optical instrument. Reflecting optics are significantly more sensitive to applied forces than refracting optics because the change in the localized optical path in a reflected wave front resulting from a given surface deformation is twice the magnitude of that deformation. Further, the angular deflection of a reflected ray due to a surface deformation is twice the magnitude of the local surface tilt. On the other hand, small linear displacements or angular misalignments of such optics in particular ways relative to other system components may not affect the overall system performance. Two good examples are a truly flat mirror that moves on its own plane without changing orientation and a Porro mirror subassembly that tilts about an axis perpendicular to the plane of reflection.

Although they have many properties in common with mirrors, diffraction gratings do not enjoy quite as much freedom of motion as mirrors because they are sensitive to the orientation of their grooves. Mirrors with curved optical surfaces are also sensitive to translations and tilts because of their image-forming properties. In general, the locations and orientations of all mirrors must be carefully maintained. Mountings for these components can then become complex optomechanical design problems. Typical examples of mountings for small mirrors are presented quantitatively to illustrate a variety of techniques that have been used successfully.

9.2 GENERAL CONSIDERATIONS

9.2.1 Defining Small Mirrors

According to US MIL-STD-1472D, *Human Engineering Design Criteria for Military Systems*, an average man using both hands can safely pick up an object from the floor, carry it, and place it on a horizontal surface about 1.5 m high if it weighs no more than 25 kg. Assuming a cylindrical plane-parallel solid mirror of the classic 6:1 aspect ratio made of ultralow expansion (ULE) glass, this mass corresponds to a diameter of about 440 mm. This sets one limit on the size of a mirror, since when mechanical equipment is required for handling the cost and complexity of working with the mirror increases dramatically. Note that the mass of small mirrors is estimated using the same techniques as that for individual lenses.

An alternate definition of small versus large mirrors is that a small mirror is rigid, capable of maintaining its optical surface figure when supported at hard points. Danjon and Couder defined a rigid glass mirror as one where $d^4/h^2 < 5\ \text{m}^2$, where d is the mirror diameter and h is the thickness. Using this criterion and again assuming a 6:1 right circular cylinder mirror, the largest "rigid" mirror is about 370 mm in diameter. Larger mirrors require more complex mountings, increasing cost.

These two criteria suggest that a small mirror is one below about 400 mm in diameter. In this chapter, a diameter of 500 mm is assumed to define the size of a small mirror. The overwhelming majority of mirrors in optical systems are below 500 mm in diameter, so this is a conservative definition.

9.2.2 Mirror Applications

Flat mirrors, used singly or as combinations of two or more, serve useful purposes in optical instruments, but do not contribute optical power and, hence, cannot form images by themselves. The principal uses of these mirrors in optical instruments are the following:

- To bend (deviate) light around corners
- To fold an optical system into a given shape or package size
- To provide proper image orientation
- To displace the optical axis laterally
- To divide or combine beams by intensity or aperture sharing at a pupil
- To divide or combine images by reflection at an image plane
- To provide dynamic scanning of beams
- To disperse light spectrally (with gratings)

Most of these functions are the same as those mentioned in the chapter on prisms (Chapter 7). Curved mirrors can do a few of these things, but their most common applications involve image formation, as is the case in reflecting telescopes.

9.2.3 Geometric Configurations

Most small mirrors have single solid substrates in the form of right circular cylinders or rectangular parallelepipeds. Some special-purpose mirrors have other face shapes. One example is the folding flat mirror in a Newtonian telescope that is elliptical in shape so as to minimize obscuration of the input beam when inclined at 45° to the axis. Mirrors can have flat, spherical, aspherical, cylindrical, or toroidal optical surfaces. Optical curved surfaces can be convex or concave. Usually, the second or back surface of a small mirror is flat, but some are shaped to make the overall profile into a meniscus. Most mirrors used in optical instruments are of the first-surface reflecting type and have thin, metallic film, reflecting coatings, such as aluminum, that are overcoated with protective dielectric coatings (typically magnesium fluoride or silicon monoxide). Nonmetallic substrates are typically borosilicate crown glass, fused silica, or one of the low-expansion materials (such as ULE or Zerodur). The thickness of the substrate is traditionally chosen as 1/5–1/6 the largest face dimension. Thinner or thicker substrates are used as the application allows or demands.

Second-surface mirrors have their reflecting coating on the back side of the mirror; the first surface then acts as a refracting surface. An image-forming mirror of this type offers distinct advantages from an optical design viewpoint as compared with the corresponding first-surface version because it has more design variables (a glass thickness, a refractive index, and one more radius) to be used for aberration correction. The refracting surface typically receives an antireflection (A/R) coating, such as magnesium fluoride, to reduce the effects of ghost images from that surface.

Mirror mass is estimated using the same methods as for lenses discussed in the chapter on mounting individual lenses (Chapter 5). Most small mirrors are right circular cylinders with a flat back. The mass W of this type of mirror, where the diameter-to-thickness ratio is defined as ξ, is given approximately by the following equation, where f_{no} is the focal ratio of the curving optical surface:

$$W = \frac{\pi}{4}\rho d^3\left(\frac{1}{\xi} - \frac{1}{32 f_{no}}\right), \tag{9.1}$$

where ρ is the mirror material density and d is the mirror diameter.

Compared to the exact mass, the error in this approximation is less than 1% for $\xi \le 10$ and $f_{no} \ge 1.0$. For an aspect ratio of $\xi = 6$, which is common, the error in the approximation is less than 0.5% for $f_{no} = 1$ and below 0.1% when $f_{no} > 2.0$.

Meniscus shapes are occasionally used for small mirrors. With the meniscus shape, the thickness is constant, with the radius of the mirror back equal to the radius of the front surface plus the mirror thickness. The approximate mass W of a meniscus mirror with an optical focal ratio f_{no}, aspect ratio ξ, and diameter d is given approximately by

$$W = \frac{\pi}{4}\rho\frac{d^3}{\xi}\left(1 - \frac{1}{128 f_{no}}\right). \tag{9.2}$$

The error in this approximation compared to the exact volume is less than 2.5% when $\xi \le 10$ and $f_{no} \ge 1.0$; when $\xi \le 10$ and $f_{no} \ge 2.0$, the error is less than 1.05%.

9.2.4 Reflected Image Orientation

Reflection from a single mirror results in an image that is left handed. This means that the object appears reversed (or reverted) in the plane of reflection. Figure 9.1 shows this reversal for an arrow-shaped object *A–B*. If the observer at the eye point looks directly at the object, the point (*B*) appears on the right. It appears on the left in the image *A′–B′*. Note that the entire image can be observed with one eye by using the portion of the mirror extending from *P* to *P′*. If the object is a word, it can

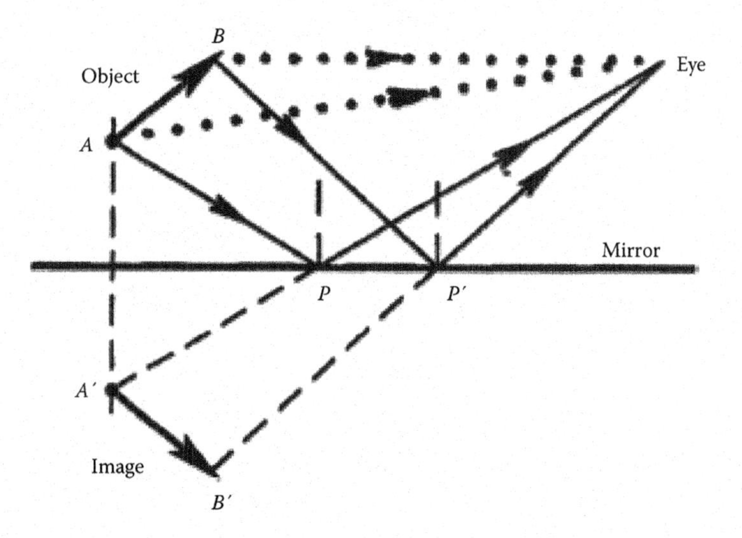

FIGURE 9.1 Reflection of an arrow-shaped object by a flat mirror as seen by an observer's eye. Note the apparent inversion of the image relative to the object as seen directly and the limited portion of the surface of the mirror actually used by the light rays.

be easily read directly, but the left-handed image will appear backward and cannot be read easily without careful thought. The view in Figure 9.2a shows the letter "P" as right handed. It can be read, even if upside down. In the view in Figure 9.2b, it appears left handed and is not as easy to read, regardless of how the page is turned.

With systems of mirrors, the orientation of the image becomes a little more complex. Each reflection reverts the image. An odd number of reflections create a left-handed image, while an even number creates a right-handed image. In optical systems where an erect and unreverted image is needed, such as a terrestrial telescope, careful consideration must be given to the number of reflections occurring in each meridian. If the planes of reflection of multiple mirrors are not oriented orthogonally, the image may appear rotated about the axis. An image rotator/derotator (such as a Pechan prism) might be needed to correct this potential orientation problem.

Each reflection at oblique incidence deviates a ray by some angle δ_i. With reflections on the same plane from multiple mirrors, deviations add algebraically. This is shown in Figure 9.3 for two mirrors. The total deviation is $\delta_1 + \delta_2$. This principle is applied in the layouts of two periscopes shown in Figure 9.4. In the view in Figure 9.4a, mirrors M_1 and M_2 are parallel and inclined to the X-axis by 45° angles. Since the mirror normals at the reflecting surfaces are opposed, the deviations have opposite signs. Hence, $\delta = \delta_1 + \delta_2 = 0$. The output ray is then parallel to the input ray. The intermediate ray path is vertical, so the X separation of the incidence points on the mirrors is zero.

The view in Figure 9.4b shows the more general case of a periscope in which the intermediate ray between the two mirrors is traveling at an angle σ to the Y-axis and the output ray travels at an angle with respect to the input ray direction. Now there are both X and Y separations (Δx and Δy) of the points of incidence. Once again, the total deviation is the sum of the individual deviations of the two mirrors; the second deviation is taken as negative for the reason stated previously. Signs assigned to other angles are as noted in the figure. Typically, the design of such a periscope starts with desired

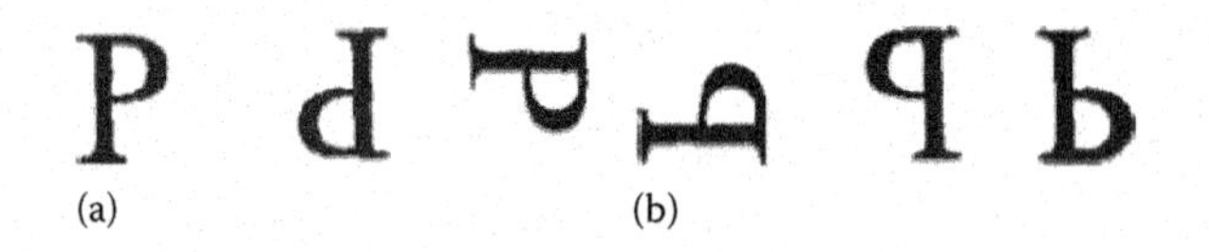

FIGURE 9.2 (a) Right- and (b) left-handed images. The images in (a) can be oriented for easy reading by rotating about the line of sight, but this is not true for the images in (b).

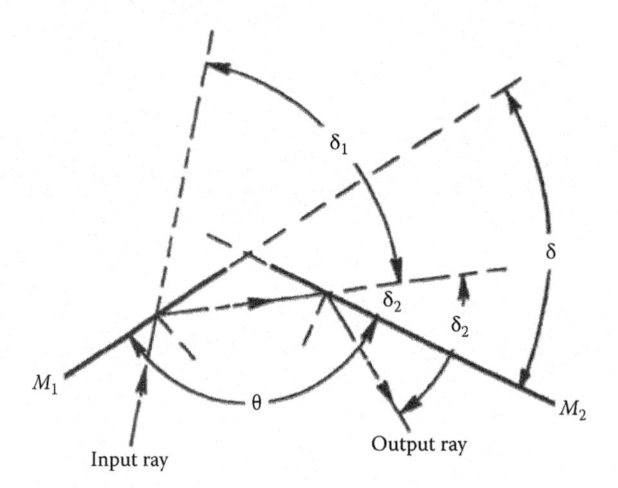

FIGURE 9.3 Deviations of a light ray upon intersecting two flat mirrors oriented at an angle δ to each other. The total deviation is the sum of δ_1 and δ_2.

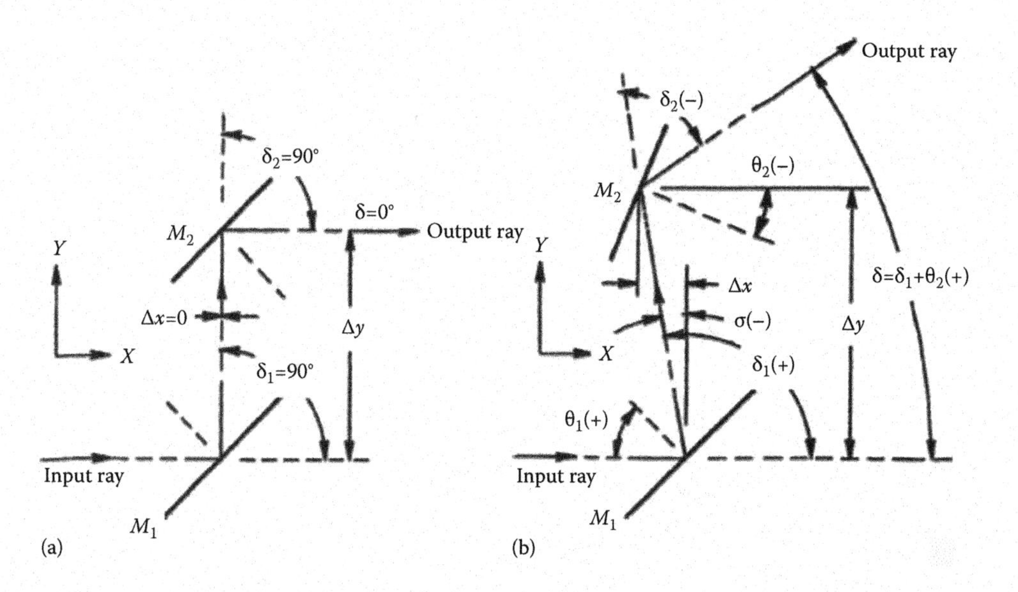

FIGURE 9.4 Deviation and lateral displacement of a ray by two mirrors arranged as a periscope. In view (a), the mirrors are parallel and oriented for 45° incidence. The more general case is shown in view (b). The algebraic signs of the angles are indicated in parentheses in the latter view.

vertical and horizontal offsets Δx and Δy plus a desired deviation δ. Equations that can be used to determine the other parameters are as follows:

$$\sigma = \arctan\left(\frac{\Delta x}{y}\right), \tag{9.3}$$

$$\theta_1 = \frac{\sigma + 90°}{2}, \tag{9.4}$$

$$\theta_2 = \frac{\delta - \sigma - 90°}{2}, \tag{9.5}$$

$$\delta_1 = 180° - 2\theta_1, \tag{9.6}$$

$$\delta_2 = \delta + \sigma - 90°. \tag{9.7}$$

This rather simple design becomes much more complex if the light path through the periscope is to be three dimensional, i.e., involving out-of-plane angles. Then, one might resort to surface-by-surface ray tracing in a lens design program or a vector analysis technique. The packaging of systems with many reflections, such as that illustrated in Figure 9.5, yields well to such techniques. The process of laying out a convoluted optical path is sometimes called "optical plumbing."

Another important aspect of the use of multiple mirror systems is the orientation of the intermediate and final images. A simple technique is to sketch the system in isometric form and to visualize the changes that take place at the mirror when a "pencil crossed with a drumstick" is "bounced"

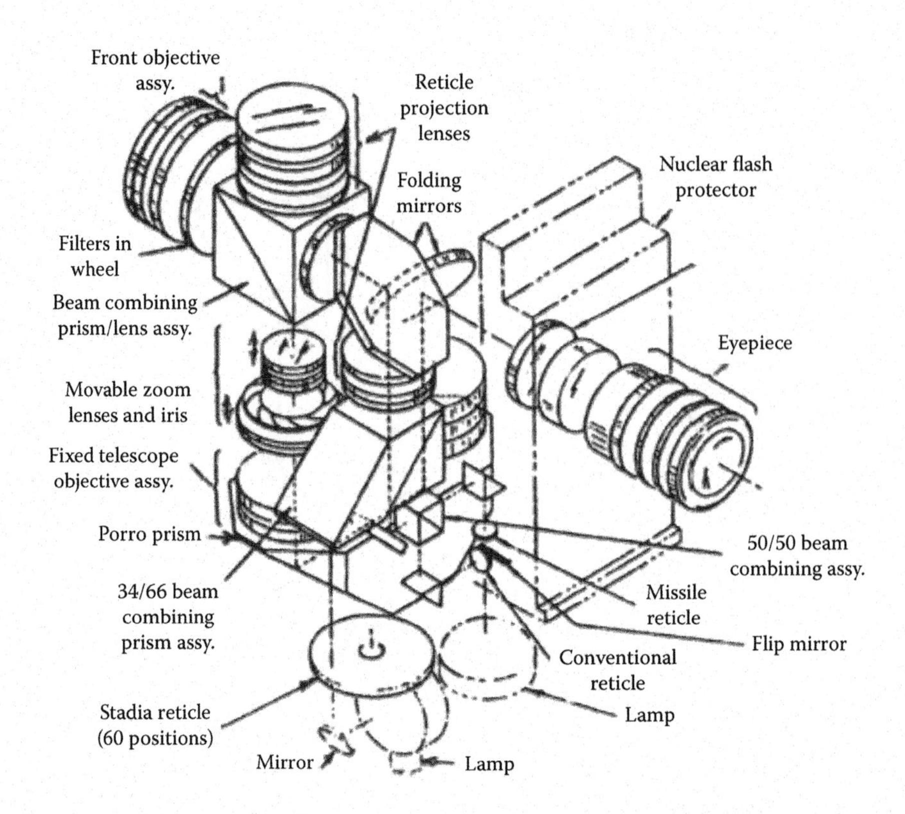

FIGURE 9.5 Optical system schematic of a zoom telescope intended for use in the periscope of a gunner for an armored vehicle. Layout of a convoluted light path such as this is facilitated by use of vector representations of mirror normal and optic axis paths. (Courtesy of the US Army, Washington, DC.)

from the reflecting surfaces. This is illustrated in Figure 9.6. The view in Figure 9.6a applies on the plane of reflection, while the view in Figure 9.6b shows how the change occurs in both meridians. The inversion that naturally occurs at an objective lens or relay lens can be included as shown in Figure 9.7a. In this three-dimensional figure, the object at *A* is projected by a lens *B* onto a screen at *S*. The center of the image is located at distances Δx and Δy from the lens. Figure 9.7b shows another three-dimensional system with optics.

Example 9.1

A periscope as in Figure 9.4b displaces the optical axis vertically by $\Delta y = 600$ mm and deviates the optical axis by $\delta = 30°$. The horizontal displacement Δx is 50 mm. What are the (1) tilt σ of the axis between the mirrors, (2) the tilts θ_1 and θ_2 of the mirror normal, and (3) the individual axis deviations δ_1 and δ_2 occurring at the mirrors?

From Equation 9.3,

$$\sigma = \arctan\left(\frac{-50\ \text{mm}}{600\ \text{mm}}\right) = -4.764°.$$

From Equation 9.4,

$$\theta_1 = \frac{-4.764° + 90°}{2} = 42.681°.$$

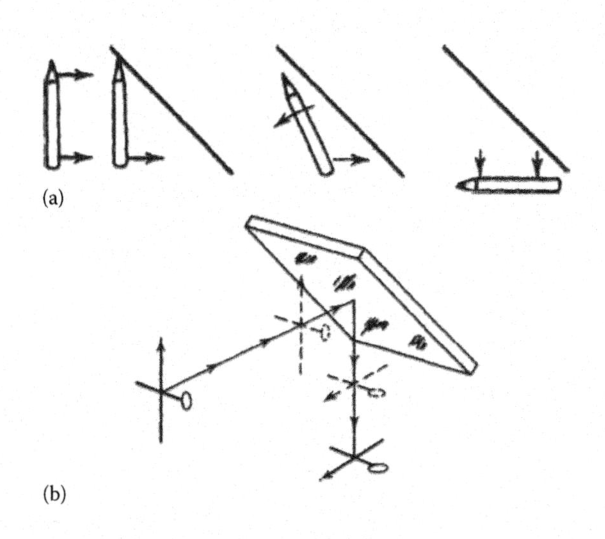

FIGURE 9.6 (a) Visualization of in-plane changes in image orientation upon reflection at a flat mirror through use of a bouncing pencil. (b) Two-dimensional image orientation changes can be visualized with an object configured as an arrow crossed by a drumstick. (Adapted from Smith, W.J., *Modern Optical Engineering*, 4th ed., McGraw-Hill, New York, 2008.)

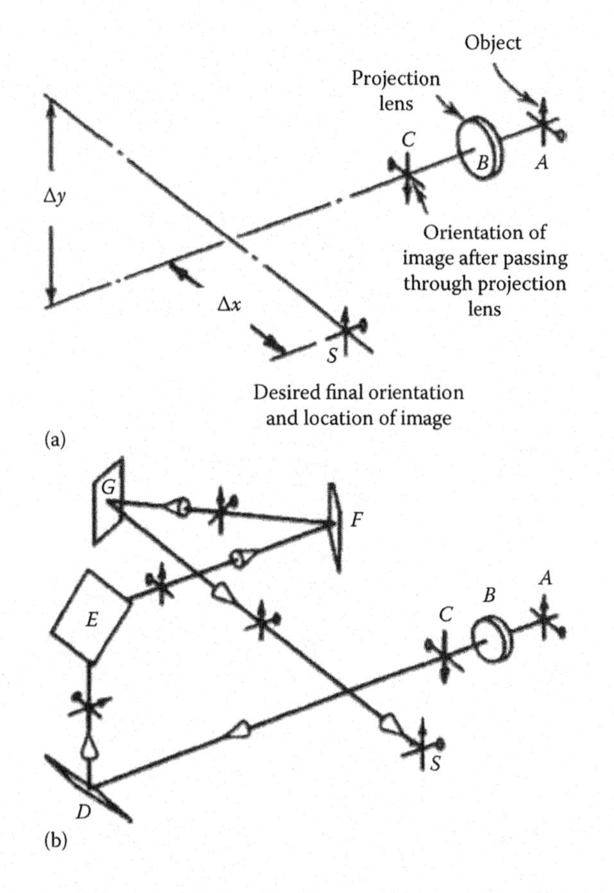

FIGURE 9.7 (a) Representation of a typical mirror system design problem in which an object at A is to be imaged at a particular location on a screen S with a particular orientation. (b) One of many possible mirror arrangements that could be designed for this purpose. (Adapted from Smith, W.J., *Modern Optical Engineering*, 4th ed., McGraw-Hill, New York, 2008.)

From Equation 9.5,

$$\theta_2 = \frac{30° - (-42.618°) - 90°}{2} = -27.618°.$$

From Equation 9.6,

$$\delta_{\cdot} = 180° - 2(42.618°) = 94.764°$$

From Equation 9.7,

$$\delta_2 = 30° + (-4.764°) - 90° = -64.764°.$$

9.2.5 Beam Prints on Optical Surfaces

The physical size of the front surface mirror is determined primarily by the size and shape of the irradiated area (called the beam print) of the light beam on the reflecting surface plus any allowances considered appropriate for mounting provisions, misalignment, and beam motion during use. This beam print can be determined from a scaled layout of the optical system showing the extreme rays of the light beam in at least two meridians. This method is rather time consuming to use and often inaccurate owing to compounded minor drafting errors. Computer-aided design programs that have the capability of representing light beams can be used with ease for this purpose. The most precise method is to have the lens designer ray trace the optical system and determine the extreme ray intercepts on mathematical representations of the reflecting surfaces.

The beam intercept contour of a symmetrical beam on a tilted flat mirror is expressed in terms of the major and minor axes of an ellipse oriented and located properly with respect to the intercept, on the surface, of the optical axis of the beam. This is shown in Figure 9.8. The following are equations for finding the beam intercept contour:

$$W = D + 2L \tan\alpha, \tag{9.8}$$

$$E = \frac{W \cos\alpha}{2\sin(\theta - \alpha)}, \tag{9.9}$$

$$F = \frac{W \cos\alpha}{2\sin(\theta + \alpha)}, \tag{9.10}$$

$$A = E + F, \tag{9.11}$$

$$G = (A/2) - F, \tag{9.12}$$

$$B = \frac{AW}{(A^2 - 4G^2)^{1/2}}, \tag{9.13}$$

where W is the width of the beam print at the mirror/axis intercept; D is the beam diameter at a reference plane perpendicular to the axis and located at an axial distance L from the mirror/axis

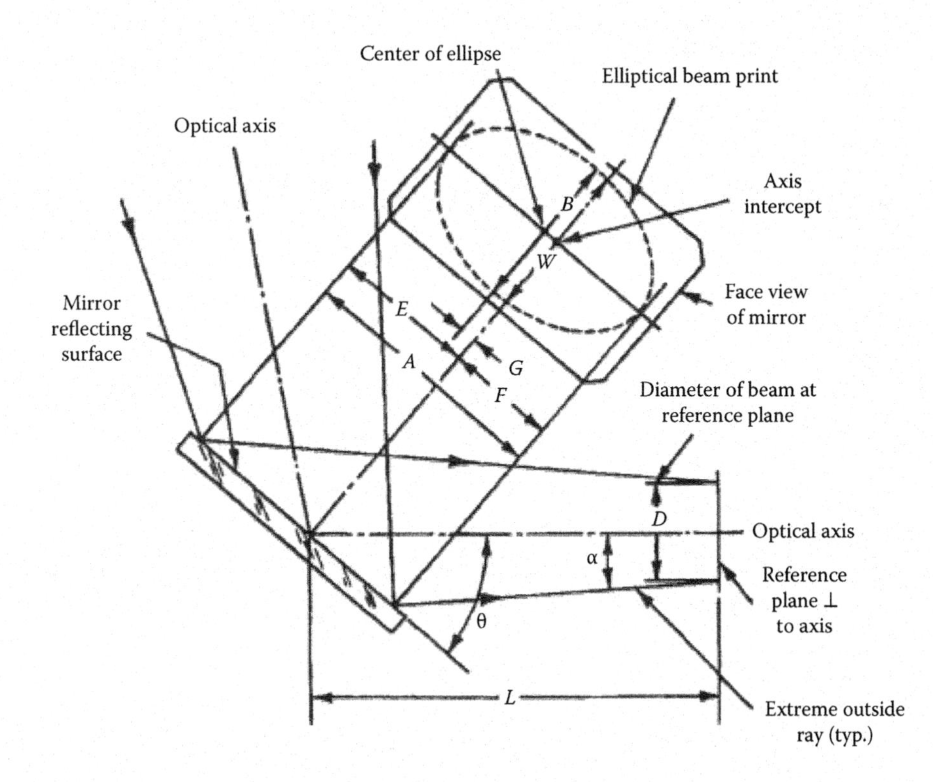

FIGURE 9.8 Geometric relationships used to define the beam print of a rotationally symmetric beam on an inclined mirror. (Adapted from Schubert, F., Determining optical mirror size, *Machine Design*, 51, 128, 1979.)

intercept; α is the beam divergence angle of the extreme off-axis reflected ray; E is the distance from the top edge of the beam print to the mirror/axis intercept; θ is the mirror surface tilt relative to the axis (or 90° minus the tilt of the mirror normal); F is the distance from the bottom edge of the beam print to the mirror/axis intercept; A is the major axis of the ellipse; G is the offset of the beam print center from the mirror/axis intercept; and B is the minor axis of the beam print.

These equations apply for either propagation direction of the beam as long as the reference plane is located where D is smaller than W. For a collimated beam propagating parallel to the axis, α and G are zero, and the preceding equations reduce to the symmetrical case where

$$B = W = D, \tag{9.14}$$

$$E = F = \frac{D}{2\sin\theta}, \tag{9.15}$$

$$A = D/\sin\theta. \tag{9.16}$$

Example 9.2

A circular beam of diameter $D = 50$ mm on a reference plane perpendicular to the axis is located at $L = 50$ mm following a flat mirror tilted 30° to the axis. The geometry is that of Figure 9.8. (1) If the divergence α of the beam is ±0.6°, what are the dimensions of the beam print on the mirror? (2) If the beam is collimated, what are the beam print dimensions?

For (1), from Equation 9.8,

$$W = 50\times10^{-3}\,\mathrm{m} + 2(50\times10^{-3}\,\mathrm{m})\tan(0.6°) = 51.047\times10^{-3}\,\mathrm{m}.$$

From Equation 9.9,

$$E = \frac{(51.047\times10^{-3}\,\mathrm{m})\cos(0.6°)}{2\sin(30°-0.6°)} = 51.990\times10^{-3}\,\mathrm{m}.$$

From Equation 9.10,

$$F = \frac{(51.047\times10^{-3}\,\mathrm{m})\cos(0.6°)}{2\sin(30°+0.6°)} = 50.138\times10^{-3}\,\mathrm{m}.$$

From Equation 9.11,

$$A = 50.990\times10^{-3} + 50.138\times10^{-3}\,\mathrm{m} = 101.128\times10^{-3}\,\mathrm{m}.$$

From Equation 9.12,

$$G = \frac{101.128\times10^{-3}\,\mathrm{m}}{2} - 50.138\times10^{-3}\,\mathrm{m} = 426\times10^{-6}\,\mathrm{m}.$$

From Equation 9.13,

$$B = \frac{(101.128\times10^{-3}\,\mathrm{m})(51.047\times10^{-3}\,\mathrm{m})}{\left[(101.128\times10^{-3}\,\mathrm{m})^2 - 4(426\times10^{-6}\,\mathrm{m})^2\right]^{\frac{1}{2}}} = 51.049\times10^{-3}\,\mathrm{m}.$$

For (2), for a collimated beam, $\alpha = 0°$, so from Equation 9.14,

$$B = W = 50\times10^{-3}\,\mathrm{m}.$$

From Equation 9.15.

$$E = F = \frac{50\times10^{-3}\,\mathrm{m}}{2\sin(30°)} = 50\times10^{-3}\,\mathrm{m}.$$

From Equation 9.16.

$$A = \frac{50\times10^{-3}\,\mathrm{m}}{\sin(30°)} = 100\times10^{-3}\,\mathrm{m}.$$

9.2.6 Mirror Coatings

Most mirrors used in optical instruments have light-reflective coatings made of metallic or non-metallic thin films. Metals commonly used as coatings are aluminum, silver, and gold because of their high reflectivities in the ultraviolet (UV), visible, and infrared (IR) spectral regions. Protective coatings such as silicon monoxide (SiO) or magnesium fluoride (MgF_2) may be placed over metallic coatings to increase their durability. Nonmetallic films consist of single layers or multilayer stacks of dielectric films. The stacks are combinations of materials with high and low indices of refraction. Dielectric reflecting films function over narrower spectral bands than the metals, but have very high reflectivities at specific wavelengths. They are especially useful in monochromatic systems such as those using laser radiation.

Figure 9.9 shows typical visible light reflectance versus wavelength curves for protected aluminum (Figure 9.9a) and UV-enhanced aluminum (Figure 9.9b) at normal incidence and for two polarization states.[1] Figure 9.10a shows the IR reflectance versus wavelength for first-surface-protected gold at normal incidence, while Figure 9.10b shows the IR reflectance of first-surface-protected silver. The IR reflectance of a typical first-surface, multilayer dielectric film at 45° incidence for two polarization states appears in Figure 9.11a. That of a second-surface mirror coating of silver at normal incidence is shown in Figure 9.11b. The latter location for the coating is advantageous from an environmental durability viewpoint because the reflecting side of the film is then better protected from the outside environment and physical damage due to handling or use than a first-surface coating. The exposed back of such a thin-film coating is typically given a protective coating such as electroplated copper plus enamel for this same purpose.

Uncoated glass surfaces are sometimes used as partial reflecting beam-splitting surfaces in applications requiring a reflectance of about 4%. The actual reflectance varies with the angle of incidence and

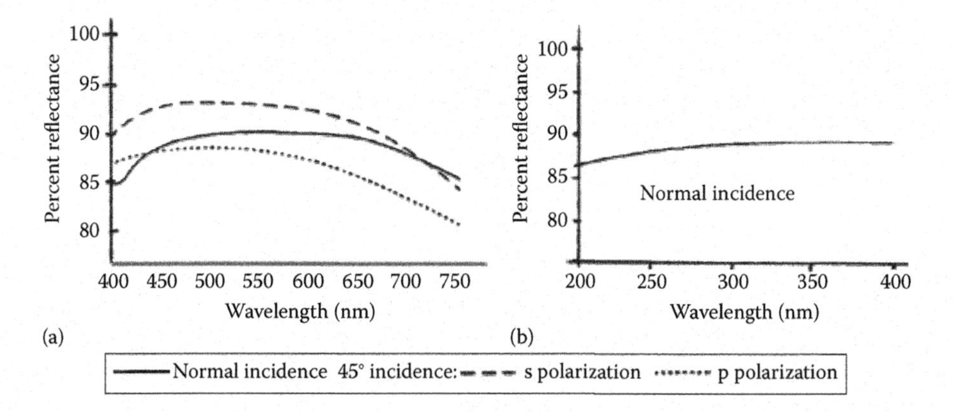

FIGURE 9.9 Reflectance versus wavelength for first-surface metallic coatings of (a) protected aluminum and (b) UV-enhanced aluminum. (From Yoder, P.R. Jr., *Mounting Optics in Optical Instruments*, 2nd ed., SPIE Press, Bellingham, WA, 2008.)

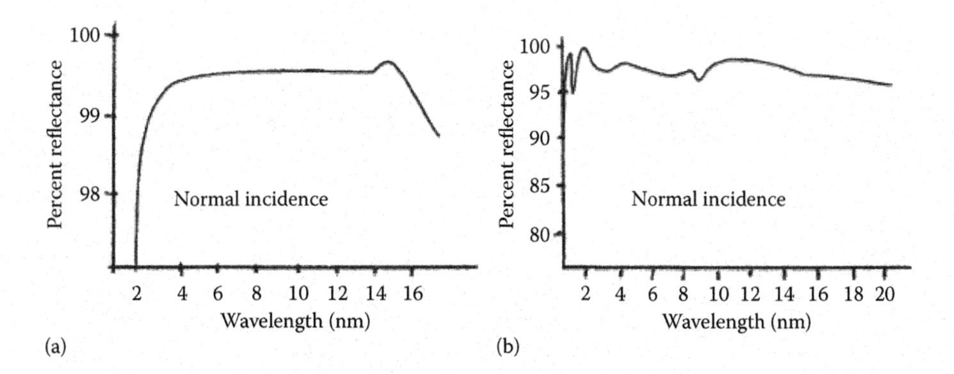

FIGURE 9.10 Reflectance versus wavelength for first-surface thin films of (a) protected gold and (b) protected silver. (From Yoder, P.R. Jr., *Mounting Optics in Optical Instruments*, 2nd ed., SPIE Press, Bellingham, WA, 2008.)

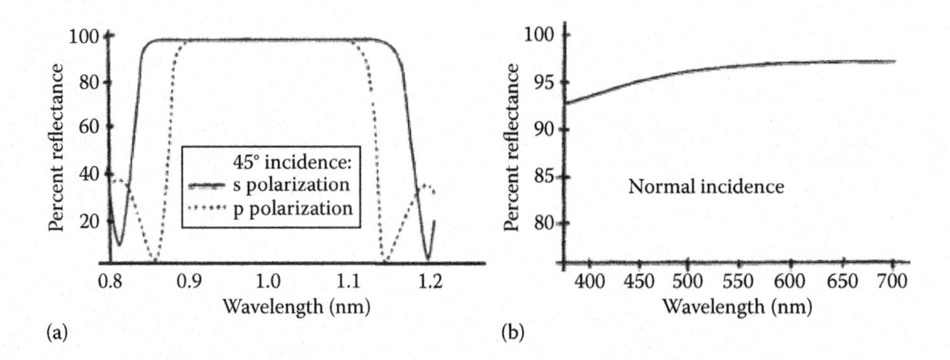

FIGURE 9.11 Reflectance versus wavelength for (a) a first-surface multilayer dielectric thin film and (b) a second-surface thin film of silver. (From Yoder, P.R. Jr., *Mounting Optics in Optical Instruments*, 2nd ed., SPIE Press, Bellingham, WA, 2008.)

the state of polarization of the incident radiation. These effects may be noted from the multiple curves in Figure 9.12, which applies to glass with an index of 1.523 in air. The solid line represents the reflectance of unpolarized light, the line of short dashes is for p-polarized light with the electric vector parallel to the plane of incidence, and the line of long dashes is for the s-polarized beam with the ***E*** vector perpendicular to the same plane.* At the polarizing angle, the p-polarized component disappears.

At normal incidence, the reflectance $(R_S)_\lambda$ of the interface between two refracting materials with refractive indices n_1 and n_2 is obtained from the Fresnel equation as follows:

$$(R_S)_\lambda = \frac{(n_2 - n_1)_\lambda^2}{(n_2 + n_1)_\lambda^2}. \tag{9.17}$$

For instance, the monochromatic reflectance of an uncoated surface on glass with index n_D = 1.523 in air at normal incidence is $R_S = (1.523 - 1.000)^2/(1.523 + 1.000)^2 = 0.043$ (4.3%). Note that this is the value plotted in Figure 9.12 at zero angle of incidence.

The Fresnel reflectance at a given wavelength of a dielectric surface such as glass at other than normal incidence is given by the following equation:

$$R_S = \frac{1}{2}\left[\frac{\sin^2(I - I')}{\sin^2(I + I')} + \frac{\tan^2(I - I')}{\tan^2(I + I')}\right], \tag{9.18}$$

where I and I' are the angles of incidence and refraction, respectively.

The first term inside the brackets refers to the s-polarized component of the radiation, and the second term inside the brackets refers to the p-polarized component. At normal incidence, this equation simplifies to Equation 9.17.

Applying Equation 9.18 to the surface of an optic with n of 1.523 and I of 70°, I' is 38.097° and $R_S = (1/2)[(\sin^2 31.902°/\sin^2 108.098°) + (\tan^2 31.902°/\tan^2 108.098°)] = 0.175$. This is approximately the value shown for that glass in Figure 9.12.

An A/R coating is usually applied to refracting optical surfaces to reduce the surface reflectance and enhance the intensities of beams reflected from those surfaces. These coatings are single-layer thin-film coatings or a stack of multiple thin-film coatings. The simplest case of a single-layer coating causes destructive interference between a first beam reflecting from the air/film interface and a second beam reflecting from the film/glass interface. Destructive interference occurs when these beams are

* To minimize confusion from the fact that both English words *parallel* and *perpendicular* begin with *p*, the name for perpendicularly polarized light comes from the German word for *normal*, which is *senkrecht*.

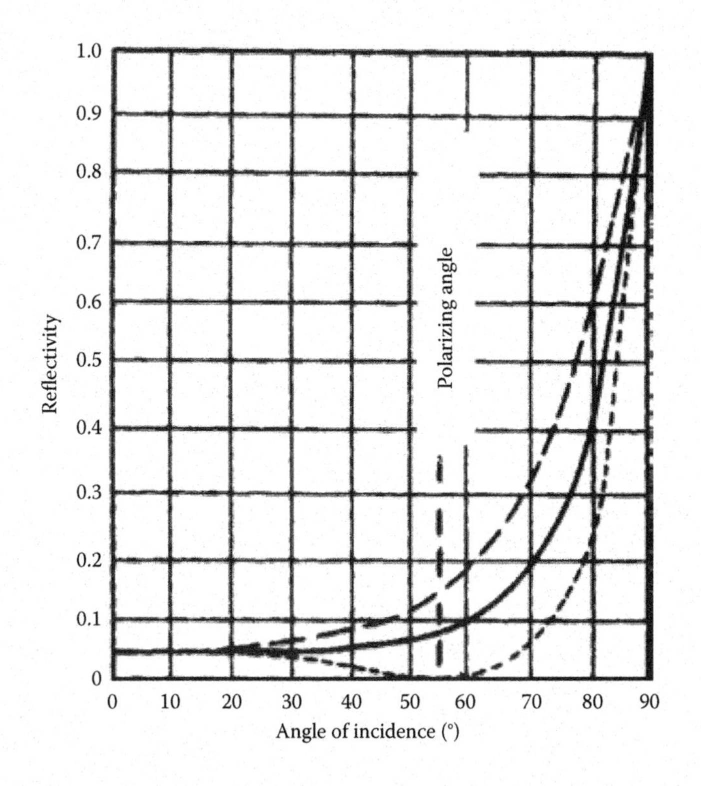

FIGURE 9.12 Reflectivity of an uncoated single air-to-glass interface at various angles of incidence. The glass refractive index is 1.523. The solid line represents unpolarized light, while the short- and long-dashed lines represent the p-polarized and s-polarized components, respectively. (Adapted from Jenkins, F.A., and White, H.E., *Fundamentals of Optics*, 3rd ed., McGraw-Hill, New York, 1957.)

exactly 180° (or one-half wavelength) out of phase. Since the second beam passes through the film twice, the desired $\lambda/2$ phase shift results if the optical thickness $(n)(\lambda)$ of the film is $\lambda/4$. The combined intensity of the two out-of-phase reflected beams is then zero, because their wave amplitudes subtract. Note that complete destructive interference occurs only at one specific wavelength and then only if the beam amplitudes are equal. The latter condition occurs if the following equation is satisfied:

$$n_2 = (n_1 n_3)^{1/2}, \tag{9.19}$$

where n_2 is the index of refraction of the thin film, n_1 is the index of refraction of the surrounding medium (typically air with $n = 1$), and n_3 is the index of refraction of the glass. All indices are at a specific wavelength λ.

If a thin-film material with exactly the right index to A/R coat for a given type of glass is not available, imperfect cancellation of the two reflected beams occurs. The resultant surface reflectance R_S is

$$R_S = \left[(R_{1,2})^{1/2} - (R_{2,3})^{1/2}\right]^2 \tag{9.20}$$

where $R_{1,2}$ is the reflectance of the air/film interface and $R_{2,3}$ is the reflectance of the film/glass interface.

High-efficiency, multilayer, dielectric A/R coatings can be designed to have zero reflectivity at a specific wavelength or significantly reduced variations in reflectivity with wavelength. Figure 9.13 shows plots of reflectance versus wavelength for a single-layer (MgF_2) coating, a "broadband" multilayer coating with low reflectivity over the entire visible spectrum, and two multilayer coatings

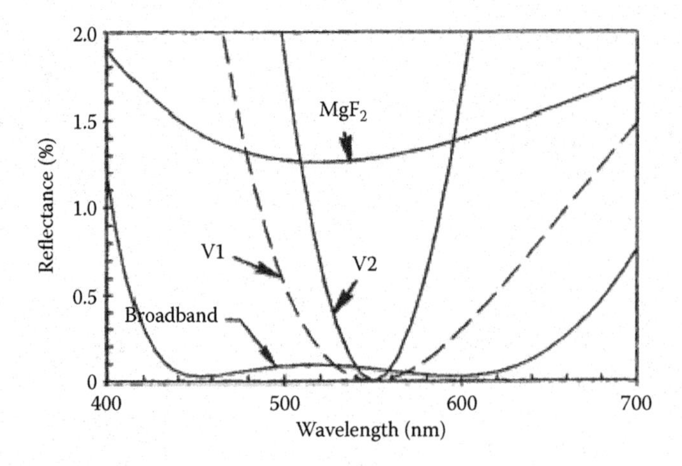

FIGURE 9.13 Variations in reflectance within the visible spectrum for several common A/R coatings. (From Yoder, P.R. Jr., *Mounting Optics in Optical Instruments*, 2nd ed., SPIE Press, Bellingham, WA, 2008.)

with zero reflectivity at $\lambda = 550$ nm. All these coatings are applied to crown glass. Coatings V1 and V2 are called "V-coats" because of their characteristic downward-pointing triangular shapes.

Example 9.3

A plane-parallel N-BK7 glass plate with a refractive index of $n_D = 1.517$ is to be coated with a single-film A/R coating designed for normal incidence. (1) What would be the ideal film index for zero reflectivity? (2) If coated with MgF_2 with an index n_D of 1.38, what is the reflectivity? (3) Compare the last result to the reflectivity without the coating.

From Equation 9.19,

$$n_2 = [(1.000)(1.517)]^{0.5} = 1.232.$$

From Equation 9.17,

$$R_S = \frac{(1.380 - 1.000)^2}{(1.380 + 1.517)^2} = 17.2 \times 10^{-3}, \quad \text{with coating.}$$

From Equation 9.17,

$$R_S = \frac{(1.517 - 1.000)^2}{(1.517 + 1.000)^2} = 42.2 \times 10^{-3}, \quad \text{uncoated.}$$

The uncoated reflectance is 0.0422/0.0172 = 2.45 or almost 2.5 times as large as the coated value.

9.2.7 Ghost Image Formation by Second-Surface Mirrors

An obvious difference between first- and second-surface mirrors is that a transparent substrate is needed for the latter, but not for the former. Appendices 6 and 7 list the mechanical properties and figures of merit of common nonmetallic and metallic mirror substrate materials. Of these, fused silica has the best refractive properties, and most of the remaining candidates have very poor or no refractive characteristics. Second-surface mirrors are usually made of optical glasses (see Appendix 6) or crystals (see Appendices 2 through 5). The plastic materials listed in Table 3.4 are not very good for mirror applications from a mechanical viewpoint. A distinct advantage of the second-surface mirror for image-forming mirrors is that an additional surface radius, asphericity, axial thickness, and index are available for controlling aberrations.

Figure 9.14 shows what happens when a converging beam reflects from a plane-parallel second-surface mirror that is tilted at 45° to the axis. Two equal-sized images are formed. The main image comes from the silvered or beam-splitting back surface, while a ghost image comes from the front surface. These images do not coincide; they are separated axially by d_A and laterally by d_L. At normal incidence, only axial separation would be observed.

The second-surface image-forming mirror configuration is most frequently used in primary or secondary mirrors in catadioptric systems for photographic and moderate-sized astronomical telescope applications. Second-surface designs obviously do not work in mirrors that do not have opaque substrates or ones with tapered or arched back surfaces.

Figure 9.15 illustrates a concave second-surface mirror and its function in forming a normal or main image of a distant object. Since the light to be reflected by the second-surface mirror must pass

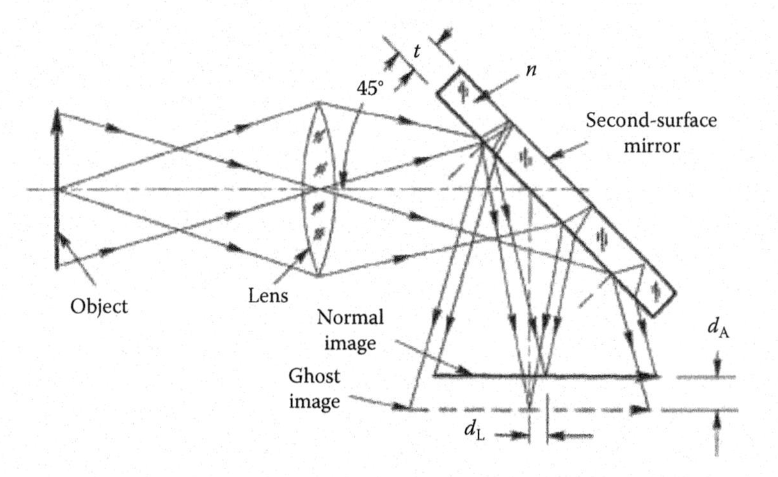

FIGURE 9.14 Ghost image formation from the first surface of a second-surface flat mirror inclined at 45° to the axis. (Adapted from Kaspereit, O.K., *Design of Fire Control Optics ORDM*2-1, Vol. 1, US Army, Washington, DC, 1952.)

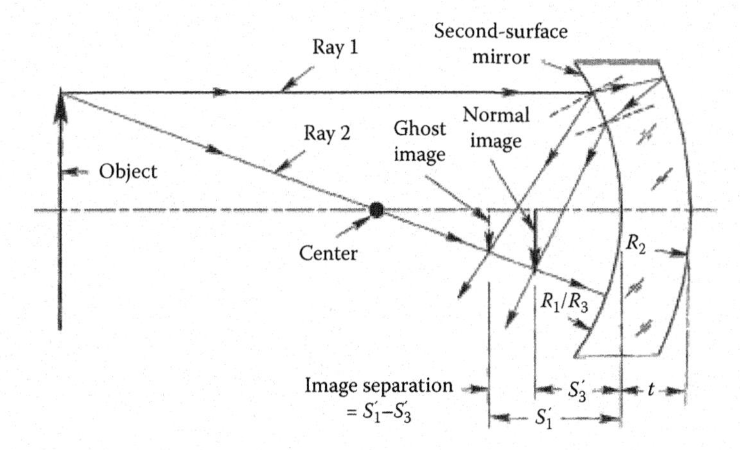

FIGURE 9.15 Ghost image formation from the first surface of a second-surface meniscus mirror with concentric spherical surfaces. (Adapted from Kaspereit, O.K., *Design of Fire Control Optics ORDM*2-1, Vol. 1, US Army, Washington, DC, 1952.)

through the front (refracting) mirror surface to get to the reflecting surface, a ghost image is formed by the front surface. If the depth of focus of the optical system includes both the normal and ghost images and the intensity of the ghost relative to that of the main image is great enough, a double image will be observed.

9.3 SEMIKINEMATIC MOUNTINGS FOR SMALL MIRRORS

When choosing mounting configurations for small mirrors, it is best to observe the basic principles of kinematics. Mirrors must be considered to be flexible plates unless they are quite thick compared with their other dimensions. The factors for the appropriateness of kinematic mounting are the inherent rigidity of the mirror; the tolerable movements and distortions of the reflecting surfaces; the magnitudes, locations, and orientations of the steady-state forces holding the mirror against its mounting surfaces during operation; the transient forces driving the mirror against or away from the mounting surfaces during exposure to extreme shock and vibration; thermal effects; the shape of the mounting surface on the mirror; the sizes, shapes, and orientations of the mounting surfaces (pads) on the mount; and the rigidity and long-term stability of the mount. The design must be compatible with assembly, adjustment, maintenance, package size, weight, and configuration constraints and be affordable.

Figure 9.16 shows a relatively simple means for attaching a glass-type mirror configured as a plane-parallel plate to a metal surface. The reflecting surface is pressed against three flat, coplanar, machined (lapped) pads by three spring clips. The spring contacts are directly opposite the pads so as to minimize bending moments. This design constrains one translation and two tilts. The spacers that position the clips are machined to the proper thickness for the clips to exert clamping forces (preload) of controlled magnitude normal to the mirror.

The spring clips should be stiff enough to restrain the mirror against the worst-case shock and vibration to which it may be subjected. They are designed as cantilevered beams with a free length equal to the distance between the edge of the restraining means (the screw in Figure 9.16) and the nearest edge of the contact area of the clip on the mirror. A safety factor (SF) of about 2 beyond the force needed to overcome imposed dynamic loads is frequently employed to place an upper limit on preload. The total preload F_A required of all the clips:

$$F_A = W\,\mathrm{SF}\,a_G, \tag{9.21}$$

where W is the mass of the mirror.

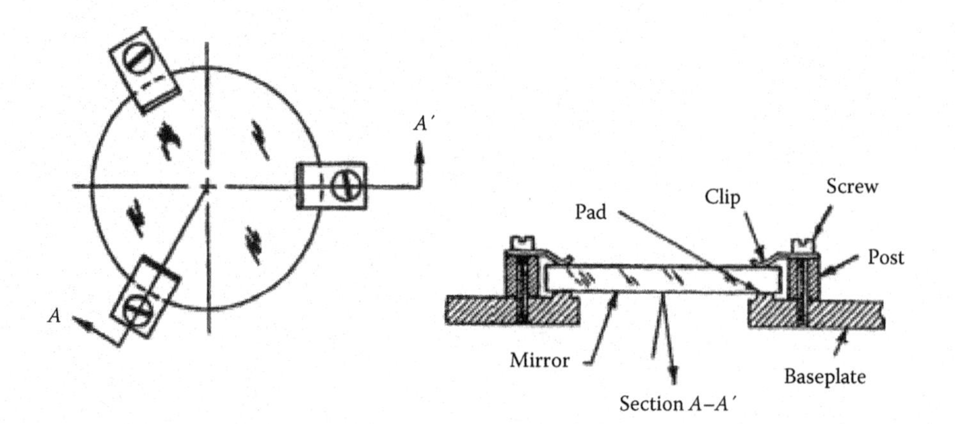

FIGURE 9.16 Concept for mounting a flat mirror to coplanar pads on a baseplate with three spring clips. (Adapted from Durie, D.S.L., Stability of optical mounts, *Machine Design*, 40, 184, 1968.)

Each of N clips should then provide a force of F_A/N units. The required deflection Δ of a spring clip to provide a particular preload is given by

$$\Delta = \frac{4F_A L^3 (1 - \nu_M^2)}{E_M b t^3 N}, \tag{9.22}$$

where ν_M is the Poisson's ratio of the spring material, F_A is the preload, L is the free (cantilevered) length of the spring, E_M is the Young's modulus of the spring material, b is the width of the spring, t is the thickness of the spring, and N is the number of springs employed.

The stress σ_B within the cantilevered spring created by the imposed bending is calculated by

$$\sigma_B = \frac{6F_A L}{bt^2 N}, \tag{9.23}$$

where all parameters are as defined previously. Note that if the spring were attached to the mount by some means that did not require it to be perforated, σ_B would be reduced by a factor of about 3.

Lateral motions of the mirror on the pads and rotation about its normal are not constrained other than by friction as shown in the design of Figure 9.16. This may be allowable because the performance of a flat mirror is insensitive to these motions. Excessive lateral movement of the optic can be prevented by adding stops or, if the mirror is round, by sizing the supports to provide a minimal clearance to the mirror. Thermal expansion differences must be taken into consideration if the mirror just touches the stops at assembly.

Figure 9.17 illustrates a less desirable mounting design in which the mirror rim rests directly on a ring-shaped supporting surface machined into the plate. Spring clips provide the clamping forces, as in Figure 9.17, but unless the supporting surface is as flat as the mirror surface, minor irregularities can occur anywhere on the metal surface. Hence, bending moments can be introduced and the reflecting surface may be deformed. Similar irregularities could result from foreign matter (such as dust) trapped between the mirror and the mounting surface. The likelihood of this happening with localized small pads is significantly less than with a continuous optic-to-mount contact.

An arrangement sometimes used when mounting flat first-surface mirrors to an unperforated baseplate is illustrated in Figure 9.18. Here, the clips are solid, so they do not bend. Resilience is

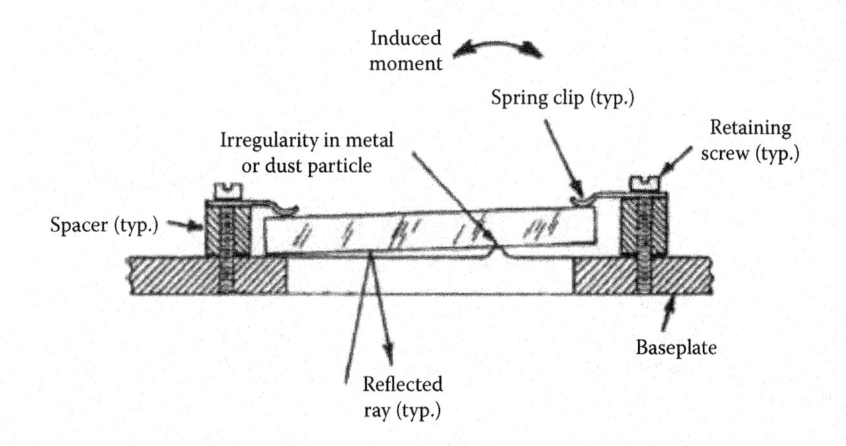

FIGURE 9.17 Less desirable design for mounting a flat mirror to a baseplate without pads. The effect of a dirt particle in the interface is depicted. (Adapted from Durie, D.S.L., Stability of optical mounts, *Machine Design*, 40, 184, 1968.)

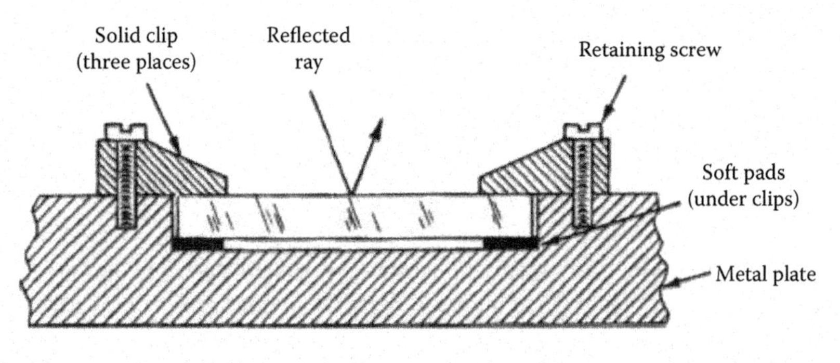

FIGURE 9.18 Common, but potentially unsatisfactory, technique for mounting a flat mirror to a baseplate by using resilient pads and solid clips.

built into the mount by inserting three small pads of soft material such as Neoprene under the mirror at the clips. This gives a form of three-point suspension around the mirror periphery. Compression of the soft pads tends to accommodate thickness variations within the mirror. The soft pads can be located between the mirror surface and clips, but then the supporting surface on the plate must be very flat or have raised pads in order not to overly distort the mirror owing to misaligned contact areas. Thin strips of plastic such as Mylar tape are sometimes used as pads although they will not be very elastic. A disadvantage of this mounting is the fact that the resilient material may, over time, become either permanently deformed or increasingly stiff so that the preload is changed.

Circular, rectangular, or nonsymmetrically shaped mirrors can often be mounted in the same manner as a lens. Circular mirrors up to perhaps 6 in. (15.2 cm) in diameter can be held with threaded retaining rings. The outside diameter (OD) limit for a threaded mount is set primarily by the increasing difficulty of machining thin, circular, retaining rings with sufficient quality in larger diameters. Larger circular mirrors can be held with continuous (i.e., annular) flanges. Some can be held with three or more cantilevered spring clips. The number of clips obviously depends upon the size of the mirror because more required total preload demands more clips to keep the bending stress in each clip within reason. Drawbacks of this type of mounting are the same as for that of the one in Figure 9.17; small defects in the mounting surface may produce unacceptable deformation of the mirror optical surface.

Figure 9.19 shows one retaining ring mounting concept that can be applied to a small, circular, first-surface mirror. The mirror is shown as a convex sphere, although a mirror with a concave surface could be mounted similarly. The reflecting surface in this example contacts a shoulder in the cell and is constrained after centering by tightening the retainer. The retainer typically has a loose fit in the internal thread of the cell, so it can align itself with the mirror if that mirror has a significant wedge. Tangential contact occurs on the polished surface of the mirror. If that surface has a sufficiently short radius, the mirror may self-center to the mechanical axis of the mount as a result of balancing of radial components of the axial force. However, as in the case of lens surfaces, ones with long radii will not self-center. A long radius is assumed for the mirror in Figure 9.19. Here, three temporary radial locating screws are used to center the mirror to the mount OD. Curved spacers are then custom-made to fit into the existing spaces at three symmetrically located points around the periphery of the mirror. These spacers should be centered axially at or near the plane of the center of gravity (CG) of the mirror. Preload is provided when the retainer is tightened. The centering screws are then removed. To prevent the spacers from shifting under shock or vibration, three holes in the cell wall nominally at the spacer centers are filled with epoxy once the assembly is complete. The spacer shown in the figure has a shallow circular recess machined into its OD to provide space for the epoxy to lock those spacers in place. Note that close tolerances on the OD of the mirror are not required here because the radial spacers are custom-fitted into whatever spaces exist at the selected locations after centering the lens.

Figure 9.20 shows a more conventional concept for mounting a concave first-surface mirror. A tangential (conical) interface is provided at the convex second surface of the mirror as recommended

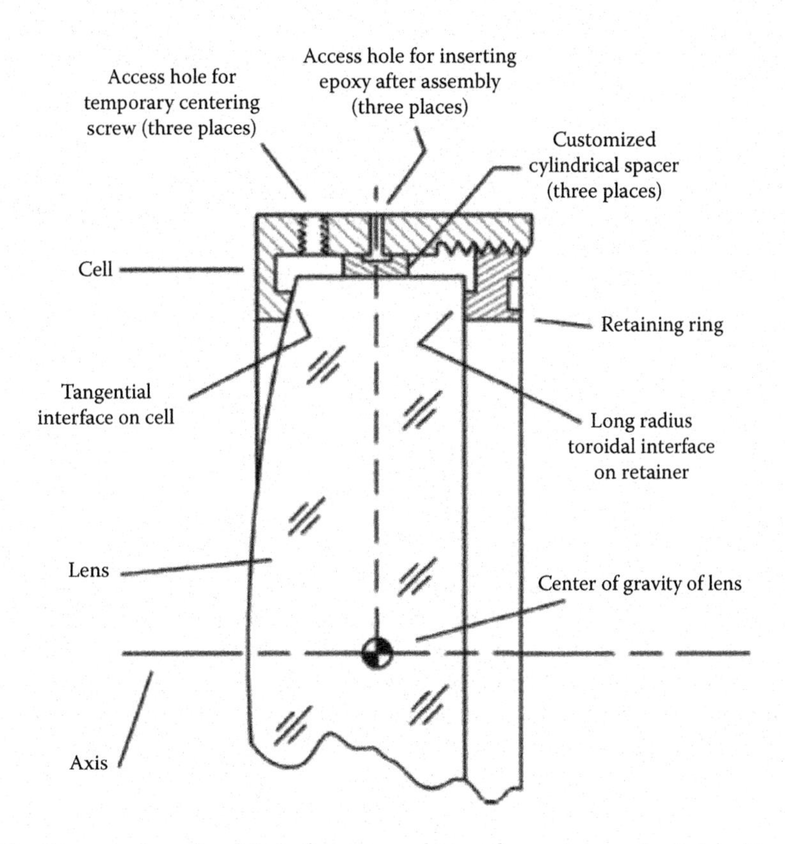

FIGURE 9.19 Conceptual configuration of a convex first-surface mirror preloaded in its mount with a threaded retaining ring. Custom-made spacers are inserted at three locations around the mirror rim after centering the optical surface with three temporary radial push screws.

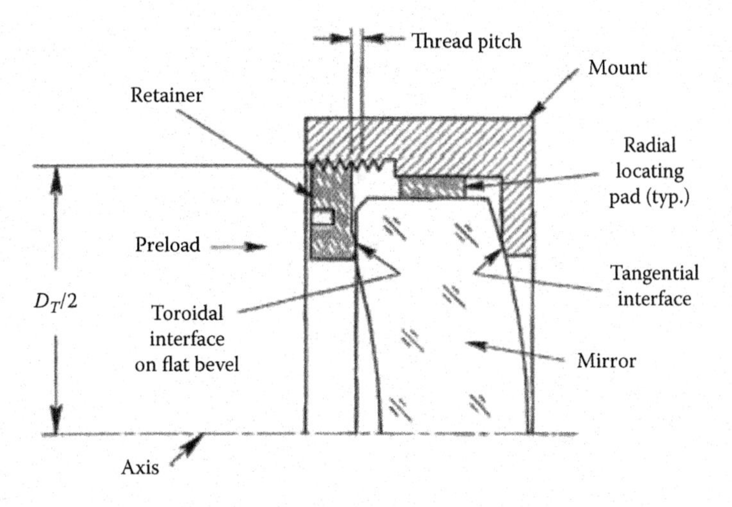

FIGURE 9.20 Conceptual configuration of a threaded retaining ring mounting for a second-surface meniscus-shaped mirror. (From Yoder, P.R. Jr., *Mounting Optics in Optical Instruments*, 2nd ed., SPIE Press, Bellingham, WA, 2008.)

in Section 5.6.2. The retainer is given a convex toroidal shape to interface with the flat bevel of the mirror.

Note that temperature changes can create problems with regard to the fit of the radial locating pads and the constancy of axial preload in this and all the other mirror mounts discussed here because of differential expansion or contraction of the optical and mechanical parts. Obviously, this problem can be reduced in severity if the metal parts are chosen to have coefficients of thermal expansion (CTEs) close to that of the mirror.

As in lens mounting, the magnitude of the nominal total preload (F_A) developed in a threaded retainer mirror mounting design with a specific torque (M) applied to the ring at a fixed temperature can be estimated by

$$F_A = \frac{5M}{D_T}, \tag{9.24}$$

where D_T is the pitch diameter of the thread.

Another mounting for a small circular mirror, in this case a concave second-surface type, is shown in Figure 9.21. Here, the mirror surface registers against a tangential interface, while the flat bevel on the front of the mirror touches a toroidal interface on the retainer. Contacts occur at the same height on both sides. The choice of these interface shapes, the dimensions, and a "loose" fit in the retainer threads ensure minimal contact stress as well as minimal tendency to bend the mirror by mount-induced moments.

The function of the continuous flange in the design shown in Figure 9.21 is essentially the same as that of the threaded retainer described earlier for lens mountings. The magnitude of the total

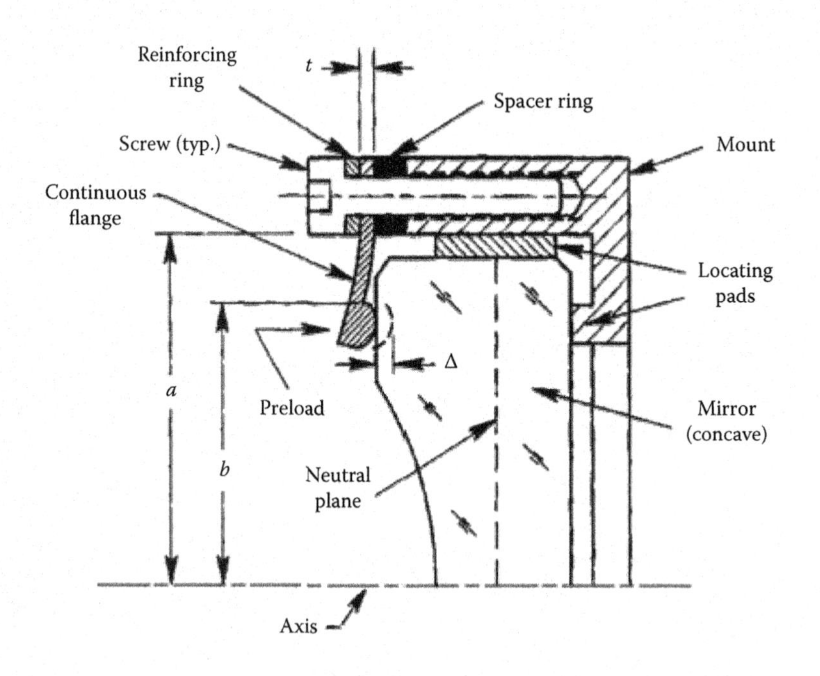

FIGURE 9.21 Conceptual configuration for a concave first-surface mirror preloaded with a continuous ring flange. (From Yoder, P.R. Jr., *Mounting Optics in Optical Instruments*, 2nd, ed., SPIE Press, Bellingham, WA, 2008.)

preload exerted for a given flange deflection Δ can be estimated using the same equations as for an individual lens. These equations are repeated here for convenience:

$$\Delta = (K_A - K_B)\frac{F_A}{t^3}, \quad (9.25)$$

$$K_A = \frac{3(m^2-1)\left[a^4 - b^4 - 4a^2b^2 \ln\left(\frac{a}{b}\right)\right]}{4\pi m^2 E_M a^2}, \quad (9.26)$$

$$K_B = \frac{3b^2(m^2-1)(m+1)\left[2\ln\left(\frac{a}{b}\right) + \frac{b^2}{a^2} - 1\right]\left\{b^2 + a^2\left[2\ln\left(\frac{a}{b}\right) - 1\right]\right\}}{4\pi m^2 E_M[b^2(m+1) + a^2(m-1)]}, \quad (9.27)$$

where t is the thickness of the cantilevered section of the flange, a is the outer radius of the cantilevered section, b is the inner radius of the cantilevered section, m is the reciprocal of Poisson's ratio (ν_m) of the flange material, and E_m is the Young's modulus of the flange material.

The spacer between the mount and the flange can be ground at assembly to the particular thickness that produces the predetermined flange deflection when firm metal-to-metal contact is achieved by tightening the clamping screws. Customizing the spacer accommodates variations in as-manufactured lens thicknesses. The flange material and thickness are the prime design variables. The dimensions a and b, and hence the annular width $(a - b)$, can also be varied, but these are usually set primarily by the mirror diameter, mount wall thickness, and overall dimensional requirements.

The stress σ_B built up in the bent portion of the flange must not exceed the yield strength σ_{ys} of the material. The following equations apply:

$$\sigma_B = \frac{K_C F_A}{t^2} = \frac{\sigma_{ys}}{SF}, \quad (9.28)$$

$$K_C = \frac{3}{2\pi}\left\{1 - \frac{2b^2\left[m - (m+1)\ln\left(\frac{a}{b}\right)\right]}{a^2(m-1) + b^2(m+1)}\right\}. \quad (9.29)$$

As in the case of lenses, the localized deflections Δ of the flange as measured between the attachment screws should be essentially the same as those existing at the screws. This ensures uniform preload around the edge of the mirror. This can be accomplished by machining the flexing portion of the flange as a thinned, annular region in a thicker ring, thereby providing extra stiffness at the clamped annular zone of the flange. Stiffening the flange with a backup or reinforcing ring as shown schematically in Figure 9.21 could also do this.

Increasing the number of screws also tends to reduce the possibility of nonuniform preload around the edge of the mirror. Using guidance from the design of high-pressure chambers to the mirror-mounting case, the number of screws N should be

$$3 \le \frac{\pi D_B}{Nd} \le 6, \quad (9.30)$$

where D_B is the diameter of the bolt circle passing through the centers of the screws and d is the diameter of the screw heads.*

This criterion is probably overly conservative in an optical instrument application, especially if a stiff backup ring is employed or if the flange is thickened in the region where it is clamped. The application of good engineering judgment and possibly experimentation is suggested.

Example 9.4

A 300 mm diameter mirror with a mass of 7.9 kg is mounted with 1 mm of radial clearance between mirror and cell (a = 300 mm + 2(1 mm) = 302 mm). The mirror is retained with a continuous flexible flange made of 7075-T6 aluminum, with elastic modulus E_M = 71.1 GPa, yield stress σ_{YS} = 490 MPa, and Poisson's ratio ν = 0.33. The flange retains the mirror against an axial acceleration $a_G = 75g$ with an SF = 2. The flange bears against a diameter that is 6 mm inside the edge, so b = 300 mm − 2(6 mm) = 288 mm. Here the flange thickness t and deflection Δ are estimated.

The flange thickness t is estimated using Equations 9.28 and 9.29. First, calculate the parameter K_C from Equation 9.29:

$$K_C = \frac{3}{2\pi}\left[1 - \frac{2\left(\frac{1}{0.33}\right)(288\,\text{mm})^2 - 2(288\,\text{mm})^2\left(\frac{1}{0.33}+1\right)\ln\left(\frac{302\,\text{mm}}{288\,\text{mm}}\right)}{(302\,\text{mm})^2\left(\frac{1}{0.33}-1\right)+(288\,\text{mm})^2\left(\frac{1}{0.33}+1\right)}\right] = 44.584\times10^{-3}.$$

Then, solving Equation 9.28 for t,

$$t = \left[\frac{(44.584\times10^{-3})(7.9\,\text{kg})\,2(75)(9.81\,\text{m/s}^2)}{490\times10^6\,\text{Pa}}\right]^{\frac{1}{2}} = 1.03\times10^{-3}\,\text{m}.$$

Calculating K_A and K_B from Equations 9.26 and 9.27,

$$K_A = \frac{3[(0.33)^{-2}-1]\left[(302\,\text{mm})^4 - (288\,\text{mm})^4 - 4(302\,\text{mm})^2(288\,\text{mm})^2\ln\left(\frac{302\,\text{mm}}{288\,\text{mm}}\right)\right]}{4\pi(0.33)^{-2}(71\times10^9\,\text{Pa})(302\,\text{mm})^2}$$

$$= 17.7\times10^{-18}\,\text{m}^2/\text{Pa},$$

$$K_B = 3[(0.33)^{-2}-1][(0.33)^{-2}+1]\left[2\ln\left(\frac{302\,\text{mm}}{288\,\text{mm}}\right) - \frac{(288\,\text{mm})^2}{(302\,\text{mm})^2} - 1\right]$$

$$\times\left[(302\,\text{mm})^4 + 2(302\,\text{mm})^2(288\,\text{mm})^2\ln\left(\frac{302\,\text{mm}}{288\,\text{mm}}\right) - (302\,\text{mm})^2(288\,\text{mm})^2\right]$$

$$\times\left\{4\pi(0.33)^{-2}(71\times10^9\,\text{Pa})\left[(288\,\text{mm})^2\left(\frac{1}{0.33}+1\right)+(302\,\text{mm})^2\left(\frac{1}{0.33}-1\right)\right]\right\}$$

$$= 838\times10^{-21}\,\text{m}^2/\text{Pa}.$$

* For instance, if D_B = 17.0 in. (431.8 mm) and d = 0.375 in. (9.525 mm), $24 \le N \le 48$.

The deflection Δ is found using Equation 9.25:

$$\Delta = (17.7\times10^{-18}\ \text{m}^2/\text{Pa} - 838\times10^{-21}\ \text{m}^2/\text{Pa})\frac{(7.9\ \text{kg})(75)2(9.81\ \text{m/s}^2)}{(1.03\times10^{-3}\ \text{m})^3} = 180\times10^{-6}\ \text{m}.$$

The required deflection could be achieved by grinding a spacer; the dimension of 180 μm (≈ 0.007 in.) is relatively easy to measure and make.

Figure 9.22 shows a semikinematic mounting for a partially reflecting mirror used as a plate-type beam splitter. The beam-splitting coating is on the front face. That face registers against small-area fixed pads and is spring loaded directly opposite these points. Here, and in any design with hard contacts against the reflecting side of the mirror, the location and orientation of that surface do not change with the temperature of the optic. Displacements of the constraints caused by temperature changes may, of course, affect the location and orientation of that surface.

The optics are constrained laterally only by friction in many of the designs presented earlier. This may be acceptable because the performance of a flat mirror is generally insensitive to these motions. Excessive lateral movement of the optic can be prevented by spring loading it against fixed stops. CTE differences must be considered if the mirror touches hard stops without any resiliency.

A mirror mounting concept that utilizes one spring-loaded mechanism and two fixed constraints at 120° from each other on a plane parallel to the reflecting surface, and three spring-loaded mechanisms to preload the mirror against coplanar pads in the orthogonal direction, is illustrated in Figure 9.23. While compression coil springs are shown, cantilevered clips could be employed. This mount is semikinematic since all six degrees of freedom (DOFs) are constrained by the spring loads and the contacts are small areas instead of points.

The interfaces between the mirror and the pads are self-aligning in this design because of the flexibilities of the springs. Intimate contacts over the pad areas are assumed. In a design without that flexibility, localized line contact between the glass and the edges of the pads could occur if the

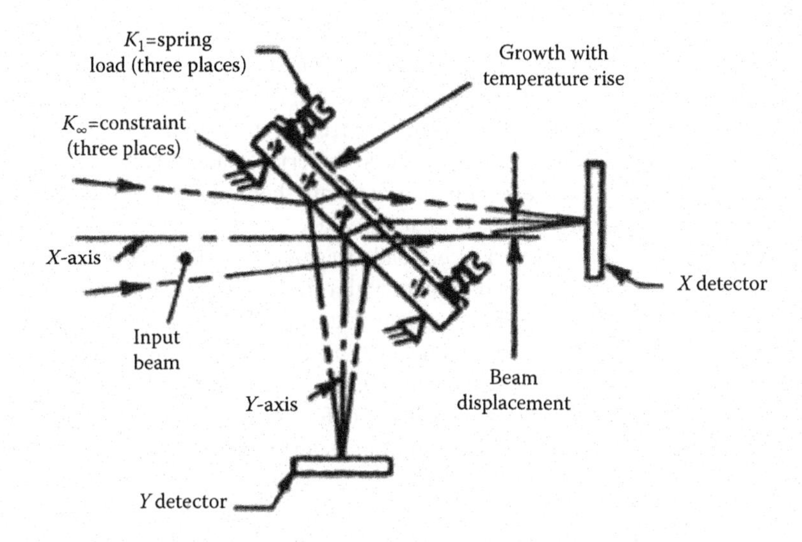

FIGURE 9.22 Mounting for a front surface beam splitter plate. (Adapted from Lipshutz, M.L., Optomechanical considerations for optical beam splitters, *Appl. Opt.*, 7, 2326, 1968.)

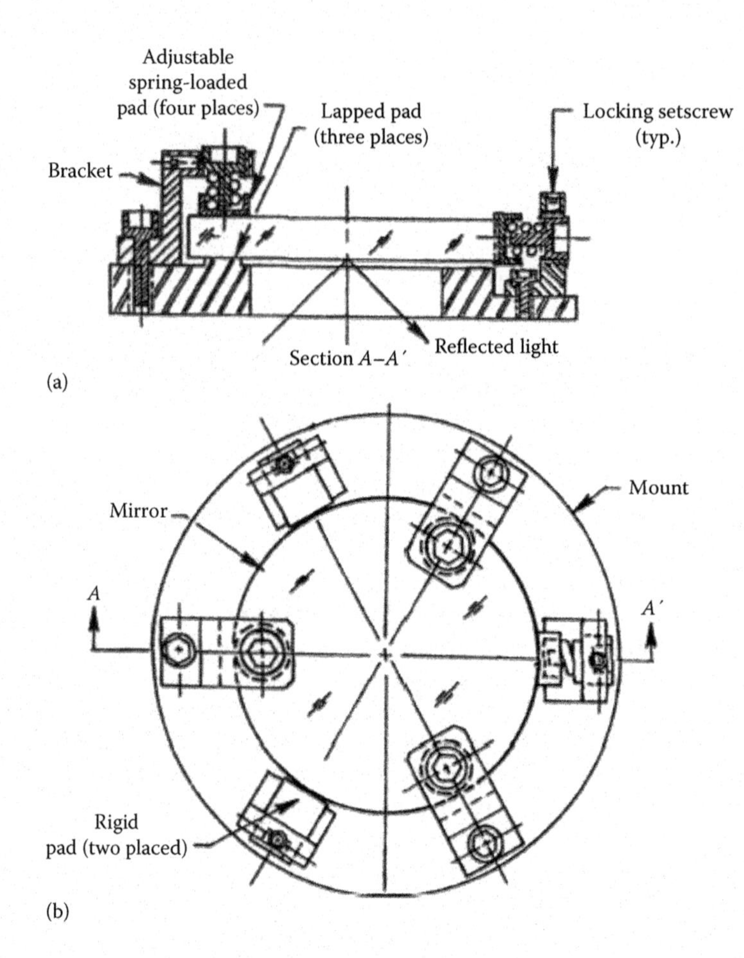

FIGURE 9.23 Concept for a spring-loaded semi-kinematic mirror mounting: (a) is a side view while (b) is a top view. (Adapted from Yoder, P.R. Jr., in *Handbook of Optomechanical Engineering*, Chap. 6 pp. 151–210, Ahmad, A., Ed., CRC Press, Boca Raton, FL, 1997.)

pads were not perfectly aligned with the mirror surface. The stress concentration that might then occur could be detrimental to the optical figure quality of the mirror. This potential problem can be eliminated and the design made more deterministic from a stress buildup viewpoint if the pads were to be provided with curved interfacing surfaces. Spherical pads on cantilevered springs are illustrated in Figure 9.23a. The interfaces with the mirror are then small circles formed by localized elastic deformation of the metal and the glass under preload.

Whenever an optic (such as a mirror) is preloaded against an opposing reference surface or surfaces, the preload force must be applied perpendicular to the faces of the optic and arranged so the line of force passes directly through the optical material toward the centers of appropriately shaped pads touching the other side of that optic. A typical configuration of this type in which a mirror is constrained axially is illustrated in Figure 9.24.[2] In the view in Figure 9.24a, the mirror interfaces are properly arranged and the mirror is not deformed by the applied forces. In the view in Figure 9.24b, the pads are not aligned with the forces and moments are applied to the mirror, tending to bend it as shown.

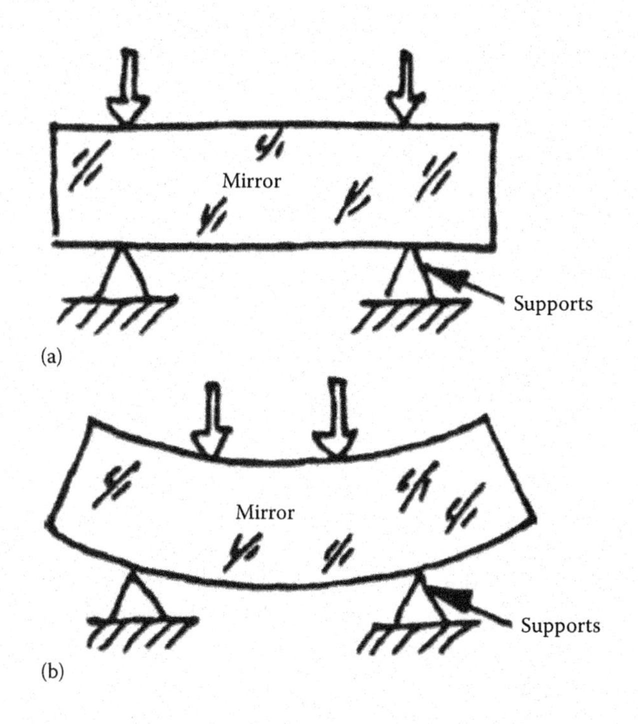

FIGURE 9.24 Schematic representations of (a) a mirror constrained by directly opposing forces and supports and (b) the same mirror subjected to a bending moment by forces directed between the supports. (From Vukobratovich, D., *Introduction to Optomechanical Design*, SPIE short course SC014, 2014.)

9.4 MOUNTING MIRRORS BY BONDING

9.4.1 Single and Multiple Bonds on Mirror Backs

Small mirrors can be mounted with glass-to-metal bonds by using adhesives. This design technique is simple and compact while also providing mechanical strength adequate for withstanding the severe shock, vibration, and temperature changes characteristic of most military and aerospace applications. The technique is also used frequently in less rigorous applications because it is easy to apply and is reliable when executed properly.

The critical aspects of a glass-to-metal bond are the inherent characteristics of the adhesive, thickness of the adhesive layer, cleanliness of the surfaces to be bonded, dissimilarity of CTEs for the materials bonded, dimensions of the bond, environment that the bonded assembly will experience, and care with which the bonding operation is performed. Usually, thin layers of epoxies or thicker layers of polyurethane adhesives are used to bond mirrors. The manufacturer's recommended procedures for applying and curing these adhesives should be followed unless special requirements of the application dictate otherwise. The means described for achieving the appropriate bond layer thickness for prism mounts may also be applied to mirror mountings. Experimental verifications of the choice of adhesive and methods of application and curing are advisable in critical applications.

The backs of first-surface mirrors with dimensions of up to ~150 mm can be bonded directly to a mechanical support. The ratio of the largest face dimension to the thickness of the mirror should be less than 6:1 so that mounting forces, acceleration effects, and adhesive shrinkage during curing or differential expansion or contraction of the mount at extreme temperatures does not excessively distort the optical surface. Figure 9.25 illustrates such a design.[3] The mirror is made of Schott BK7 glass, 51 mm in diameter and 8.4 cm in thickness (6:1 ratio). Its mass is about 40 g. The mounting base is stainless steel type 416 and has a single circular flat raised surface (or land) to which the mirror is to be bonded. The design task is to determine how big this bond should be to withstand some specified level of shock or vibration acceleration with a specified SF.

Guidelines for determining the appropriate adhesive area for bonding prisms to mechanical mounts are provided in the chapter on prism mounting (Chapter 7). These guidelines apply to mirrors as well. For the convenience of the reader, the important equations for calculating the minimum bond area are given here from that section.

The minimum area of the bond, A_{min}, is determined by

$$A_{min} = \frac{W a_G \, SF}{\sigma_{eYS}}, \tag{9.31}$$

where W is the mass of the optic, a_G is the worst-case expected acceleration factor, SF is the safety factor, and σ_{eYS} is the shear or tensile strength of the adhesive joint.

The SF should be at least 2 and possibly as large as 4 to allow for some unplanned, nonoptimum conditions, such as inadequate cleaning during processing. Note that the thickness of the adhesive layer (which is always small compared to the thickness of the mirror), the Young's modulus E of the adhesive and its Poisson's ratio ν do not appear in Equation 9.31. These parameters can affect the performance of the mirror if the bond is significantly thicker.

Because the dimensional changes of the adhesive bond during curing (shrinkage) and during temperature changes (expansion or shrinkage depending upon the sign of the change) are proportional to the in-plane dimensions of the bond, the bond area should not be too large. If a large total bond area is necessary to hold a heavy mirror, the bond should be divided into a group of smaller

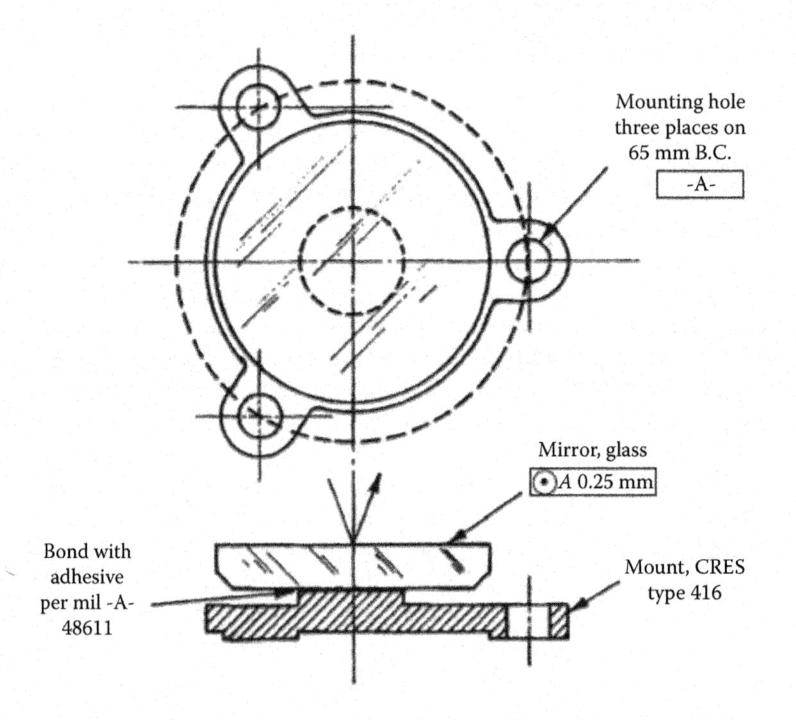

FIGURE 9.25 First-surface flat mirror subassembly with the optic bonded on its back to a flat pad on the mounting flange. (From Yoder, P.R. Jr., Non-image-forming optical components, *Proc. SPIE*, 531, 206, 1985.)

areas such as a triangular or ring pattern of circles or spots of any shape. Figure 9.26 shows some examples of multiple bond areas.

The maximum size of any individual bond is determined by thermal stress induced by temperature change. Bond size must be kept small enough to limit the stress to the yield of the adhesive, with the appropriate SF, or the yield stress of the mirror substrate material, whichever is lower. The "spot bond" equations for thermal stress in adhesive bonds are discussed in the chapter on prism mounting (Chapter 7); again, for convenience, these equations are given in the following:

$$\tau = \frac{(\alpha_1 - \alpha_2)\,\Delta T\, G \tanh(\beta r)}{\beta h_e}, \tag{9.32}$$

$$\beta = \left\{ \frac{G_e}{h_e} \left[\frac{(1-\nu_1^2)}{E_1 h_1} + \frac{(1-\nu_2^2)}{E_2 h_2} \right] \right\}^{\frac{1}{2}}, \tag{9.33}$$

where α_1 and α_2 are the CTEs of the prism and mount materials; E_1 and E_2 are the elastic moduli of the prism and mount materials; ν_1 and ν_2 are the Poisson's ratios of the prism and mount materials; h_1 and h_2 are the thicknesses of the prism and its mount; ΔT is the change in temperature; G_e is the shear modulus of the adhesive; and h_e is the adhesive bond thickness.

The pointing stability of a bonded mirror with temperature is a concern, since the CTE of adhesives is an order of magnitude higher than that of metals. If there is a variation in thickness between multiple adhesive bonds, this variation can cause the mirror to tilt when the temperature is changed. Consider the adhesive bond geometry in Figure 9.26b, with three bonds equally spaced on a common radius r. If one pad varies in thickness by Δh_e from the thickness of the other pads, the tilt θ_T of the mirror in radians for a temperature change ΔT is

$$\theta_T = \frac{2}{3} \frac{\Delta h_e \alpha_e \Delta T}{r}, \tag{9.34}$$

where α_e is the CTE of the adhesive and r is the radius of the circle for the bonds.

Equation 9.34 can be used to set the tolerance for bond thickness; similar equations can be developed for other bond geometries. Bond thickness can be controlled using methods discussed in the

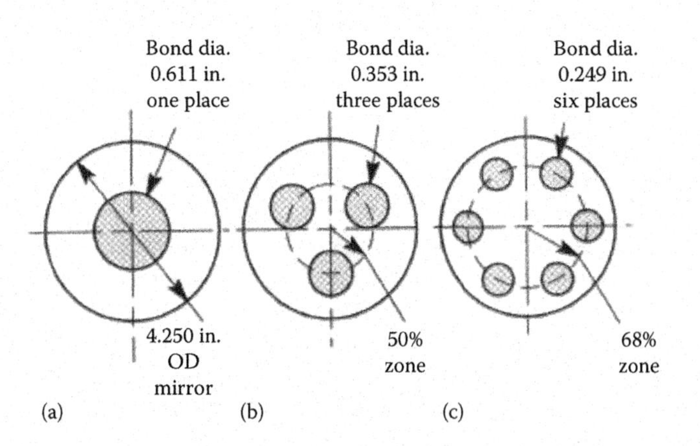

FIGURE 9.26 Arrangement of equal total area bonds from the design in Example 9.4 drawn to the same scale on the back of a first-surface mirror of diameter D: (a) single centered bond, (b) three bonds in an equilateral triangle pattern at the 50% zone, and (c) six bonds in a ring at the 68% radial zone.

chapter on prism mounting (Chapter 7). For example, incorporating glass microspheres of constant size into the adhesive is one way of controlling bond thickness.

Example 9.5

A flat circular Zerodur mirror of diameter 4.250 in. (107.95 mm) and thickness 0.708 in. (17.983 mm) is bonded to a metal base with 3M EC-2216 B/A adhesive. The bonded assembly is to withstand shocks of $a_G = 200g$ with an SF = 4. The density of Zerodur is 0.091 lb/in.^{3} (2530 kg/m^3). Assume that the bond strength is 2500 psi (17.24 MPa). What are the minimum bond area A_{min} and diameter D (1) of a single bond, (2) of each of three equally sized bonds arranged in a triangular pattern, and (3) of six equally sized bonds arranged in a circle at the 68% radial zone of the mirror?

The mirror mass is

$$W = \frac{\pi(2530\ \text{kg/m}^3)(17.983\times10^{-3}\,\text{m})(107.950\times10^{-3}\,\text{m})^2}{4} = 416\times10^{-3}\,\text{kg}.$$

From Equation 9.31,

$$A_{min} = \frac{(416\times10^{-3}\,\text{kg})(200)(9.81\,\text{m/s}^2)4}{17.24\times10^6\,\text{Pa}} = 189.4\times10^{-6}\,\text{m}^2.$$

Then the single bond diameter D_1 is

$$D_1 = 2\sqrt{\frac{189.4\times10^{-6}\,\text{m}^2}{\pi}} = 15.53\times10^{-3}\,\text{m (0.611in.)}.$$

The diameter of the three bonds D_3 is

$$D_3 = 2\sqrt{\frac{189.4\times10^{-6}\,\text{m}^2}{3\pi}} = 8.966\times10^{-3}\,\text{m (0.353 in.)}.$$

The diameter of the six bonds D_6 is

$$D_6 = 2\sqrt{\frac{189.4\times10^{-6}\,\text{m}^2}{6\pi}} = 6.340\times10^{-3}\,\text{m(0.249 in.)}.$$

The spot bonds are shown in Figure 9.26.

9.4.2 Annular Ring Mounts

Small mirrors can be mounted by annular bonding between the mirror edge and cell. This mounting technique is similar to adhesive bonding discussed in the chapter on mounting individual lenses (Chapter 5). This method can be applied to small and moderate-sized mirrors just as well as lenses. Figure 9.27 illustrates such a mounting. The elastomer fills the entire gap of thickness t_E between the mirror and cell inside diameter (ID). The same equations used for adhesive bonding of lenses apply. Rectangular mirrors can also be secured using this technique.

Figure 9.28a shows another technique for elastomerically potting a mirror into a mount. The mirror is attached to its mount with 12 discrete segments or pads of elastomer applied within the gap between the mirror OD and the cell ID. In this case, the mirror is fused silica ($\alpha_G = 0.5 \times 10^{-6}$), the cell is Kovar ($\alpha_M = 5.5 \times 10^{-6}$), and the elastomer is Dow Corning 6-1104 silicone ($\alpha_E = 261 \times 10^{-6}$).

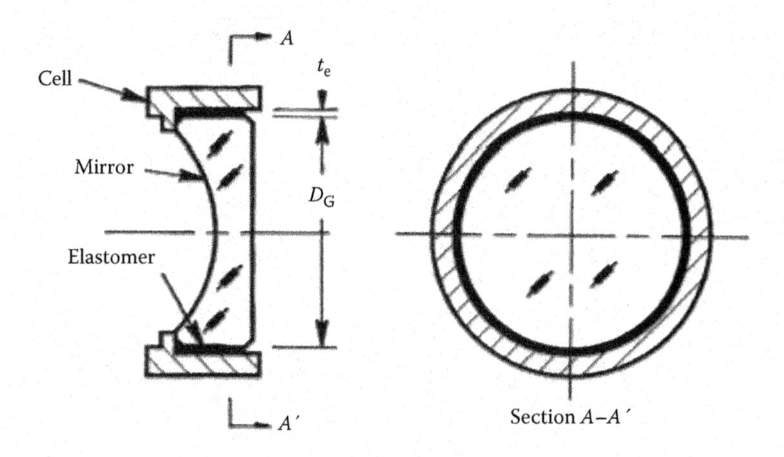

FIGURE 9.27 Schematic configuration of a concave first-surface mirror potted with an elastomeric ring into a cell.

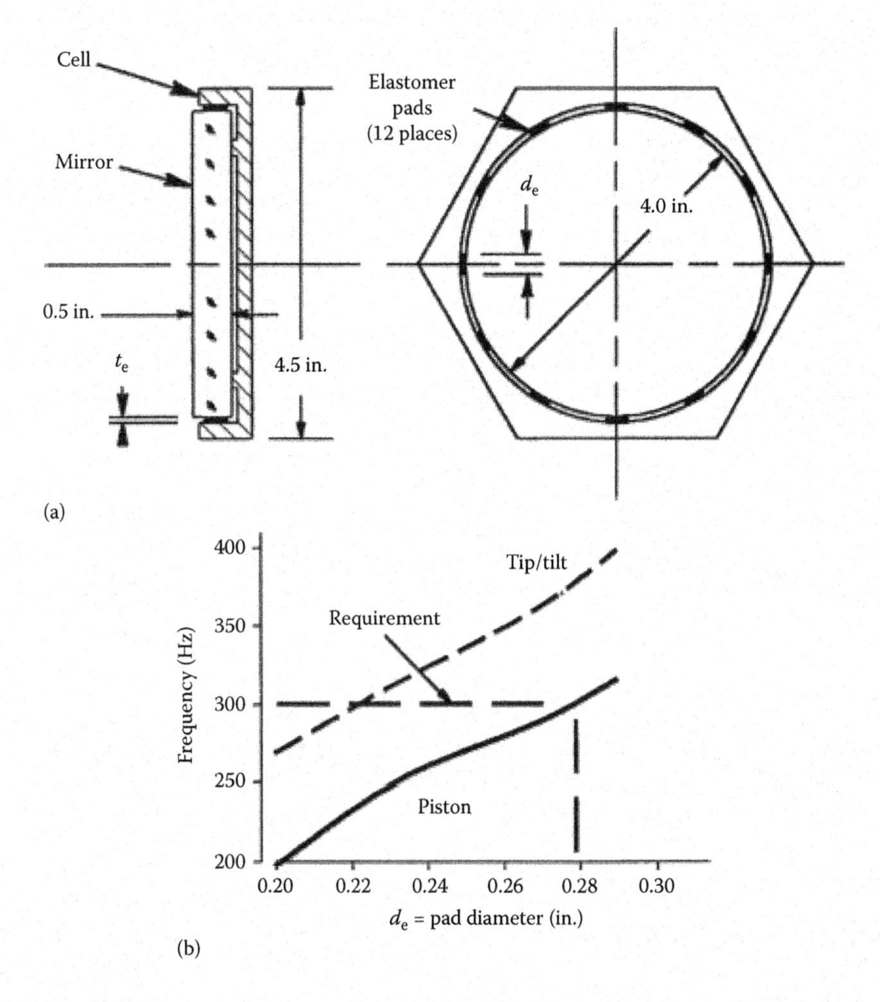

FIGURE 9.28 (a) Flat mirror rim mounted on multiple (12) discrete elastomer pads of dimension d_E and thickness t_E. (b) Plots of fundamental frequency for piston and tip/tilt vibrational modes for this mirror mounting. (Adapted from Mammini, P. et al., Sensitivity evaluation of mounting optics using elastomer and bipod flexures, *Proc. SPIE*, 5176, 26, 2003.)

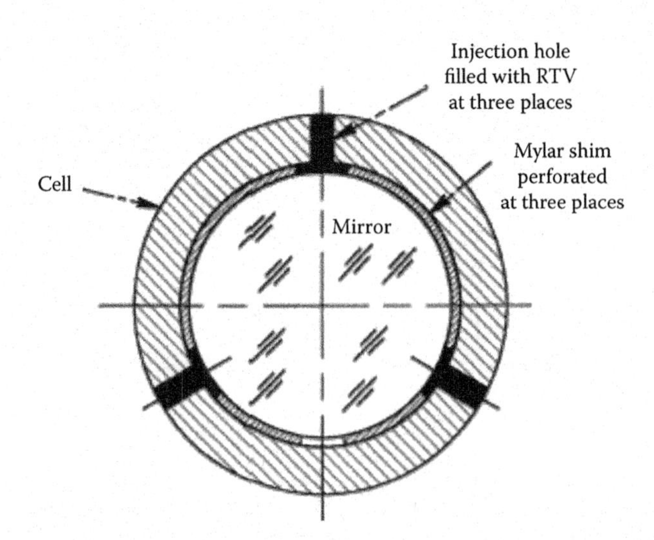

FIGURE 9.29 Mirror mounting concept in which three pads of an elastomer (room-temperature vulcanizing [RTV]) are injected through three holes in the cell wall and through three corresponding holes in a Mylar radial shim. (Adapted from Vukobratovich, D., *Introduction to Optomechanical Design*, SPIE short course SC014, 2014.)

The nominal "athermal" thickness of the elastomer pads according is 914 µm. The pad edge dimension (if square) or diameter (if circular) is d_E. This dimension is a design parameter.

Results of a finite-element analysis of vibrational modes for this design are shown in Figure 9.28b. The fundamental frequencies of the piston and tip/tilt modes varied with t_E. This is expected since changing the adhesive thickness alters the stiffness of the bond, with a resulting change in frequency. Figure 9.28b shows these variations as spline fits through the data points listed in the referenced paper. The application required that these frequencies be at least 300 Hz. The long vertical line shows that d_E should then be at least 7.11 mm. The actual dimension used was 7.34 mm. Thermal analysis showed that a 10°C temperature change would cause an out-of-plane surface distortion of less than 1/300 wavelength at 633 nm over the entire mirror surface.

Another technique is to radially constrain a circular lens or mirror with a single shim strip of Mylar located between the OD of the optic and the ID of the cell as shown in Figure 9.29. The shim is perforated at three places such that the holes line up with radially directed holes through the cell wall. An RTV compound is injected through the holes to reach the rim of the mirror. Pads of the RTV compound that have formed after curing hold the mirror from rotating about its axis (clocking) and constrain the optic radially.

9.5 FLEXURE MOUNTS FOR SMALL MIRRORS

In this section, a few typical examples of ways in which small mirrors can be mounted on flexures are discussed. There are two prime reasons for using flexure mounting. One is to prevent deforming the mirror surface due to differential thermal expansion of the mirror and its mount, and the other is to keep the vertex of a curved optical surface centered to an optical axis in an essentially stress-free manner.

Figure 9.30 illustrates the principle of one type of in-plane flexure mounting for a circular mirror in a cell. The cell is supported from three flexure blades. The curved arrows indicate the directions of free motion for each flexure acting alone. Ideally, these lines of freedom should intersect at a point (such as the CG of the mirror substrate), the flexure lengths should be equal, and the fixed ends of the three flexures should form an equilateral triangle. The function of this system of flexures may be explained as follows. In the absence of *C*, the combination of flexures *A* and *B* will permit rotation only about point *O*, which is the intersection of flexure *B* with a line extending flexure *A*. With *C* in place, rotation about *O* is prevented since flexure *C* is stiff in that direction.

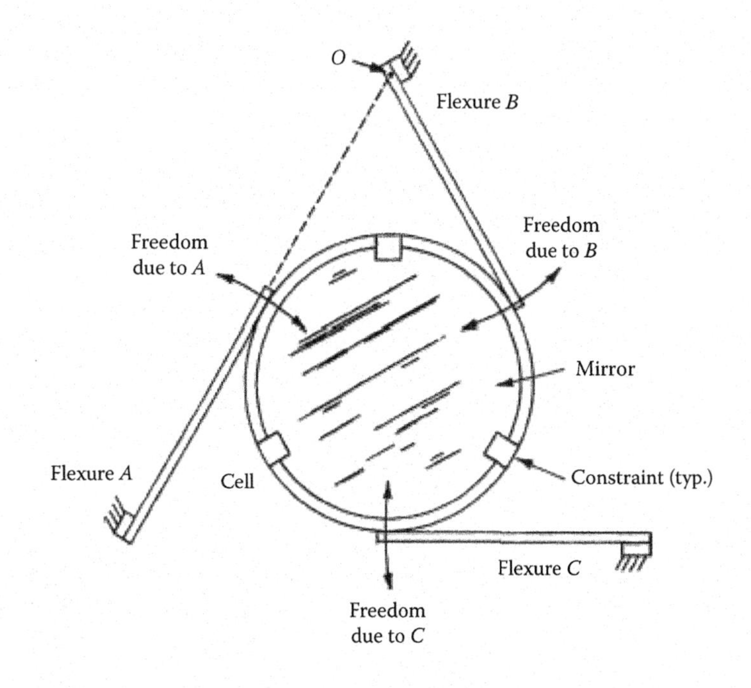

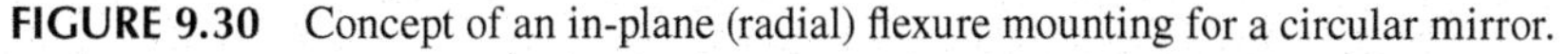
FIGURE 9.30 Concept of an in-plane (radial) flexure mounting for a circular mirror.

Although not apparent in the frontal view of the figure, the flexure blades may have sufficient depth perpendicular to the page to provide beam strength to prevent the mirror from translating along its normal. If not, axial constraints are needed. Uniform temperature changes cause thermal expansion of all parts of the subassembly, but radial motion of the optic will be impeded without stressing the mirror. The only permitted motion due to expansion or contraction is a small rotation about the mirror normal through the intersection of the lines of freedom. This occurs because of slight changes in the lengths of the flexures. The magnitude θ of this rotation in radians may be approximated by the expression $\theta = 3\alpha\Delta T$, where α is the CTE of the flexure material (typically in parts per million per degree [ppm/°]) and ΔT is the temperature change. If the flexure is beryllium copper with α of 17.8 ppm/K and ΔT is ±20 K, θ is only ±3.7 arcmin. This rotation would be inconsequential in most applications.

A quite different technique for bonding a mirror to a mechanical support is sketched in Figure 9.31. Here, a circular mirror is bonded to three flat flexure blades that are, in turn, attached mechanically

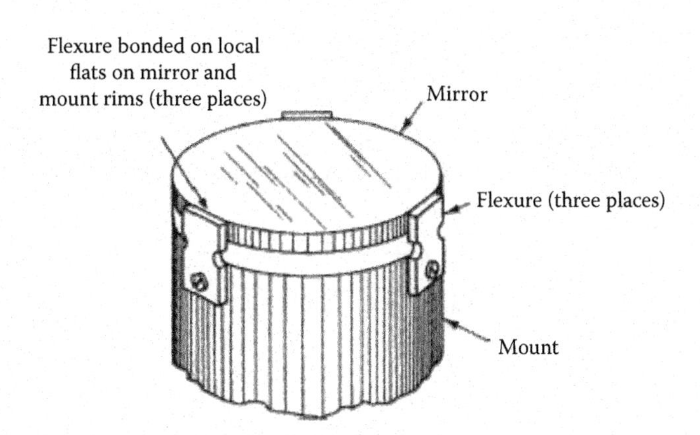

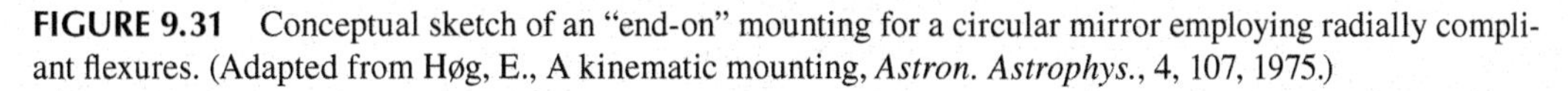
FIGURE 9.31 Conceptual sketch of an "end-on" mounting for a circular mirror employing radially compliant flexures. (Adapted from Høg, E., A kinematic mounting, *Astron. Astrophys.*, 4, 107, 1975.)

by screws, rivets, or adhesive to a circular post or tube of essentially the same diameter as the mirror. The flexures are flat, so they can bend radially to accommodate differences in thermal expansion. They are of the same free length and material so that thermally induced tilts are minimized. The local areas on both the mirror and mount where the springs are attached are flattened to obtain an adequate contact area for bonding and to prevent cupping of the springs. The flexures should be as light and flexible as is consistent with vibration and shock requirements. This mounting arrangement can be used to support image-forming mirrors as well as flat ones. The design tends to keep the optic centered despite temperature changes.

Flexure mounts are analyzed using the principle from the chapter on mirror performance (Chapter 8); fundamental frequencies are given by Equations 8.33 through 8.37. In some designs, flexure stress may be a concern. If the individual flexures are simple constant cross-sectional beams in bending, fixed at both ends, and not loaded axially, the stress in the flexure σ_f is given by

$$\sigma_f = \frac{3E_f \delta h_f}{L_f^2}, \tag{9.35}$$

where E_f is the flexure elastic modulus, δ is the maximum end deflection of the flexure, h_f is the flexure thickness in the bending direction (rectangular cross section), and L_f is the flexure length.

Example 9.6

The mass of a 120 mm diameter fused silica mirror is 0.5 kg. The mirror is subjected to a temperature change of 75 K and is mounted to a 6061 aluminum cell by three tangential flexures equally spaced around the circumference; each flexure is made of 6Al–4V titanium. A minimum radial frequency of 1000 Hz is required. A flexure thickness of h_f = 1.5 mm is desirable. Find the flexure length L_f and width b_f.

The deflection of the end of the flexure is equivalent to the difference in contraction between the mirror and aluminum cell. This is found by multiplying the mirror radius (120 mm/2 = 60 mm) by the difference in CTE times the temperature change, or

$$\delta = (23.4\times10^{-6}\,\text{K}^{-1} - 580\times10^{-9}\,\text{K}^{-1})(60\times10^{-3}\,\text{m})(75\text{K}) = 1.041\times10^{-3}\,\text{m}.$$

For 6Al–4V titanium, the elastic modulus E_f = 114 GPa and the yield stress is σ_{YS} = 102 MPa. From Equation 9.35, the flexure length L_f is given by

$$L_f = \left[\frac{3(114\times10^9\,\text{Pa})(1.041\times10^{-3}\,\text{m})(1.5\times10^{-3}\,\text{m})}{102\times10^6\,\text{Pa}}\right]^{\frac{1}{2}} = 72.3\times10^{-3}\,\text{m}.$$

Note that the flexure length is independent of the width with respect to bending stress. A conservative estimate of the radial frequency is given by Equation 8.36, and the tangential stiffness k_T is given by Equation 8.35; then,

$$f_n \cong \frac{1}{2\pi}\sqrt{\frac{3}{2}\frac{k_T}{m}} = \frac{1}{2\pi}\sqrt{\frac{3}{2}\frac{E_f b_f h_f}{L_f W}}.$$

Solving for the width of the flexure b_f,

$$b_f = \frac{8\pi^2(1000\text{Hz})^2(0.5\text{kg})(72.3\times10^{-3}\,\text{m})}{3(114\times10^9\,\text{Pa})(1.5\times10^{-3}\,\text{m})} = 5.57\times10^{-3}\,\text{m}.$$

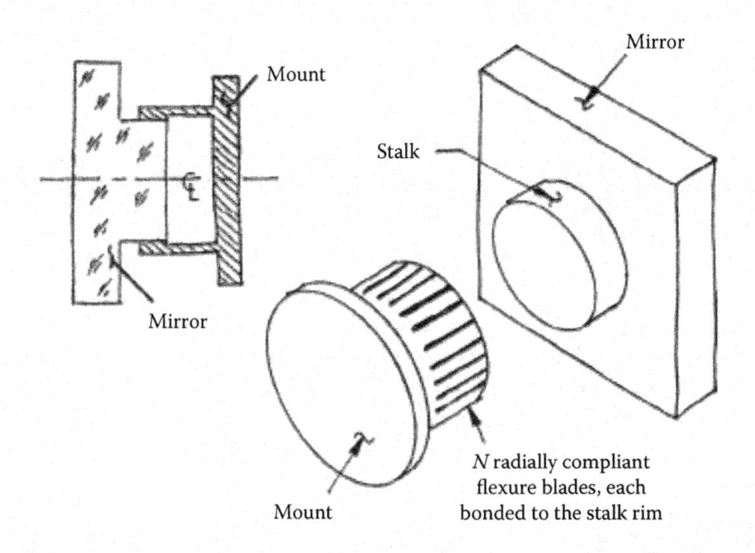

FIGURE 9.32 Concept for a radially compliant mounting for a square mirror featuring an integral central stalk. The flexure blades machined into the mount are bonded to the stalk. (Adapted from Vukobratovich, D., *Introduction to Optomechanical Design*, SPIE short course SC014, 2014.)

Figure 9.32 illustrates a mounting for a so-called *mushroom mirror.* The mirror has a cylindrical protrusion or stalk centrally located on its back face. This stalk is an integral part of the mirror or bonded in place with an adhesive. The intersection of the integral stalk and the back face of the mirror should have a fillet for ease of fabrication and for minimizing residual stress rather than the sharp corner indicated in the figure. The stalk is inserted into and bonded to the IDs of a series of N flexure blades machined into a cylindrical portion of the mounting flange. The flexures are radially compliant and act in the same manner as the flexures in Figure 9.31 to keep the mirror centered when the temperature changes. Mirror deformations also are minimized.

Figure 9.33a shows a flexure mounting concept in which the cylindrical rim of a circular mirror is bonded to the centers of three flexures oriented tangentially to that rim and attached to the structure at both ends (see Figure 9.33b). Some tangential compliance might well be provided in the interface to the mount at one end of each flexure to allow for temperature changes if the flexure is not made of the same material as the mount. Special fixtures are needed to align the mirror with the flexures at the time of bonding.

Figure 9.34 shows the mechanical design of a boss bonded to the rim of a 381 mm diameter mirror to attach a flexure. The material of such a boss should be selected to match its CTE with that of the mirror as closely as possible. For example, a boss made of Invar 36 might be used with a ULE or Zerodur mirror. An epoxy such as 3M EC-2216 B/A might be chosen for use with these materials. The adhesive would be injected through the small access hole at the center of each boss. The appropriate bond thickness could be obtained by the use of shims or microspheres added to the adhesive. Care should be exercised to align the sides of the square holes approximately with the plane of the mirror surface. Flexures such as that shown in Figure 9.35 would then be epoxied to this boss and attached with screws (or epoxied) to the mount structure.

Figure 9.36 shows schematically some other types of bosses, threaded studs, and flexures that have been successfully bonded to mirrors to allow them to be attached to the optical instrument structure. Those in the view in Figure 9.36a are bonded into recesses or notches ground into the mirror substrate, while those in the view in Figure 9.36b are bonded externally to the mirror surfaces.

A mirror of square or rectangular shape can be supported in a cell attached to three deep cantilevered flexure blades as shown in Figure 9.37. The dashed lines indicate the directions of freedom (approximated as straight lines). In this case, the intersection of these lines, which is the stationary thermal center, does not coincide with the geometric center of the mirror or the CG of this particular mirror and cell combination. By changing the angles of the corner bevels and relocating the flexures, the intersection point could be centralized and the design improved from

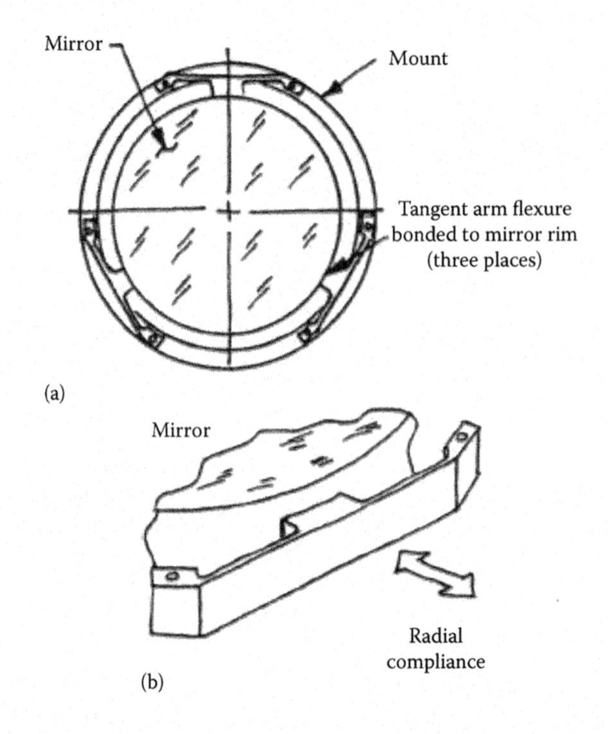

FIGURE 9.33 (a) Conceptual sketch of the mounting of a circular mirror by using three tangentially oriented flexures supported at both ends as shown in view (b). (Adapted from Vukobratovich, D., and Richard, R., Flexure mounts for high-resolution optical elements, *Proc. SPIE*, 959, 1988.)

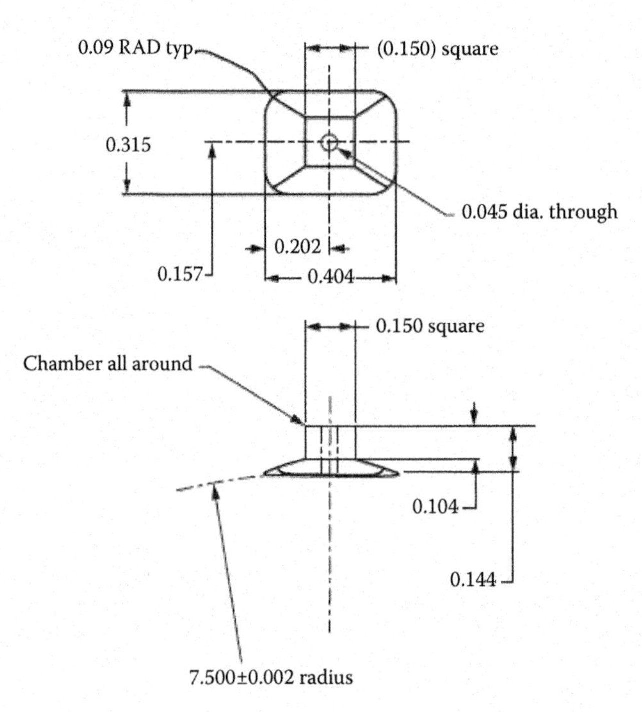

FIGURE 9.34 Design of a boss suitable for bonding to the rim of a 15.0 in. (38.1 cm) diameter mirror for flexure mounting in the fashion of Figure 9.30. Dimensions are inches. (From Yoder, P.R. Jr., *Mounting Optics in Optical Instruments*, 2nd ed., SPIE Press, Bellingham, WA, 2008.)

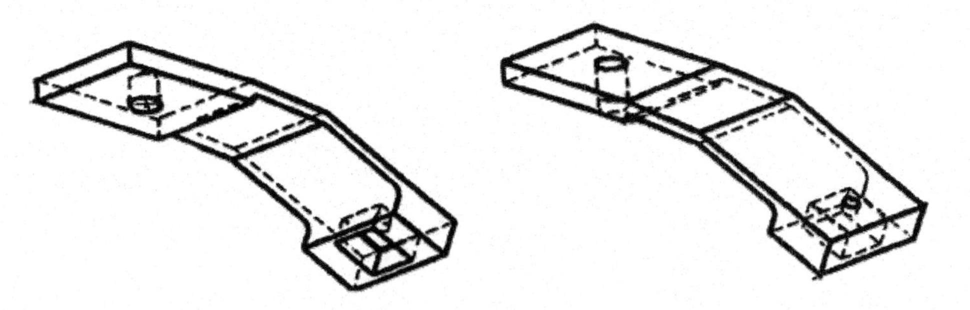

FIGURE 9.35 Top and bottom views of a cantilevered flexure shaped to interface with the boss in Figure 9.34 and the mount for a circular mirror. (From Yoder, P.R., Jr., *Mounting Optics in Optical Instruments*, 2nd ed., SPIE Press, Bellingham, WA, 2008.)

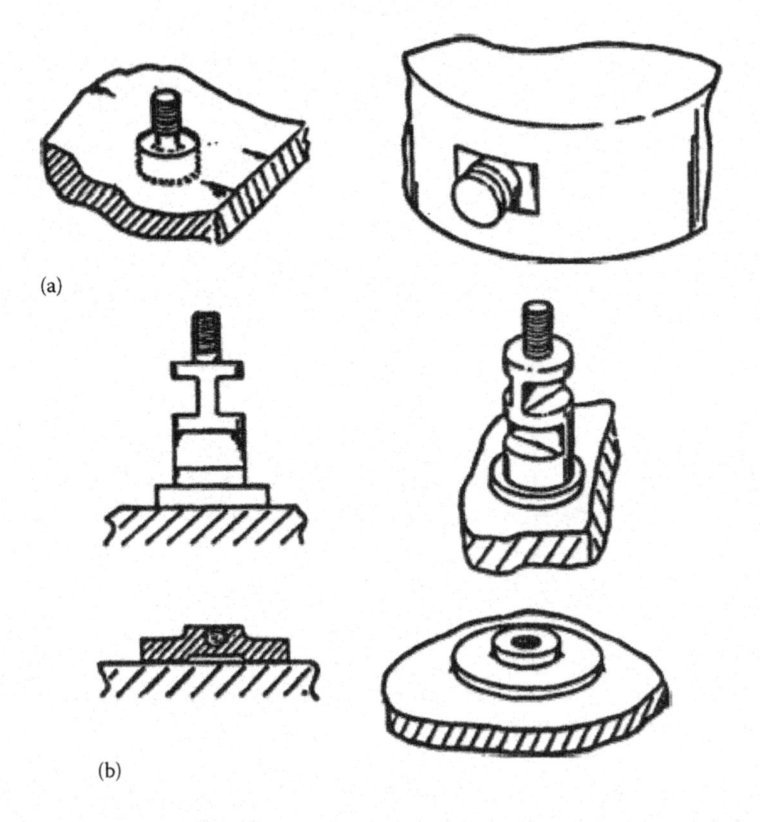

FIGURE 9.36 Schematic illustrations of a variety of bosses, threaded studs, universal-joint flexure studs, and pads that can be bonded to the rims or backs of mirrors for attaching flexures. Those in view (a) are bonded into recesses or notches in the mirror substrate, while those in view (b) are bonded externally to the mirror surface. (From Yoder, P.R. Jr., *Mounting Optics in Optical Instruments*, 2nd ed., SPIE Press, Bellingham, WA, 2008.)

a dynamic viewpoint. Differential thermal expansion across the mount-to-structure interface can then occur without stressing the mirror. Axial movement of the mirror is prevented by the high stiffness of the blades in that direction.

If the rectangular mirror is to be mounted without a cell, bosses configured as shown in Figure 9.36, but with flat bonding surfaces might be attached directly onto the rim of the mirror. They would be attached in the same manner described for the circular mirror with bosses. A straight version of the flexure shown in Figure 9.35 would then be used at three places to interface the mirror with the mount structure.

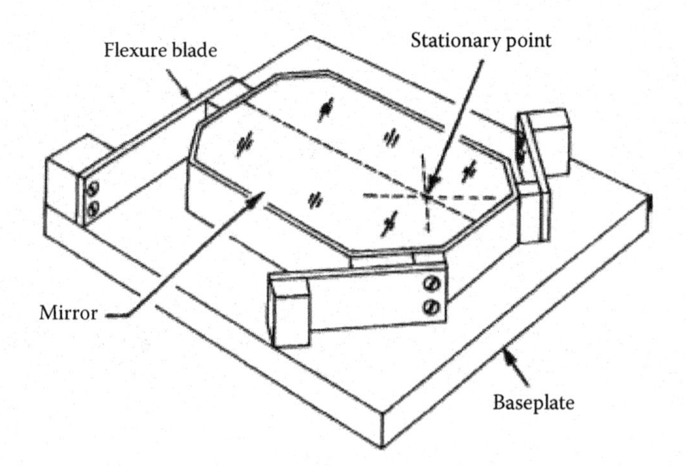

FIGURE 9.37 Concept of flexure mounting of a cell-mounted rectangular mirror. (From Yoder, P.R. Jr., *Mounting Optics in Optical Instruments*, 2nd ed., SPIE Press, Bellingham, WA, 2008.)

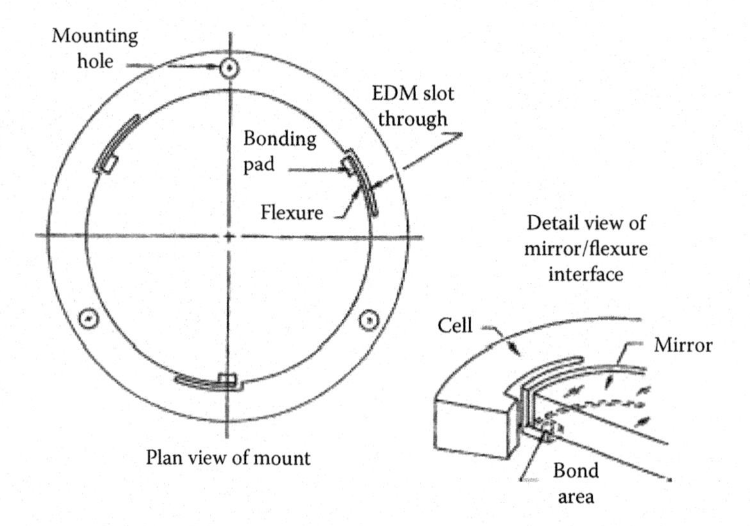

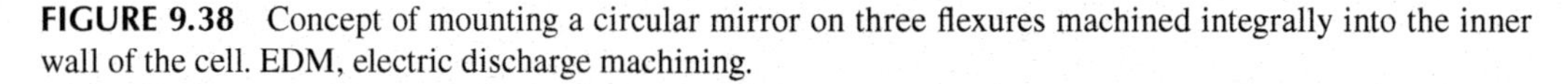

FIGURE 9.38 Concept of mounting a circular mirror on three flexures machined integrally into the inner wall of the cell. EDM, electric discharge machining.

Another concept of mounting a circular mirror with cantilevered tangential flexures is depicted in Figure 9.38. Here the flexures are integral to the body of the ring-shaped mounting. The flexures typically would be created by machining narrow slots into the ID of the mount by using an electric discharge machining process. Once again, the blades are stiff in the tangential and axial directions and compliant radially as would be appropriate to negate decentrations that are due to temperature changes. Mounts of this general type are frequently used in complex projection optical systems for creating patterns on computer chips by microlithography, where highly accurate optical component centration is mandatory.

Mirrors used in high-precision, high-performance applications may benefit from being mounted in the manner shown conceptually in Figure 9.39. Here a circular mirror with three bonded-on bosses is attached to tangentially oriented arms containing dual sets of universal-joint-type flexures. Three adjustable axial metering rod-type supports that include flexures are also attached to the bosses. Such a mounting is essentially insensitive radially to temperature changes because of the action of the tangent arm flexures. The thermal compensation mechanisms shown in the axial supports make the design insensitive in that direction to temperature changes. The

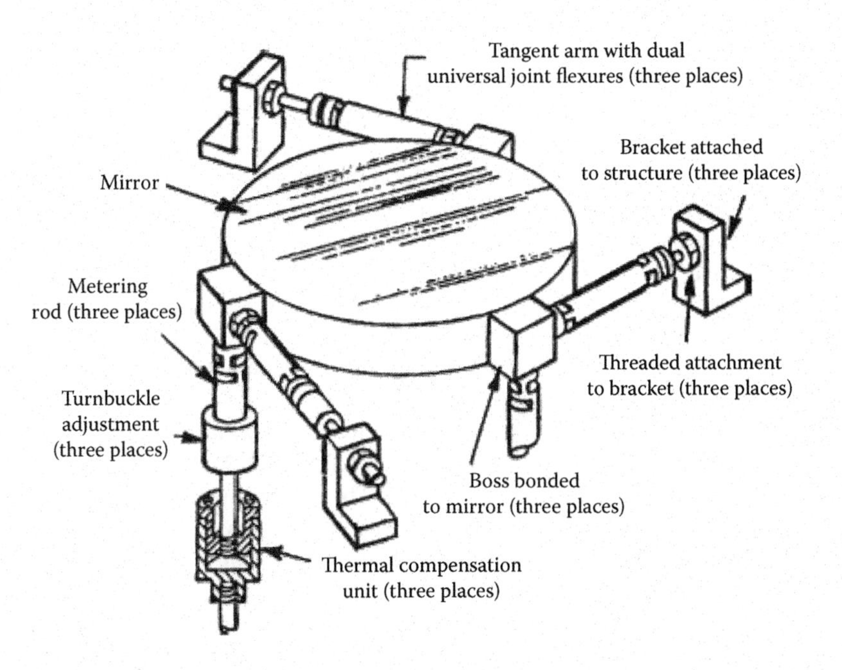

FIGURE 9.39 System of flexures configured to minimize displacement, tilt, despace, and/or distortion of the optical surface of a circular mirror caused by temperature changes or mounting forces. The mount also provides adjustments for all six DOFs. (From Yoder, P.R. Jr., *Mounting Optics in Optical Instruments*, 2nd ed., SPIE Press, Bellingham, WA, 2008.)

latter mechanism consists of selected lengths of dissimilar metals arranged in a reentrant manner. Differential screws might be employed to advantage as the means for attaching the fixed ends of the tangent arms to the brackets in some applications. This would provide fine adjustment of the lengths of the tangent flexures. The turnbuckle mechanisms shown in the metering rods would facilitate axial adjustment. These could also be differential screws. Two-axis tilts of the mirror can be adjusted by differential motion of these axial mechanisms. These mechanisms allow control and stabilization of all six DOFs of the optic.

PROBLEMS

1. A mirror with a diameter of d = 200 mm and thickness of h = 25 is made of Corning Pyrex glass, with a concave radius of R = 1.2 m. The density of Pyrex is ρ = 2230 kg/m^3. Compare the masses given by the exact and approximate methods. *Answer:* The exact mass is 1.60515 kg, the approximate mass is 1.60548 kg, and the error is 211×10^{-6}.
2. A meniscus mirror with a diameter of d = 200 mm and constant thickness of h = 30 mm is made of 6061 aluminum, with a concave radius of 1.0 m. The density of 6061 aluminum is ρ = 2790 kg/m^3. Compare the masses given by exact and approximate methods. *Answer:* The exact mass is 2.623 kg, the approximate mass is 2.638 kg, and the error is about 5.59×10^{-3}.
3. The primary mirror of a small Cassegrain telescope is made in the form of a meniscus, with a diameter of d = 350 mm and thickness of 40 mm. There is a 100 mm diameter hole through the center of the mirror. The mirror focal ratio is f_{no} = 2.0. The mirror is made of Corning Pyrex, with a density of ρ = 2230 kg/m^3. Estimate the mirror mass by using approximate methods. *Answer:* The approximate mass of the mirror is 7.92 kg.
4. A scanning periscope displaces the optical axis by Δy = 300 mm and displaces the axis horizontally by Δx = 40 mm. The output beam deviation is $\delta = \pm 15°$. What are the tilts θ_1 and θ_2 of the mirror normals? *Answer:* The angle of mirror M_1 is $\theta_1 = 48.8°$; when the

output beam angle is +15°, the angle of mirror M_2 is $\theta_2 = -41.3°$, and when the output beam angle is −15°, the angle of mirror M_2 is $\theta_2 = -56.3°$.

5. The primary mirror of a small Newtonian telescope has a diameter of $d = 150$ mm and a focal ratio of $f_{no} = 8$. The diameter of the image at the focus is 25 mm. A folding flat tilted at $\theta = 45°$ brings the image outside the telescope tube, with $L = 150$ mm. What are the elliptical dimensions W and A of the minimum size folding flat and what is its offset G from the optical axis? *Answer:* The elliptical mirror size is $A = 57.6$ mm and $B = 40.7$ mm; the required offset is 1.5 mm.
6. A collimated laser beam of diameter $W = 25$ mm is deflected through an angle of 120°. What is the minimum elliptical folding mirror size? (Hint: Remember that the mirror angle is half of the angle of deflection.) *Answer:* The minimum width $B = 25$ mm, and the minimum length of the ellipse $A = 50$ mm.
7. What is the uncoated reflectance of zinc sulfide in air at a wavelength of 10.6 µm, where the index of refraction is $n = 2.192$? *Answer:* The reflectance $R_S = 0.139$.
8. The mass of a small flat mirror is 154×10^{-3} kg. The mirror is subjected to an acceleration of $a_G = 75g$ normal to the optical surface. Three spring clips made of 17-7 PH stainless steel in condition C retain the mirror; the thickness of the clips is $t = 1.27$ mm (stock steel thickness), and their width is $b = 6$ mm. For 17-7 PH stainless steel in condition C, the yield strength $\sigma_{YS} = 1.21$ GPa, the elastic modulus $E_M = 200$ GPa, and Poisson's ratio $\nu_M = 0.3$. With an SF = 2, what is the length L of the spring and the deflection Δ? *Answer:* The spring length $L = 25.8$ mm, and the deflection $\Delta = 1.91$ mm.
9. The mass of a small right circular cylinder mirror is 0.972 kg, and its diameter is 150 mm. The mirror is subjected to an acceleration of $a_G = 50g$ normal to the circular optical surface. A continuous flange made of AISI 302b stainless steel retains the mirror. There is 1 mm of radial clearance all around the mirror, and the flange contacts the mirror at a distance of 3 mm radially from the circumference. The properties of AISI 302b stainless steel are an elastic modulus of $E_M = 193$ GPa, a yield stress of $\sigma_{YS} = 703$ MPa, and a Poisson's ratio of $\nu_M = 0.25$. The stainless steel is available in a thickness of 381×10^{-6} m (0.015 in.), and this thickness is used in the flange. For this thickness, what is the SF? What is the deflection Δ of the flange to ensure that the preload force does not go to zero with the calculated SF? *Answer:* SF = 4.236; with this SF, the flange deflection is $\Delta = 352 \times 10^{-6}$ m.
10. The mass of a small folding flat is 21.5×10^{-3} kg. The mirror is subjected to an acceleration of $a_G = 75g$ in any direction; an SF = 4 is required. A single circular spot bond is used to attach the mirror to its mount. The adhesive shear strength is $\sigma_{eYS} = 14$ MPa. What is the diameter of the minimum size spot bond on back of the mirror? *Answer:* The spot bond diameter is 2.4 mm.
11. The diameter of a fused silica mirror is 300 mm, its thickness is 50 mm, and its mass is 7.776 kg. The mirror is bonded to a 6061 aluminum plate; the aluminum plate is 25 mm thick. The adhesive bond thickness is 75×10^{-6} m. The mirror is subjected to an acceleration of $a_G = 20g$, and an SF = 2 is required. The mirror is bonded at 25°C and must survive at a low temperature of −51°C. For fused silica, the elastic modulus is $E_G = 73$ GPa. Poisson's ratio is $\nu_G = 0.17$, and the CTE is $\alpha_G = 580 \times 10^{-9}$ K^{-1}. For 6061 aluminum, the elastic modulus is $E_M = 69$ GPa, Poisson's ratio is $\nu_M = 0.33$, and the CTE is $\alpha_M = 23.3 \times 10^{-6}$ K^{-1}. For the adhesive, the shear modulus $G_e = 207$ MPa and the yield stress in shear is $\sigma_{eYS} = 14.5$ MPa. What is the minimum adhesive area for the acceleration load? What is the largest bond radius, assuming a circular bond, based on thermal stress? Using the largest bond radius, how many bond areas are required, rounding up to the next highest integer? (Hint: Use Equations 9.32 and 9.33 to find the largest bond area.) *Answer:* The minimum bond area set by the acceleration is $A_{min} = 210.4 \times 10^{-6}$ m^2, and the radius of the largest bond area set by thermal stress is $R = 3.05$ mm. Dividing the largest bond area set by thermal stress into the minimum bond area required by acceleration gives $N = 8.854$; this is rounded up to $N = 9$.

12. The mass of a 150 mm diameter mirror is 0.972 kg. The mirror is bonded to a mount by using three circular bond areas, equally spaced on a 100 diameter centered on the mirror axis. The diameter of each bond area is 10 mm. The yield strength of the adhesive is $\sigma_{eYS} = 17.2$ and the CTE is $\alpha_e = 102 \times 10^{-6}$ K^{-1}. What is the maximum acceleration a_G set by the bond yield, assuming an SF = 4? If the maximum tilt tolerance of the mirror is 5×10^{-6} radian for a temperature change of 75 K, what is the bond thickness tolerance Δh_e? *Answer:* The maximum acceleration is $a_G = 89.6g$; and the bond thickness tolerance is $\Delta h_e = 49 \times 10^{-6}$ m.
13. A mirror similar to the one in Figure 9.32 is mounted by flexures attached to a stalk on its back. The mirror mass is 3 kg, and the stalk diameter is 60 mm. The mirror is made of fused silica, with a CTE of $\alpha_G = 580 \times 10^{-9}$ K^{-1}. There are 12 flexures bonded to the central stalk; each flexure is $b_f = 12$ mm wide and $L_f = 15$ mm long. The flexures and mount are made of 7075-T6 aluminum, with an elastic modulus of $E_f = 71.7$ GPa, a yield strength of $\sigma_{YS} = 503$ MPa, and a CTE of $\sigma_f = 23.6 \times 10^{-6}$ K^{-1}. With an SF = 4, what flexure thickness h_f is necessary for a temperature change of $\Delta T = 75$ K? What is the approximate radial frequency of the mirror on its flexure mount? (Hint: Use Equations 8.35 and 8.36 to estimate the frequency.) *Answer:* The flexure thickness is $h_f = 2.54 \times 10^{-3}$ m, and the natural frequency in the radial direction $f_r \approx 2717$ Hz.
14. A 150 mm diameter right circular Zerodur mirror with a mass of 1.12 kg is mounted by three flexures that are tangential to its edge and equally spaced around the mirror circumference, as in Figure 9.30. The length of each flexure from point of attachment to mirror is $L_f = 44$ mm; the flexure thickness in the bending direction is $h_f = 2.5$ mm, and the flexure depth or width is $b_f = 16$ mm. The flexures are made of 17-PH stainless steel in condition H1025, where the elastic modulus is $E_f = 196.5$ GPa and the yield strength is $\sigma_{YS} = 1$ GPa, and are fixed at both ends. The mirror mount is made of 6061 aluminum with a CTE of $\alpha_M = 23.4 \times 10^{-6}$ K^{-1}, and the CTE of Zerodur is $\alpha_G = 50 \times 10^{-9}$ K^{-1}. Due to uncertainty about the environmental conditions, a large SF = 10 is required. Using the SF, what is the maximum temperature change without exceeding yield in the flexures? What is the approximate radial natural frequency of the mirror on its flexure mount? (Hint: Use Equations 8.35 and 8.36 to estimate the frequency.) What is the axial natural frequency in the direction normal to the mirror optical surface? *Answer:* The maximum temperature without exceeding yield of the flexures is $\Delta T = 75$ K; the radial natural frequency $f_R = 2500$ Hz, and the axial natural frequency $f_A = 662$ Hz.

REFERENCES

1. Yoder, P.R. Jr., *Mounting Optics in Optical Instruments*, 2nd ed., SPIE Press, Bellingham, WA, 2008.
2. Vukobratovich, D., *Introduction to Optomechanical Design*, SPIE short course SC014, 2014.
3. Yoder, P.R. Jr., Non-image-forming optical components, *Proc. SPIE*, 531, 206, 1985.

FURTHER READING

Hatheway, A.E., Designing elastomeric mirror mountings, *Proc. SPIE*, 6665, 666504, 2007.

Horn, W., The Aurora Project square mirror arrays, *Proc. SPIE*, 817, 218, 1987.

Mammini. P., Holmes, B., Nordt, A., and Stubbs, D., Sensitivity evaluation of mounting optics using elastomer and bipod flexures, *Proc. SPIE*, 5176, 26, 2003.

Schubert, F., Determining optical mirror size, *Mach. Des.*, 51, 128, 1979.

Vukobratovich, D., and Richard, R., Flexure mounts for high-resolution optical elements, *Proc. SPIE*, 959, 18, 1988.

Willis, C.L., and Petrie, S.P., Bonding glass to metal with plastic for stability over temperature II, *Proc. SPIE*, 4771, 38, 2002.

Yoder, P. R. Jr., and Vukobratovich, D., *Opto-mechanical Systems Design*, 4th ed., CRC Press, Boca Raton, FL, 2015.

10 Design and Mounting of Metallic Mirrors

10.1 INTRODUCTION AND SUMMARY

The advantages of metal mirrors compared to conventional glass mirrors are lower manufacturing cost, simplicity of mounting, high thermal conductivity, higher tolerable stress, and suitability for the same type of material athermalization. The higher specific stiffness of some metals such as beryllium provides better performance for dynamic applications such as scanning and stabilization of a mirror. The disadvantages of metal mirrors are increased surface scattering and dimensional instability.

The discussion in this chapter begins in Section 10.2 with some general considerations of metal mirrors as distinguished from nonmetallic ones. A variety of examples of metal mirror design and fabrication techniques are then discussed. Mirrors made of various materials are considered in Sections 10.3 through 10.5. Because of the inherent porosity of some metal surfaces, it is advantageous to plate them with an amorphous material. Electrolytic nickel (EN) and electroless nickel (ELN) are very common types of plating for such mirrors. In Section 10.6, the characteristics of these types of plating and descriptions of their applications are summarized.

Some, but not all, of the metals discussed here are compatible with material removal and final contouring to optical-quality surfaces by precision turning with single-point, gem-quality natural diamond tools. This process, commonly called single-point diamond turning (SPDT), is discussed in Section 10.7.

In Sections 10.8 and 10.9, some of the techniques most commonly used to mount metal mirrors in optical instruments are discussed. It is quite feasible to interface many smaller metallic substrates with the various conventional mounts discussed earlier in this book for nonmetallic mirrors. There are design differences in such mountings because the mechanical properties of metals differ from those of nonmetals. Highly successful methods for supporting small to moderate-sized metal mirrors involve mounting provisions built into the mirrors themselves. Integral mounting techniques are reviewed, and some precautions to be observed in such designs are summarized.

The chapter closes in Section 10.10 with brief considerations of multiple metallic optical and mechanical component interfaces using SPDT techniques to facilitate assembly and optical alignment.

10.2 GENERAL CONSIDERATIONS OF METAL MIRRORS

The mirror designer has a number of metallic materials from which to choose. Table 10.1 lists key candidates. Table 10.2 lists properties generally considered in a material trade-off analysis for a state-of-the-art mirror application. From the fabrication viewpoint, the metal mirror designer should consider the available process options for each step and select those apparently best suited for the particular application. The typical fabrication cycle includes most, if not all, of the following steps: formation of the substrate, geometric shaping, stress relieving, plating (typically with ELN), optical finishing, testing, and coating. The stress-relieving step is especially important as a means of minimizing the internal stress within the substrate that will tend to relieve itself with time or with changes in temperature, thereby causing the mirror surface to distort.

Selection of a mirror material is based on performance, cost, schedule, and risk. The accuracy of the contours of the optical surface of the mirror is the most important performance parameter, and it is influenced by material, geometry, mounting, environment, and manufacturing process. Cost,

TABLE 10.1
Typical Metallic and Metal Matrix Mirror Material Types

Aluminum	Beryllium	Copper	Molybdenum	Stainless Steel
356	I-70-H	101 (OFHC)	Low carbon	304
024	220-H	Glidcop™	TZM	316
5083 and 5086	I-250			416
6061	I-400			17-4PH
Tenzalloy	S-200-FH			
SXA	O-50 and O-30			

Source: Paquin, R.A., Selection of materials and processes for metal optics, *Proc. SPIE*, 65, 12, 1975; Paquin, R.A., Metal mirrors, Chapter 4 in *Handbook of Optomechanical Engineering*, pp. 69–110, CRC Press, Boca Raton, FL, 1997; Howells, M.R., and Paquin, R.A., Optical substrate materials for synchrotron radiation beam lines, *Proc. SPIE*, CR67, 339, 1997.

Note: OFHC, oxygen-free high-thermal conductivity. TZM, titanium–zirconium–molybdenum.

schedule, and risk are determined by an identical list of factors. In evaluating a material for use in a metal mirror, all of these factors must be considered. Discussions of cost and risk are given in the sections on each material later in this chapter.

Selection of mirror materials for performance is based on figures of merit (FOMs). Deflection under acceleration and fundamental frequency are proportional to the specific stiffness, which is the ratio of the elastic or Young's modulus E to the density ρ or E/ρ. The overall change in size after a change in temperature is proportional to the coefficient of thermal expansion (CTE). In the presence of a steady-state temperature gradient, the distortion is proportional to the thermal distortion coefficient, which is the ratio of the CTE α to thermal conductivity k. The time to reach thermal equilibrium after a change in temperature is proportional to the thermal diffusivity, which is the ratio of the thermal conductivity to the product of density and heat capacity. For cryogenic applications, enthalpy is important; this is the amount of heat that must be removed to lower the temperature of the material. Also important for cryogenic applications is the uniformity of the CTE.

The allowable stress for mirror materials is set by the deformation tolerance of the optical surface. Glassy materials are perfectly elastic up until fracture, so the allowable stress is the fracture stress associated with some probability of failure at a given lifetime. At the low stresses associated with typical mirror applications, the yielding of metals is nonlinear with applied stress. A common approximation is the microyield stress as the allowable stress in a metal. Microyield is defined as the amount of stress necessary for a permanent deformation of 1 part per million (ppm) or strain of $\varepsilon = 10^{-6}$.

Table 10.3 compares the FOMs of two common glassy mirror materials, Pyrex and Zerodur, with those of the most common metal mirror materials, copper, aluminum, and beryllium. For many materials, the properties change with temperature; Table 10.3 gives values for room temperature, nominally 300 K. Further discussion of material FOMs are given in Chapter 3 on materials and Chapter 8 on mirror performance.

A review of the material properties in Table 10.3 provides some of the reasons for selecting metals over glassy materials for some applications. Beryllium provides better specific stiffness and hence improvements in weight and vibration frequency compared to any glassy material. Thermal equilibrium times after a change in temperature are about two orders of magnitude shorter for metals in comparison with glassy materials. The thermal distortion indices for beryllium and copper are comparable to that of Schott Zerodur and are 55 and 67 times better, respectively, than that of Pyrex. Aluminum is superior to Pyrex with respect to both thermal distortion and thermal diffusivity, although the thermal distortion index of Zerodur is better than that of aluminum.

When cost is considered, aluminum becomes attractive for many mirror applications including infrared (IR) systems and scanning applications. Copper is used for low-cost high-energy laser

TABLE 10.2
Key Material Properties Influencing the Behavior of Mirrors

Mechanical	Physical	Optical	Metallurgical	Fabrication	Others
Young's modulus	CTE	Reflectivity	Crystal structure	Machinability	Temperature sensitivity of the listed parameters
Yield strength	Density	Absorption	Phases present	Polishability	Availability (including size and cost)
Microyield strength	Thermal conductivity	Complex refractive index	Voids and inclusions	Plateability	
Fracture toughness	Thermal diffusivity		Grain size	Forgability	
Modulus of rupture	Specific heat		Recrystallization temperature	Weldability	
	Radiation resistance		Stress relief temperature	Brazeability	
	Vapor pressure		Annealing temperature	Solderability	
	Electrical conductivity		Dimensional stability	Heat treatability	
	Corrosion potential				

Source: Paquin, R.A., Selection of materials and processes for metal optics, *Proc. SPIE*, 65, 12, 1975; Howells, M.R., and Paquin, R.A., Optical substrate materials for synchrotron radiation beam lines, *Proc. SPIE*, CR67, 339, 1997.

Note: Entries are not necessarily in order of significance.

TABLE 10.3
Comparison of Properties of Selected Glassy and Metal Mirror Materials

	Material				
Property	**Pyrex Corning 7740**	**Zerodur**	**Aluminum 6061-T6**	**Copper OFHC**	**Beryllium S-200FH**
Specific stiffness E/ρ (m^2/s^2)	28.3×10^6	35.7×10^6	25.5×10^6	12.9×10^6	164×10^6
CTE α (K^{-1})	3.3×10^{-6}	100×10^{-9}	23.6×10^{-6}	17×10^{-6}	11.4×10^{-6}
Thermal distortion index α/k (m/W)	2.92×10^{-6}	68.5×10^{-9}	141×10^{-9}	43.4×10^{-9}	52.8×10^{-9}
Thermal diffusivity $k/(\rho - c\rho$ (m^2/s)	483×10^{-9}	721×10^{-9}	69×10^{-6}	114×10^{-6}	60.5×10^{-6}
Allowable stress (MPa)	10	10	140	12	40

Note: All properties are at 300 K. The allowable stress for glassy materials is the fracture stress and that for metal materials is the microyield stress.

mirrors since it can be diamond turned and is superior in thermal distortion and thermal diffusivity. Beryllium is expensive and is used where distortion under acceleration is important, or weight is critical. Applications for beryllium include high-speed scanners, stabilized optics, and space systems.

Quantitative data on the physical characteristics of most of the material types listed in Table 10.3 are in Chapter 3. Because there is keen interest in athermalized optomechanical designs for various applications, many published evaluations are comparisons of key material properties as functions of temperature. For example, Figure 10.1 shows variations in the CTEs of two metals, aluminum and beryllium, compared with those of some glass and ceramic-type mirror materials.[1] Obviously, these two metals are much more sensitive to temperature change than the nonmetals—except at very low temperatures. Table 10.4 lists the measured values of the CTEs of a series of metals as functions of temperature over an extended range. High conductivities enhance performance at cryogenic temperatures.

In general, all materials approach zero-expansion properties at cryogenic temperatures. ULE has a very small CTE at or near 300 K. Fused silica has a zero CTE near 190 K; beryllium and

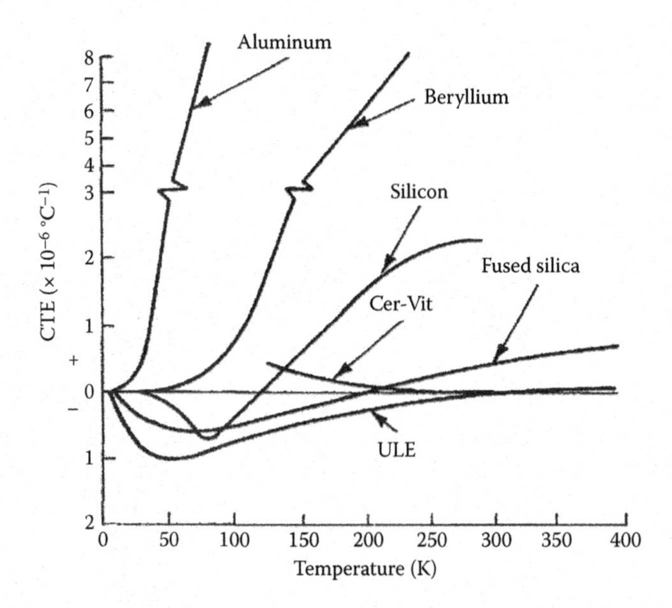

FIGURE 10.1 CTE variations with temperature for some mirror materials. (From Paquin, R.A., Beryllium Mirror Technology: State-of-the-Art Report, and Goggin, W., Perkin-Elmer Report IS 11693, Norwalk, CT, 1972.)

TABLE 10.4
Temperature Dependence of the CTEs (× 10^{-6} K^{-1}) of Selected Materials

Temperature (K)	6061 Al	Be	Cu	Au	Fe	304 CRES	416 CRES	Mo	Ni	Ag	Si	α-SiC	β-SiC
5		0.0003	0.005	0.03	0.01				0.02	0.015		0.01	
10		0.001										0.02	
20		0.005				9.8	4.3	0.3			0		
25		0.009	0.63	2.8	0.2			0.4	0.25	1.9	0	0.03	
50		0.096	3.87	7.7	1.3	10.5	4.9	1	1.5	8.2	−0.2	0.06	
75		0.47							4.3		−0.5	0.09	
100	12.2	1.32	10.3	11.8	5.6	11.4	6	2.8	6.6	14.2	−0.4	0.14	
125	18.7	2.55											
150	19.3	4.01				12.4	7				0.5	0.4	
175	20.3	5.54											
200	20.9	7.00	15.2	13.7	10.1	13.2	7.9	4.6	11.3	17.8	1.5	1.5	
225	21.5	8.32											
250	21.5	9.50				14.1	8.8				2.2	2.8	
293	22.5	11.3	16.5	14.2	11.8	14.7	9.5	4.8	13.4	18.9	2.6	3.3	3.26
300		11.5										3.4	3.29
350	23.8												3.46
400	25.0	13.6	17.6	14.8	13.4	16.3	10.9	4.9	14.5	19.7	3.2	4	3.62
450	26.3												3.77
500	27.5	15.1	18.3	15.4	14.4	17.5	12.1	5.1	15.3	20.6	3.5	4.2	3.92
600	30.1	16.6	18.9	15.9	15.1	13.6	12.9	5.3	15.9	21.5	3.7	4.5	4.19
700		17.8	19.5	16.4	15.7	19.5	13.5	5.5	16.4	22.6	3.9	4.7	4.42

Source: Paquin, R.A., Properties of metals, Chapter 35 in *Handbook of Optics*, 2nd ed., Vol. II, Bass, M. et al., Eds., Optical Society of America, Washington, DC, 1994. With permission of Optical Society of America.

aluminum reach this condition at ~40 and ~15 K, respectively. Although a low value of CTE in the temperature region of interest is important in mirror design, it is also important for that property to be uniform throughout the substrate. Nonuniformities in CTE cause bumps, depressions, or more complex changes in surface geometries to appear at temperatures other than that at which the mirror was fabricated. Materials that have low cryogenic CTEs generally have much higher CTEs at room temperature where final polishing and figuring take place. This makes achievement of a specific surface figure and high smoothness difficult. Common practice is to iteratively test at low temperature and correct the surface at room temperature until the desired result is obtained or the process ceases to improve the low-temperature performance. This practice is called "cryonull figuring."

Minimizing spatial temperature gradients requires a high thermal conductivity k to dissipate the effects of surface thermal loads on the mirror. This is especially true for mirrors exposed to high thermal irradiation. Similarly, specific heat c_ρ is of importance. These properties and their general temperature dependences for temperatures below 300 K are shown graphically in Figure 10.2. Glasses have low values of k and c_ρ from 300 K to absolute zero. Metals have considerably higher values for these parameters at these temperatures. Room-temperature values for these parameters may be found in Table 3.6.

Aluminum, copper, and molybdenum have cubic crystal structures that serve best as mirror substrates in wrought forms. Castings of these materials tend to be more porous, but are less expensive than forgings. Cast beryllium is thermally anisotropic owing to its hexagonal crystal structure. If pulverized and processed by powder metallurgy techniques, it forms highly successful mirror

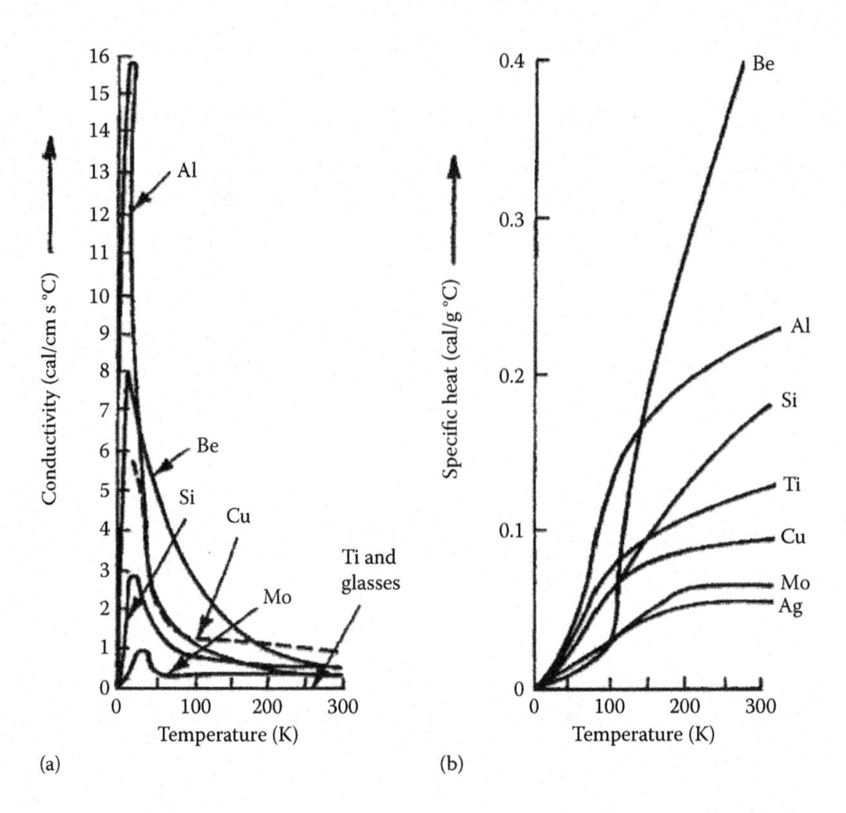

FIGURE 10.2 Low-temperature properties of selected candidate mirror materials: (a) thermal conductivity; (b) specific heat. (From Paquin, R.A., and Goggin, W., Beryllium Mirror Technology: State-of-the-Art Report, Perkin-Elmer Report IS 11693, Norwalk, CT, 1972.)

substrates. Molybdenum and copper are frequently used in water-cooled mirrors for high-energy applications. Thoriated tungsten alloy and stainless steel have also been employed in mirrors.

Appreciable levels of residual internal stress can develop in metal substrates as a result of casting, forging, machining, and some heat-treating processes. Such stresses cause dimensional instabilities and usually reach maximum levels near the surfaces. Each metal substrate must therefore undergo stress relieving during manufacture. The nature of this operation required for a given optic is dictated by the application and the fabrication stage at which the stress relieving is performed. Chemical etching, heat-treating, or combinations of these processes are usually used for this purpose.

10.3 ALUMINUM MIRRORS

Aluminum and its alloys are lightweight, strong, relatively inexpensive, and readily machinable materials widely used in optical instrument components and structures. The costs of aluminum mirrors are lower by a factor of 1.5–2.0 compared to those of glass optics with the same optical specifications. An additional advantage is ease of mounting by bolting. Aluminum is often used in "same-material" athermalized systems where both optics and support structure are made of the same material. Aluminum is common in cryogenic applications because its high thermal conductivity simplifies attaining and maintaining low temperatures.

The choice of the type of aluminum used in a mirror substrate is determined by the combination of performance and cost. Strength properties are not important for mirror substrates, with the important exception of the microyield strength, which does not correlate with conventional 0.2% yield or tensile strengths. The specific stiffness values and thermal properties of aluminum alloys are nearly identical. Selection of aluminum alloys for mirror applications is based on the ease of fabrication, quality of surface finish when diamond turned, dimensional stability, and cost.

The most common aluminum alloy for mirrors is 6061, in the T6 condition. This alloy is relatively inexpensive and easy to machine. Optical fabrication processes for 6061 are well established. Two issues with 6061 are dimensional stability and surface finish.

The long-term dimensional stability of aluminum mirrors is controversial. Selection of alloy and process control is critical for good long-term stability. Typically, special heat treatments in addition to those given in standards such as AMS 2770 are necessary for best dimensional stability. The goal of these additional heat treatments is to stabilize the alloy by removing residual stress. Any residual stress left from fabrication can change the optical figure of the mirror with time. Tables 10.5 and 10.6 give the typical thermal cycling treatments for selected forms of aluminum alloys.

A three-step heat treatment process is used for 6061, with the objective of removing stress from machining and optical finishing. After rough machining, the mirror is left about 1.3 mm (0.05 in.) oversized with respect to final dimensions. Material removal should be symmetrical about the neutral plane of the mirror to minimize distortion from machining. After rough machining, solution heat treatment at 530°C (980°F) followed by quench produces a T4 condition. Rapid cooling by quenching in plain water can produce a nonuniform stress distribution through the thickness of the mirror. Further machining of the mirror can change the surface stress, causing surface distortion. Use of a glycol/water quench solution slows down the surface cooling rate during quenching, producing a more uniform stress distribution.

After solution heat treatment, the mirror is age hardened to produce the final T6 condition. Typically, the AMS 2770 age hardening heat treatment is employed, with the mirror heated to 175°C (350°F) for 8 hours. An additional benefit is stress relief during age hardening.

Alternate processes after solution heat treatment are uphill quenching and overaging. An additional means of reducing residual stress after solution heat treatment is uphill quenching. Uphill quenching consists of reducing the mirror temperature to that of liquid nitrogen (77 K) and then rapidly quenching in a boiling water bath. The efficiency of the uphill quench with respect to removing residual stress and stabilizing the mirror is controversial. Overaging subjects the mirror to 260°C

TABLE 10.5
Thermal Cycling Treatments for Stabilizing Various Aluminum Mirror Substrates

Material	Treatment	Cycle Duration at High Temp.	Cycle Duration at Low Temp.	Total No. of Cycles	Temperature Change Rate
6061	Solution treat at 530°C, quench in PG,[a] age at 175°C in stages	30 minutes at 150°C	30 minutes at <40°C	>3	<3°C/minute or <1°C/minute or during polishing
SXA (after finish machining)	Solution treat at 495°C, quench in PG,[a] age at 190°C for 12 h	30 minutes	Quench to −195°C	Until no dimension change	
SXA (after ELN plating)	Solution treat at 495°C, quench in PG,[a] age at 190°C for 12 h	4 hours at 170°C	30 minutes at −75°C	>5 or until no dimension change	<5°C/minute
5000 series alloys	Anneal at 350°C, slow cool	60 minutes at 150°C	30 minutes at less than −40°C	>3	<3°C/minute or <1°C/minute during polishing

Source: Howells, M.R., and Paquin, R.A., Optical substrate materials for synchrotron radiation beam lines, *Proc. SPIE*, CR67, 339, 1997.

[a] PG is 20% polyalkylene glycol solution in water.

TABLE 10.6
Candidate Stress Relief Methods for Mirrors Made from Al 6061-T651 Plate Stock

Step	SR1 Control–No Heat Treatment	SR2 Simple Heat Treatment	SR3 SR2 + Uphill Quench	SR4 Solution Treatment with 28% Glycol Quench	SR5 SR4 Treatment with Uphill Quench	SR6 Solution Treatment with H_2O Quench and Uphill Quench
1	Rough machine	Rough machine	Rough machine	Rough machine	Rough machine	Rough machine
2		Heat treat at 260°C for 2 hours	Heat treat at 260°C for 2 hours			
3				Solution treat at 530°C	Solution treat at 530°C	Solution treat at 530°C
4				Quench within 15 s in UCON Quenchant A at 29–35°C	Quench within 15 s in UCON Quenchant A at 29–35°C	Quench within 15 s in H_2O at 18–24°C
5			Uphill quench: allow to reach 23°C, slowly place in LN_2, and dunk in boiling H_2O		Uphill quench: allow to reach 23°C, slowly place in LN_2, and dunk in boiling H_2O	Uphill quench: allow to reach 23°C, slowly place in LN_2, and dunk in boiling H_2O
6				Age at 175°C, 8 hours	Age at 175°C, 8 hours	Age at 175°C, 8 hours
7	Finish machine	Finish machine	Finish machine	Finish machine	Finish machine	Finish machine
8				Age at 175°C, 8 hours	Age at 175°C, 8 hours	Age at 175°C, 8 hours
9				Thermal cycle 3× at rate <1.7°C/minute, cool to 83 K, hold 30 minutes, heat to 23°C, hold 15 minutes, heat to 150°C, hold 30 minutes, and cool to 23°C	Thermal cycle 3× at rate <1.7°C/minute, cool to 83 K, hold 30 minutes, heat to 23°C, hold 15 minutes, heat to 150°C, hold 30 minutes, and cool to 23°C	Thermal cycle 3× at rate <1.7°C/minute, cool to 83 K, hold 30 minutes, heat to 23°C, hold 15 minutes, heat to 150°C, hold 30 minutes, and cool to 23°C
10	Figure test at 293 and 80 K 3×	Figure test at 293 and 80 K 3×	Figure test at 293 and 80 K 3×	Figure test at 293 and 80 K 3×	Figure test at 293 and 80 K 3×	Figure test at 293 and 80 K 3×

Source: Ohl, R. et al., Comparison of stress relief procedures for cryogenic aluminum mirrors, *Proc. SPIE*, 4822, 51, 2002.

(500°F) for 2 hours, which is a higher temperature than normally prescribed for the T4 condition. Overaging lowers strength and hardness; the advantage is improvement in dimensional stability by reduction in residual stress, although effectiveness is also controversial.

After age hardening, the mirror is finish machined to within 50–150 μm (0.002–0.006 in.) of the final mirror contour before diamond turning. A conventional stress relief of 175°C (350°F) for 2–8 hours reduces residual stress from this machining operation.

Finally, a series of thermal cycles further lowers residual stress and "stabilizes" the mirror against long-term changes in dimension. Each cycle begins by cooling the mirror to about 85 K, near that of liquid nitrogen; then, the mirror is heated to about 150–160°C (300–325°F). Dwell times at hot and cold temperatures are about 30 minutes to ensure that thermal equilibrium is reached. The rate of temperature change is around 8°C/minute (15°F/minute). A minimum of three cycles is performed. Stress release during thermal cycling is an exponential process, with the majority of stress removed during the first two cycles; increasing the number of cycles beyond three normally does not significantly decrease stress while adding cost. Increasing the lower temperature of the cycle to about −130°C (143 K or −202°F) to permit use of a commercial programmable furnace does not greatly change the amount of residual stress and is common practice, particularly in high-volume production.

The optical surfaces of aluminum mirrors are produced by diamond turning. The surface roughness of a diamond-turned aluminum mirror surface is between 40 and 100 Å root-mean-square (RMS). Although the surface finish produced by diamond turning varies with different aluminum alloys, for 6061, which is the most common aluminum alloy used for mirrors, the surface finish ranges from 60 to 80 Å RMS at best. This surface roughness is acceptable for IR applications, but produces excessive scatter for most visible applications.

One index of surface scatter is the Strehl ratio, which is the ratio of the intensity of the central peak of the diffraction disk maximum to the actual intensity. When the Strehl ratio is greater than 0.8, the system performance is limited only by diffraction. For a mirror surface of roughness δ_{RMS}, the reduction in the Strehl ratio S_R is given by

$$S_R = \frac{1 - e^{-\left(\frac{4\pi\delta_{RMS}}{\lambda}\right)^2}}{2}, \tag{10.1}$$

where λ is the wavelength.

Example 10.1

The surface roughness of a diamond-turned mirror is 60 Å. Is the optical system diffraction limited if the wavelength is 1.06 μm?

The reduction in the Strehl ratio at 1.06 μm is

$$S_R = \frac{1 - e^{-\left(\frac{4\pi\delta_{RMS}}{\lambda}\right)^2}}{2} = \frac{1 - e^{-\left[\frac{4\pi(60\times10^{-10}\,\mathrm{m})}{1.06\times10^{-6}\,\mathrm{m}}\right]^2}}{2}$$

$$S_R = 4.71\times10^{-3}$$

$$1 - S_R = 0.995 \geq 0.8.$$

Since $1 - S_R = 0.995$ is greater than the 0.8 necessary for diffraction-limited performance, the mirror is diffraction limited.

One method for improving surface finish is to plate the aluminum substrate with ELN and then polish the plating after diamond turning to improve surface roughness. ELN plating is further discussed in the section on plating of mirrors (Section 10.6). A second method is to plate the surface with an amorphous layer of aluminum and then to diamond turn this layer. Finally, methods for directly polishing the bare aluminum surface to improve surface roughness after diamond turning are coming into use.

A commercial process called AlumiPlate™ plates a layer of amorphous, high-purity aluminum on the mirror substrate. Bimetallic bending is not a concern since the CTE of the plating is almost identical to that of the aluminum mirror substrate. The surface finish of a diamond-turned AlumiPlate mirror is about 40 Å RMS, which is better than that of bare aluminum. AlumiPlate is relatively expensive and is soft, making cleaning difficult.

Postpolishing of bare aluminum surfaces after diamond turning is a way of avoiding the problems associated with plated substrates. Until relatively recently, postpolishing of bare aluminum was considered impractical, but a variety of techniques are now being used with success. Cabot Electronics developed a chemical mechanical polishing method that produced a 20 Å RMS surface finish on test mirrors. Direct polishing of bare aluminum by using a method developed by Astron produces a surface finish of about 20 Å RMS in mirrors up to 300 mm diameter. The visible quality (VQ) process developed by the ELCAN division of Raytheon produces an average surface roughness of about 20 Å RMS on diamond-turned 6061 aluminum. By 2011, several hundred aluminum mirrors with diameters of up to 600 mm were produced using the VQ process. Another bare aluminum polishing process was developed by the Tinsley division of L-3; this also produces an average surface roughness of 20 Å RMS.

Further improvement in surface finish of bare aluminum substrates is possible by the use of rapid-solidification-processed 6061 aluminum. New alloys suitable for mirror applications made using this process include RSA-6061 and RSA-905 (RSA stands for rapidly solidified aluminum). Other unconventional processes used in the RSA alloys include spin casting. A surface roughness of 20 Å RMS is readily achieved by postpolishing of RSA alloys, with surface roughness being as good as 10 Å RMS on some test mirrors.

10.4 BERYLLIUM MIRRORS

Beryllium has several unique properties: low density (70% that of aluminum), high stiffness-to-weight ratio (its elastic modulus is 1.5 times higher than that of steel), high specific heat (highest of any metal, 1825 J/kg K), and high conductivity. Owing to its hexagonal crystalline structure, beryllium is highly anisotropic. Its CTE is typically 7.7×10^{-6} °C^{-1} along the axis perpendicular to the hexagonal face axis and 10.6×10^{-6} °C^{-1} along the orthogonal axes.

One important factor associated with beryllium is cost. The historical cost of a beryllium mirror made to the same optical surface specification is about three times higher than a bare aluminum and four times higher than an ELN-plated aluminum mirror. A more recent estimate lowers the cost ratio for beryllium to about 1.5 times more expensive than a comparable aluminum optic, while another puts the cost of beryllium about twice that of aluminum. Related to cost is the higher technical risk associated with beryllium, leading to possible delays in schedule. System cost considerations may involve more than just the cost of the optics. For example, in space systems where launch costs per unit mass are high, the lower mass of beryllium may produce a cost saving that more than offsets the higher cost of the material.

To develop an isotropic structure and increase strength, beryllium parts are usually fabricated by powder metallurgical methods. Isotropy of macroparticles is approached statistically with a large number of very small particle grains with random relative orientations. Figure 10.3 illustrates schematically the vacuum hot-pressing (VHP) method for making a blank such as one that might be used as a mirror substrate. The powder is fed into a cylindrical graphite or steel die and, with mechanical vibration, is compacted to approximately 50% of bulk density. The two end rams are then driven together to consolidate the material to better than 99% of bulk density. The process takes place under vacuum and at an elevated temperature to promote sintering.

A more uniform material comes from a process involving hot sintering followed by hot isostatic pressing (HIP), shown in Figure 10.4. The HIP process is more expensive, but it yields blanks

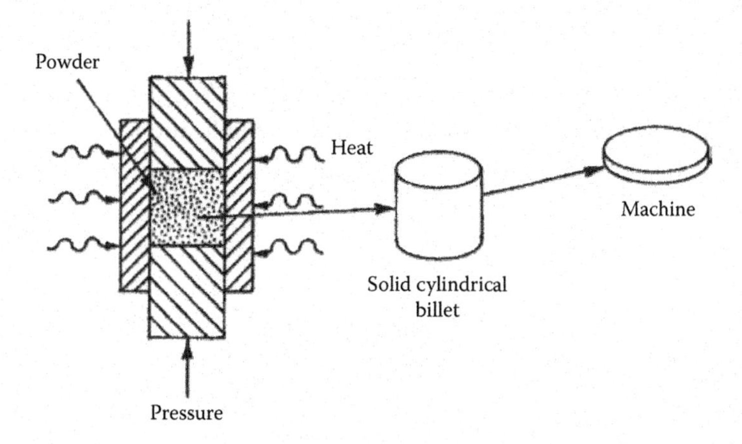

FIGURE 10.3 Schematic of the VHP process for making a blank substrate for a mirror. (From Paquin, R.A., and Goggin, W., Beryllium Mirror Technology: State-of-the-Art Report, Perkin-Elmer Report IS 11693, Norwalk, CT, 1972.)

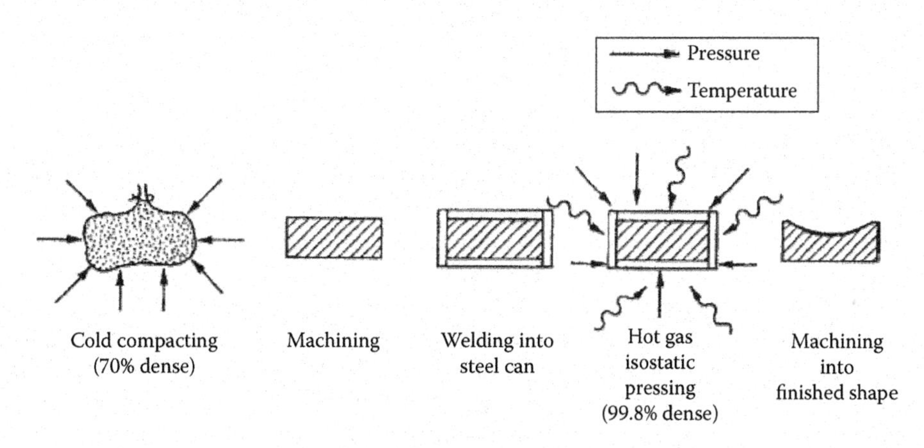

FIGURE 10.4 Beryllium mirror blank preparation by the cold compacting plus HIP process. (From Paquin, R.A., and Goggin, W., Beryllium Mirror Technology: State-of-the-Art Report, Perkin-Elmer Report IS 11693, Norwalk, CT, 1972.)

with low porosity and few inclusions. Some grain boundary concentrations of metallic impurities and beryllium oxide are characteristically observed. These concentrations do not affect microyield strength, machinability, or growth structure (rib print-through) in the polished overcoating. Yield strength varies as $d^{-1/2}$, where d is the grain size according to the Hall–Petch relationship, hence the trend toward smaller grain sizes and related higher consolidation in beryllium production.

Machining beryllium requires special techniques to minimize subsurface damage and residual stress (Table 10.7).[2] The machinability of beryllium is about 55% (with 1113 steel as 100%); the metal machines like a lightweight cast iron. Typically, beryllium is machined by a series of progressively lighter and lighter cuts in a three-step process (rough, intermediate, and finishing), with heat treatment and acid etching to control subsurface damage between each level of material removal. Rough machining is to about 1.25 mm of final dimensions, while intermediate machining is to within 125 μm of the final dimensions. As a final step after finish machining, thermal cycling provides good long-term dimensional stability.

Machining of beryllium induces microcracks in the surface of the material extending to a depth of about 100 μm or more; in addition, cracks called twins may form. Twins and subsurface damage lower the ultimate tensile strength by up to 30% and reduce ductility, with elongation below 1%. Acid

TABLE 10.7
Typical Sequence of In-Process Fabrication, Annealing, and Thermal Cycling Stress Relieving Steps for Be Mirrors to Be Used in Cryogenic Applications

Rough machine
Acid etch
Anneal at 790°C
Finish machine
Acid etch
Thermal cycle three to five times (limits determined by application, but at least −40 to + 100°C)
Grind, etch, and thermal cycle
Figure and thermal cycle
Final polish and thermal cycle

Source: Paquin, R.A., Chapter 4: Metal Mirrors, *Handbook of Optomechanical Engineering*, A. Ahmad, ed., pp. 89–110, CRC Press, 1997.

Note: When ELN or Be coatings are used, they should be applied after the grind/etch/thermal cycle steps. For thin Be coatings, deposition should be after the figure/thermal cycle step.

etching to a depth of 50–125 μm removes the subsurface damage. Heat treatment is usually performed in addition to acid etching; typical heat treatment is at 570–800°C for between 1 and 2 hours, with heating and cooling rates below 150°C/hour. Heat treatment is performed in a vacuum furnace at a pressure below 100 mTorr to avoid oxidation or discoloration of the beryllium; alternately, a controlled environment furnace can be used with a dry gas at a dew point lower than −60°C.

Beryllium is vulnerable to corrosion and must be protected during fabrication and in use. Exposure to halogens can cause pitting of beryllium. Only about 2 ppm chlorine in water will attack and pit beryllium. Household tap water can contain up to 4 ppm of chlorine and hence is likely to pit beryllium if used for cleaning. Human saliva is highly corrosive to beryllium, and protective masks are mandatory when polishing bare beryllium. Bare beryllium is almost never used; the surface of beryllium is usually protected against corrosion by plating, painting, or chemical conversion coatings.

The fastening of separate pieces of beryllium (such as plates made by VHP) by welding, brazing, diffusion bonding, or adhesive bonding to make a mirror substrate is quite difficult and generally gives poor results. Early experiments with brazing beryllium mirrors used silver lithium and aluminum silicon alloys as brazing materials. A 20 in. (50.8 cm) diameter spherical mirror 1.0 in. (2.54 cm) thick with a brazed eggcrate core and brazed-on front and back face sheets weighing less than 6 lb (2.7 kg) was fabricated and tested by Paquin and Goggin in 1972.[1] Sensitivity to changes in ambient temperature of about λ/25 per degree Fahrenheit for visible light was measured. Open-back machining, chemical milling, or advanced powder metallurgy methods are preferable techniques for achieving weight reduction.

The very smooth, low-scatter surface quality of some large glass mirrors—typically measured as low as 5 Å RMS at very high spatial frequencies—has yet to be achieved on polished bare beryllium mirrors (the average surface roughness of the James Webb Space Telescope (JWST) primary is 32 Å RMS). This characteristic of the surface is frequently called microroughness. Current polishing techniques typically achieve 15–25 Å microroughness on bare beryllium surfaces of moderate aperture. Experiments have indicated that a thin film of beryllium sputtered onto a beryllium substrate can be polished to optical surfaces with ~5 Å microroughness.

For long-wavelength IR applications, it is unnecessary to coat beryllium; its reflectivity at various wavelengths is typically as that shown in Figure 10.5.[3] At shorter wavelengths, applying a thin film of Be or a thicker (sputtered) layer of that material to the Be substrate improves the reflectance.

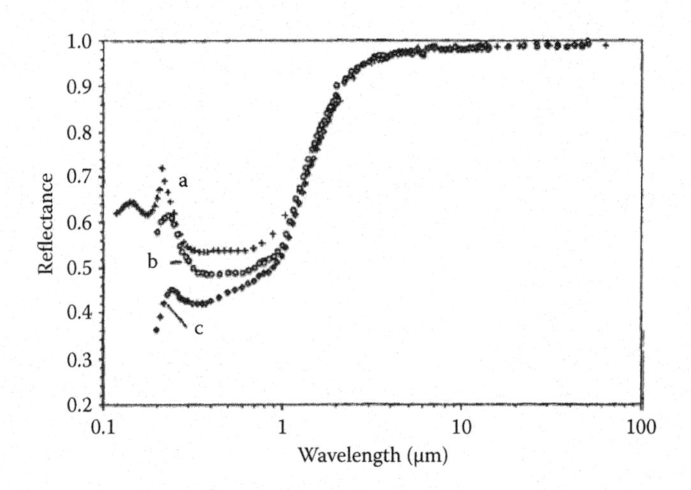

FIGURE 10.5 Reflectance versus wavelength of polished surfaces of evaporated high-purity thin-film Be (curve a), high-purity thick Be cladding (curve b), and bulk HIP'ed Be (2% BeO) (curve c). (From Paquin, R.A., Properties of metals, Chapter 35 in *Handbook of Optics,* 2nd ed., Vol. II, Bass, M. et al., Eds., pp. 35.1–35.78, Optical Society of America, Washington, DC, 1994. With permission of Optical Society of America.)

All-beryllium mirrors have the following three advantages over those with ELN-plated surfaces: they are not susceptible to bimetallic effects when the temperature changes, they are slightly lighter, and they are less vulnerable to radiation damage due to natural or nuclear radiation such as what might be encountered in some types of space missions.

Mirrors intended for use at cryogenic temperatures should be temperature cycled and tested at or near the operating temperature. Compensation for changes at low temperature with or without plating can then be built into the mirror figure. Figure 10.6 shows a computer plot of in-process

FIGURE 10.6 In-process computer plot of equivalent double-pass surface error contours needed in the Be primary mirror for the IRAS Telescope at room temperature to compensate for instrumental errors and changes anticipated for operation at 2 K. (From Harned, N. et al., *Opt. Eng.*, 20, 195, 1981.)

surface errors for the bare beryllium primary mirror for the Infrared Astronomical Satellite (IRAS) Telescope showing the required room-temperature correction needed on the reflecting surface to compensate for changes anticipated at 2 K based on in-process interferometric tests at the latter temperature as well as known instrumental errors.[4]

Further improvement in the cryogenic performance of beryllium mirrors was obtained by a special grade of I-70-H. This material was used in the Spitzer Space Telescope, after successful cryogenic testing of 0.5 and 0.85 m mirrors. The I-70 beryllium powder was chemically cleaned prior to consolidation to ensure uniformity. To minimize residual stress, the test mirrors were annealed and acid etched at each intermediate step of grinding and polishing, as well as being subjected to frequent thermal cycles. As a result of these precautions, the cryogenic distortion of 0.5 m mirror was the lowest of any beryllium mirror tested at the National Aeronautics and Space Administration (NASA) Ames Research Facility. The thermal distortion of the optical surface between room temperature and 4.4 K was 0.096λ RMS (λ = 633 nm). Similar results were achieved on the 0.85 m mirror of the Infrared Telescope Technology Testbed, with an optical figure change of about 0.5λ RMS between 295 and 5 K and less than 0.02λ RMS optical figure hysteresis.

Cryogenic optical figure stability is improved with a new formulation of beryllium called O-30 that is created by a gas atomization process. This process involves vacuum melting of solid high-purity beryllium and pouring it through a small orifice where it encounters a stream of high-velocity gas that breaks the stream into small spherical droplets as it cools under ambient atmospheric pressure. The resulting material has the lowest beryllium oxide content ever produced by powder metallurgy technology, is highly isotropic, and behaves well during HIP'ing. The mechanical characteristics of HIP'ed O-30 are compared to those of other forms of this material in Table A7.1 in Appendix 7.

The low scatter and isotropy of HIP'ed O-30 beryllium makes it ideal for optical mirrors, especially ones intended for cryogenic applications. Tests conducted under the Subscale Beryllium Mirror Demonstrator and Advanced Mirror System Demonstrator (AMSD) programs on 0.5 m diameter and 1.4 m hexagonal mirrors, respectively, demonstrated that O-30 Be would be better than ULE or SiC for making the primary, secondary, and tertiary mirrors of the JWST. Figure 10.7

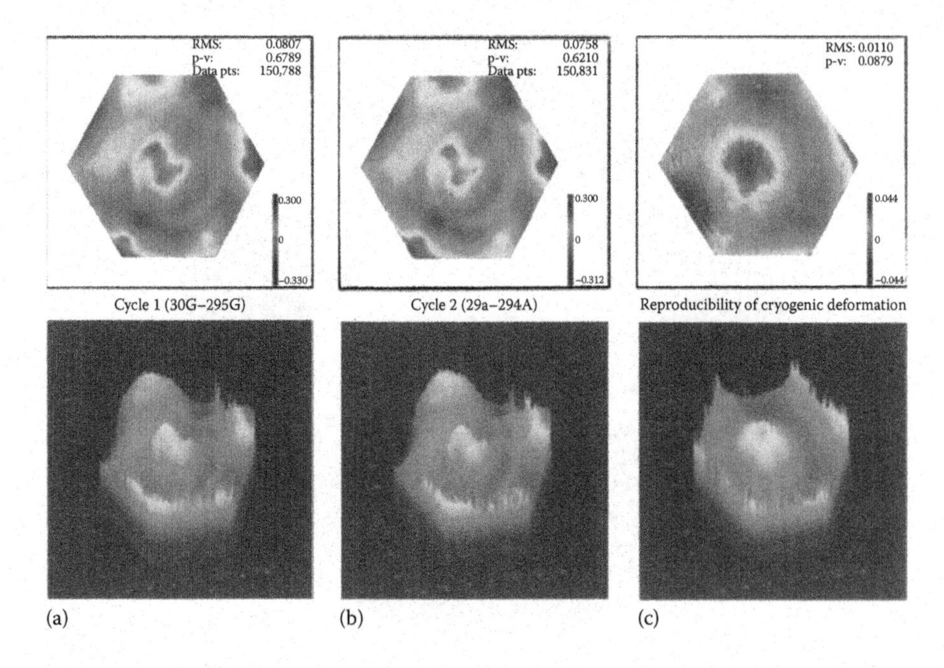

FIGURE 10.7 Interferometric test results for the AMSD beryllium mirror: (a) cycle 1, (b) cycle 2, and (c) reproducibility (difference) for two cycles. Figure errors are in micrometers. (From Parsonage, T., JWST beryllium telescope: Material and substrate fabrication, *Proc. SPIE*, 5494, 39, 2004.)

shows the results of the interferometric cryogenic testing of the AMSD mirror.[5] Reproducibility from one 300–30 K cryocycle to the next was 0.0110 μm RMS. The aerial density of the mirror was <15 kg/m², and its first mode frequency was at least 220 Hz. These results support expectations that the JWST mirrors will meet all specifications.

When launched, the JWST will be state of the art with respect to both lightweight and beryllium optics. The JWST is a 6.5 m aperture three-mirror anastigmat telescope with a focal ratio of *f*/20. It is intended to be diffraction limited at a wavelength of 2 μm, with a resolution of about 100 marcsec.

The primary mirror consists of 18 lightweight open-back hexagonal beryllium mirrors; each mirror is about 1.32 m across from flat to flat. (See Figure 10.8.[6]) The total collecting area of the composite mirror is about 25 m². The areal density of the mirror is 13.8 kg/m², although the mirror mounting hardware raises the areal density of the assembly to 26.2 kg/m². Because the diameter of the primary mirror is larger than the diameter of the shroud on the launch vehicle, it will be necessary to fold the segments before launch and then deploy them on orbit. This requires a reliable system for detecting and adjusting the position of the individual mirrors.

Each individual mirror segment is attached to a triangular "delta plate" by means of three bipod actuators. (See Figure 10.9.) The actuators form a hexapod providing six degrees of freedom (DOFs) in adjustment. The range of travel is 3.3 mm in piston, 1.7 mm in decenter, 1.22 mrad in tilt, and 1.7 mrad in clocking. The resolution of the actuator is 7.5 nm. Each bipod actuator is attached to the mirror via a "whiffle" plate; the latter is attached to the mirror at three points by using titanium flexures.

There are six more flexures at the perimeter corners of the mirror segments, each of which is attached to a radius of curvature (ROC) strut. All six of the ROC struts are tied together at the center of the segment, forming a six-sided pyramid. At the apex of the pyramid is a seventh actuator, pushing against the center of the mirror segment. This seventh actuator is used to change the ROC of the mirror, with a stroke of −2.102 to +5.528 μm.

Polishing of the JWST flight mirrors, including four spares, was completed by L-3 Integrated Optical Systems—Tinsley in 2012. All mirrors met the specifications for optical performance; the overall figure error for the composite primary mirror was 13.5 nm RMS, while the tolerance was 21.2 nm RMS. Average midspatial frequency errors, where the frequency is defined as clear aperture to a length of 222 mm, was 12.3 nm RMS. The average high-spatial frequency error,

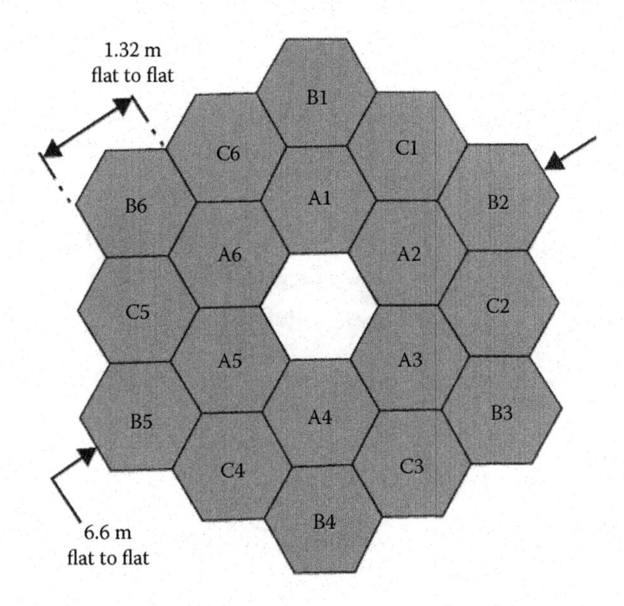

FIGURE 10.8 Array of 18 individual hexagonal beryllium mirror segments that make up the JWST primary. (From Wells, C. et al., Assembly integration and ambient testing of the James Webb Space Telescope primary mirror, *Proc. SPIE*, 5484, 859, 2004.)

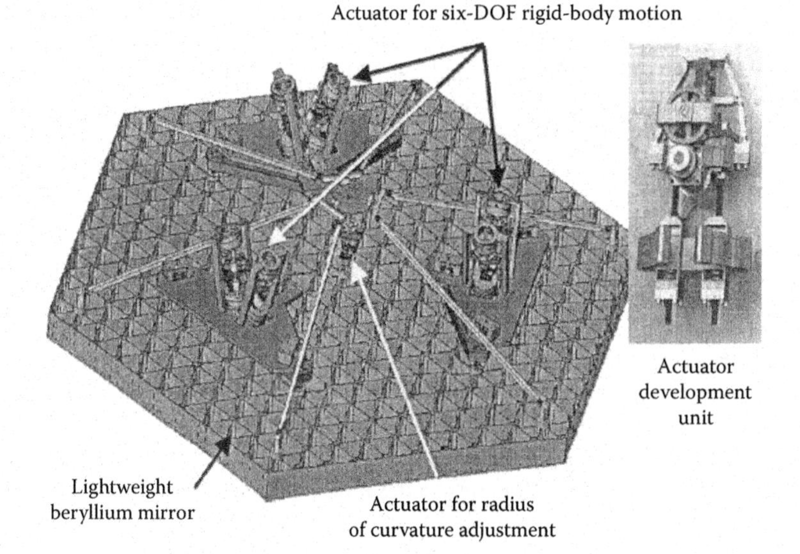

FIGURE 10.9 Rear view of JWST primary segment showing the seven actuators provided for optical figure and radius of curvature adjustment in orbit. (From Wells, C. et al., Assembly integration and ambient testing of the James Webb Space Telescope primary mirror, *Proc. SPIE*, 5484, 859, 2004.)

where the frequency is defined as a length of 222 mm–80 μm, was 6.1 nm RMS. The average surface roughness of the segments was 3.2 nm RMS. Figure 10.10 shows one segment undergoing visual examination at NASA Goddard Space Flight Center.

Beryllium has earned an unfortunate reputation as a hazardous material. Inhalation of beryllium particulate may lead to a debilitating lung disease; this condition is called chronic beryllium

FIGURE 10.10 One finished JWST primary segment being inspected at NASA Goddard Space Flight Center in 2013. (Courtesy of NASA, Washington, DC, http://www.flickr.com/photos/nasawebbtelescope/89603389001.)

disease (CBD). CBD is a lung disease associated with tumorlike granulations and scarring of the lung tissue; there is no cure. The greatest hazard is inhalation of dust particles with a diameter below 10 µm. Currently the US exposure limit set by the Occupational Safety and Health Administration is 2 µg/m^3 in an 8-hour period; there is discussion about lowering this limit to 0.2 µg/m^3 in an 8-hour period. Medical data show that only about 3% of workers who work with Be appear to be susceptible to CBD. An additional hazard is development of dermatitis from handling beryllium; this condition is avoided by wearing gloves when handling bare beryllium (there is no hazard when handling plated or painted beryllium). With proper precautions used in manufacturing facilities, the health risk from machining bare Be is minimal. Simple exhaust systems at the source of dust creation with efficient filters are very effective. In loose abrasive grinding of optical surfaces, there is no hazard if the slurry is captured and not allowed to dry on exposed surfaces. The contaminated waste is not considered hazardous under US laws and can be disposed of by a licensed disposal firm without problems. Fine dust of other optical materials such as SiO_2 and SiC could also cause respiratory disease if inhaled, so they should be handled and machined very carefully.

10.5 MIRRORS MADE FROM OTHER METALS

10.5.1 General Considerations

Metal mirrors are frequently used in applications involving high-energy electromagnetic beam irradiance of the reflecting surface and requiring active cooling by flowing pressurized fluid through channels machined into the substrate or indirect cooling by way of contact with cooled pressure plates. Aluminum, copper, molybdenum, tungsten, and silicon carbide are popular materials for such mirrors. Stainless steel has also been employed in some instances, particularly for synchrotron beamline mirrors. Copper, molybdenum, and silicon carbide are discussed in this section.

10.5.2 Copper Mirrors

Bare copper mirrors perform well in the IR because they are highly reflective at these wavelengths. Reflectance of 98.6% is typical at 10.6 µm wavelength. At visible wavelengths, they are much less efficient; ~60% reflectance at 0.5 µm is typical. These mirrors have good thermal properties and are economical to produce. A visible oxide layer forms on a freshly polished bare copper surface within 120 hours, but the reflectivity at 10.6 µm does not degrade by more than ~0.5%, nor does the reflectivity at 0.633 µm degrade by more than ~2% in that period. The scattering effect of the polished surface increases with age.

Copper mirrors are usually made of oxygen-free, high-thermal conductivity (OFHC) base material with characteristics as listed in Table 3.6. A lower-cost mirror material with slightly reduced performance characteristics has been called "laboratory copper." Mirrors made from either material are typically configured as solid disks or rectangular plates with thicknesses of about one-sixth the diameter or diagonal of the face. OFHC copper mirrors as large as 30 in. (76 cm) in major face dimension have been made to a surface figure accuracy of $\lambda/20$ for visible light. There is reported success in ion-polishing copper mirrors to increase smoothness and decrease absorptivity.

A form of copper called Glidcop AL-15, UNS C15715, with 0.3 wt.% $A1_2O_3$, has been used successfully for cooled beam folding and focusing mirrors in synchrotron beamlines. The material can be ELN plated, ruled to make gratings, and brazed if special precautions are followed. Its characteristics are given in Table A7.1 in Appendix 7.

10.5.3 Molybdenum Mirrors

Molybdenum mirrors have favorable thermal and mechanical properties for use in high-thermal irradiance applications and are generally cooled. The faceplate thickness of cooled high-energy

TABLE 10.8
Recommended Etching Processes for Some Metals

Substrate	Etchant Acid	Temperature (°C)	Material Removed (mm)	Process Time (Minutes)
Be–Cu	50% acetic 30% nitric 10% hydrochloric 10% orthophosphoric	70	0.008	~15
Mo and TZM	50% sulfuric 50% nitric	78	0.003	~10

Source: Kurdock, J. et al., *Appl. Opt.*, 14, 1808, 1975.

laser mirrors must be thin to maximize heat transfer; however, the pressurized coolant distorts the thin faceplate. The thermal expansion of the faceplate also induces distortion. An FOM for cooled mirror materials is the ratio of the elastic modulus E to the CTE α, or E/α. Molybdenum is about nine times better than copper with respect to this FOM.

The reflectivity of bare, polished Mo is typically greater than 90% for wavelengths longer than 1.8 μm. Low-carbon, vacuum-arc-cast plate is the preferred material for polished Mo mirrors. This is produced by vacuum-arc melting of powdered Mo in a water-cooled copper mold. It is generally of high purity and has a minimum of second-phase particles. Other forms used with varying degrees of success are powder metallurgy Mo produced from compacted and sintered Mo powder and TZM alloy* produced by vacuum-arc melting of carbon-deoxidized Mo with titanium and zirconium in a water-cooled mold. Powder metallurgy Mo typically has voids and oxide particles as impurities. The additives in TZM promote a dispersion of fine carbon particles that improve the strength and high-temperature stability of that material.

Superpolishing Mo, TZM, and beryllium–copper mirrors to microroughness less than 20 Å RMS requires special techniques. The important features of the recommended process are controlled grinding, i.e., the use of progressively finer abrasives, with total material removal equal to at least three times the size of the previously used (coarser) grit, to minimize subsurface damage and acid etching to eliminate work-hardened layers. Table 10.8 defines the etching treatments found to be successful.

10.5.4 Silicon Carbide Mirrors

Silicon carbide mirror blanks are made primarily by either of two processes: chemical vapor deposition (CVD) or reaction bonding (RB). The physical characteristics of both types of SiC are summarized in Table 3.6. The CVD process involves pyrolysis of vapor methyltrichlorosilane in excess hydrogen within a subatmospheric-pressure CVD reactor. The resulting SiC is a theoretically dense, polycrystalline material, free of voids and microcracks. The CVD process is also compatible with replication of high-precision optical surfaces onto substrates. RB SiC is made by casting a slurry of α-SiC† into a mold, drying, prefiring, firing, and "siliconizing" to fill pores. A cladding of silicon

* TZM is an alloy of molybdenum, titanium, zirconium, and carbon. It has a higher recrystallization temperature, higher creep strength, and higher tensile strength than pure molybdenum.

† α-SiC is the hexagonal form, while β-SiC is the cubic isotropic form of single-crystal material.

may then be added to facilitate polishing. Both materials are generated or ground with high-speed diamond wheels or diamond grit on hard laps.

Measurements of SiC support the following conclusions: (1) elastic modulus degrades by only 10% when the temperature of the SiC is raised from room temperature to 1650°C (2462°F); (2) SiC can be polished to a surface roughness of less than 0.1 nm RMS; (3) the bidirectional reflectance distribution function of a highly polished SiC sample is 1×10^{-5} at 10.6 and 0.633 μm wavelengths for angles of 3–18° from specular; (4) minimal figure change ($\lambda/125 = 0.005$ μm RMS) occurs when a SiC sample is cooled from room temperature to −190°C (−310°F); (5) no appreciable surface degradation occurs when thermal atomic oxygen flux of 2×10^{18} atom/cm^2 s is directed onto the sample for 6 hours; (6) small changes in surface reflectivity in the wavelength range of $200 < \lambda < 600$ nm occur when an atomic oxygen beam of broad energy spectrum (0–10 eV) with a peak at 5 eV is directed onto CVD SiC; and (7) electron beam irradiation of SiC yields a surface damage threshold of 0.5 cal/cm^2.

Reaction-bonded SiC is a two-phase material manufactured (cast) by well-established techniques developed for nonoptical applications. This is a desirable material for mirrors because of the flexibility of the fabrication process (allowing adjustment in physical, thermal, and mechanical properties), low cost relative to other forms of SiC, compatibility with ion beam sputtering to obtain a smooth polishable surface, and scalability to large sizes. Areal densities as low as 10 kg/m^2 and microroughness of 8–15 Å on small flat and spherical SiC mirrors are possible.

The proprietary Ceraform process developed by United Technologies Corporation is used to produce RB SiC mirrors. Plots of CTE (or α) and κ for the material in comparison with the same parameters for 6061 Al, I-70A Be, and ULE as functions of temperature are shown in Figures 10.11 and 10.12.[7] The inverse thermal distortion index k/α or thermal stability FOM versus temperature for those same materials is shown in Figure 10.13. This FOM was defined as k/α. It should be noted that using room-temperature data from Table 3.6, this FOM would be essentially identical for RB SiC (30% Si) and CVD SiC.

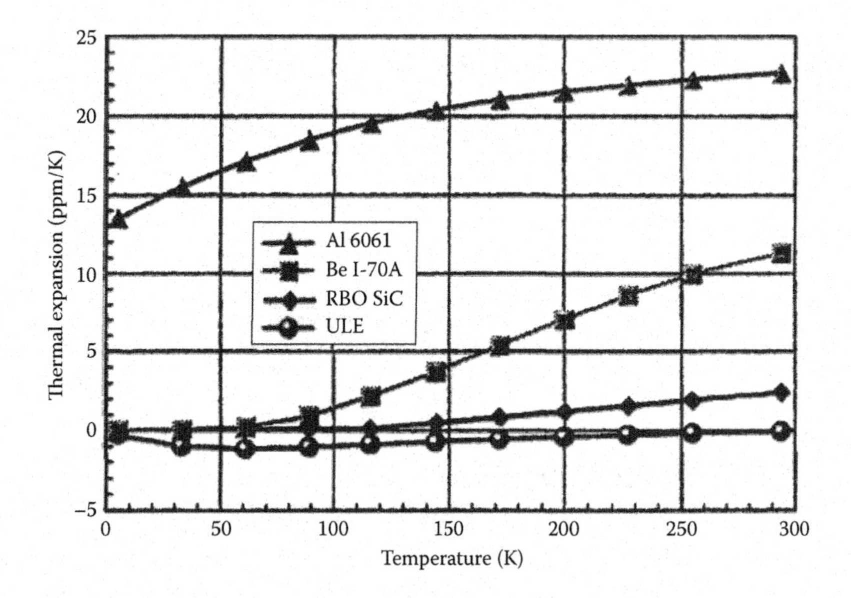

FIGURE 10.11 Plots of CTE versus temperature for four types of mirror materials. (From Ealy, M.A. et al., CERAFORM SiC: Roadmap to 2 meters and 2 kg/m^2 areal density, *Proc. SPIE*, CR67, 53, 1997.)

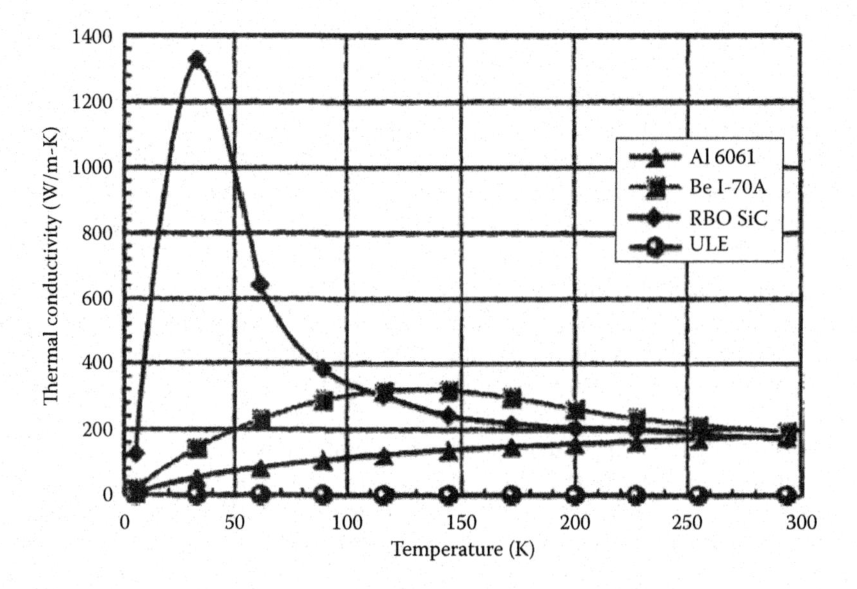

FIGURE 10.12 Plots of thermal conductivity versus temperature for four types of mirror materials. (From Ealy, M.A. et al., CERAFORM SiC: Roadmap to 2 meters and 2 kg/m^2 areal density *Proc. SPIE*, CR67, 53, 1997.)

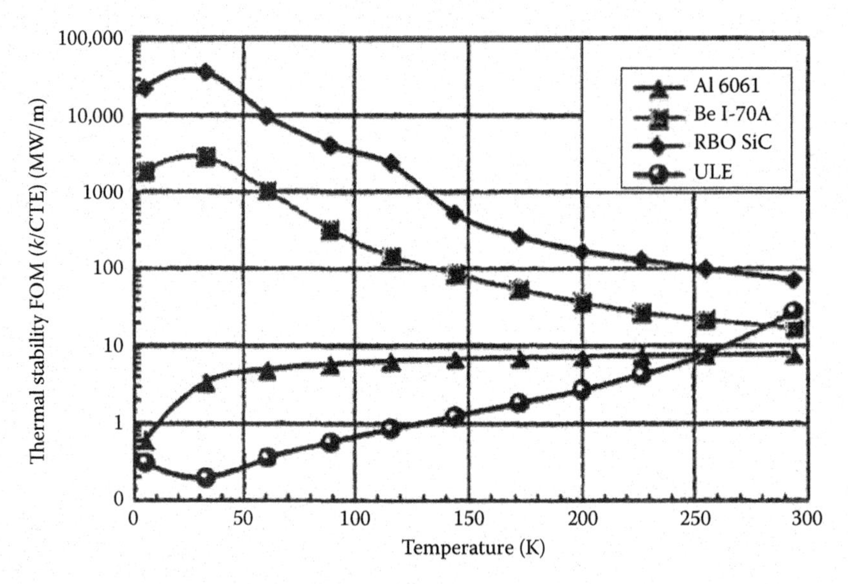

FIGURE 10.13 Plot of thermal stability FOM (k/α) versus temperature for four different types of mirror materials. (From Ealy, M.A. et al., CERAFORM SiC: Roadmap to 2 meters and 2 kg/m^2 areal density, *Proc. SPIE*, CR67, 53, 1997.)

10.6 PLATING OF METAL MIRRORS

Because of the inherent crystalline structure, softness, and ductility of some metals, it is virtually impossible to achieve a high-quality optical finish directly on the base metal. Aluminum and beryllium benefit by having thin layers of a metal such as nickel plated onto the base metal and the optical surface single-point diamond turned and polished into these layers to achieve a smooth finish. The smoothness of optical surfaces on substrates made of CVD or RB silicon carbide can be improved with a thin layer of vapor-deposited pure copper added. Gold can be plated onto various types of metal substrates to form good IR-reflecting mirrors.

Similar improvements in surface smoothness can be achieved in some cases with plated layers of the same material as the substrate. The proprietary AlumiPlate process for plating aluminum on aluminum is a prime example. The smoothness and thermal damage threshold of copper and molybdenum mirrors can be improved if a thin amorphous layer of the base metal is deposited onto the substrates before polishing.

The most frequently used plating material is nickel. Two basic processes are available for plating nickel layers onto mirror substrates. These are electrolytic and electroless plating. EN can be plated to a thickness of 760 μm (0.03 in.) or more. It has a Rockwell hardness of 50–58. The process is simple but slow and does not require precise temperature control. Typically, 60 ± 8°C (140 ± 15°F) is adequate. Uniformity of coating thickness is not easily attained with this process. ELN is an amorphous material with phosphorus content in the range of 11–16%. It can be plated more evenly and is more corrosion resistant, and its application process is less complex mechanically and electrically than that of EN. On the negative side, the maximum practical thickness of the ELN layer is about 200 μm (0.008 in.), so the substrate must have nearly the proper contour before plating. The process temperature of 93°C (~200°F) for ELN plating is higher than that for EN and must be controlled to ±3°C. The Rockwell hardness of ELN plating is typically 49–55, but this can be increased by heat treatment.

The CTE of ELN varies with phosphorus content. A rule of thumb is that a change of 0.5% in phosphorus content changes the CTE by about 1 ppm. The CTE α_P of ELN as a function of percentage of phosphorus content P, for phosphorus contents between 7% and 12%, is given approximately by

$$\alpha_p = (36 - 2.3\,P) \times 10^{-6}\,K^{-1}. \quad (10.2)$$

The surface finish of diamond-turned ELN plated mirrors is dependent on the phosphorus content of the plating. The best surface roughness after diamond turning is produced with a phosphorus content of 13%. Reducing the phosphorus content to 11.6% degrades surface roughness by a factor of about 1.5. Below a phosphorus content of 10.8%, there *is* noticeable damage to the SPDT tool and an increase in roughness by a factor of 2.

A phosphorus content of 11.5% is often used as a compromise between matching the CTE between plating and substrate, reducing residual stress, and minimizing surface roughness after diamond. Normally, the tolerance for phosphorus content is ±0.5%. Use of an 11.5% phosphorus content means that there will be a difference in CTE between the ELN plating and both aluminum and beryllium substrates. In addition to phosphorus content, the CTE of ELN is dependent on annealing temperature and thickness.

Mismatch between thermal expansion characteristics of the plated layer and the base metal of the mirror substrate is a cause of dimensional instability in the completed optic. For nickel and beryllium, this mismatch is about 2 ppm/K, whereas that for nickel and aluminum is about five times larger. The resultant bimetallic effect may be quite significant in high-performance systems.

Bimetallic bending effects are minimized by plating both sides of a mirror and contouring the back. Shown in Figure 10.14[8] are a series of ELN-plated aluminum concave mirrors with a diameter of 180 mm (7.09 in.). Design variations from the baseline plano-concave shape (view in Figure 10.14a) are (1) increasing the substrate thickness to resist bending (view in Figure 10.14c); (2) meniscus shape (view in Figure 10.14d); (3) designing the substrate with a symmetric cross section to produce equal and opposite bending effects (view in Figure 10.14e); and, (4) for all configurations, plating both sides of the substrate with equal and unequal thicknesses of nickel. The plano-plano configuration of the view in Figure 10.14b is included for general information with regard to the effect of front-to-back surface-plating thickness differences.

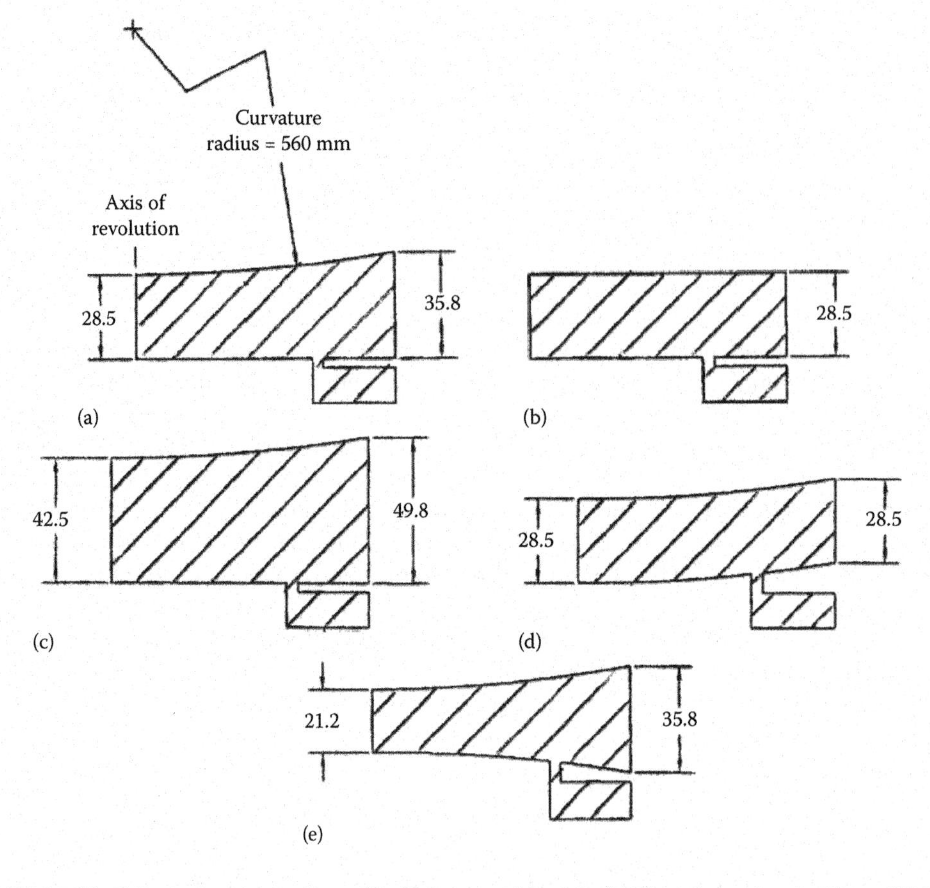

FIGURE 10.14 Mirror shapes investigated by Vukobratovich et al. (1997) to determine the bimetallic bending effects of various ELN claddings. (a) Curved front-flat back, (b) plane-parallel, (c) deep section curved front-flat back, (d) meniscus, and (e) double concave.

Studies of these mirror shapes showed that (1) the closed-form results do not correlate well with the finite-element analysis (FEA) results, the FEA results being considered more accurate; (2) surface deformations comprising both correctable aberrations (piston and focus) and uncorrectable aberrations need to be determined rather than just surface departure from nominal; (3) because of mounting constraints on the back of the mirror, equal thickness platings front and back may increase rather than decrease bimetallic bending; (4) increasing mirror thickness does not help; and (5) symmetric shaping of the substrate significantly reduces bending even if the back surface is not plated.

Referring to Figure 10.14, ELN plating on aluminum surfaces works best with only the front surface plated for the baseline and double-concave mirror configurations. The best plating arrangement for the thick and meniscus substrates is equal thicknesses front and back. With ELN plating on beryllium substrate configurations, the best arrangement would be plating on the front surface only. This applies to all mirror configurations.

A factor contributing to the long-term stability of ELN-coated metallic mirrors is the stress inherent in the coating. Internal stress is dependent on the phosphorus content of the deposited nickel. In most cases, it is possible to specify a phosphorus content (typically <12%) for zero stress when annealed. Varying the chemical composition and heat treatment applied allows the zero residual stress level to be located at the center of a given operational temperature range.

Bimetallic bending stress between an ELN-plated mirror and substrate is greatest at the edge of the plated surface. During a change in temperature, the plating may lift off the substrate at the edge, eventually leading to overall failure. One way to minimize this problem is to machine a "nickel-lock" feature into the mirror. This is a shallow groove in the substrate, nominally 125 μm deep and wide at or slightly before the edge of the plating.

The stress in plated coatings is conveniently measured by plating one side of thin metal strips as witness samples of the base material to be coated. Generally, these strips measure 102 mm × 10.2 mm × 76 μm, with the last dimension depending upon the particular material. The opposite faces of the strips are ground flat and parallel with inherent bending not exceeding 25 μm. Release of residual stress due to the plating bends the strip. The magnitude of the bending, and hence the stress, is determined by placing the strips on edge under a microscope and measuring contour departure from a straight line along the long dimension. Typical bending magnitudes for ELN coatings on aluminum are in the 250–380 μm range; thus, these are easily measured with a reasonable accuracy.

10.7 SINGLE-POINT DIAMOND TURNING OF METAL MIRRORS

The use of single-crystal diamond cutting tools and specialized machinery to produce precision surfaces on selected materials by very accurately cutting away a thin portion of the surface is called "single-point diamond turning," "precision machining," and "precision diamond turning" by various people. The first terminology is used here and is abbreviated as SPDT hereafter in this chapter. The technique has developed from crude experiments to fully qualified production processes since the early 1960s.

The basic SPDT process generally involves the following steps: (1) preform or conventionally machine the part to a rough shape with ~100 μm (0.004 in.) excess material left on all surfaces to be processed; (2) heat-treat the part to relieve stress; (3) mount the part with minimal induced stress in an appropriate chuck or fixture on the SPDT machine; (4) select, mount, and align the diamond tool on the machine; (5) finish machine the part to final shape and surface quality with multiple light cuts under computer control; (6) inspect the part (in situ, if possible); and (7) clean the part to remove cutting oils and solvents. For some applications of the optic, plating the surfaces following step 2 is required to provide an amorphous layer to be diamond turned. In some cases, step 7 is followed by polishing the optical surface or surfaces to smooth it or them and applying the appropriate optical coating.

The machine used for SPDT operations can legitimately be termed an *instrument* since it unquestionably meets the classical definition of Whitehead, who states, "an instrument may be defined as any mechanism whose function is directly dependent on the accuracy with which the component parts achieve their required relationships" (see ref. 9, p. 2). In this case, that accuracy is achieved, in part, from inherent mechanical rigidity and freedom from self-generated and external vibration and thermal influences. The predictability and high resolution of rotary and linear motions and low wear characteristics of its mechanisms are inherent attributes of a good SPDT design.

Inherent in the SPDT technique is the production of a periodic grooved surface that scatters and absorbs incident radiation. Figure 10.15a illustrates schematically a highly magnified view of the localized contour of the turned surface in a machine functioning in a facing operation as shown in Figures 10.16 and 10.17. The inherent roughness of the surface is depicted in Figure 10.18. The diamond tool has a small curved nose of radius R. The motion of the tool across the surface creates parallel grooves as indicated on the right. The theoretical peak to valley height h of the resulting cusps is given by the following simple equation involving the parameters designated in Figure 10.15b:

$$h = \frac{f^2}{8R}, \tag{10.3}$$

where f is the transverse linear feed of the tool per revolution of the surface. For example, if the spindle speed is 360 revolutions per minute (rpm), the feed rate is 8.0 mm/minute and the tool radius

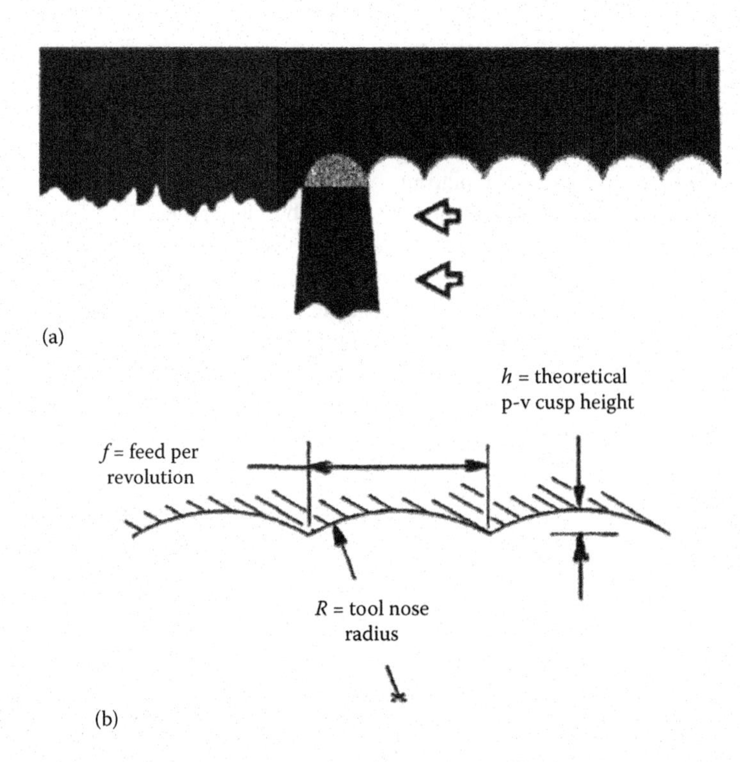

FIGURE 10.15 Schematic illustrations of (a) a single-point diamond tool advancing from right to left over the surface of a substrate and (b) the geometry of the cusped surface resulting from the SPDT process. (Adapted from Rhorer, R.L., and Evans, C.J., Fabrication of optics by diamond turning, Chapter 41 in *Handbook of Optics,* 2nd ed., Vol. II, Bass, M. et al., Eds., pp. 41.1–41.12 Optical Society of America, Washington, DC, 1995. With permission of Optical Society of America.)

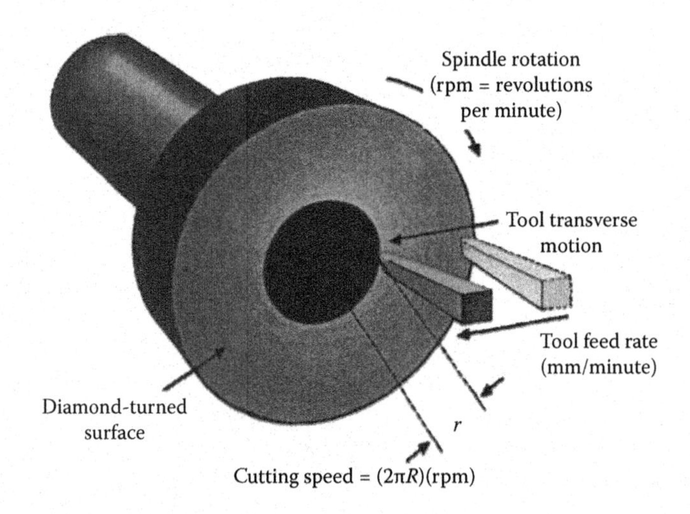

FIGURE 10.16 Schematic of an SPDT facing operation. (Adapted from Rhorer, R.L., and Evans, C.J., Fabrication of optics by diamond turning, Chapter 41 in *Handbook of Optics,* 2nd ed., Vol. II, Bass, M. et al., Eds., pp. 41.1–41.12 Optical Society of America, Washington, DC, 1995. With permission of Optical Society of America.)

FIGURE 10.17 Photograph of a facing operation on an SPDT instrument. (Courtesy of Rank, Taylor Hobson, Inc., Keene, NH.)

is 6 mm, the value for $f = 8.0/360 = 22.2$ μm per revolution. Hence, $h = (0.0222)^2/[(8)(6)] = 1.03 \times 10^{-5}$ mm or 1.03 Å. Note that the width of each cusp equals f. The RMS value of the height δ_{RMS} is about $h/3.325$ or $\delta_{RMS} \approx 0.3h$.

The actual surface roughness of a diamond-turned surface is often much worse than predicted from the preceding equation. Lubrication of the machined surface, tool edge geometry and deterioration, and vibration are machining parameters affecting surface roughness. Low-order vibration is one of the biggest sources of surface finish degradation, and elaborate vibration isolation systems are incorporated in some diamond-turning machines.

The tool of an SPDT machine must follow an extremely accurate path relative to the surface being cut throughout the procedure. There are several sources of errors in the machined surface contour beyond the cusped structure just described. These are (1) waviness from inaccuracies of

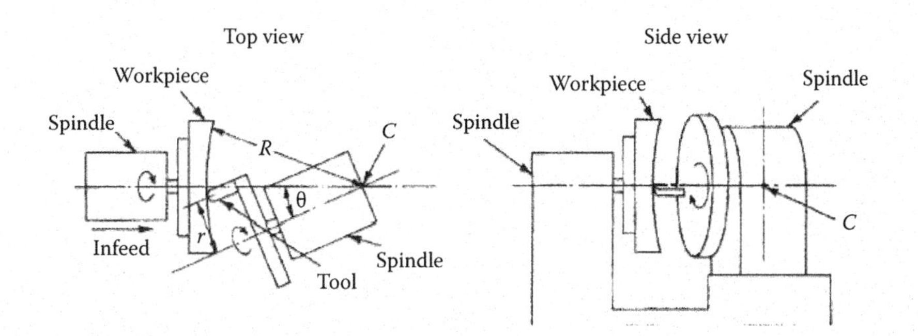

FIGURE 10.18 Schematic of a dual rotary axis fly-cutter SPDT instrument configured to machine concave spherical surfaces. Here, $R = r/\sin\theta$. (From Parks, R.E., Introduction to diamond-turning, SPIE short course, 1982.)

travel of slide mechanisms providing tool motions; (2) nonrepeatability of axial, radial, and tilt motions relative to the spindle rotation; (3) external and self-generated vibrations; (4) effects within the turned material wherein differential elastic recovery of adjacent grains and impurities cause contour "steps" or "orange peel" appearance in the surface; and (5) repeated structure within the cusps because of irregularities in contour of the tool cutting edge.

The earliest applications of SPDT were to windows, lenses, and mirrors for use in the IR region because such optics could have rougher and less accurate surfaces than ones for use with shorter wavelengths. With advances in SPDT technology, diamond-machined optics can today be made smooth enough to be used in visual and lower-performance ultraviolet instruments. Surface microroughness typically achieved in quantity production on bare 6061 aluminum mirror surfaces is 80–120 Å RMS, while that achieved on plated surfaces is about 40 Å RMS.

Materials that can be machined by SPDT techniques with more or less success are listed in Table 10.9. Compatibility with the process is not an intrinsic property of the material. Rather, it is an expression of practicality. Some materials, such as ferrous metals, ELN, and silicon, *can* be diamond turned but wear the cutting tools rapidly, so this technique is not generally considered economical for machining them. Some alloys of listed generic metals can be processed with great success, while others are doomed to failure. For example, good surfaces can be cut on 6061 aluminum, while 2024 aluminum typically ends up with poor surfaces. In general, ductile materials (those hard to polish by traditional methods) are compatible with SPDT, whereas hard, brittle materials polish easily, but are not suitable for SPDT. In some cases, brittle materials can be processed to high precision on an SPDT machine by substituting a grinding head for the diamond cutter.

Plate, rolled, extruded, or forged wrought forms of metals are most commonly used for SPDT, but considerable success has been achieved in the diamond turning of carefully prepared near-net-shape castings of types 201-T7, 716-T5, and 771-T52 aluminum alloys. To present a homogeneous substrate to the diamond, a casting must be made of virgin, metallurgical pure raw material (<0.1% impurity), its hydrogen content must be <0.3 ppm, the material handling equipment and gating

TABLE 10.9
Materials That Usually Can Be Diamond Turned

Aluminum	Calcium fluoride	Mercury cadmium telluride
Brass	Magnesium fluoride	Chalcogenide glasses
Copper	Cadmium fluoride	Silicon (?)
Beryllium copper	Zinc selenide	Poly(methyl methacrylate)
Bronze	Zinc sulfide	Polycarbonate
Gold	Gallium arsenide	Nylon
Silver	Sodium chloride	Polypropylene
Lead	Calcium chloride	Polystyrene
Platinum	Germanium	Polysulfone
Tin	Strontium fluoride	Polyamide
Zinc	Sodium fluoride	Ferrous metals (?)
ELN (K >10%)	potassium dihydrogen phosphate (KDP)	
EN (?)	potassium titanyl phosphate (KTP)	

Source: Rhorer, R.L. and Evans, C.J., Fabrication of optics by diamond turning, in *Handbook of Optics*, 2nd ed., Vol. II, Bass, M. et al., Eds., pp. 41.1–41.12 Optical Society of America, Washington, DC, 1995. With permission of Optical Society of America; Gerchman, M., Specifications and manufacturing considerations of diamond-machined optical component, *Proc. SPIE*, 607, 36, 1986.

Note: Materials with (?) cause rapid diamond wear.

system must not increase the impurity level of the raw material, and the rate of casting solidification must be carefully controlled so that cooling occurs isotropically from the optical surface inward.

Polycrystalline materials may not machine very well because their grain boundaries may be emphasized by the cutting action of the tool. However, this problem must be evaluated on a case-by-case basis. As an example, for IR applications, the differences in SPDT results when diamond turning single and polycrystalline germanium are insignificant. The presence of grain boundaries does not seem to cause brittle fracture to surfaces.

There are two basic types of SPDT instruments: the *lathe* type, in which the workpiece rotates and the diamond tool translates, and the *fly-cutter* type, in which the tool rotates and the workpiece translates. There are a variety of different geometries of SPDT instruments designed to create cylinders, exterior and interior cones, flats, spheres, toroids, and aspheres. Here only five of these configurations are considered.

Figure 10.19 shows a lathe-type SPDT instrument with the linear tool axis parallel to the spindle axis.[10] This resembles the conventional machinist's lathe. The workpiece is mounted between live and dead centers (as shown) or cantilevered from a faceplate on the spindle. By appropriate fixturing, the diamond tool can also be positioned to turn the inside diameter (ID) of a hollow cylinder. If the linear tool slide is rotated about a vertical axis so as to lie at a horizontal angle to the spindle axis, this instrument can be used to machine external or internal cones.

Figure 10.20 illustrates a fly-cutting SPDT instrument. Here the workpiece is a flat generated by a single or double series of slightly displaced, parallel, arcuate tool cuts. If the spindle axis is not

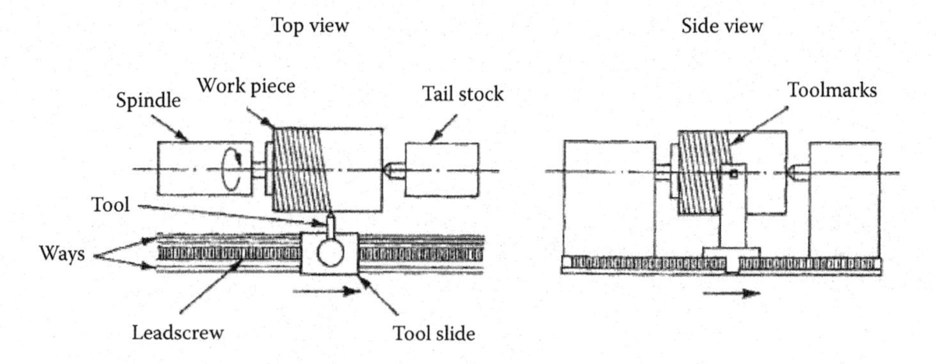

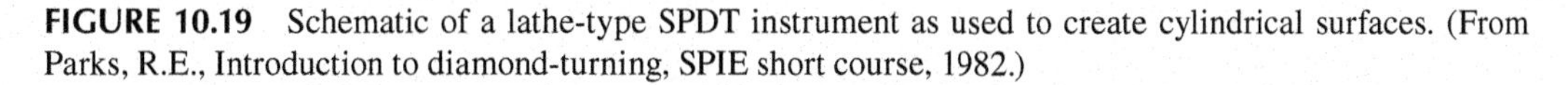
FIGURE 10.19 Schematic of a lathe-type SPDT instrument as used to create cylindrical surfaces. (From Parks, R.E., Introduction to diamond-turning, SPIE short course, 1982.)

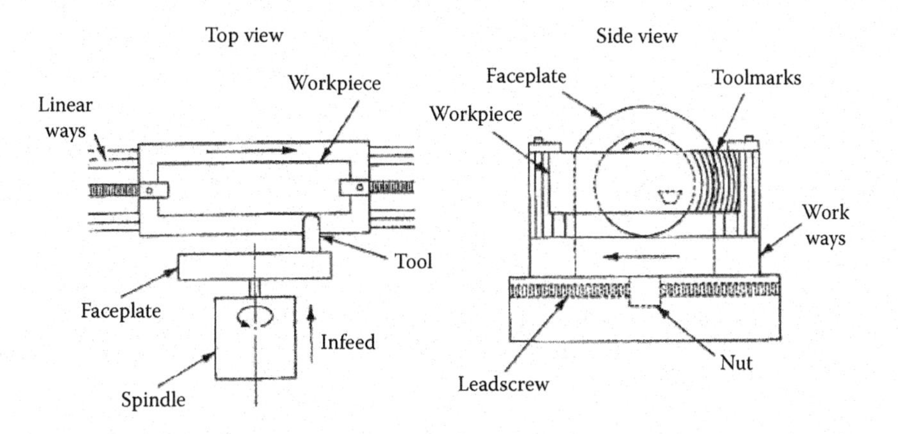

FIGURE 10.20 Schematic of a fly-cutter SPDT instrument as used to create flat surfaces. (From Parks, R.E., Introduction to diamond-turning, SPIE short course, 1982.)

accurately perpendicular horizontally to the linear axis, the surface becomes cylindrical. Faceted scanner mirrors are machined by a variation of this geometry in which the workpiece is indexed about an axis inclined with respect to both the spindle and linear axis. The only practical method of machining scanner mirrors directly into a substrate with closely spaced, flat, internally reflecting facets is through SPDT techniques.

Another type of fly-cutting SPDT instrument designed to create spherical surfaces is illustrated in Figure 10.18. Here, two rotary motions about coplanar and intersecting axes carry both the workpiece and the tool. The radius R created equals $r/\sin\theta$, where these parameters are as depicted in the figure. The function is similar to that of a diamond-cup surface-generating machine as used in mechanical and optical shops. As shown, a concave surface is machined; a convex one is created if the point C representing the intersection of the two axes is moved behind (i.e., to the left of) the workpiece. This is generally accomplished by supporting the tool on a yoke with arms passing on both sides of the workpiece spindle. If the tool is mounted on a linear feed rotating about the axis through point C of this figure, the configuration is called an "R–θ" instrument. It cuts aspherical as well as spherical surfaces.

Figure 10.21 shows an instrument with four axes, two rotary and two linear, the latter being stacked vertically. The Z-axis of this system is a limited, rapid linear motion of the diamond tool. By coordinating the motion of the tool with the rotational position of the workpiece, a nonaxisymmetric surface can be generated. Rotation of the tool occurs about the center of the circular cutting edge through action of a stepper motor. This motor is indexed after individual or every few rotations of the workpiece spindle. This keeps the tool normal to the workpiece and eliminates error due to variation in radius, hardness, and finish along the tool cutting edge. It also allows a shorter-radius tool to be used, thus allowing greater slope variation on the machined surface. With this number of motions, convex and concave aspheric surfaces can be easily machined. Figure 10.22 shows a way in which three rectangular aperture, off-axis parabolas can be mounted for simultaneous precision diamond turning on such an instrument. Centering pins and reference flats on the mirrors control orientation.

A variety of single- and multiple-axis SPDT machines with different workpiece size capacities and surface contour capabilities are available from various manufacturers worldwide. For example, Figure 10.23 shows a typical commercial five-axis SPDT instrument. Although inclusion here does

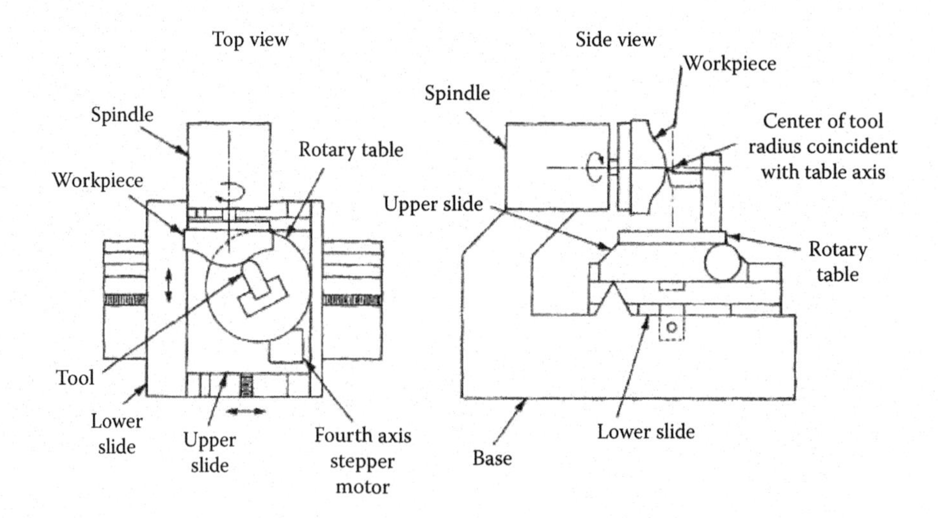

FIGURE 10.21 Schematic of a four-axis SPDT instrument with stacked linear slides and stepwise control of the orientation of the diamond tool relative to the workpiece. (From Parks, R.E., Introduction to diamond-turning, SPIE short course, 1982.)

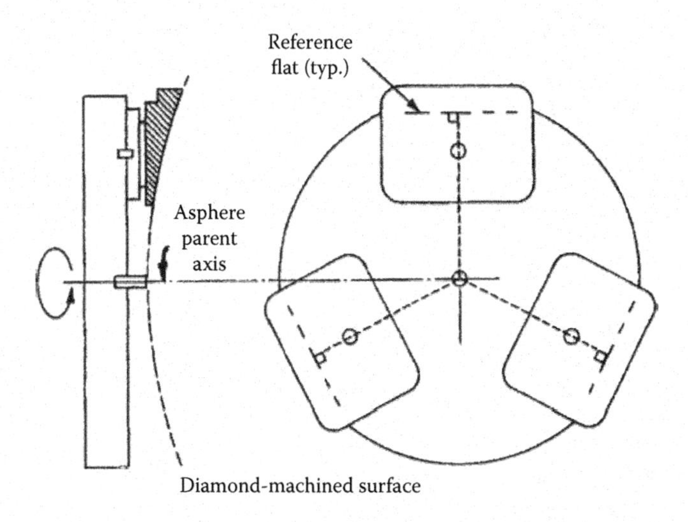

FIGURE 10.22 Typical fixturing for SPDT of multiple off-axis aspheric mirrors. (Adapted from Curcio, M.E., Precision-machined optics for reducing system complexity, *Proc. SPIE*, 226, 91, 1980.)

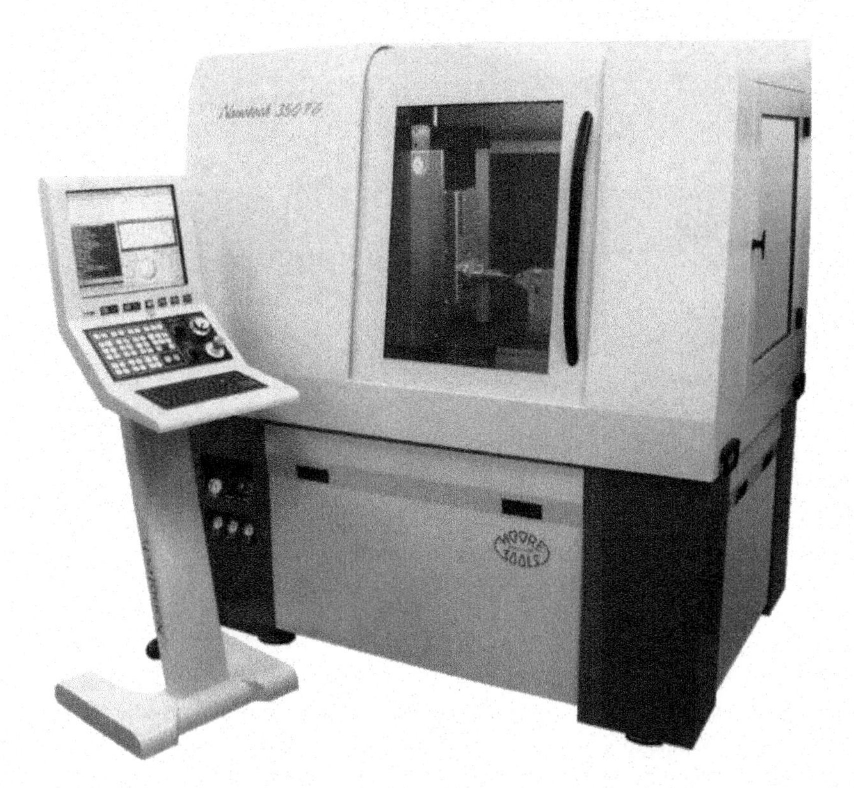

FIGURE 10.23 Photograph of an ultraprecision, free-form five-axis SPDT/free-form generator, the Moore Nanotech® model 350FG. (Courtesy of Moore Nanotechnology Systems, Swanzey, NH.)

not constitute endorsement of this product, it does represent the state of the art in commercially available production instruments in early 2014. The specifications and capability of this instrument, the Nanotech 350FG ultraprecision free-form generator produced by Moore Nanotechnology Systems, of Swanzey, New Hampshire, are summarized in Table 10.10. Figure 10.24 shows an off-axis parabolic mirror machined into 6061 aluminum substrate with the surface generator of

TABLE 10.10
Key Features and Capabilities of the Nanotech 350FG Computer Numerically Controlled Machining System Shown in Figure 10.23

System capabilities	Three- to five-axis turning of aspherical and toroidal surfaces, rotary ruling of freeform surfaces, and raster fly cutting of freeforms, linear diffractives, and prismatic optical structures
Workpiece capacity	500 mm diameter × 300 mm long
Configuration	Monolithic composite epoxy–granite structure. Dimensions (m), 1.42 × 1.57 × 0.46
Form accuracy	≤150 nm
Surface finish	≤3.0 nm
Performance	Linear axis position feedback resolution, 0.034 nm
	Rotary axis encoder angular resolution, 0.25 arcsec

Source: Moore Nanotechnology Systems, Swanzey, NH.

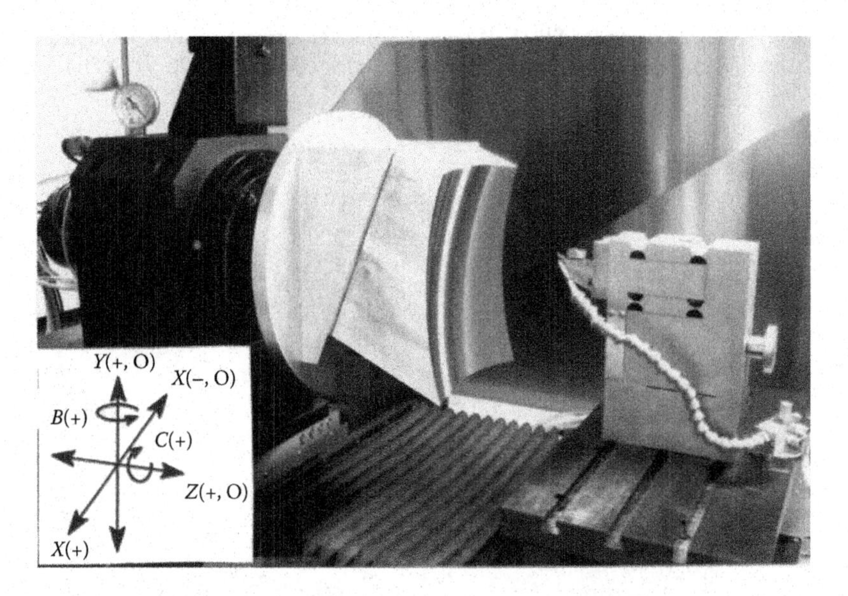

FIGURE 10.24 Photograph of a large (14 in. × 14 in. aperture) off-axis parabolic mirror surface being SPDT finished on a 6061 aluminum substrate in the five-axis machine shown in Figure 10.23. The axes of motion are indicated in the insert diagram. (Courtesy of Moore Nanotechnology Systems, Swanzey, NH.)

Figure 10.23. The process used is called slow-slide-servo machining, where the *Z*-axis (carrying the cutting tool) undulates in and out in coordination with the rotation of the spindle (*C*-axis) as defined in the insert detail view. The tool is fed from edge to center as it moves along the *X*-axis. The size of the optical workspace shown is 350 mm^2. The combined weight of the mounting fixture and mirror is ~18.2 kg.

Surfaces routinely created by modern SPDT generators are perfectly adequate for IR applications and some visible-light applications. With postprocess polishing, most surfaces can be brought to high visible-light quality standards.

Single-crystal diamond chips have unique characteristics that make them ideal for SPDT applications. They are very hard (when properly oriented), have low contact friction, are very stiff mechanically, have good thermal properties, and take an edge sharp to atomic dimensions. They can also be resharpened when wear becomes excessive. The cutting-edge radius may vary for the particular application from infinity to less than 760 μm. Typical maximum defect depths in properly sharpened diamond tools are <8 nm as indicated by scanning electron microscope measurements. The radius can be held constant to better than 150 nm for typical (short) radii. The diamond chips may be brazed to standard lathe tool bits as shown in Figure 10.25 for physical support. When so mounted, they can be easily handled and attached to the SPDT instrument. Cubic boron nitride tools have also proven effective for SPDT machining of bare beryllium substrates.

Figure 10.26 shows a multiple fly-cutter head with three diamond tools installed so that cuts can be made at different locations on the workpiece, at different depths, or by differently shaped tools during each pass so as to reduce the number of passes required to finish the surface.

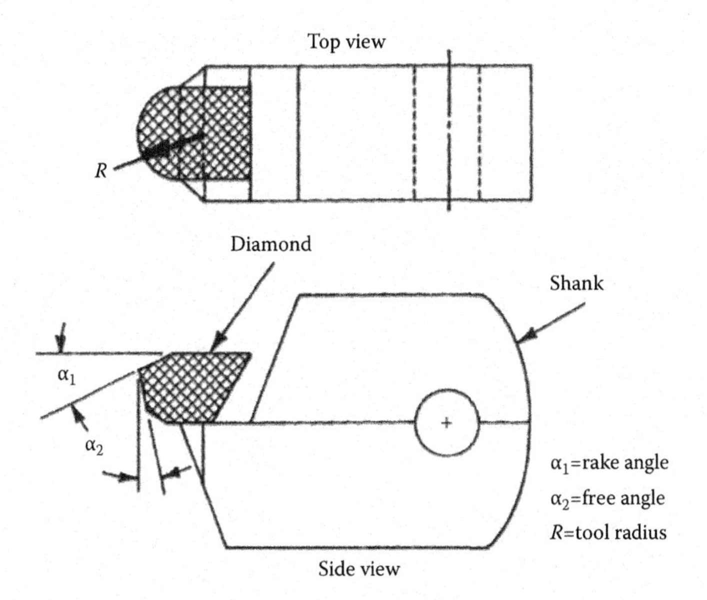

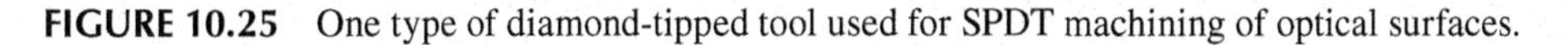

FIGURE 10.25 One type of diamond-tipped tool used for SPDT machining of optical surfaces.

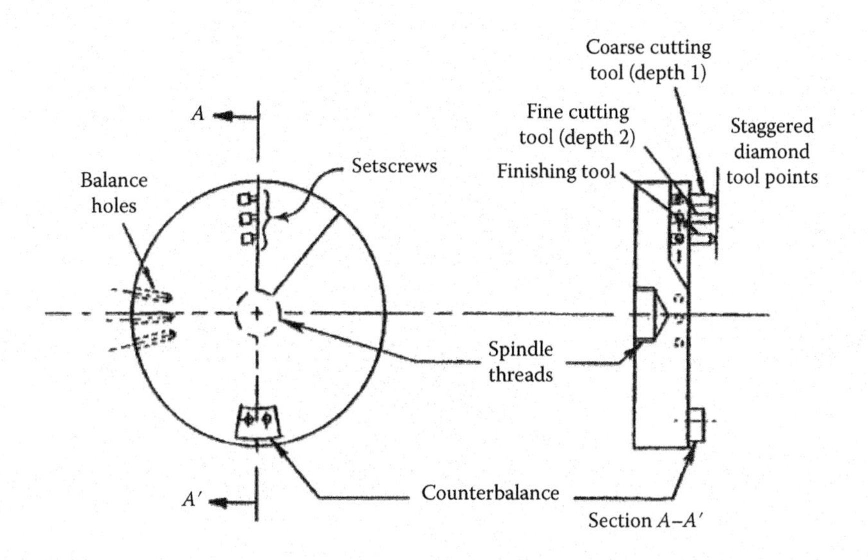

FIGURE 10.26 Schematic of a multiple-diamond fly-cutter head. (Adapted from Sanger, G.M., The precision machining of optics, Chapter 6 in *Applied Optics and Optical Engineering*, Vol. 10, Shannon, R.R., and Wyant, J.C., Eds., pp. 251–390, Academic Press, San Diego, CA, 1987.)

Stress-free mounting of the workpiece on the SPDT instrument is vital to achieving true surface contours and accuracy of machined dimensions. Techniques employed to ensure minimum stress include vacuum chucking, potting, and flexure mounts. Figure 10.27 illustrates the vacuum technique as applied to a thin germanium lens element, whereas Figure 10.28 illustrates a potting technique as applied to a circular mirror substrate, and Figure 10.29 shows a flexure technique as applied to an axicon element. The use of centering chucks to facilitate SPDT operations on crystalline lenses and on optomechanical subassemblies is discussed in Section 6.6 of this book.

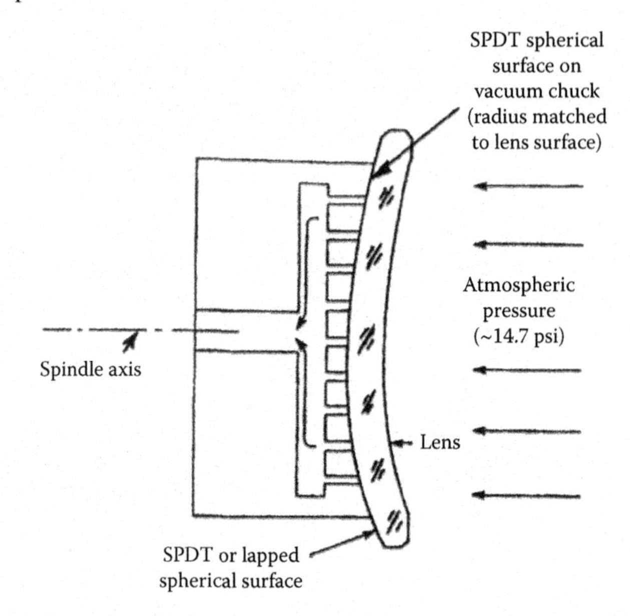

FIGURE 10.27 Schematic diagram of a vacuum-chuck technique for supporting a thin germanium lens on the spindle of an SPDT instrument. (Adapted from Hedges, A.R., and Parker, R.A., Low stress, vacuum-chuck mounting techniques for the diamond machining of thin substrates, *Proc. SPIE*, 966, 16, 1988.)

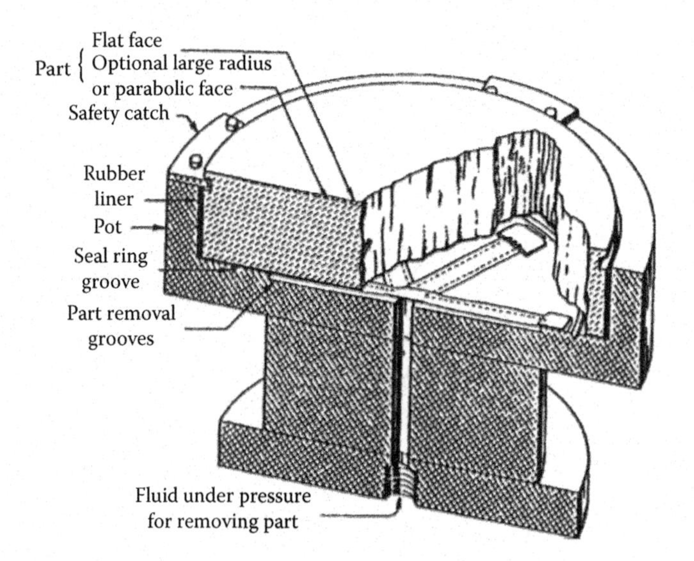

FIGURE 10.28 Schematic diagram of a potting technique for supporting a circular mirror substrate on the spindle of an SPDT instrument. (From Sanger, G.M., The precision machining of optics, Chapter 6 in *Applied Optics and Optical Engineering,* Vol. 10, Shannon, R.R., and Wyant, J.C., Eds., pp. 251–390, Academic Press, San Diego, CA, 1987.)

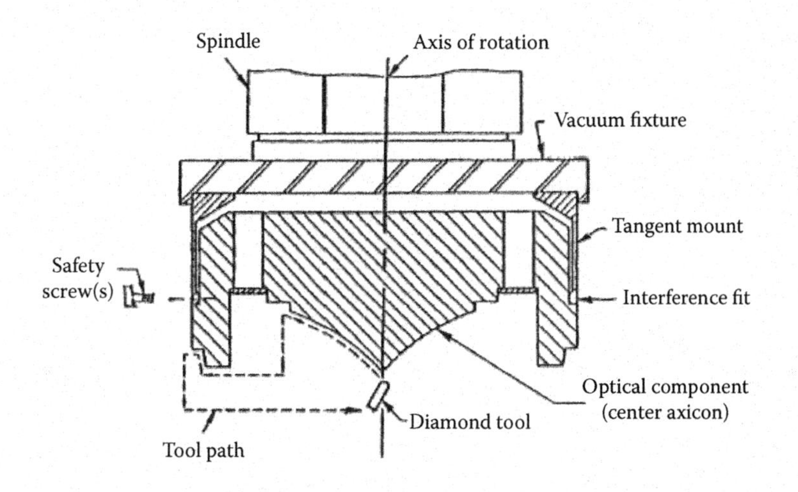

FIGURE 10.29 Schematic diagram of a flexure technique for supporting an axicon substrate on the spindle of an SPDT instrument. (From Sanger, G.M., The precision machining of optics, Chapter 6 in *Applied Optics and Optical Engineering,* Vol. 10, Shannon, R.R., and Wyant, J.C., Eds., pp. 251–390, Academic Press, San Diego, CA, 1987.)

A simple vacuum chuck uses a compliant seal around the edge of the mirror, such as an O-ring. If a full vacuum is present on one side of the mirror ($\approx 10^{-3}$ Torr or 10^{-6} bar), the load per unit area on the mirror holding it to the chuck is about 100 kPa. With a friction coefficient of about 0.5 between seal and mirror and safety factor (SF) of 4, this is enough to hold a mirror made of aluminum that is over 450 mm thick. Some chucks use direct metal-to-metal seals for better stability, but even with a friction coefficient as low as 0.1, a vacuum chuck can hold a 150 mm thick mirror. For a compliant (worst-case) edge seal, the force F_c needed to move a mirror of surface area A laterally on a vacuum chuck is $F_c = \mu \Delta P$, where μ is the friction coefficient between the chuck and mirror and ΔP is the pressure differential acting on the mirror surface (about 100 kPa assuming a vacuum behind the mirror).

The pressure differential in a vacuum chuck induces deformation in the surface of the mirror. It may be necessary to reduce the pressure differential to minimize this distortion. Alternately, a stiff internal backup structure can both stabilize the mirror position and minimize distortion. However, the mirror surface is likely to follow the contours of the backup structure; this effect is called "print-through." To minimize print-through, a rule of thumb is that the backup structure must be flat to the same tolerance as the mirror optical surface; hence, diamond turning is often used to create such supports.

If a circular mirror of diameter d and thickness h is supported around its edge and a pressure differential ΔP holds the mirror against the chuck, the maximum surface deflection δ_P is

$$\delta_P = \frac{3}{256}\frac{\Delta P}{E}(1-\nu)(5+\nu)\frac{d^4}{h^3}, \tag{10.4}$$

where E is the elastic modulus of the mirror material and ν is the Poisson's ratio of the mirror material.

For a rectangular mirror of length L and width b that is edge mounted on a vacuum chuck and subjected to a pressure differential ΔP, the maximum surface deflection δ_P is

$$\delta_P = \frac{3}{4\pi^4}\frac{\Delta P}{E}(1-\nu)(5+\nu)\frac{L^4 b^4}{h^3(L^2+b^2)^2}. \tag{10.5}$$

Example 10.2

A 150 mm diameter aluminum mirror is 20 mm thick; it is supported by a compliant vacuum chuck around its edge. If the maximum tolerable surface deflection is 0.125 wave at 1.06 μm, what is the maximum pressure differential used to hold the mirror in place?

Solving for the pressure differential ΔP,

$$\Delta P = \frac{256}{3}\delta_P E \frac{h^3}{d^4}\frac{1}{(1-\nu)(5+\nu)}.$$

Now 0.125 wave at 1.06 μm is 0.125(1.06 μm) = 132.5 nm. For aluminum, E = 69 MPa and ν = 0.33; then,

$$\Delta P = \frac{256}{3}(69\times10^9\,\text{Pa})(132.5\times10^{-9}\,\text{m})\frac{(20\times10^{-3}\,\text{m})^3}{(150\times10^{-3}\,\text{m})^4}\frac{1}{(1-0.33)(5+0.33)}$$

$$\Delta P = 3.45\times10^3\,\text{Pa}.$$

A pressure differential of just 3.45 kPa would suffice to hold the mirror in place. Since atmospheric pressure is about 101 kPa, this is 3.45/101, or a reduction of just 3.4% in pressure.

Diamond-turned surfaces may not have sufficient smoothness for some purposes, so postturning polishing may be needed. Surface microroughness is especially important in the case of SPDT optics because of the cyclic nature of the process. The grooves scatter and diffract light more than a smooth surface. Figure 10.30 defines the useful ranges of some of the methods used to measure this parameter.[11]

Typical surface roughness values for different surfaces when diamond turned are given in Table 10.11. The surface roughness in diamond-turned mirrors is determined by material properties, tool

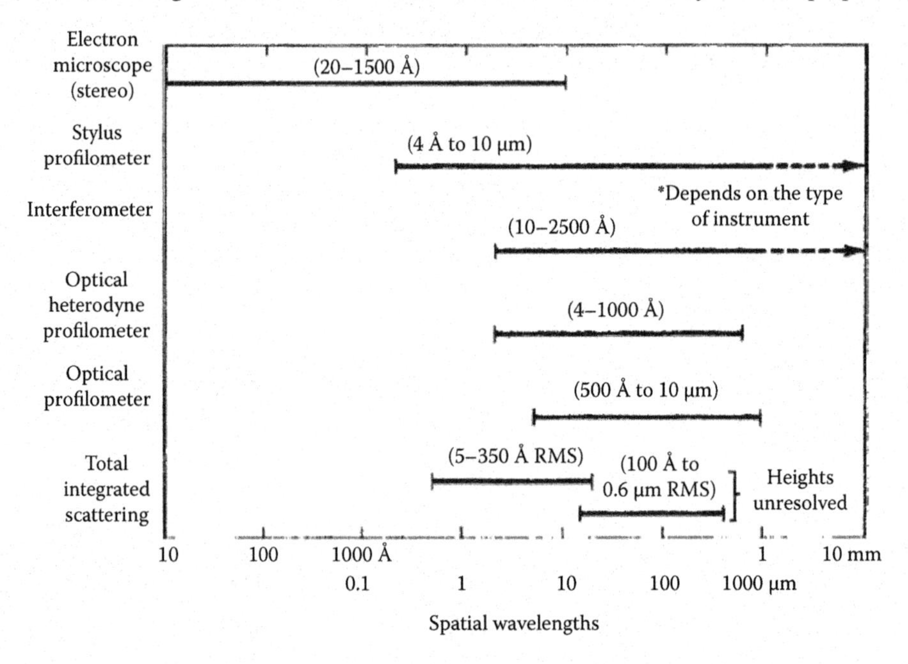

FIGURE 10.30 Capabilities of various techniques for assessing surface microroughness. The bars represent the applicable ranges of spatial wavelengths, while the numbers in parentheses give the approximate limits of surface height measurements. (From Bennett, J.M., and Decker, D.L., Surface characterization of diamond-turned metal optics, *Proc. SPIE*, 288, 534, 1981.)

TABLE 10.11
Typical Surface Roughness Values for Optical Surfaces Produced by Diamond Turning

Surface Material	Surface Finish (Å RMS)
Aluminum, 6061	60–100
AlumiPlate	20–40
ELN	20–40

condition, and cutting process. Material properties influencing surface finish include anisotropy, grain size, and inclusions. Tool condition is set by the initial geometry of the diamond tool and wear during use. The cutting process is the selection of speeds and feed rates, and it can be controlled during fabrication.

Variation in grain size, as well as grain orientation, influences surface roughness. Generally, the smaller the grain size, the better the surface roughness. Similarly, close-packed grain structures found in longitudinally processed bar, or forgings, improve surface finish. Hard inclusions, typically in the form of silicon for aluminum alloys, degrade surface finish. During diamond turning, these inclusions are ripped out of the alloy matrix, scratching the surface and blunting the edge of the tool. It is common to specify "no remelt" when purchasing aluminum for diamond turning; since recycled alloy is thought to be inferior in purity, especially with respect to silicon inclusions. Better surface finishes are obtained with high-purity aluminum alloys, with fine-grained forged material often preferred. This explains why amorphous, high-purity materials lacking a grain structure such as AlumiPlate and ELN improve surface finish when compared to conventional engineering alloys.

10.8 CONVENTIONAL MOUNTINGS FOR METAL MIRRORS

Small and moderate-sized metal mirrors can be mounted in the same ways discussed in Chapter 9 for nonmetallic mirrors if there are no unusual requirements inherent in the application. Unusual requirements include extreme temperatures, exposure to high-energy radiation (such as that from lasers or solar simulators), and extreme acceleration, shock, or vibration. Because metal mirrors can have optimally located attachment interfaces machined directly into the backs of the substrates, their mountings can be simpler, and the mirrors would experience smaller self-weight deflections than if mounted by clamping near their rims against annular or localized pads by using retaining rings or multiple springs.

The primary differences between the designs of mountings for metal mirrors and those made of glass-type materials lie in the values of key mechanical parameters such as density, modulus of elasticity, Poisson's ratio, thermal conductivity, CTE, and specific heat. Table A7.1 in Appendix 7 lists these and some other parameters for several important metallic materials as well as nonmetallic ones.

10.9 INTEGRAL MOUNTINGS FOR METAL MIRRORS

An attribute of most types of metal mirrors that facilitates mounting is their compatibility with machining. This includes conventional and SPDT shaping and surfacing and drilling and threading of holes to accept fasteners such as screws. If the latter operations take place in regions of the mirror sufficiently removed from the optical surface(s), little or no optical performance degradation results. Although attractive because of its simplicity, direct mounting by screwing a mirror to a structural interface can lead to problems if care is not exercised in the design and execution of both surfaces involved in the interface. As a rule of thumb, the mounting surfaces must be flat to the same tolerances and coplanar (or, in some designs, simply parallel) as the related optical surface(s) if mirror distortion is not to be introduced when the screws are tightened. The use of kinematic mounting techniques or provision of rotational compliance at the interfaces tends to reduce distortion, but may result in complex designs. The thermal properties of the mirror and mount should match as nearly as possible in order not to introduce distortions when the temperature changes.

A simple approach unique to metal mirrors uses SPDT processes to establish mounting surfaces integral to the mirror in the same machine setup as used to create the optical surface. Figure 10.31a illustrates schematically a design that uses slots to create mounting ears that isolate the optical surface from the bending stresses that might be introduced as the mirror is attached to its interface.[12] The multiple mounting surfaces are, in this case, diamond turned on the back of the same piece of material that has the reflecting surface diamond turned into its front surface. Figure 10.31b is a photograph of ears on such a mirror. In this design, the mirror's edges are heavily beveled all around. The slots are cut with a cylindrical "core cutter" whose axis is parallel to the mounting surface. Threaded holes for attachment screws can be seen.

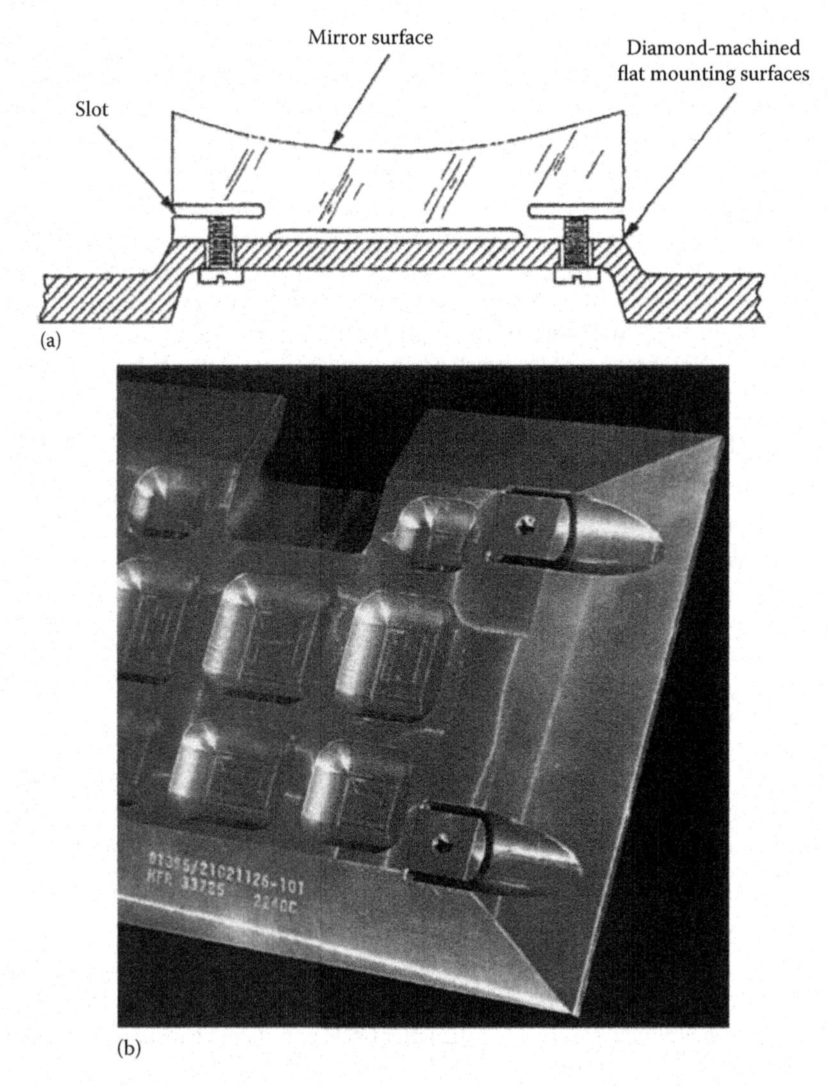

FIGURE 10.31 (a) Schematic diagram of one type of strain-free mounting for a metal mirror. (b) Photograph of mounting ears (flexures) formed by a diamond-tipped core cutter moving parallel to the reflecting surface of a metal mirror and cutting into the heavily beveled edges. Also shown are machined recesses that reduce mirror weight. (From Zimmerman, J., *Opt. Eng.*, 20, 187, 1981.)

Figure 10.32 shows a flat mirror that has its mounting pads on the front side of the mounting flange to facilitate machining them parallel to the reflecting surface in the same SPDT instrument setup. This simplifies alignment of the mirror at the time of installation by essentially eliminating the angle error (wedge) between these critical surfaces.

The integral flexure arm mount shown in Figure 10.33 is a stress-free design that has been successfully applied to several beryllium mirrors. The mating surfaces of both the arms and the surface to which they were attached were precision lapped to minimize distortion of the mirror surface when clamped in place. The mirror supports were not sufficiently stiff to hold the mirror during rough machining or grinding. The substrate was held by the cylindrical ring on the back of the mirror during these operations. It was later transferred to the flexure arms for final figuring when the forces exerted would be smaller.

Figure 10.34a and b shows mounting flexure tab features machined into the rear surfaces of two metal mirrors intended for use in the Infrared Multi-object Spectrograph developed as a facility for the 3.8 m Mayall Telescope of the Kitt Peak National Observatory and as a pathfinder for

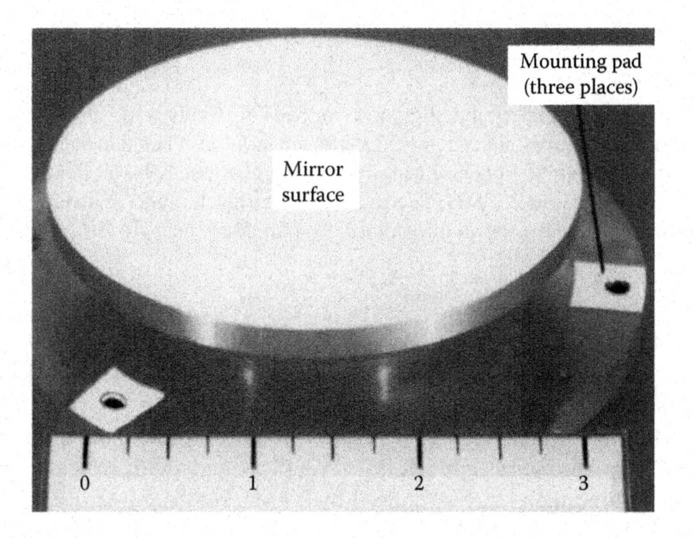

FIGURE 10.32 Example of a metal mirror with front mounting pads created in the same SPDT setup as the reflecting surface. (From Zimmerman, J., *Opt. Eng.*, 20, 187, 1981.)

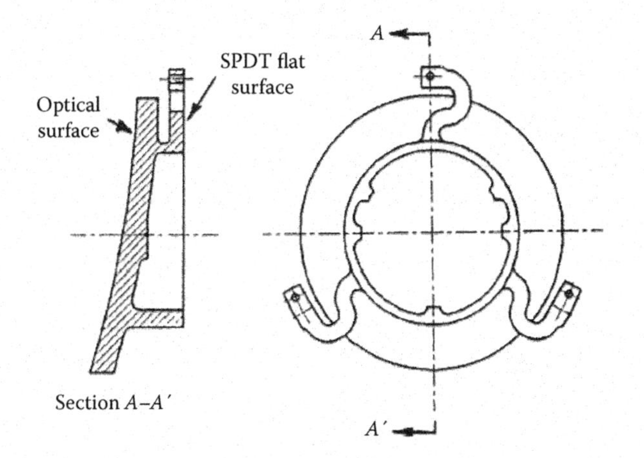

FIGURE 10.33 Be mirror with integral flexure-arm supports. (Adapted from Sweeney, M.M., Manufacture of fast, aspheric, bare beryllium optics for radiation hard, space borne systems, *Proc. SPIE*, 1485, 116, 1991.)

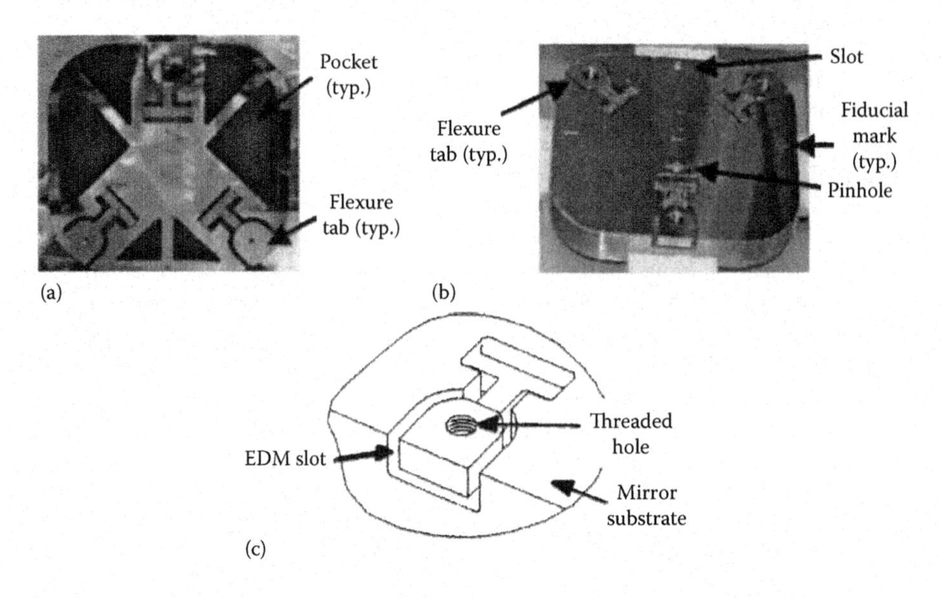

FIGURE 10.34 (a, b) Rear mounting and alignment surfaces of two aspheric aluminum mirrors featuring flexure tabs to minimize mounting-induced optical surface distortions. The fiducial markings, pinhole, and slot shown in view (b) facilitate alignment during machining and assembly. (c) Schematic diagram of one flexure tab with a threaded screw hole. EDM, electrical discharge machining. (From Ohl, R. et al., Design and fabrication of diamond machined aspheric mirrors for ground-based, near-IR astronomy, *Proc. SPIE*, 4841, 677, 2003.)

the planned multi-object spectrograph in the JWST.[11] The mirror in the view in Figure 10.34a is an off-axis section of a concave prolate ellipsoid measuring 264 mm × 284 mm (10.39 in. × 11.18 in.). Pockets are machined into the rear surface of this mirror to reduce its weight. The mirror in the view in Figure 10.34b is an off-axis section of a convex oblate ellipsoid measuring 90 mm × 104 mm (3.54 in. × 4.09 in.). It is not lightweight. Both mirrors are made with a 6:1 diameter-to-thickness ratio of 6061-T651 aluminum and stress relieved by a preferred heat-treating method as indicated in the column headed "SR5" in Table 10.6.[13] All mounting and optical surfaces of both mirrors are finish machined by SPDT processing.

To facilitate assembly and alignment, the flexure tabs are cut into the mounting surfaces by a plunge electrical discharge machining process to form the shapes indicated schematically in Figure 10.34c. The flexures minimize optical surface deformation due to mounting forces by bending to translate as much as ±25 μm (0.001 in.) and tilt by ±0.1° A threaded hole for attaching screws is provided in each tab.

The spectrograph is to operate at 80 K. Its structure is fabricated from the same type of material (Al 6061-T651) as the mirrors so as to form an athermal assembly. Alignment of the system is facilitated by several cross-hair fiducial marks scribed into the mirror rear and side surfaces during SPDT machining. One of these is pointed out in Figure 10.34b. The slot and pinhole indicated in that same view serve as alignment references when the substrate is attached to the SPDT machine.

The general design principles underlying all these mirror examples are the following: (1) Mounting stresses are isolated from the mirror surface by incorporating flexure arms, geometric undercuts, or slots that create a form of flexure mounting. (2) The mirror should be designed to be a stiffer spring than the interfacing mounting structure. Deformations then occur in the mount rather than in the mirror substrate. (3) The mirror should, if possible, be held during machining in precisely the same manner as it will be held during operation. Then, mounting strains will be the same in both conditions. (4) The mounting surfaces should be machined flat and parallel to the same degree of precision as the optical surface(s).

It should be noted that compliance with the second of these principles might result in a design in which rigid-body displacement or tilt of the optical surface can occur during mounting. Provision should, in this case, be made for alignment adjustment subsequent to installation. A further possible consequence of adhering to this principle is elimination of the need for adherence to the final principle.

In Section 10.7, the bimetallic effect of plating metal mirrors with dissimilar material is discussed. For the mirror shapes in Figure 10.14, the best configuration for a 180 mm (7.09 in.) diameter beryllium mirror is a plano-concave one in which only the front (optical) surface is ELN plated. Figure 10.35 shows the optomechanical design for that mirror. The view in Figure 10.35a is a section through the mirror showing the axial interface pads at three places and a pilot diameter that engaged the ID of a hole in the telescope backplate for centering purposes. The pilot diameter is shown in

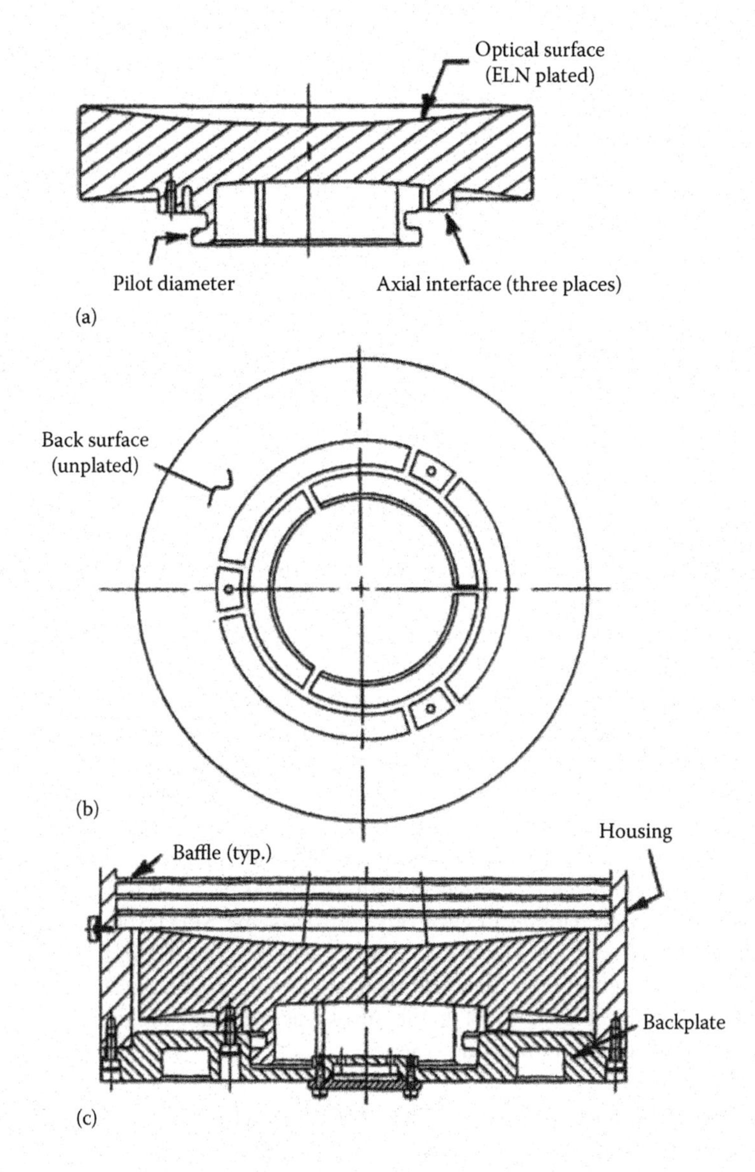

FIGURE 10.35 Optomechanical design for mounting a binconcave aluminum mirror incorporating the results from an investigation of bimetallic effects of ELN plating. Mirror cross section shown in (a), and plan in (b). Cross section of mounted mirror shown in (c). (Adapted from Vukobratovich, D. et al., Therm-Optic analysis of bi-metallic mirrors, *Proc. SPIE* 3162, 12, 1997.)

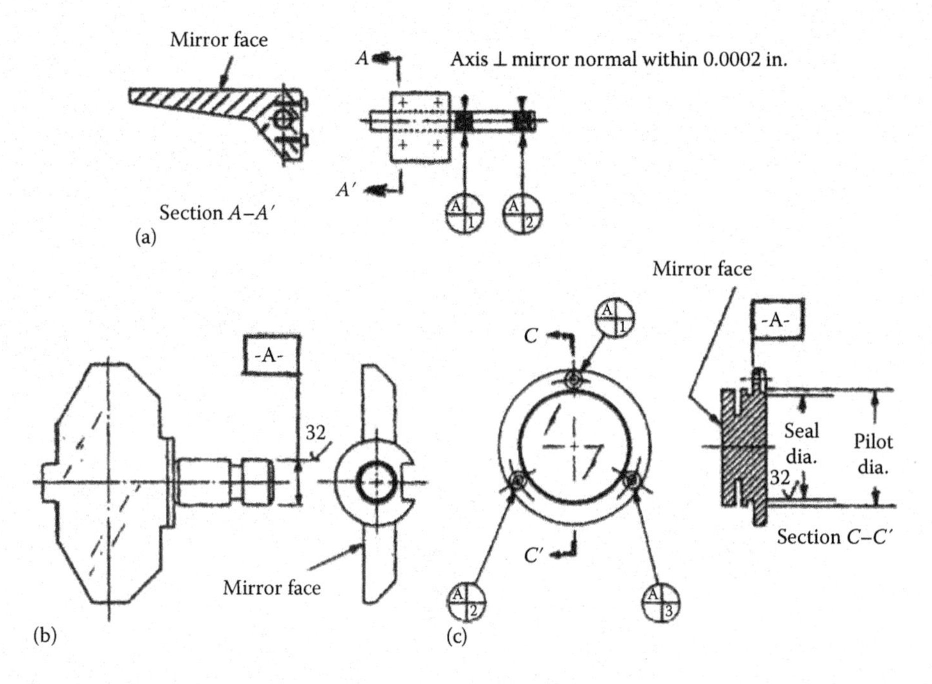

FIGURE 10.36 Three configurations of mirrors fabricated by SPDT techniques to accurately orient the optical surfaces to mechanical mounting interfaces. Dimensions are in inches. (Adapted from Addis, E.C., Value engineering additives in optical sighting devices, *Proc. SPIE*, 389, 36, 1983.)

the view in Figure 10.35b, and its interface with the backplate is shown in the view in Figure 10.35c. The mirror is attached with three screws as shown. By SPDT machining the axial pads perpendicular to the axis of the optical surface, the tilt alignment and axial location of the mirror were automatically established.

Additional examples of SPDT-machined optics that were designed to establish a high degree of accuracy between the optical surface and mechanical interface by using the latter as the datum for machining are shown in Figure 10.36a through c. These designs were intended for military applications, so they were ruggedized for high shock and vibration levels.

In the view in Figure 10.36a, a 6061-T6 aluminum mirror is clamped to a closely toleranced centerless ground stainless steel shaft. The bearing surfaces (shown shaded) ground on the shaft were used as the part orientation reference during SPDT machining of the mirror face. When pivoted in high-class bearings, the subassembly provided a precision optical tracking mechanism. Another aluminum scan mirror is shown in the view in Figure 10.36b. Here, the cylindrical surface -A- provides the machining datum. The mirror normal is perpendicular to -A-. The mirror shown in the view in Figure 10.36c is a static component with its optical face and its mounting interfaces at -A- accessible for machining parallel to each other in the same SPDT setup. The pilot diameter indicated in the view Figure 10.36c interfaces with a corresponding ID in the optical device. An O-ring seal is accommodated in the groove adjacent to the pilot diameter. This aluminum component resembles that shown in Figure 10.34 and features a necked-down region to serve as a flexure to isolate the optical surface from mounting stresses.

10.10 INTERFACING MULTIPLE SPDT COMPONENTS TO FACILITATE ASSEMBLY AND ALIGNMENT

A major advantage of SPDT as a means for manufacturing precision optical components is the ability to integrate locating and optical surfaces directly into each workpiece of multiple-component systems during fabrication, frequently without removing the workpiece from the SPDT instrument.

TABLE 10.12
Representative Mechanical Tolerances Produced by Diamond Turning

Feature	Tolerance
Parallelism	±0.0002 in. (5 μm)
Perpendicularity	±0.0001 in. (2.5 μm)
Flatness	±0.0001 in. (2.5 μm)
Optical surface run out with respect to data	±0.0005 in. (12.5 μm)
Minimum radial assembly clearance	0.0004 in. (10 μm)
Diameter	±0.0008 in. (20 μm)
Part location via pinning	±0.0004 in. (10 μm)

This ensures maximum alignment accuracy of the optical surfaces to other portions of the overall system.

Much tighter mechanical tolerances are possible with diamond turning when compared to conventional machine tool practice. This reduction in tolerance greatly simplifies the assembly process. In some cases, no optical alignment or adjustment is necessary. Table 10.12 is a list of tolerances representative of diamond turning practice.

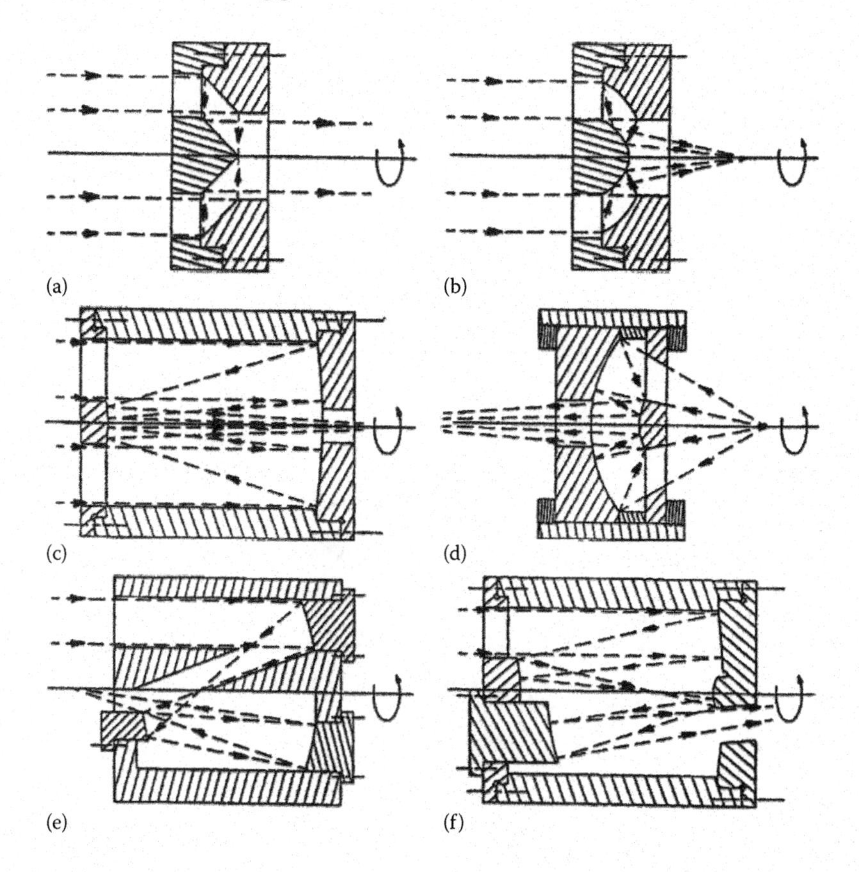

FIGURE 10.37 Six types of SPDT-machined multiple-component optical systems configured with integral locating surfaces to facilitate alignment and simplify assembly. The axis of symmetry is indicated for each system. (a) Reflaxicon beam expander. (b) Fast Cassegrain telescope. (c) Slow Cassegrain telescope. (d) Schwarzschild microscope objective. (e) Inserted three-mirror system. (f) Combination four-mirror system. (Adapted from Gerchman, M., Diamond-turning applications to multimirror systems, *Proc. SPIE*, 751, 116, 1987.)

Figure 10.37 schematically shows six types of optical assemblies featuring this type of construction. Each system shown has at least one SPDT mechanical interface between separate optomechanical components. Each of the latter components is SPDT machined to accurately align its optical surface(s) with the mechanical interfaces. The axis of symmetry of each system is indicated by a curved arrow. All but one system (Figure 10.37e) has a mirror machined integrally with a spider support. The system in the view in Figure 10.37a involves two centered conical (axicon) reflecting optical surfaces, hence the name *reflaxicon*. The fast Cassegrain telescope shown in the view in Figure 10.37b requires only two components because it is so short, while the longer, slow Cassegrain telescope in the view in Figure 10.37c is most conveniently made with three components. The central component is essentially a spacer. The reflecting microscope objective system in the view in Figure 10.37d also has a spacer that allows focus to be established. It is assembled with two threaded retaining rings. Note that the light path is from right to left in this view. The three-mirror system in the view in Figure 10.37e has separate, off-axis optical components that need to be machined with reference mechanical surfaces or locating pins to facilitate rotational alignment. It also has integral stray light baffling provisions. The relatively complex four-mirror system in the view in Figure 10.37f embodies features of all the other systems.

The mechanical interfaces are configured generally as shown in Figure 10.38 to provide centering and axial positioning. Minimal stress is introduced into the components if the radial interface involves close sliding contact, all surfaces providing the axial interface are coplanar flats and accurately normal to the component axes or a toroidal surface contacting a flat, and the bolt constraints are centralized in the contacting pads.

Figure 10.39 shows a sectional view of an unobscured aperture, 10-power, afocal telescope assembly comprising two parabolic mirrors (one of which is off-axis) and a housing with two integral stray light baffles. Figure 10.40 is a drawing of the primary mirror, while Figure 10.41 is a drawing of the secondary mirror.

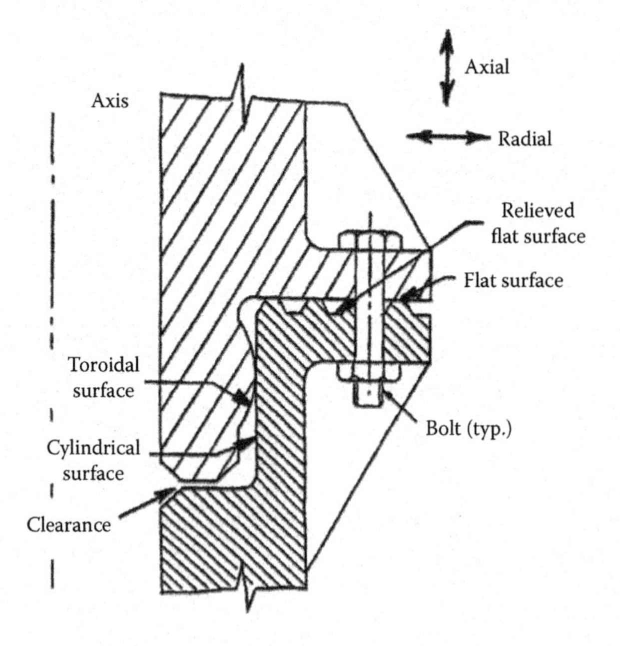

FIGURE 10.38 Typical interface between SPDT optical components to ensure axial and radial alignment and to minimize stress buildup due to mounting forces. (Adapted from Sanger, G.M., The precision machining of optics, Chapter 6 in *Applied Optics and Optical Engineering*, Vol. 10, Shannon, R.R., and Wyant, J.C., Eds., pp. 251–390, Academic Press, San Diego, CA, 1987.)

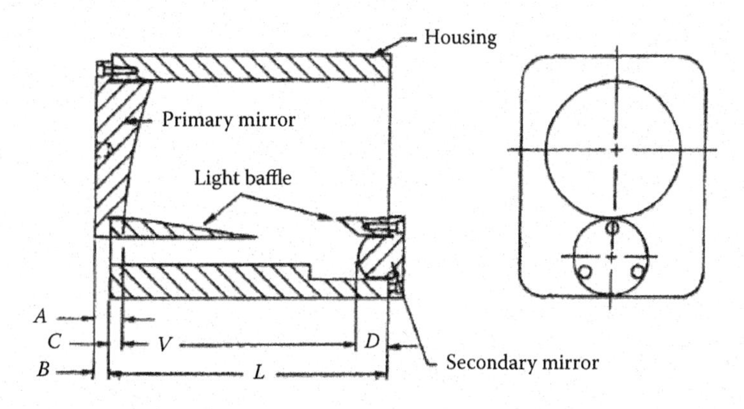

FIGURE 10.39 Example of a 10-power, afocal telescope comprising primary and secondary mirrors and a telescope housing with two light baffles. The mechanical interfaces are SPDT machined for accurate alignment at assembly. (Adapted from Morrison, D., Design and manufacturing considerations for the integration of mounting and alignment surfaces with diamond turned optics, *Proc. SPIE*, 966, 219, 1988.)

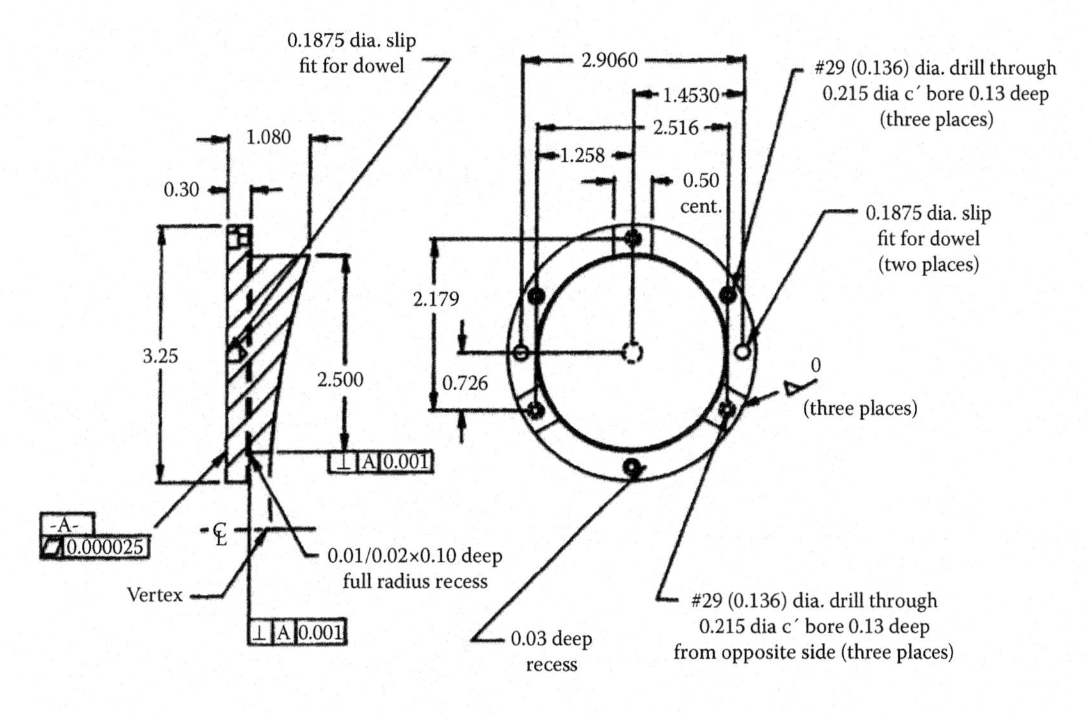

FIGURE 10.40 Drawings of the primary mirror for the telescope shown in Figure 10.39. (Adapted from Morrison, D., Design and manufacturing considerations for the integration of mounting and alignment surfaces with diamond turned optics, *Proc. SPIE*, 966, 219, 1988.)

Each mirror has a flat flange on the reflecting surface side of the substrate that interfaces with the parallel ends of the housing. The interfacing surfaces and the optical surfaces are SPDT machined to high accuracy with regard to location and minimal tilt with respect to the optical axes. The flat surfaces also serve as alignment references during setup for testing. The length of the housing controls the vertex-to-vertex separation of the mirrors. The end surfaces are flat to $\lambda/2$ at 0.633 μm, parallel to 0.5 arcsec, and separated by the nominal length ±127 μm. The actual length is measured to ±250 nm, and the part is serialized for identification.

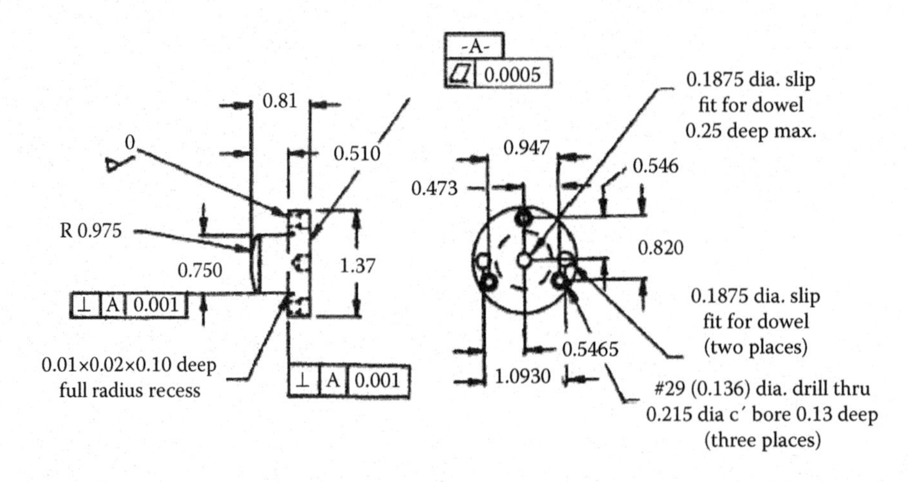

FIGURE 10.41 Drawings of the secondary mirror for the telescope shown in Figure 10.39. (Adapted from Morrison, D., Design and manufacturing considerations for the integration of mounting and alignment surfaces with diamond turned optics, *Proc. SPIE*, 966, 219, 1988.)

The primary mirror is attached to a subplate (fixture) for diamond turning. This subplate is vacuum chucked to the SPDT instrument and diamond-turned flat to $<\lambda/2$ at 0.633 μm. The rim of the subplate is then turned to ±127 nm roundness to provide an accurate reference for centering six precision, jig-bored, dowel holes in pairs 51.692 mm apart on a 50.800 mm bolt circle to match the dowel pinholes on the primary mirror. A central dowel hole also is bored at this time. Three mirror blanks are mounted on the subplate as shown in Figure 10.42 for simultaneous machining.

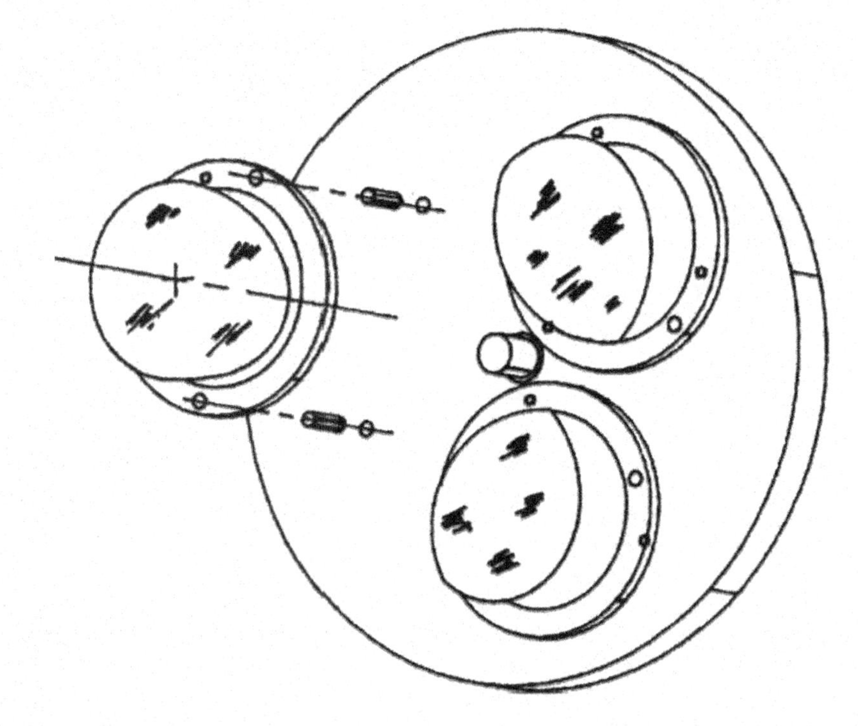

FIGURE 10.42 Schematic fixturing arrangement for simultaneous SPDT machining of three off-axis secondary mirrors. This is functionally similar to that shown in Figure 10.22. (Adapted from Morrison, D., Design and manufacturing considerations for the integration of mounting and alignment surfaces with diamond turned optics, *Proc. SPIE*, 966, 219, 1988.)

After the optical surfaces of a set of primary mirrors are completed, their actual axial thicknesses are measured to 25 nm and recorded as dimension A of Figure 10.39. Nominally, this dimension is 16.970 mm. The mirrors are then individually mounted on a vacuum chuck by using a central dowel for centering. The mounting flange is turned until $A - B = 6.350 \pm 0.051$ mm. The actual dimension C, to the nearest microinch, is then recorded. The length L of the housing is also measured to the nearest microinch.

The secondary mirror is mounted individually on a vacuum chuck by using a central dowel for centering, and the optical surface is diamond turned. Its flange is then machined in the same setup so that its dimension D (see Figure 10.39) equals $L - V - C$. All mirrors are machined in the same manner so that they are automatically positioned correctly upon installation.

Because all critical dimensions have been machined to close tolerances, no adjustments are needed during assembly. The fabrication process has determined, in each system, the optical alignment. This telescope could be completely assembled within 30 minutes.

Another telescope featuring similar diamond-turned features for facilitating alignment is shown in Figure 10.43 schematically. All components were made of 6061 aluminum for thermal stability. All surfaces marked SPDT were diamond turned as described in the following. The 203.2 mm diameter primary mirror had an integral mounting flange isolated from the optical surface by a necked-down flexure region similar in principle to that of Figure 10.36. The primary and secondary mirrors had integral spherical reference surfaces diamond-turned concentric with the nominal telescope focal point as indicated in Figure 10.43.

After conventional machining to near net shape and size, the secondary mirror was mounted on an SPDT instrument for diamond turning the back (nonoptical) surface. The substrate was then turned over and attached to a vacuum chuck for diamond turning the outside diameter (OD) and ID

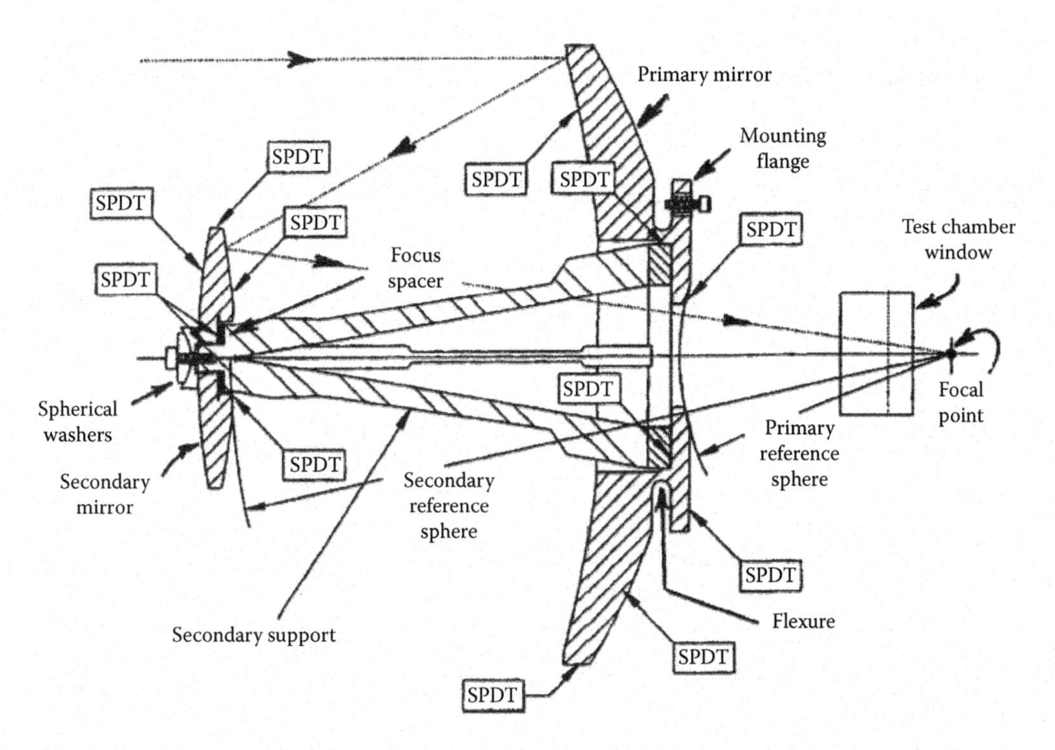

FIGURE 10.43 Schematic diagram of an all-aluminum telescope with optical surfaces, mechanical interfaces, and alignment reference surfaces SPDT machined for ease of assembly. (Adapted from Erickson, D.J. et al., Optimization of the optomechanical interface employing diamond machining in a concurrent engineering environment, *Proc. SPIE*, CR43, 329, 1992.)

of the mirror, the convex aspheric optical surface, the concave spherical reference surface, and the interface for the focus spacer.

The following surfaces on the primary mirror were diamond turned in a single machine setup: the flat mounting flange surface, the concave spherical reference surface, the convex spherical mirror back surface, and the OD and ID of the mirror. It was then turned over and mounted by its flange to the SPDT faceplate. The substrate was centered to the spindle axis by minimizing runout of the precision OD. The concave aspheric optical surface and the axial interface for the secondary support were then turned, and the ID of the mirror was matched to the conventionally machined OD of the secondary support. Without removing the primary mirror from the spindle, the secondary support was attached with screws (not shown in Figure 10.43) and the OD and axial interface for the secondary mirror were turned. This ensured accurate alignment of the mirror axes.

After removal of the primary/secondary support subassembly from the spindle, the focus spacer was ground to thickness and parallelism and the secondary mirror was installed. When the axial separation of the optical surfaces was correct, fringes could be observed between an auxiliary reference surface concentric with the focal point and both diamond-turned reference surfaces on the mirrors. The authors indicated that no subsequent alignment was needed to achieve $<\lambda/4$ reflected wave front accuracy at $\lambda = 0.633$ μm from production telescopes.

Elimination of optical alignment and adjustment during assembly by incorporation of mounting features by diamond turning is now called "snap-together assembly." A snap-together assembly is suited for high-volume production of optics, particularly IR systems with looser tolerance requirements than in the visible. A snap-together assembly often makes use of precision dowel pins to provide the locational interface, with the pins assembled into diamond-turned holes in the optics. One limitation on the accuracy of a snap-together assembly is distortion of the optics relative to the mounting features by angular acceleration ("centrifugal force") during diamond turning. Postpolishing of the optical surfaces, to improve surface roughness and optical figure after diamond turning, may shift the location of the optical axis relative to the alignment features. These errors can be compensated by using test fixtures with mounting features identical to those used in assembling the system.

Off-axis multimirror optical systems have a well-earned reputation for difficulties in fabrication, particularly during alignment. As discussed earlier, diamond turning can be used to fabricate the optical surfaces, although this does not resolve the alignment problem. One approach is to diamond-turn reference surfaces outside the clear aperture during the same setup as for the optical surfaces. For example, there was an annular flat reference mirror outside the clear aperture of the axicon optic shown in Figure 10.29. During alignment, this flat was used in autocollimation to measure the tilt of the optic relative to the optical axis.

PROBLEMS

1. The surface roughness of a diamond-turned mirror is 120 Å RMS. Is the mirror diffraction limited at a wavelength of 1.54 μm? *Answer:* Since $1 - S_R = 0.995 > 0.8$, the mirror is diffraction limited at 1.54 μm.
2. What is the difference in maximum deflection between a square and circular mirror of the same size ($d = L$) if both are edge mounted and subjected to the same pressure differential on a vacuum chuck? Assume that materials and thickness are the same. *Answer:* At constant thickness and material, a circular mirror with a diameter that is the same as the length of a square mirror deflects 0.66 as much.
3. The diameter of a circular aluminum mirror is 150 mm, and the thickness is 20 mm. A vacuum chuck with backup support is used to hold the mirror for diamond turning. There are seven circular holes in the chuck, with a full vacuum in each hole ($\Delta P = 1$ atm = 101.3 kPa), and friction coefficient $\mu = 0.15$ between chuck and mirror. The SF

for the chuck is SF = 4. If the density of aluminum is 2710 kg/m^3, what is the diameter of each hole in the chuck? *Answer:* The diameter of each hole in the vacuum chuck is about 21.3 mm.
4. At constant thickness, what is the ratio of thermal equilibrium times for aluminum (6061-T6) and Pyrex (Corning 7740) mirrors? *Answer:* It takes about 143 times longer for a Pyrex mirror to reach thermal equilibrium than one made of aluminum.
5. The primary mirror of the Russian 6 m BTA telescope is made of a borosilicate glass similar to that of Corning Pyrex 7740. There are problems with the long thermal equilibrium time of the primary, which is about 72 hours. One idea for solving the thermal equilibrium time was an aluminum primary. The mirror is about 600 mm thick. What are the deflection and thermal equilibrium time ratios for mirrors made of 6061-T6 aluminum and Pyrex 7740, assuming that the mirror thickness is the same? Neglect shear effects and use specific stiffness data from Table 8.3. *Answer:* The deflection of the aluminum replacement mirror is about 1.04 times worse than that of the glass mirror, and its thermal equilibrium time is 30.24 minutes, or a ratio of 143 times shorter than that of the Pyrex mirror.
6. Compare the focal stability of mirrors made of aluminum and beryllium in a cryogenic optical system operating at a temperature of 100 K when subjected to the same linear axial temperature gradient. Assume the same thickness for the mirrors. *Answer:* The focal shift of the beryllium mirror at 100 K is about 11% of that of the aluminum mirror.
7. At a temperature of 293 K, what is the difference in CTE between ELN with 11.5% phosphorus content and beryllium? *Answer:* The difference in CTE $\Delta\alpha = 1.75 \times 10^{-6}\ K^{-1}$.
8. The spindle speed of a diamond-turning machine is 250 rpm, and the feed rate is 7 mm/minute. If the tool radius is 6 mm, what is the RMS roughness of the diamond-turning surface? Is the optical surface diffraction limited at a wavelength of 1.06 μm if this surface roughness is increased by a factor of 10? *Answer:* The RMS surface roughness is 23.6 Å, and the Strehl ratio is 0.962; the system is diffraction limited.

REFERENCES

1. Paquin, R.A., and Goggin, W., Beryllium Mirror Technology: State-of-the-Art Report, Perkin-Elmer Report IS 11693, Norwalk, CT, 1972.
2. Paquin, R.A., Metal mirrors, Chapter 4 in *Handbook of Optomechanical Engineering,* pp. 69–110, CRC Press, Boca Raton, FL, 1997.
3. Paquin, R.A., Properties of metals, Chapter 35 in *Handbook of Optics*, 2nd ed., Vol. II, Bass, M., Van Stryland, E.W., Williams, D.R., and Wolfe, W.L., Eds., Optical Society of America, Washington, DC, 1994.
4. Harned, N. et al., Alignment and evaluation of the cryogenic corrected Infrared Astronomical Satellite (IRAS) telescope, *Opt. Eng.*, 20, 195, 1981.
5. Parsonage, T., JWST beryllium telescope: Material and substrate fabrication, *Proc. SPIE*, 5494, 39, 2004.
6. Wells, C., Whitman, T., Hannon, J., and Jensen, A., Assembly integration and ambient testing of the James Webb Space Telescope primary mirror, *Proc. SPIE*, 5484, 859, 2004.
7. Ealy, M.A. et al., CERAFORM SiC: Roadmap to 2 meters and 2 kg/m^2 areal density, *Proc. SPIE*, CR67, 53, 1997.
8. Vukobratovich, D. et al., Therm-optic analysis of bi-metallic mirrors, *Proc. SPIE*, 3162, 12, 1997.
9. Whitehead, T.N., *The Design and Use of Instruments and Accurate Mechanisms, Underlying Principles*, Dover, New York, 1954.
10. Parks, R.E., Introduction to diamond-turning, SPIE short course, 1982.
11. Bennett, J.M., and Decker, D.L., Surface characterization of diamond-turned metal optics, *Proc. SPIE*, 288, 534, 1981.
12. Zimmerman, J., Strain-free mounting techniques for metal mirrors, *Opt. Eng.*, 20, 187, 1981.
13. Ohl, R. et al., Design and fabrication of diamond machined aspheric mirrors for ground-based, near-IR astronomy, *Proc. SPIE*, 4822, 57, 2002.

FURTHER READING

Addis, E.C., Value engineering additives in optical sighting devices, *Proc. SPIE*, 389, 36, 1983.

ASME (American Society of Mechanical Engineers), *Heat Treatment of Wrought Aluminum Alloy Parts*, AMS2770J, ASME, New York, 2011.

Erickson, D.J., Johnston, R.A., and Hull, A.B., Optimization of the optomechanical interface employing diamond machining in a concurrent engineering environment, *Proc. SPIE*, CR43, 329, 1992.

Gerchman, M., Specifications and manufacturing considerations of diamond-machined optical components, *Proc. SPIE*, 607, 36, 1986.

Janeczko, D. Metal mirror review, *Proc. SPIE CR38*, 258, 1991.

Klein, C.A., Mirror figure-of-merit and material index-of-goodness for high power laser beam reflectors, *Proc. SPIE*, 288, 69, 1981.

Morrison, D., Design and manufacturing considerations for the integration of mounting and alignment surfaces with diamond turned optics, *Proc. SPIE*, 966, 219, 1988.

Sanger, G.M., The precision machining of optics, Chapter 6 in *Applied Optics and Optical Engineering*, Vol. 10, Shannon, R.R., and Wyant, J.C., Eds., pp. 251–390, Academic Press, San Diego, CA, 1987.

Vukobratovich, D., and Schaefer, J.P., Large stable aluminum optics for aerospace applications, *Proc. SPIE*, 8125, 81250T, 2011.

Yoder, P.R. Jr., and Vukobratovich, D., *Opto-mechanical Systems Design*, 4th ed., CRC Press, Boca Raton, FL, 2015.

Zimmerman, J., Strain-free mounting techniques for metal mirrors, *Opt. Eng.*, 20, 187, 1981.

11 Kinematic Design and Applications of Flexures

Contributed by Jan Nijenhuis

11.1 INTRODUCTION AND SUMMARY

Optical components (such as mirrors, lenses, and prisms), mounts, brackets, and housings are a few examples of bodies that make up an optical system or instrument. Each of these bodies must be mounted to or within the structure of the instrument. Common mounting methods include bolts, clamps, retaining rings, and adhesives. Usually, these are sufficient to guarantee that the various bodies stay in place when subjected to operational and environmental loads. However, for optical systems, these often are not adequate because of stringent requirements for positional accuracy, stability, and freedom from mechanical strain. Then, it becomes evident that much skill and artisanship are needed to realize design objectives. Ways to do so are described throughout this book. This chapter concentrates on the basic principles for mounting various bodies. Each is kinematically mounted, and all of their six degrees of freedom (DOFs) are independently constrained *once* or intentionally left free for adjustment purposes. These bodies are assumed to be rigid and infinitely stiff. Overconstraints usually result in loads on the components that will tend to deform them, thereby degrading the performance of the instrument.

Each rigid body has three DOFs concerning translations and three DOFs concerning rotations. These are usually defined along the axes of a Cartesian coordinate system because any translation or rotation can be defined as a linear combination of these motions. Constraining these six DOFs means that the body cannot move—at least not kinematically. Applying external loads may then introduce mechanical deflections and dimensional changes caused by the limited stiffnesses of the mechanical supports. This becomes important when the actual stiffness of a support has to be calculated and evaluated in terms of the specification requirements. The latter process is illustrated by worked-out examples.

This chapter first discusses, in a systematic way, various ways of constraining the DOFs of a simple rigid body, starting by constraining one DOF and adding additional constraints until the body cannot move. Typical structural elements called struts and various forms of leaf springs are introduced. Each of these is capable of restraining one or more DOFs. Collectively, these elements are commonly known as *flexures*.

The invaluable benefit of thinking in terms of controlling DOFs is that it allows the user to lay out support structures (and mechanisms) without having to do any calculations. Just by reasoning, support structures can be defined. Having done that, the structural elements are dimensioned by taking into account the body loads, the performance requirements, and environmental inputs. A satisfactory solution is not always found the first time. Iterations may be necessary. With experience, the user will quickly arrive at the best solution, i.e., one using minimum material to achieve maximum performance.

The hardware described or referenced in this chapter ranges from simple optomechanical subassemblies to high-level systems. Application of the rules given here for properly constraining each DOF applies at all levels of complexity. Many examples of simple flexure mountings are given here. Additional examples are included elsewhere in this book.

11.2 CONTROLLING DEGREES OF FREEDOM

11.2.1 Beauty of Static Determined Designs

When each of the six DOFs of a rigid body is constrained, that body is said to be static determined. Using six equations for force and moment equilibrium, one can calculate the reaction forces and provide the required strength/stiffness in the rigid body support. Adding more constraints to the body will make it overdetermined. A question that is often asked about static determined structures is why they are so important. The argument that is usually given is that classical overdetermined structures work perfectly, so why make things difficult? At first glance, this is true and one should always use existing designs or standard products when they meet the requirements. However, a problem arises when new designs have to work better than old ones. Micrometer or milliradian accuracies are often not good enough. In such cases, thinking in terms of DOFs provides an advantage by bringing logic to the design process. At the same time, we need to be aware that thinking in DOFs has proven to be more difficult than most people expect. The reward for doing so is that the predictability and understanding of structures under mechanical and thermal loads become much more comprehensive. Also, the absolute accuracies of, for example, positioning devices, are improved.

There is a strong tendency today for proposed mechanical structure designs to be analyzed prematurely using finite-element method analysis. The result may indicate no problem—even in the case of an overdetermined case. However, that result is only one of many that could be obtained about the design. It provides very limited knowledge. The analyst needs to make many more calculations to uncover the interrelated dependencies of all the design parameters. Usually, this ends in a massive number of data—which are often difficult to handle. Paying attention to the DOFs of the design and knowing in advance how structural elements will work make the design process more direct. Simple examples of progressively more complex applications of constraints are described here to illustrate these points.

11.2.2 Controlling One DOF

Figure 11.1 shows a rectangular rigid body that is constrained in one direction by a single *strut*. Design equations for struts are given in Section 11.3.1. At the moment, it is sufficient to know that this could be an elongated part (perhaps a rod), with small lateral dimensions (or diameter) compared to its length. Such a part has high stiffnesses in tension and compression compared to its bending or torsional stiffnesses. Therefore, the latter can be neglected.

Another way of constraining one DOF uses *leaf springs*. A leaf spring is the same as a thin plate. It is capable of resisting loads that lie in-plane. Out-of-plane loads result either in bending or torsional deflections. Hence, a leaf spring constrains three DOFs, i.e., two in-plane translations and one in-plane rotation. In Figure 11.2a, two leaf springs are oriented perpendicular* to each other and attached along one edge. This combination is called a *folded leaf spring*. By connecting two leaf springs to each other so as to form a dihedral edge, only one DOF (translation in the Y direction) is constrained. In addition, the only way that loads can be transferred from the loaded body to the fixed world (here referred to as *ground*) is along the dihedral edge. Figure 11.2b illustrates an equivalent configuration using a very rigid single strut instead of the folded leaf spring.

Another way of constraining one DOF while connecting the body to the ground is by using four struts as shown in Figure 11.3. Two struts lie on the Y–Z plane and two on a plane parallel to the X–Y plane. All four meet at the common point A. Again, the DOF in the Y direction is constrained. The practical impact of the different ways of constraining one DOF by using elements such as struts, leaf flexures, or folded leaf flexures will be explained further in Section 11.3.

* The two leaf springs are oriented perpendicularly as a rule. However, this is not necessary. For practical reasons (e.g., limited space), it may be advantageous to deviate from this rule. This is discussed further in Section 11.2.5.

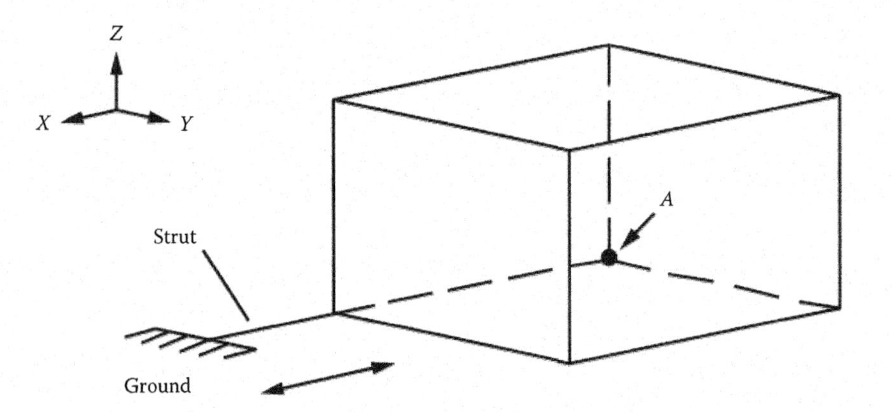

FIGURE 11.1 One DOF (translation of a rigid body in the X direction) is constrained by a single strut to the ground. The double-headed arrow indicates this constraint.

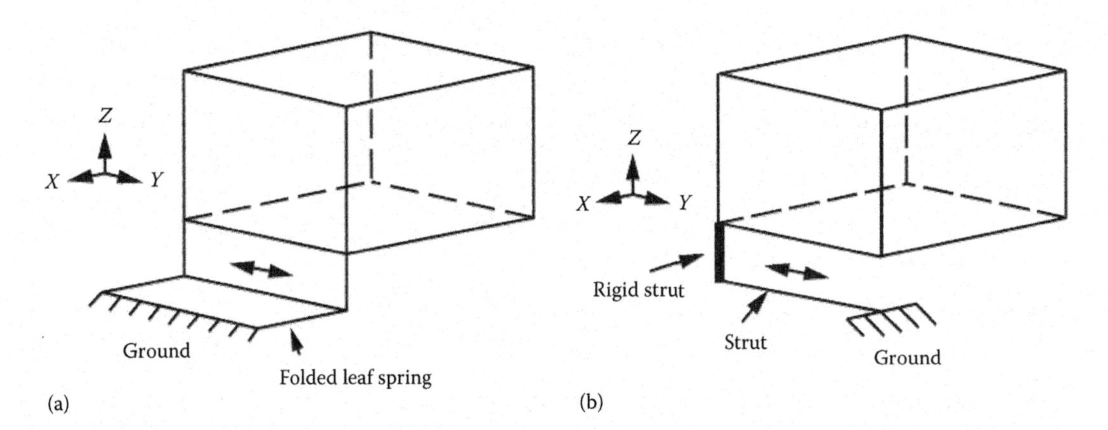

FIGURE 11.2 (a) Translation of a rigid body in the Y direction is constrained by a folded leaf spring to the ground. (b) Equivalent one DOF constraint of translation along the Z-axis using a single rigid strut.

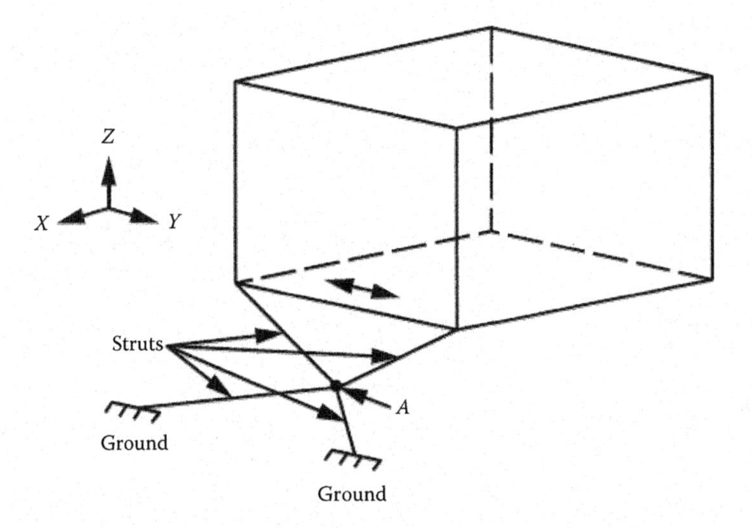

FIGURE 11.3 Configuration using four intersecting struts lying on perpendicular planes to constrain one DOF of a rigid body in the Y direction.

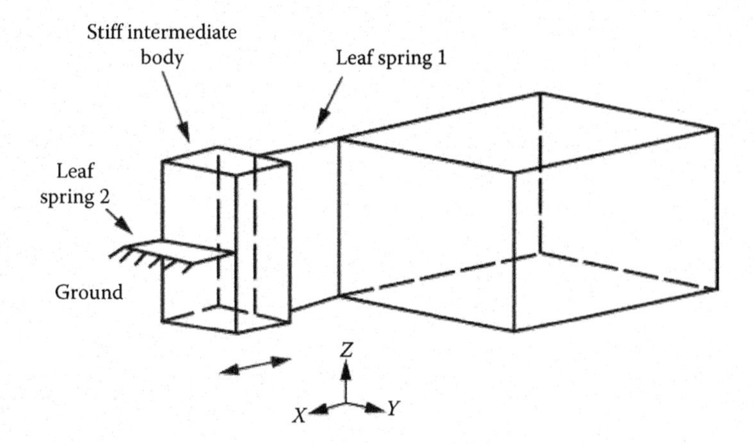

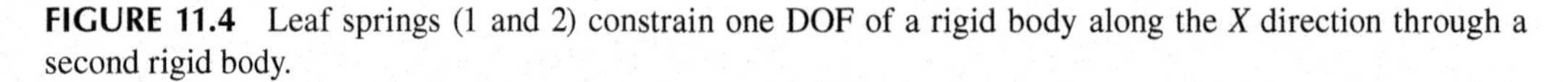

FIGURE 11.4 Leaf springs (1 and 2) constrain one DOF of a rigid body along the X direction through a second rigid body.

Figure 11.4 shows two leaf springs connecting a rigid body to the ground through an intermediate infinitely stiff body. In this case, only the single DOF in the X direction is constrained because the out-of-plane load-carrying capabilities of the leaf springs are negligible compared to their in-plane load capacities. In Section 11.3.3, this is quantified.

11.2.3 Controlling Two DOFs

Now we address how to constrain two DOFs. As one can expect, this is done by connecting more struts or leaf springs to the rigid body. This may seem rather trivial, but it is very important that

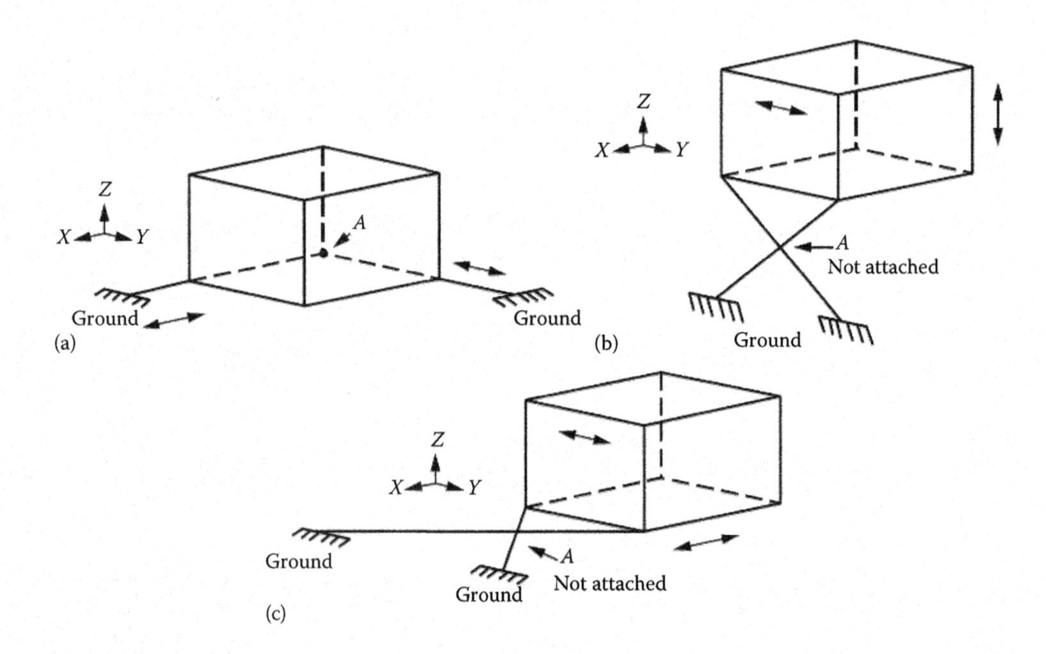

FIGURE 11.5 (a) This rigid body is constrained by two struts intersecting at point A inside the body. The body cannot translate along the X- or Y-axis, but can translate along the Z-axis and rotate about any of its axes. In (b) and (c), the body is similarly constrained in two DOFs, but the intersection points A are outside the body.

one understands the implications of adding constraints, step by step, until all the DOFs of a rigid body are constrained. In Figure 11.5a, a rigid body is constrained in the *X* and *Y* directions by the use of two perpendicular struts. These two struts intersect at point *A* inside the body. That point can translate only in the *Z* direction. All three rotations about the intersecting edges at *A* remain free. By attaching the struts as shown in Figure 11.5b, a rotation point *A* outside the rigid body and on the *Y*–*Z* plane is created. By attaching them as shown in Figure 11.5c, point *A* is moved to the *X*–*Y* plane. Note that the two struts are not physically connected at point *A*.

The two struts can also be parallel as shown in Figure 11.6. They then define a plane in which no translation in the *X* direction and no rotation around the *Y*-axis are possible. Hence, two DOFs are constrained.

If one of the two struts shown in the last figure is moved so that they do not intersect and are not parallel (i.e., they are skewed), the situation shown in Figure 11.7 results. In this case, no specific point can be identified as a rotation center. Translations along the *X*- and *Y*-axes are constrained. The other four DOFs are free.

Instead of using struts, the DOFs of a rigid body can also be constrained by the use of a constricted leaf spring. See Figure 11.8. Two holes next to each other and two collinear cuts are made in the thin plate. Between the two remaining plate sections, a small link remains at point *A*. Loads in the *X* and the *Y* directions can be transferred at *A*. The main reason for introducing this method of connection is that it is *simple* and inexpensive to make. Also, the plate section can easily be attached

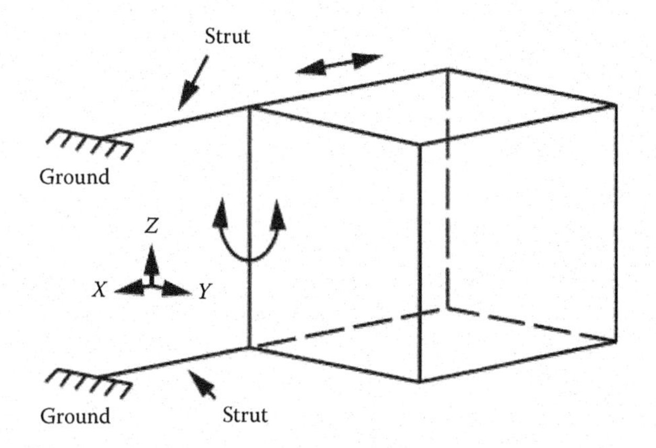

FIGURE 11.6 Two parallel and vertically separated struts constrain two DOFs of a rigid body, i.e., against translation along the *X*-axis and rotation about the *Y*-axis.

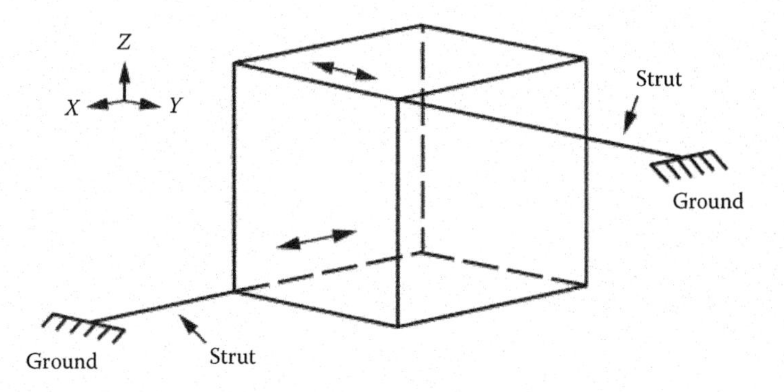

FIGURE 11.7 Two vertically separated and skewed (i.e., nonparallel and nonintersecting) struts constrain two DOFs of a rigid body (translations along the *X* and *Y* directions).

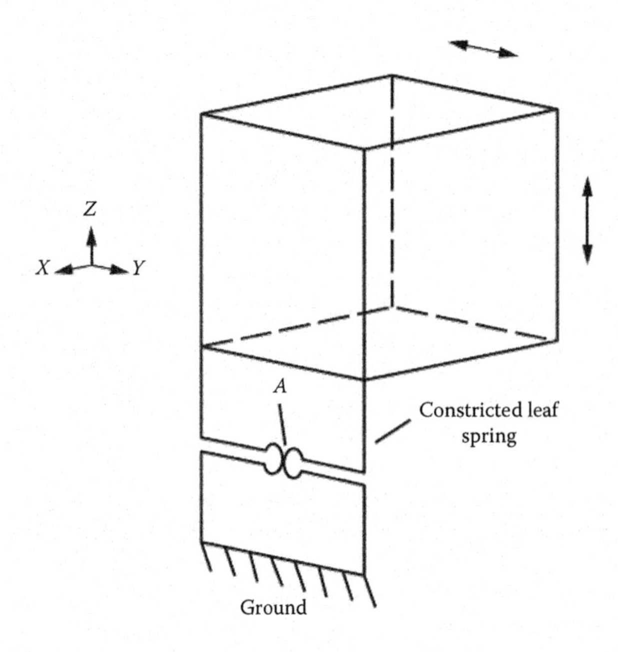

FIGURE 11.8 One constricted leaf spring is used to constrain a rigid body against translations in the *Y* and *Z* directions.

to the rigid body and to the ground because the connections have considerable lengths. The location of point *A* can be adjusted according to the needs of the designer.

11.2.4 Controlling Three DOFs

When three DOFs are constrained, three DOFs are still free. Assuming the constrained DOFs are to be translations, the configuration might be as in Figure 11.9a. Three struts are connected to the body at point *A*. The three rotations at point *A* are not constrained, so small rotations limited by the stress properties of the struts are possible. Rotation larger than a few degrees will cause small parasitic translations due to the effective shortening of the struts.

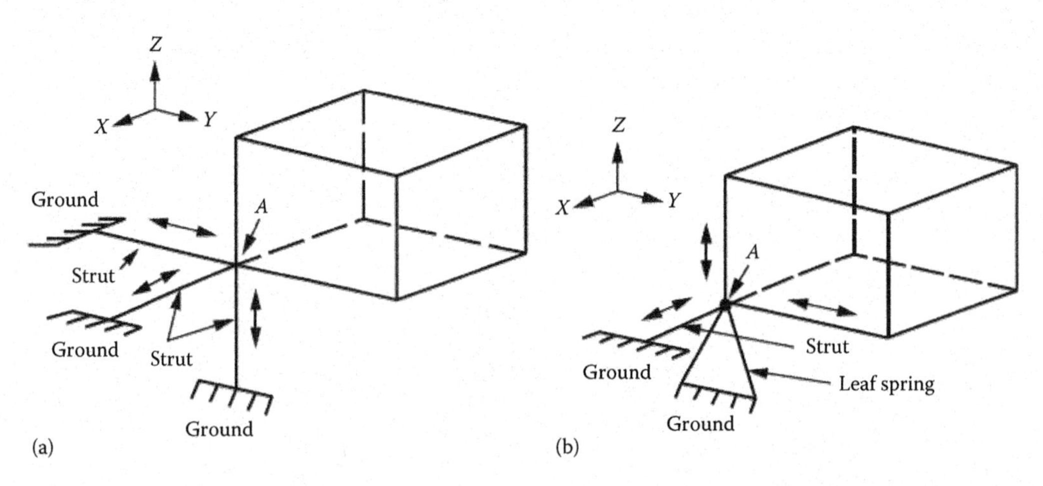

FIGURE 11.9 (a) Three mutually perpendicular struts intersecting at point *A* constrain a body in all three translations. Three rotations are free. (b) The equivalent configuration with a flexible ball joint or a two-axis elastic hinge on a leaf spring.

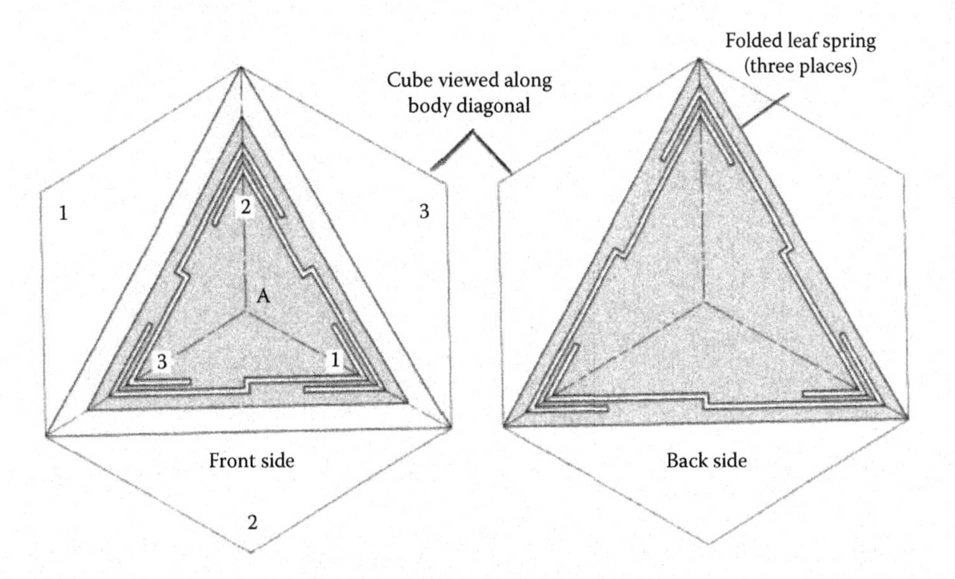

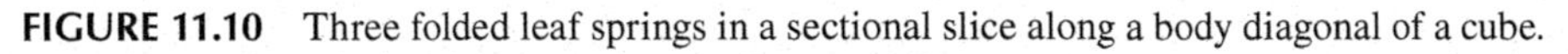
FIGURE 11.10 Three folded leaf springs in a sectional slice along a body diagonal of a cube.

Alternatively, three folded leaf springs can be used to constrain three translations. See Figure 11.10. Such a device can be made by using the wire erosion machining (electrical discharge machining [EDM]) process. A section of constant thickness is taken from a cube. The normal to that section coincides with the body diagonal of the cube. Leaf spring 1 is obtained by EDM done normal to side 1 of the cube, etc. Hence, the two sections of the leaf spring are perpendicular to each other in contrast to what the figure suggests. The hinge point is the corner of the cube. Being a monolithic structure, this hinge will show no friction or hysteresis. Each rotation is limited by the small widths of the gaps provided by the EDM cut.

By orienting three struts in different ways, other options for constraining three DOFs result. For example, if two of three coplanar struts are parallel to each other and the third perpendicular to those struts (see Figure 11.11a), rotation about the *Y*-axis and translations along the *X*- and *Y*-axes are constrained. The same results can be obtained by using a single leaf spring, as shown in the view in Figure 11.11b. From a mechanical point of view, the latter is easier to realize. Practical problems may prevent such a solution.

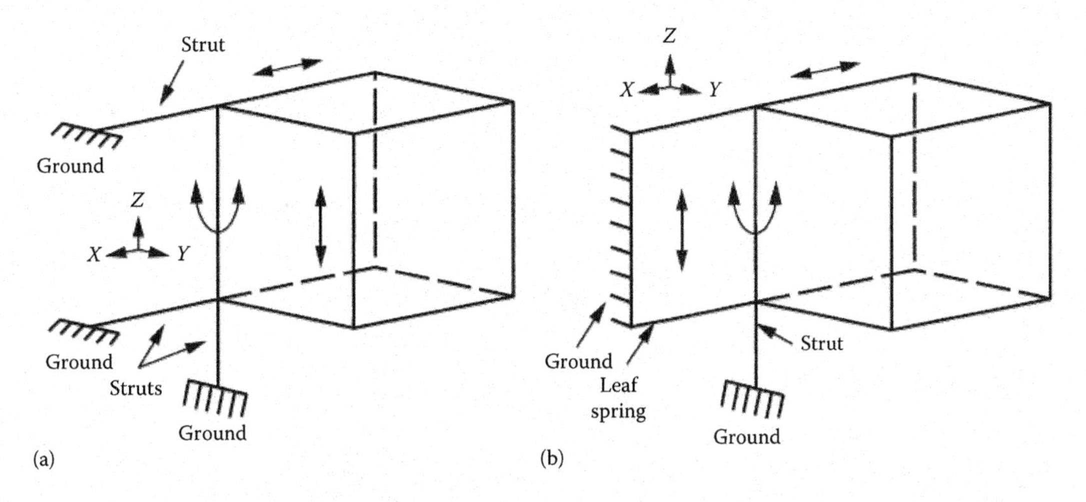

FIGURE 11.11 (a) Three coplanar struts constrain a rigid body in three DOFs. (b) Functionally equivalent configuration using a leaf spring.

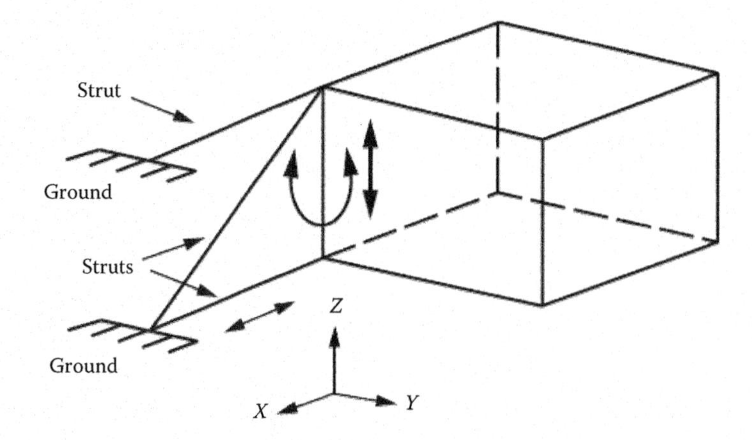

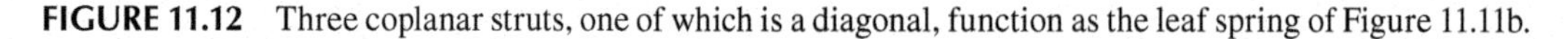

FIGURE 11.12 Three coplanar struts, one of which is a diagonal, function as the leaf spring of Figure 11.11b.

By introducing a diagonal strut between the two parallel struts of Figure 11.10a, a classical truss structure results, as shown in Figure 11.12. Two translations and one rotation are constrained. The functions of the struts are then exactly as in Figure 11.11b. Once again, there may be practical problems that prevent the use of the latter solutions.

In another approach, two of the struts are coplanar, while the third is oriented perpendicular to the first two and moved to another corner of the body. This is shown in Figure 11.13. Translation along the X direction and rotation about the Y-axis are constrained, just as is translation in the Z direction. It is not possible to identify a rotation center for the rigid body because the struts do not intersect.

In the situation shown in Figure 11.14, rotations about the X- and Z-axes as well as translation in the Z direction are possible.

An interesting situation results when all three struts are parallel, but not coplanar. This is shown in Figure 11.15. It is now impossible for the rigid body to translate in the X direction or to rotate about the Y- or Z-axis. The body is free to translate and rotate on a plane perpendicular to the three struts. This might be a very useful property in case that the rigid body is to be aligned.

In Figure 11.16, the three struts in Figure 11.15 are replaced by three folded leaf springs. Moreover, the rigid body has been configured as a cylinder. Such a cylinder can easily be

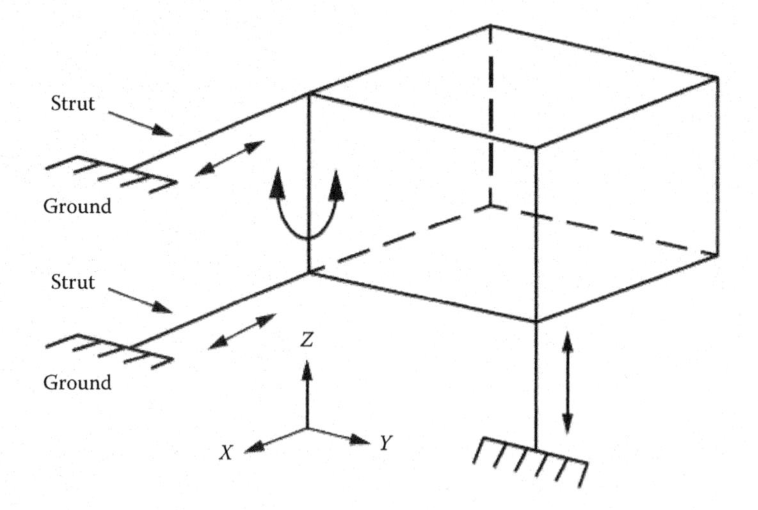

FIGURE 11.13 Three struts constrain the body against translations along the X and Z directions and rotation about the Y-axis.

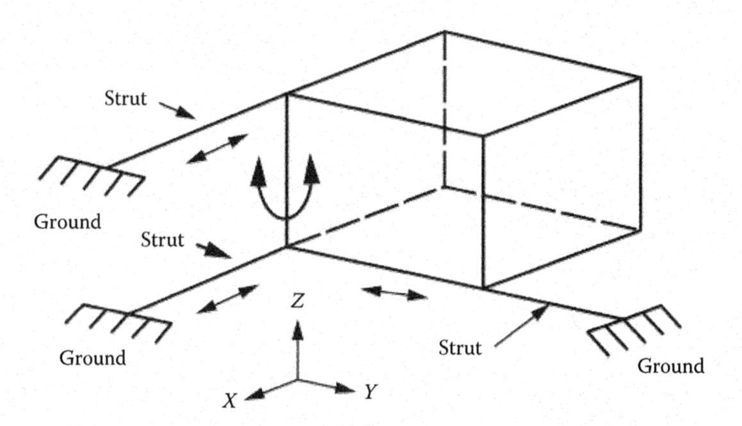

FIGURE 11.14 Three struts plus one strut lying on the $X–Y$ plane prevent translations of a rigid body on the X and Y directions, as well as rotation about the Y-axis, for a total of four DOFs constrained. This might be a very useful property in case that the body needs to be translated in one direction and tilted in two orthogonal directions with respect to that direction.

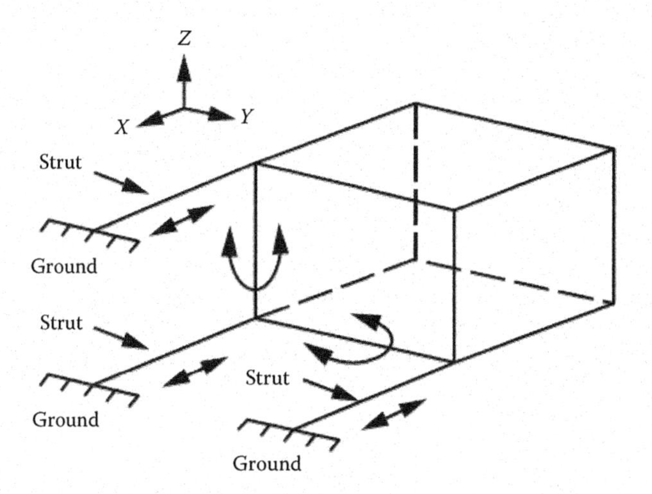

FIGURE 11.15 Three parallel struts, two on the $X–Z$ plane plus one on the $X–Y$ plane and separated in the X direction from the first two, constrain the body against translation along the X-axis and rotations about the X- and Z-axes.

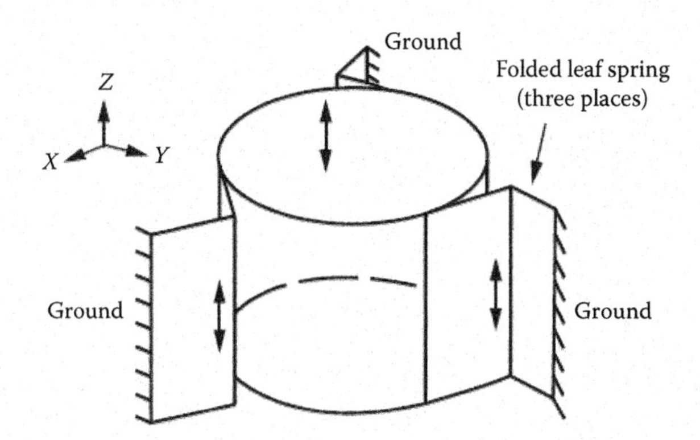

FIGURE 11.16 Three folded leaf springs attached to a cylindrical body at 120° intervals so their dihedral edges are parallel to each other and to the axis of the cylinder constrain the body against translations along all three axes. Limited rotation about the Z-axis is free.

recognized as possibly the barrel that holds a lens. The optical axis of the lens is oriented along the Z-axis. For centering purposes, it is important that lateral displacements are possible without disturbing the axial orientation. The use of folded leaf springs has another important benefit compared to the use of struts. With the latter, lateral movements on the Y–Z plane will cause small displacements in the X direction (this would be focus of the lens) because the struts are deflected laterally. For alignment of an optical component, this might be unacceptable. The solution of Figure 11.16 does not have that drawback. In addition to the desirable lateral motions, it is possible to rotate the optic around the Z-axis. This would be useful when an off-axis mirror is aligned. Because a lens usually is rotationally symmetric, rotations have no effect on its optical performance. For mechanical reasons, however, it is important that the remaining three DOFs are constrained.

It should be noted that the three struts or folding lines should not be coplanar because that means that the system becomes overconstrained in certain DOFs and not three but four DOFs remain to be constrained. The rotation about the axis normal to the folding lines and coplanar with the three folding lines would no longer be constrained if all three struts were to be located on a vertical plane.

By rotating the three leaf springs by 90°, the situation in Figure 11.17 results. Again, we have an interesting situation for mounting a lens or mirror. In-plane motions of the optic are impossible but it is possible to rotate the lens around the X- and the Y-axes. Thus, a tip/tilt motion is possible. Also, a translation in the Z direction is possible.

Again, it is possible to replace the three leaf springs by struts. That results in the configuration shown in Figure 11.18. Let us assume that the three struts have equal length and that they are made of the same material. If the temperature changes, the dimensions of the cylinder and of the struts will change slightly. However, the center of the cylinder will stay in the same place. This center is called the *thermal center* (TC). Only a slight rotation around the Z-axis will result. For a rotationally symmetric lens, this has no adverse consequences. In case that it is a problem, it can be solved by going back to the solution of Figure 11.17 because then no rotation will result. In that case also, the center of the cylinder is the TC. It should be noted that the location of the TC depends upon the forces that the struts or leaf springs exert on the lens. Hence, using nonidentical struts/leaf springs will result in a shift of the TC. In Figures 11.17 and 11.18, the struts are identical and oriented at intervals of 120°. The stiffness in any direction on the X–Y plane will be identical. Reasonable

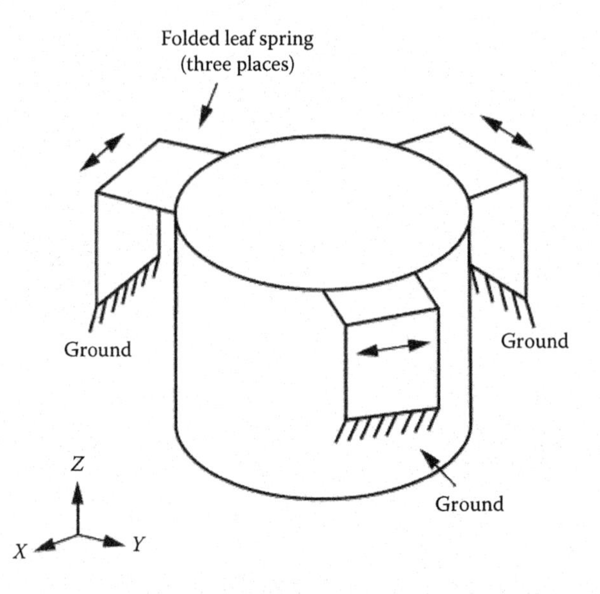

FIGURE 11.17 Three folded leaf springs attached to a cylindrical body at 120° intervals with their dihedral edges coplanar constrain the body against translations along the X- and Y-axes and rotation about the Z-axis.

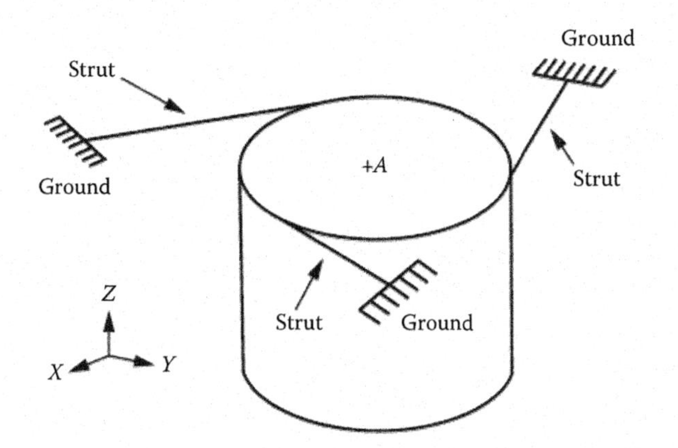

FIGURE 11.18 Three coplanar struts oriented tangentially at 120° intervals about the circular cross section of a cylindrical body constrain two translations (*X* and *Y*) and rotation about the *Z*-axis. Rotations (tilts) about the *X*- and *Y*-axes are free.

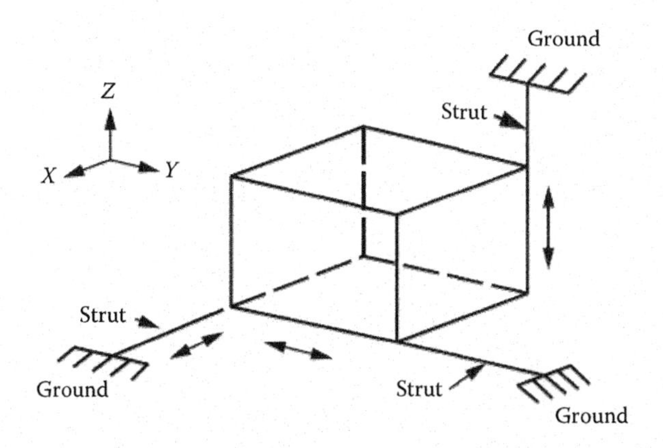

FIGURE 11.19 Three struts, each parallel to a principal axis and attached to different corners of a rigid body so two struts are coplanar and the third perpendicular to the plane of the first two, constrain translations of the body along each axis, but allow rotations about all three axes.

changes in these angles will have no effect on the location of the TC. However, such a change would affect the stiffness distribution of the design on the *X–Y* plane.

Three struts can also be used as shown in Figure 11.19. The struts have different orientations, and they do not intersect. Hence, no translations are possible. The three rotations are still free. The figure shows how these rotations are oriented. Any linear combination of these three rotations is possible.

More options exist for constraining three DOFs. However, the many examples already given should be sufficient for the reader to understand what happens when various alternatives are used.

11.2.5 Controlling Four DOFs

Constraining four DOFs means that only two DOFs remain free. They can be two rotations, two translations, or one translation and one rotation. An example of the first option is shown in Figure 11.20. It is similar to the situation shown in Figure 11.18, but now the translation in the *Z* direction has been blocked. Note that the intersection *A* of the plane containing the coplanar struts and the centerline of the fourth strut cannot move. Note also that the two remaining DOFs are two rotations parallel to

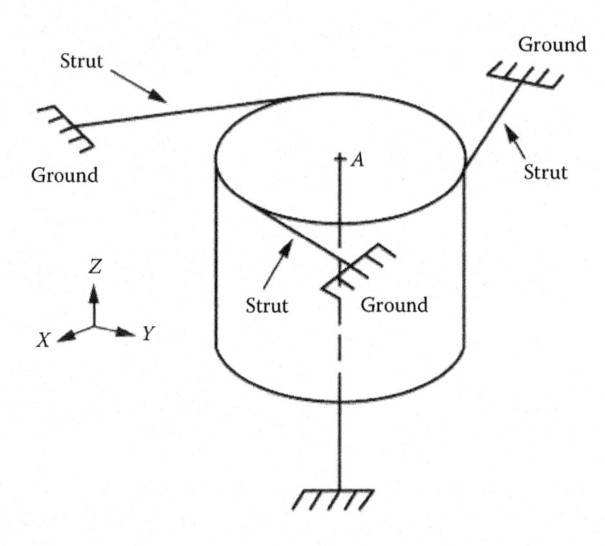

FIGURE 11.20 Three coplanar struts at 120° intervals about a cylindrical body plus one strut parallel to the *Z*-axis and passing through point *A* constrain all three translations plus rotation about the *Z*-axis.

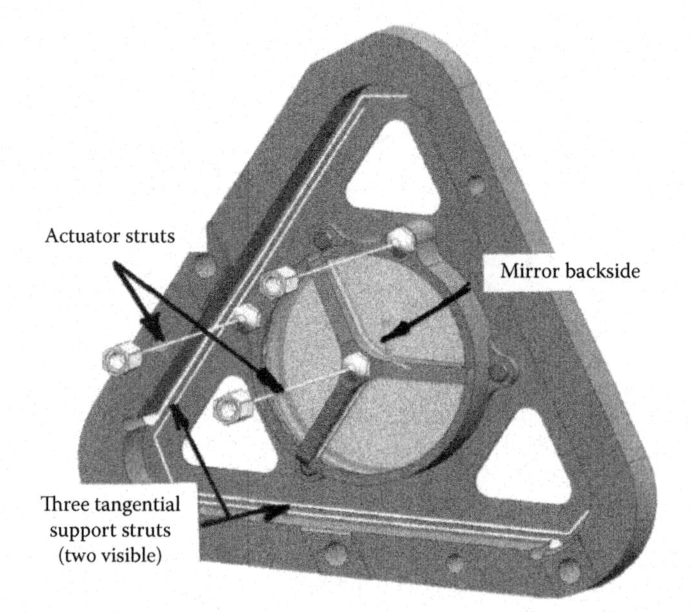

FIGURE 11.21 A tip/tilt mechanism functionally similar to that in Figure 11.20 comprises a mirror in a triangular cell on three lateral support struts (two are visible) inside a triangular ring fixed to the ground. One central support strut fixed to the ground constrains the axial position of the mirror. Two struts parallel to the mirror axis and connected to the rim of the mirror are adjustable in length to tilt the mirror normal through two orthogonal small angles.

the *X*- and *Y*-axes and centered at point *A*. Let us suppose that the cylindrical body has a mirror surface diamond turned coplanar with the three struts. The mirror surface is then mounted on a form of Cardan joint.* For scanning applications, this may be a very favorable configuration because the system pupil can then be located at the mirror and the reflected beam would rotate around the mirror vertex. Figure 11.21 is a representation of an actual mechanism of this type.

* A means for transmitting rotary motion around corners described first by Cardano (circa 1545)[1]; it was a forerunner to the now ubiquitous universal joint.

The four struts could also be arranged as two sets of parallel constraints, as shown in Figure 11.22. The remaining DOFs are one rotation around the *X*-axis and one translation in the *Z* direction. Other options for orienting the struts exist. However, one has to be careful not to overconstrain some of the DOFs. One such case is illustrated in Figure 11.23. The rotation around the *Y*-axis is here twice constrained. There are three DOFs that have not been constrained. These are the translation in the *Y* direction and the rotations around the *X*- and *Z*-axes. Rotating one of the vertical struts into the *X–Y* plane or rotating one of the horizontal struts by 90° around the *Z*-axis would correct this unwanted condition.

Another way to constrain four DOFs uses two constricted leaf springs, as shown in Figure 11.24a. The free DOFs are rotation around the hanging axis connecting the two constrictions and translation in the *X* direction. An alternate is shown in Figure 11.24b. Four struts are used. The two intersections of the struts define the rotation axis. This axis does not move even when the intersections move as result of bending of the struts. There is no need for the pair of struts to be coplanar. A small offset parallel to the direction of the rotation axis has no impact on the performance, and it avoids any physical intersection of the struts.

Figure 11.25 shows that the two pairs of struts can also have virtual intersection points *A* and *B*. Often this is very practical because the physical location of the rotation axis may interfere with other structural members. The rotation stiffness of this configuration is considerably greater than that of Figure 11.24a or b. One might expect that mechanisms exist with two free translations (two DOFs). However, this is not possible with the use of only four struts.

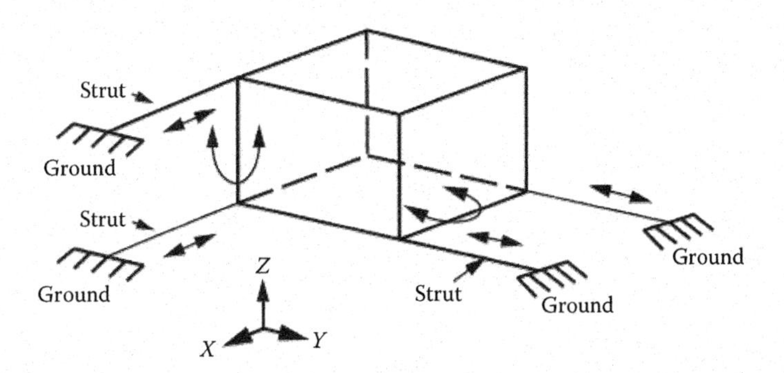

FIGURE 11.22 One set of parallel struts lying on the *X–Z* plane and attached to the ground and a second parallel set lying on the *X–Y* plane constrain a rigid body against two translations and two rotations. Two DOFs (rotation about the *X*-axis and translation along the *Z*-axis) are free.

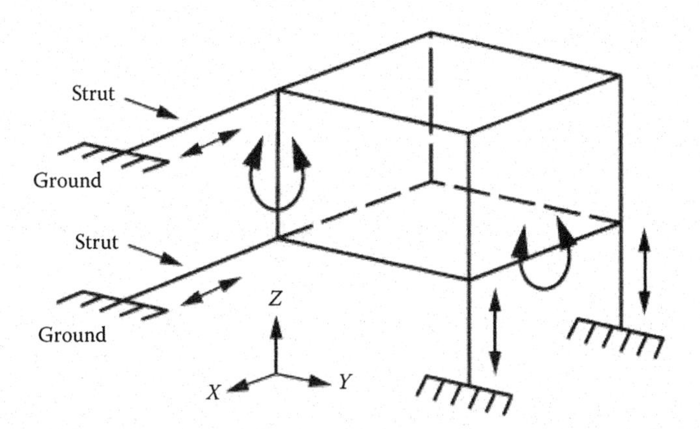

FIGURE 11.23 Two sets of parallel struts constrain a rigid body against *X* and *Z* translations plus *X* and *Y* rotations. Rotation about the *Y*-axis is overconstrained.

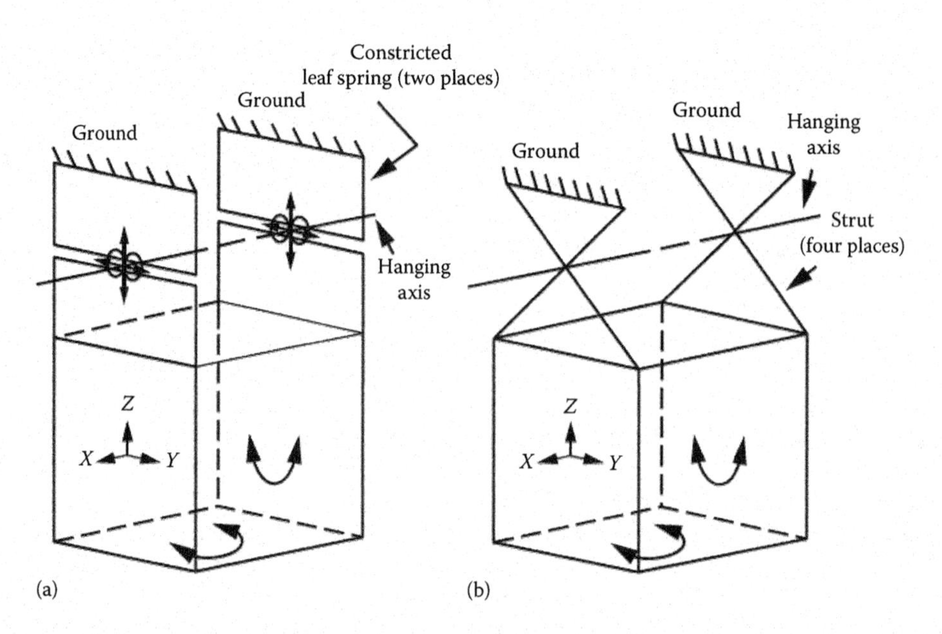

FIGURE 11.24 (a) Two constricted leaf springs constrain four DOFs. These are translations along the *Y*- and *Z*-axes and rotations about the same axes. Rotation about the hanging axis connecting the hinge centers and translation along the *X*-axis are free. (b) Alternate configuration using four diagonally oriented struts (not connected at their crossover points) to constrain the same DOF.

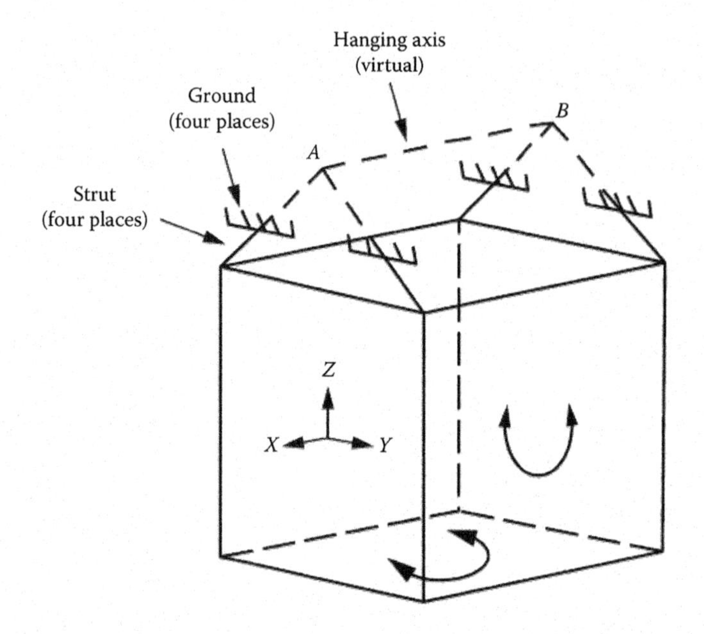

FIGURE 11.25 Two sets of struts with virtual intersections (*A* and *B*) constrain a rigid body against translations along the *Y*- and *Z*-axes and rotations about the *Y*- and *X*-axes.

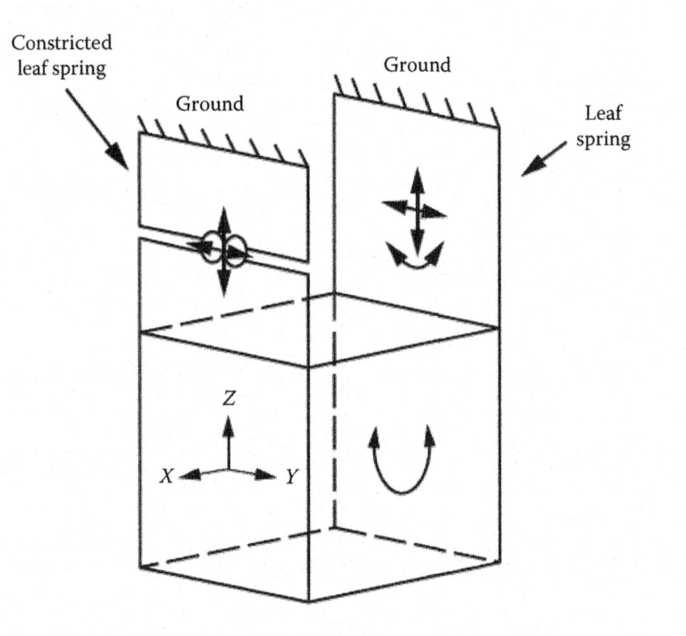

FIGURE 11.26 Two parallel leaf springs (one constricted) constrain a rigid body so as to allow only translation parallel to the *X*-axis.

11.2.6 Controlling Five DOFs

One step further is to restrain all DOFs except for one. This means that the remaining DOF will be either a translation or a rotation. An example of the former case is shown in Figure 11.26. A leaf spring and a constricted leaf spring are used. Only translation along the *X* direction is free. In practice, two normal parallel leaf springs are often used to obtain a linear guiding system based on elastic deformation. The benefits are simplicity and better dynamic performance. The latter attribute is the result of the equal stiffnesses of both leaf springs.

With two supposedly identical leaf springs, the system becomes overconstrained. Whether this is a problem or not depends on a number of aspects. If the parallel guiding system is made from a single piece of material (i.e., it is monolithic) by the use of, e.g., cutting and wire erosion, this is considered to be no problem. The manufacturing accuracies can be so high that the consequences of being overdetermined become negligible. However, when two separate leaf springs are used, problems may arise. The two leaf springs may not be perfectly parallel (because of mechanical tolerances) causing them to "fight each other," and the rigid body does not exactly follow the movement dictated by the two leaf springs. Both leaf springs are then subject to in-plane bending. As a result, the rigid body will experience a small rotation around the *X*-axis during the *X*-axis translation. Whether this is a problem depends on the application and the required accuracies.

Another issue might be that bolting separate leaf springs to the rigid body might be susceptible to microslip.* Slip inevitably means hysteresis and loss of accuracy. Mechanical solutions to reduce these effects have been developed, but are outside the scope of this chapter.

The configuration shown in Figure 11.27 uses a leaf spring and a constricted leaf spring to constrain five DOFs. The remaining DOF is a rotation around the intersection of the two neutral planes

* In the United States, the equivalent term for *microslip* is *stick-slip*.

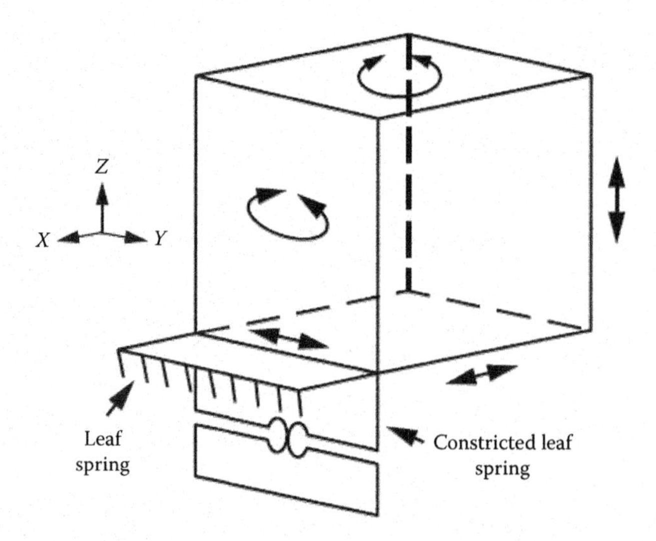

FIGURE 11.27 Two perpendicular leaf springs (one constricted) constrain a rigid body so as to allow only rotation about the *Y*-axis.

of the leaf springs. Using two leaf springs will make the design overconstrained, although the consequences will be marginal if the manufacturing tolerances are minimized.

Figure 11.28 shows two configurations with the body supported on five struts. In the view in Figure 11.28a, five struts constrain five DOFs: three translations and two rotations. Rotation about the *Y*-axis is free. The view in Figure 11.28b also has five struts. It also constrains five DOFs. All rotations are constrained. Only the translation along the *Y*-axis is free.

A particularly important variation of the one-free-DOF theme is shown in Figure 11.29. One edge of the body is supported by a leaf spring. This constrains translations in the *Z* and *Y* directions and rotation about the *X*-axis. Translation along the *X*-axis is constrained by two struts attached to the ends of the front lower edge of the body. These struts also prevent rotation about the *Z*-axis. Rotation about an axis parallel to the *Y*-axis is free. Figure 11.30 shows a type of commercially available device commonly referred to as a *flexural pivot bearing* or, a *flex pivot*, which is widely used as a precision hinge in a great variety of optomechanical instrumentation applications. This type

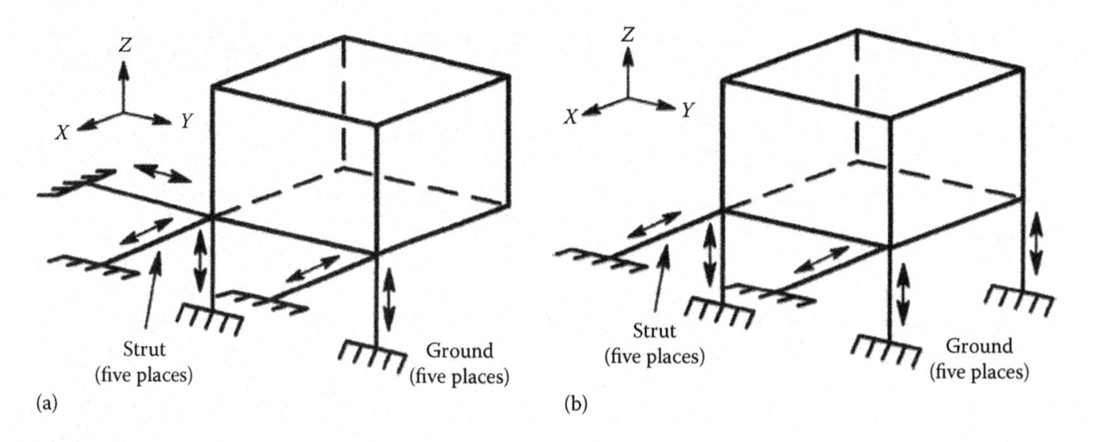

FIGURE 11.28 Arrangements of five struts that constrain five DOFs. (a) Rotation about *Y* is free. (b) Translation along *Y* is free.

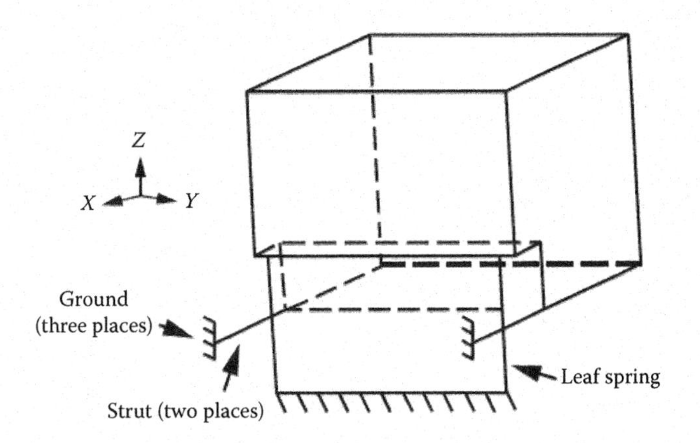

FIGURE 11.29 Two struts, both parallel to the *X*-axis, and a leaf spring parallel to the *Y–Z* plane constrain a rigid body to rotate only about the *Y*-axis.

FIGURE 11.30 Commercially available, single-ended (cantilevered) cross-spring pivot (left) that functions as in Figure 11.29. On right is a double-ended version of this device. Both types are available in a variety of sizes and load-carrying capacities. (Courtesy of Riverhawk Company, New Hartford, NY.)

of device is available in a single-ended version (left view) or double-ended version (right view). Construction typically uses corrosion-resistant steel parts brazed or welded together. Monolithic units can also be made by combining milling and wire erosion operations. This tends to be a very expensive method. Both types come in a wide range of sizes and load-carrying capabilities.

An important issue with both types of flex pivots shown in Figure 11.30 is that they are actually overdetermined. Replacing the two struts by two parallel leaf springs does this. Each leaf spring constrains three DOFs; hence, the configuration is four times overdetermined ($(3 \times 3) - 5 = 4$). Whether this is a problem depends on the manufacturing tolerances. Without doubt, it is a very practical device. However, there is a tendency for designers to use two cross spring pivots to realize a rotation axis, just as one would do using ordinary ball bearings. This is not conducive to optimal performance because of the resulting problem of aligning the pivots.

When the best possible performance is needed, one should consider using the configuration shown in Figure 11.31. The common line *AB* is the rotation axis. This is a variation of the *Haberland cartwheel hinge.* Its performance as a hinge is more accurate than the cross spring pivot, although the maximum obtainable rotation is reduced. It can best be made by wire erosion from a solid block, thereby creating a monolithic hinge.

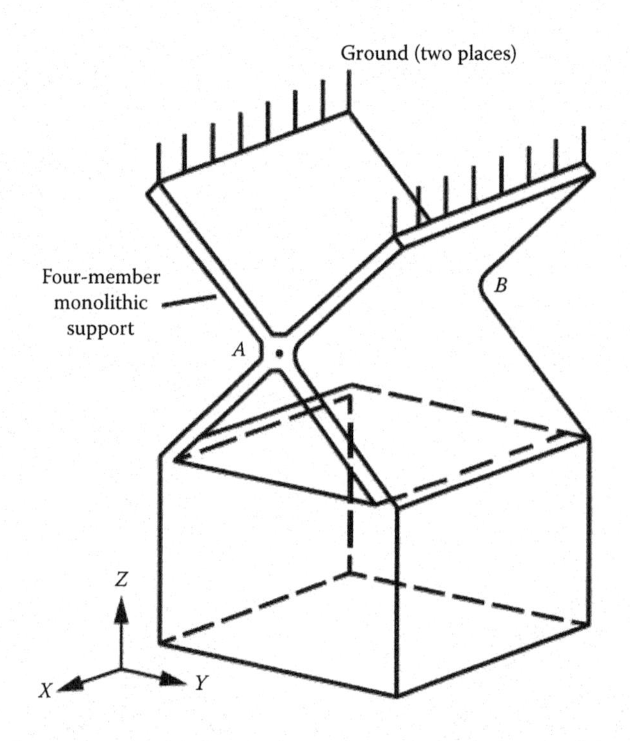

FIGURE 11.31 Two folded leaf springs (on left and right) with their dihedral edges connected at line *A*–*B* constrain a rigid body in five DOFs. Only the rotation about *AB* is free. This is based on the Haberland hinge.

11.2.7 Controlling Six DOFs

When all six DOFs of a body are constrained by struts, all three rotations and all three translations can be adjusted simply by varying the strut lengths. Figure 11.32 illustrates a body supported in this manner.

We observe that the lengths of the three vertical struts constrain translation in the *Z* direction as well as rotations around axes parallel to the *X*- and *Y*-axes. For example, lengthening strut 6 causes the rigid body to rotate about edge *AB*. Similarly, lengthening strut 3 will cause a body rotation about edge *BC*. Changing the lengths of struts 5 and 6 equally will result in a rotation about edge *AD*. Application of similar logic allows the remaining altered DOFs to be defined.

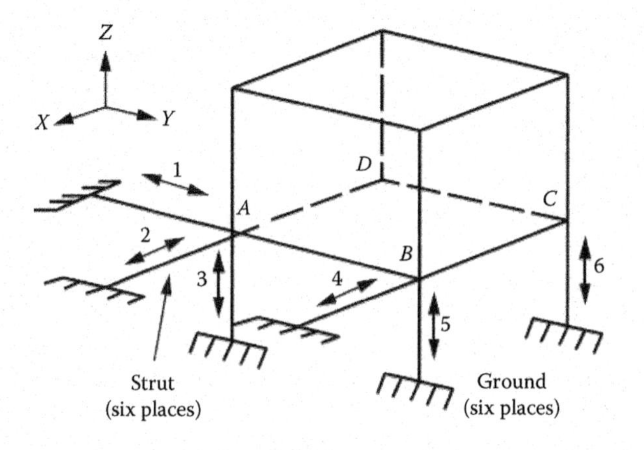

FIGURE 11.32 Six struts (numbered), three parallel to the *Z*-axis and three perpendicular to that axis, constrain a body in all six DOFs. All six DOFs can be varied by changing the lengths of the struts.

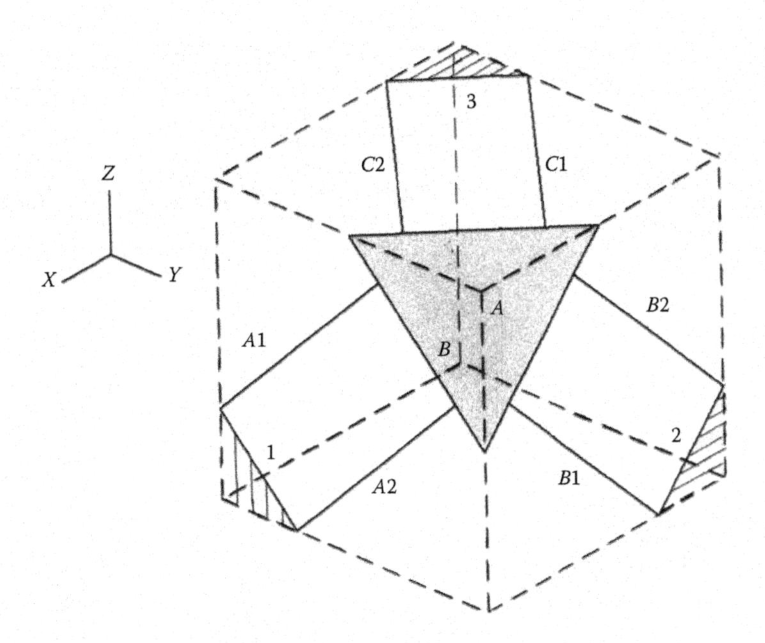

FIGURE 11.33 Three pairs of parallel struts (*A*1–*A*2), (*B*1–*B*2), and (*C*1–*C*2) constrain a rigid body (dashed cube outline) in all six DOFs with respect to the ground (shaded triangular outline).

A different way for constraining six DOFs is shown in Figure 11.33. A dashed cube is drawn to enable better orientation. Parallel strut pairs *A*1–*A*2, *B*1–*B*2, and *C*1–*C*2 are located on perpendicular planes. They are connected to a rigid body (triangular shaded area). Translating edges 1, 2, and 3 individually or in combination will result in translation along the *X*-, *Y*-, and/or *Z*-axes. For example, applying identical translations to all three edges simultaneously in the direction of cube corner *A* will result in a translation of the rigid body along the body diagonal *AB* of the cube. Rotations are not possible because of the parallelograms.

In Figure 11.34, a cube is drawn along with the diagonals of its six surfaces. The diagonals form two tetrahedrals (*A* and *B*). We assume that *B* is fixed to the ground, while *A* is attached by way of six struts to *B*. These struts are numbered 1 to 6. Their lengths are equal and the angles between adjacent struts are 90°. This structure is well known and is called a *hexapod* or, more formally, a *Stewart platform*. It is widely used in aircraft flight simulation platforms. Adjusting the struts in

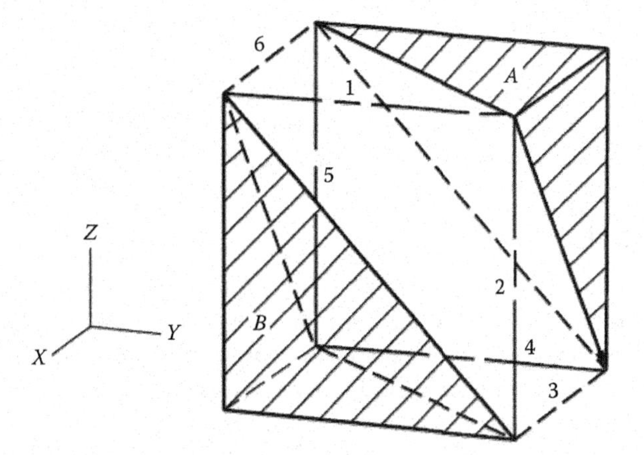

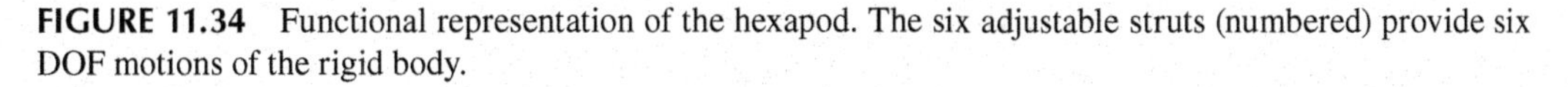

FIGURE 11.34 Functional representation of the hexapod. The six adjustable struts (numbered) provide six DOF motions of the rigid body.

FIGURE 11.35 Stewart platform for use as the mechanism for a flight simulator. (Courtesy of AXIS Flight Training Systems GmbH, Graz, Austria.)

length individually or in groups allows the operator to move *A* relative to *B* in all six DOFs. Figure 11.35 shows a typical hexapod mechanism as used in the flight simulator role.

A much smaller, commercially available hexapod is shown in Figure 11.36. This mechanism provides travel of ~220 mm (~8.7 in.) with loads to ~1000 kg (~2200 lb) and reported positional repeatability of ±0.002 mm (±8 × 10^{-5} in.). Devices such as this are typically used in test laboratories, in electro-optical sensors, in astronomical instrumentation, and in military applications.

11.2.8 Internal DOFs

The previous sections have explained how the DOFs of a rigid body are constrained—as if that condition could be absolute. In practice, the body itself may show lack of stiffness. For example, a portable table (see Figure 11.37) that is often used by campers has four supports, each with two feet. All eight feet can be in contact with the uneven ground because of the lack of torsion stiffness of the table itself. The same flexibility applies to each set of two seats.

Similar situations often occur in optomechanical hardware. A body experiencing bending or torsion loads will deflect if it lacks stiffness. In some cases, the deflections due to bending or torsion loads are large compared to those resulting from tension or shear loads.

FIGURE 11.36 Commercially available hexapod platform adjustable in all six DOFs by changing the lengths of the six independently controlled struts. (Courtesy of Physik Instrumente, Karlsruhe, Germany, http://www.pi.ws.)

FIGURE 11.37 Camping table with four legs that lacks torsional stiffness so it can rest all feet on the uneven ground.

11.2.9 Quasi-Static DOF Constraints

All structural elements constrain one or more DOFs, and for each of these, certain stiffnesses apply. The remaining DOFs are not constrained. This means that the constraint stiffnesses for these DOFs are small compared to those of the constrained DOFs and can therefore be neglected. All structural elements that have been introduced so far are based on elastic behavior. Now, two more important structural elements need to be introduced. These are the sphere and the cylinder. Depending on the ways in which they are used, they constrain a specific number of DOFs.

In Figure 11.38a through d, the circles represent a sphere or a cylinder (end view) contacting a flat, a concave cylinder, or two forms of a V-groove. The body that makes contact with a reference surface theoretically is constrained in the direction or directions defined by the line or lines connecting the contact point(s) to the centerline of the body's. However, that is true only when there is no friction. For instance, in the case of rolling contact by a sphere on a flat, this friction is very small and the sphere indeed is constrained in one DOF. In the case of sliding contact, friction is usually not negligible. Only if properly lubricated (resulting in friction coefficients $\ll 0.1$) can it be ignored. Often the sphere is fixed to a "moving body." Hence, it is not capable of rotating as is moves. It can only slide. If the body is a cylinder and it interfaces with two surfaces, as shown in Figure 11.38c or d, it is constrained in two DOFs.

The view in Figure 11.38d shows the more general case of a sphere or cylinder contacting a V-groove of some specific angle α. Depending on the value of α, different stiffnesses horizontally and vertically are obtained. The dimension x can be quantified as $x = R \sin [90 - (\alpha/2)]$. This view can also be interpreted as a section through a sphere contacting a conical reference surface. That sphere is then constrained in all three DOFs. This configuration is considered in detail in Section 11.3.6. Table 11.1 provides a summary of the DOF constraints that can be provided by a sphere or

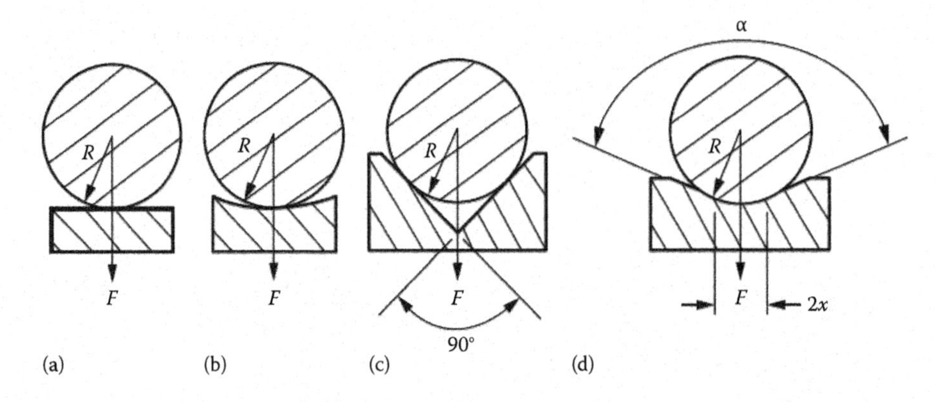

FIGURE 11.38 (a–c) Views of a sphere or cylinder of radius R interfacing with a flat plane, with a concave sphere, and with a 90° V-groove. View (d) shows the general case for the sphere or cylinder interfacing with a groove of angle α.

TABLE 11.1
DOF Limitations When Using a Sphere or Cylinder Interface with Various Structural Geometries

Structural Element	Flat Surface	V-Groove	Cone
Sphere	One point of contact, thereby constraining one translation	Two points of contact, thereby constraining two translations	Annular ring (line) of contact, thereby constraining three translations
Cylinder	One line of contact, thereby constraining one translation and one rotation	Two lines of contact, thereby constraining two translations and one rotation	Not possible

cylinder depending on the contacting body. In practice, remarkably good results can be obtained using these structural elements. However, one should realize that this is largely due to the high-precision parts that can be obtained by modern manufacturing techniques. In the following sections, further details show clearly the differences between fully elastic structural elements and elements like the sphere and cylinder.

11.3 STRUCTURAL ELEMENTS USED TO CONTROL DOFs

Structural elements, such as struts and leaf springs, have been introduced. In this section, they are examined in detail and their specific features are explained. Folded and constricted leaf springs are also discussed. It is shown that efficient use of material is a very important feature of a good structural design.

11.3.1 Strut (One DOF)

11.3.1.1 Stiffness Properties

A strut in its simplest form is a body with a constant circular or rectangular cross section in which the length is large compared to the sectional dimension(s). When cantilevered at one end, it can bend transversely under a force F, as illustrated in the views in Figure 11.39a and b, or change length axially as illustrated in the view in Figure 11.39c. The case shown in the view in Figure 11.39b represents a condition in which the right end of the strut is guided.

The axial stiffness (in tension or compression) of any strut (in newtons per meter) is, as shown in Figure 11.39c, given by

$$C_{\text{AXIAL}} = \frac{EA}{L}, \tag{11.1}$$

where E is the Young's modulus (in newtons per square meter), the cross section area A (in square meters) = bh for a rectangular strut or $\pi d^2/4$ for a circular strut, and L is the length (in meters).

The bending stiffnesses (in newton-square meters) for a circular section strut and for a rectangular section strut are as follows:

$$\text{For case a:} \quad C_{\text{BENDING}} = \frac{3EI}{L^3} \tag{11.2}$$

$$\text{For case b:} \quad C_{\text{BENDING}} = \frac{12EI}{L^3} \tag{11.3}$$

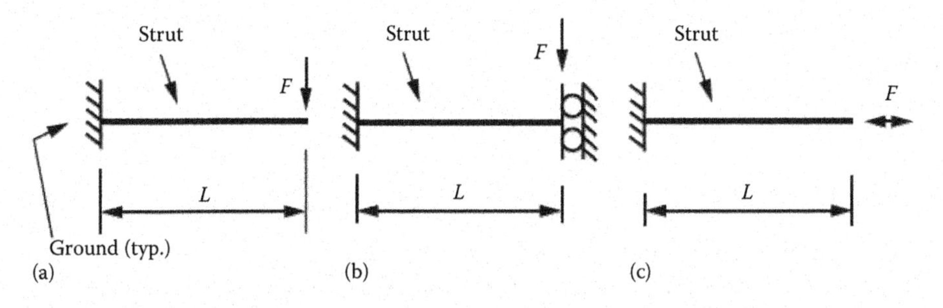

FIGURE 11.39 Simple rod-shaped strut loaded (a, b) in bending and (c) in compression or tension. In view (b), the motion is guided on the right.

In these equations, I is the moment of inertia (in quadratic meters [m^4]). For a circular section strut, $I = (\pi/64)d^4$, while for a rectangular strut with dimensions b and h, $I = (1/12)bh^3$ when $h > b$ or $I = (1/12)hb^3$ when $b > h$. The reader should note from these equations that the bending stiffness of a rectangular strut does not change when the length-to-thickness ratio does not change (provided that the width does not change).

To compare the two types of stiffness, their ratio is calculated in Example 11.1 for a typical case.

Example 11.1

(1) Calculate the stiffnesses of an aluminum strut with a diameter of 1.0 mm and a length of 50.0 mm in tension and bending modes. (2) What is the ratio of these parameters? Assume Young's modulus E to be 70 GPa (= 70×10^9 N/m).

For (1), from Equation 11.1,

$$C_{\text{AXIAL}} = (70 \times 10^9)(\pi)(1.0 \times 10^9)^2(1/4)/(50.0 \times 10^{-3}) = 1.1 \times 10^6 \text{ N/m}.$$

For (2), from Equation 11.2,

$$C_{\text{BENDING}} = (3)(70 \times 10^9)(\pi/64)(1.0 \times 10^9)^4/(50.0 \times 10^{-3}) = 82.5 \text{N/m}.$$

The ratio of these stiffnesses is $1.1 \times 10^6/82.5 = 13{,}333$.

From this simple calculation, we observe that the tension stiffness of a strut will be several orders of magnitude (OMs) greater than its bending stiffness provided that its length is much greater than its diameter.

Often, it is said that mechanical parts have a high or low stiffness without defining exactly what is meant. Therefore, the following rule of thumb is given. High stiffness typically means stiffness values $>10^6$ N/m. Low stiffness means values $<10^3$ N/m. Comparing high to low stiffness means that there are three-OM differences. This rule is often used to determine whether a rigid body is properly decoupled in all DOFs. That is also why neglecting these low stiffnesses has only a minor effect on the calculations. Furthermore, it eliminates variables when performing calculations—which make them easier.

In practice, it is not always possible to get a three-OM difference. As a compromise, a lower ratio can be accepted at the expense of larger errors when comparing theory and practice. Decoupling of only two OMs is generally considered to be insufficient. Designs capable of decouplings of four or five OMs also exist and are usually related to machines that operate in the nanometer range.

The stiffness of a strut is related to its dynamic performance. Assuming that the strut is constraining a vertical translation mode of a rigid body (see Figure 11.40a), the first-order estimates of the first natural frequencies (or eigenfrequencies) of the strut in the axial direction is given by Equation 11.4:

$$f_{\text{translation}} = \left(\frac{1}{2\pi}\right)\left(\frac{C}{W}\right)^{1/2}, \tag{11.4}$$

$$f_{\text{rotation}} = \left(\frac{1}{2\pi}\right)\left(\frac{k}{J}\right)^{1/2}, \tag{11.5}$$

where $k = Cb^2$, C is the tension stiffness (in newtons per meter), and W is the weight* of the body (in kilograms). Note that the ratio of stiffness to weight determines this first natural frequency. Hence,

* Here weight is taken as mass times acceleration, where the acceleration is 1g ($g \approx 9.81$ m/s2); elsewhere in this textbook, mass is given explicitly.

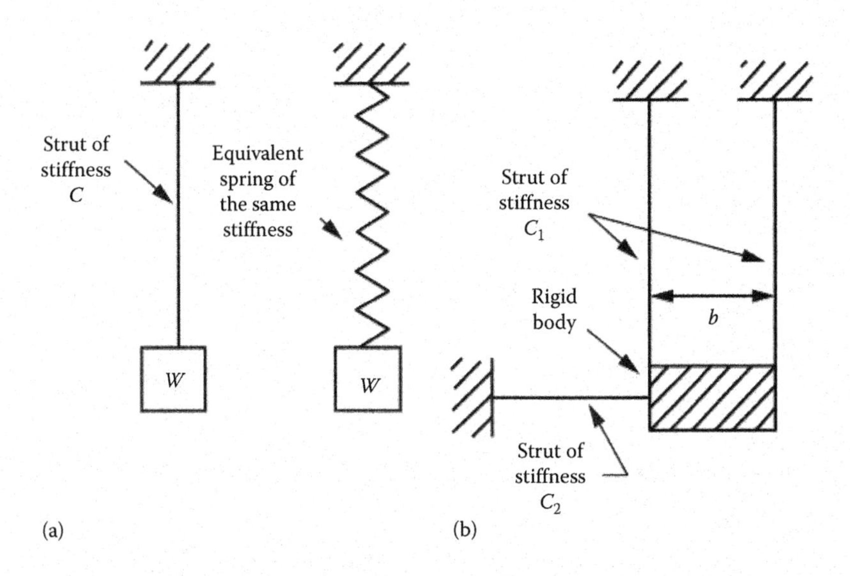

FIGURE 11.40 Point weight W supported (a) by a strut and by the equivalent spring. In each case, the support has a stiffness of C. (b) Rigid body supported by three struts, each with a specific stiffness.

with relatively low stiffness and low weight, one can still obtain a high first natural frequency f. Such a mount might be suitable for mounting a small lens or prism (with a weight of a few grams) that is supported using thin struts.

Similarly, a first-order estimate of the first rotational natural frequency can be obtained. See Figure 11.40b. This figure shows a rigid body supported by two parallel struts that prevent rotation and translation in the vertical direction. The horizontal strut prevents translation in the horizontal direction. The frequency f for the rotational mode is given by Equation 11.5, where k is the rotational stiffness (in newton-meters per radian) and J is the weight moment of inertia (in kilogram-square meters) of the body.

11.3.1.2 Reduced Buckling

The same tensile stress exists in any cross section of the strut. That means that the highest possible tension stiffness has been realized for a given amount of material. It is important that one realizes this when, for example, optimal dynamic properties of a structure are needed.

A simple rectangular section (prismatic) strut has a rather limited axial compression load capability due to buckling. To solve this, one should look at the bending moment in the strut as function of the axial position.

Usually, high bending moments occurs only at the ends of the strut. That means that the rest of the strut will deform very little and that a more massive cross section of the struts is allowed at these locations. The resulting increase in the bending stiffness of the strut will be very limited. This is illustrated in Figure 11.41.

Figure 11.41c and d has equal lateral stiffnesses provided that the thin sections are identical and the length ratios as given in the figure are realized. We make the reasonable assumption that the deformation of the thick section is negligible compared to that of the thin section. Please note that the bending stiffness of the strut increases by a factor of $(7/6)^3 = 1.6$ if the strut with a thick center section is not increased in length by a factor of 7/6 compared to the original prismatic strut.

Assuming that the tensile stiffness of the thick section is large compared to that of the thin section, it can be concluded that this stiffness has increased by a factor of $6/(1 + 1) = 3$. In reality, it will be slightly smaller due to the sudden change in the cross-sectional area. For the same reason, stress concentrations will occur at these points. This may require that a more gradual change in the cross section be provided to reduce the stress level.

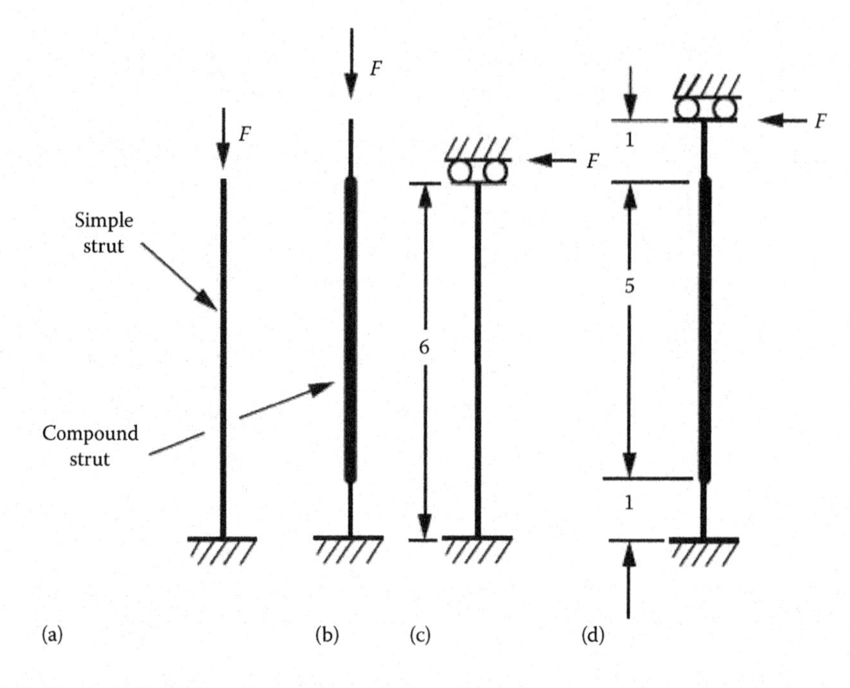

FIGURE 11.41 (a, b) Technique for increasing the buckling load capacity of a strut under compression. The strut (c) has a length of six units and a specific stiffness in bending. In (d), another strut with a total length seven units is stiffened in bending by a central section with a greater diameter and a relative length of five units.

11.3.1.3 Improving the Tension-to-Bending Stiffness Ratio

The ratio of the tension stiffness to bending stiffness of a strut is given by the following equation:

$$\frac{C_{\text{TENSION}}}{C_{\text{BENDING}}} = (\text{Q})\left(\frac{L^2}{d^2}\right). \tag{11.6}$$

Typically, the constant Q is 16/3 or 4/3 depending on the applicable hardware configuration. Often this ratio needs to be extreme, for instance, 1000. This ratio might be too extreme for a practical design. In that case, assuming $Q = 16/3$, from Equation 11.6, we find that a length-to-thickness ratio of 13.7 is needed.

A way to improve this situation is to change the configuration of the thin section while maintaining its cross-sectional area constant. In an effective design, a short localized region of the strut cross section becomes a rectangular leaf instead of a circular rod. See Figure 11.42. The cross-sectional area of the leaf is set equal to that of the rod. The tension stiffness then remains unchanged, while the bending stiffness in one direction decreases by t^2, where t is the leaf thickness. Hence, reducing the thickness by a factor of 2 improves the stiffness ratio of the thin section by a factor of 4. It also means that the decoupling ratio of the modified strut will change similarly.

A drawback of this feature is that the bending stiffness in the perpendicular lateral direction increases considerably. Increasing the width by a factor of 2 increases the bending stiffness by a factor of 4. This can be overcome by introducing a second leaf spring section that has been rotated by 90° in series with the first leaf section. Note that a small transition section is needed between the two leaf spring sections. It has little effect on performance. A sufficient total length needs to be provided to accommodate four thin leaf spring sections in combination with the thicker intermediate sections of the strut. In practice, this is hardly a problem because the required deflections are usually small. The leaf spring sections will then be short.

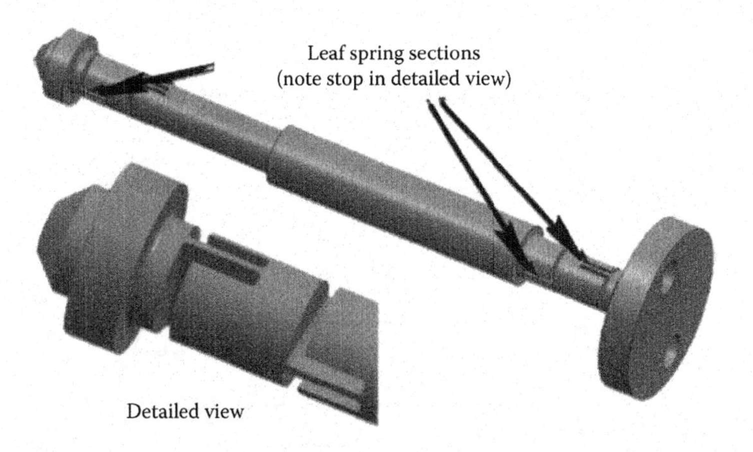

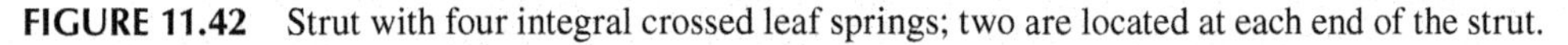

FIGURE 11.42 Strut with four integral crossed leaf springs; two are located at each end of the strut.

The thin leaf springs shown in Figure 11.42 are obtained by milling. That puts a clear limit on the section thickness. To further reduce the thickness, wire erosion is used. If the section becomes too thin, it becomes vulnerable to damage due to handling (i.e., plastic deformation or even failure). That is why a stop is often integrated into the design of the hinge. An example is shown in the detailed view of Figure 11.42. Thin slots are machined into the strut on both sides of the flexure blade. When the flexure bends, the adjacent slot closes and limits that bending. When the hinge is manufactured by use of wire erosion, such a stop is really easy to implement without increasing the manufacturing cost. Please note that "thin" is used here in a relative sense; i.e., one should always estimate the bending stress in the flexure blade to ensure that its thickness is reasonable.

Figure 11.43b shows a strut in which the two square-cornered leaf spring sections (see Figure 11.43a) are replaced by contoured elastic hinges. The bending deformation is more concentrated. The hinge line is defined by the center of the thinnest section. To obtain the same bending stiffness as the milled hinge, the thin section must be quite small. The consequences for the stress levels as function of the required deformations are discussed in Sections 11.3.3 and 11.3.5.

In some strut designs, different sections have two or more cylindrical lengths of different diameters or other cross-sectional shapes and sizes. The stiffness in tension is then computed as a set of springs in series using the following equation:

$$C_{\text{EQUIVALENT}} = \left(\sum_{i=1}^{n} c_i^{-1} \right)^{-1}. \tag{11.7}$$

Figure 11.44 illustrates three geometrical varieties of struts which provide different degrees of stiffness in tension. Progressive improvement in decoupling is demonstrated as the design evolves

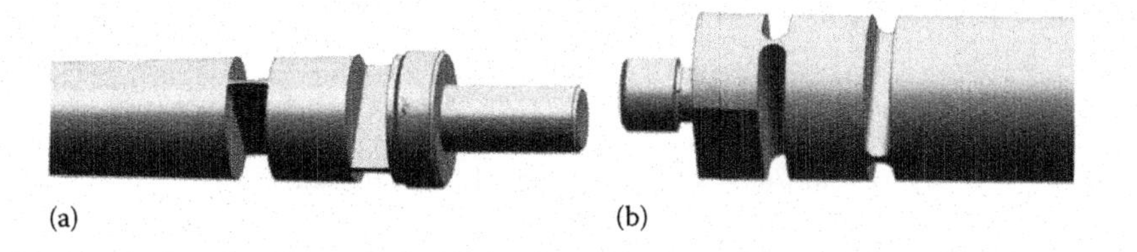

FIGURE 11.43 (a) Strut end with two crossed leaf springs created by milling. Note the sharp corners that tend to concentrate stresses. (b) Strut end with two crossed contoured elastic hinges created by wire erosion. The smoothly shaped corners tend to distribute stress and make the strut more durable.

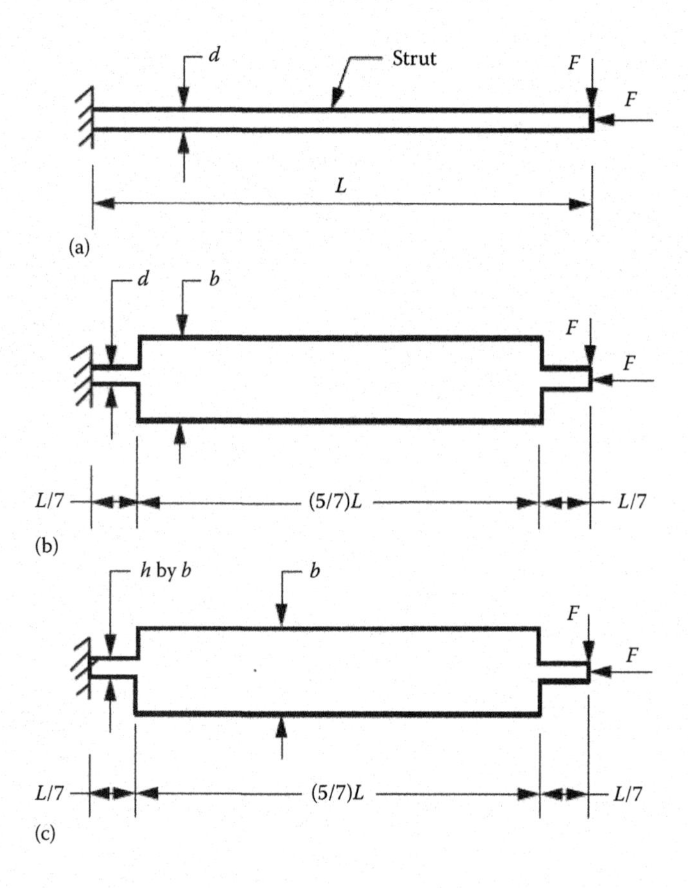

FIGURE 11.44 (a) Dimensioned cylindrical strut with a constant cross section over its length. (b) Strut of the same total length stiffened with a larger cylindrical central section. (c) Strut with parallel rectangular leaf springs at the ends.

from that of Figure 11.44a to that of Figure 11.44c. Examples 11.2 through 11.4 then show calculations of those stiffnesses and their decoupling ratios. Progressive improvement in decoupling is demonstrated as the design evolves from that of Figure 11.44a to that of Figure 11.44c.

Example 11.2

An aluminum strut with a simple cylindrical configuration (see Figure 11.44a) has the following dimensions: d = 2.000 mm (0.079 in.) and L = 50.000 mm (1.969 in.). Its Young's modulus is 70 GPa (4.827 × 10^{-4} lb/in.^{2}). (1) What are the stiffnesses in tension and bending for the strut? (2) What is its decoupling ratio?

For (1), calculate the cross-sectional area A and the moment of inertia I:

$$A = \pi d^2/4 = 3.142 \times 10^{-6} \text{m}^2 (2.027 \times 10^{-9} \text{in.}^2),$$

$$I = \pi d^4/64 = 7.85 \times 10^{-13} \text{mm}^4 (1.886 \times 10^{-6} \text{in.}^4).$$

From Equation 11.2,

$$C_{\text{BENDING}} = 3EI/L^3 = (3)(70 \times 10^9)(7.854 \times 10^{-13})/(50 \times 10^{-3})^3 = 1.32 \times 10^3 \text{N/m} \ (7.538 \text{lb/in.}).$$

From Equation 11.1,

$$C_{TENSILE} = EA/L = (70\times10^{9})(3.142\times10^{-6})/0.050 = 4.399\times10^{6}\,\text{N/m}\,(2.512\times10^{4}\,\text{lb/in.}).$$

For (2), the ratio of tensile stiffness to bending stiffness is $1.32 \times 10^3/4.399 \times 10^6 = 3333$. This is the decoupling ratio of the stiffnesses. It is reasonable, but not extreme.

Example 11.3

An aluminum strut with an enlarged portion as shown in Figure 11.44b has its total length L divided into three parts of relative lengths 1, 5, and 1. It has a triple cylindrical configuration and the following dimensions: d = 2.000 mm (0.079 in.), L = 50.000 mm (1.969 in.), and $b = 5d$ = 10.000 mm (0.394 in.). The two smaller diameter sections have diameters of d and lengths of $L/7$ = 7.143 mm (0.282 in.). Its Young's modulus is 70 GPa (4.827×10^{-4} lb/in.^{2}). (1) What are the stiffnesses of the strut in tension and in bending? (2) What is the new decoupling ratio?

For (1), calculate the cross-sectional areas of the large and small cylinders:

$$A_L = 5^2\pi b^2/4 = 7.854\times10^{-5}\,\text{m}^2\,(5.067\times10^{-8}\,\text{in.}^2),$$

$$A_S = \pi d^2/4 = 3.142\times10^{-6}\,\text{m}^2\,(2.02\times10^{-9}\,\text{in.}^2).$$

Applying Equation 11.1,

$$C_L = EA_L/(5/7)L = (70\times10^{9})(7.854\times10^{-5})/(5/7)(50\times10^{-3}) = 1.5\times10^{8}\,\text{N/m},$$

$$C_S = EA_S/(1/7)L = (70\times10^{9})(3.142\times10^{-6})/(1/7)(50\times10^{-3}) = 3.2\times10^{7}\,\text{N/m}.$$

Applying Equation 11.7,

$$\begin{aligned} C_{TENSILE} &= 1/(1/C_L + 2/C_S) = 1/\left[(1.5\times10^{8}) + (2/(3.2\times10^{7})\right] \\ &= 1.40\times10^{7}\,\text{N/m}\,(175.125\,\text{lb/in.}). \end{aligned}$$

The bending stiffness of the strut will now increase from that in Example 11.2 by the ratio $(7/6)^3$. Thus,

$$C_{BENDING} = \left[(7/6)^3\right][1.32\times10^{3}] = 2.10\times10^{3}\,\text{N/m}\,(11.991\,\text{lb/in.}^2).$$

For (2), the decoupling ratio of this strut is $1.40 \times 10^7/2.10 \times 10^3 = 6697$. It is twice that of the strut in Example 11.2.

Example 11.4

A large diameter cylindrical strut of diameter b has each end reduced to a thin rectangular section with dimensions b by h, as shown in Figure 11.44c. Here, $h = 5b$. The cross-sectional areas of these end sections equal those of the thin cylinder sections of Example 11.3.

The tensile stiffness of this strut is the same as that of Example 11.3 because the end section areas are the same. Hence,

$$C_{\text{tensile}} = 1.40 \times 10^7 \text{ N/m},$$

Dimension h = (area in Example 11.3 end section)/$b = 3.142 \times 10^{-6}/5 \times 10^{-3} = 0.31 \times 10^{-6}$ m.

The bending stiffness is decreased from that in Example 11.3 in proportion to the ratio of the moments of inertia of the two sections, i.e., of diameter d versus that of the rectangular area bh, as follows:

$$C_{\text{BENDING}} = (2.1 \times 10^{-3})(I_{\text{RECTANGLE}}/I_{\text{CIRCLE}}) = \left[(1/12)bh^3\right]/\left[(\pi/64)(d^4)\right]$$
$$= [(1/12)(5 \times 10^{-3})(0.31 \times 10^{-3})/[(\pi/64)(2 \times 10^{-3})^4 = 68.9 \text{ N/m } (0.393 \text{ lb/in.}).$$

The decoupling ratio = 1.40 × 10^7/68.9 = 203,000. This represents a dramatic improvement compared with the last example (more than five OMs!).

11.3.2 Leaf Spring (Three DOFs)

A strip of sheet material acts as a leaf spring. This type of structural element is capable of resisting loads in plane with the sheet. However, because of its relatively small thickness, it bends under loads perpendicular to the sheet.

The various parameters of the leaf spring are shown in Figure 11.45. Its tension/compression stiffness can be calculated using Equation 11.1. The load F causes shear and bending deformations. Which one is dominant depends on the layout of the leaf. It is advantageous for the length L to be smaller than the width b. This can be verified by examination of the following formulas. We see that the bending deflection is proportional to L^3, while the shear deflection is proportional to L. Hence, the bending deflection is much more sensitive to the magnitude of the length L and should

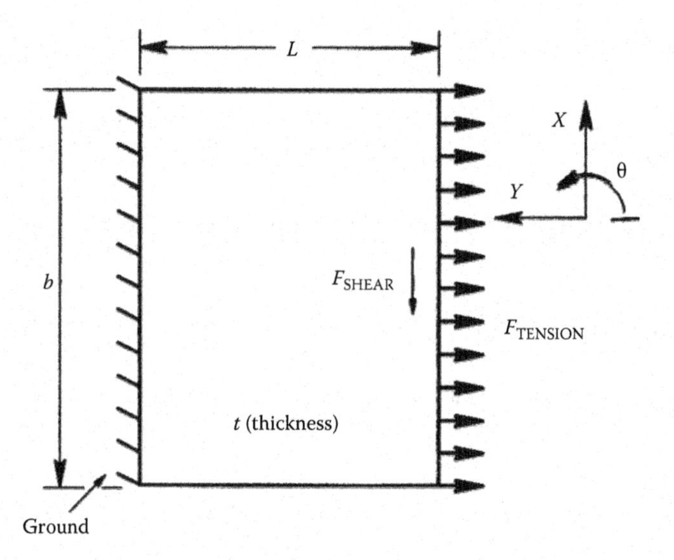

FIGURE 11.45 Leaf spring loaded by a distributed tensile load F_{TENSION} and a shear load F_{SHEAR}. The leaf thickness is t.

be minimized. The pertinent relationships, which involve the shear modulus G, Young's modulus E, and Poisson's ratio ν, are as follows.

The shear deformation due to F_{SHEAR} is

$$f_{SHEAR} = \frac{FL}{Gbt}. \tag{11.8}$$

The bending deformation due to F is

$$f_{BENDING} = \frac{FL^3}{3EI}. \tag{11.9}$$

The deformation ratio due to F is

$$\frac{f_{BENDING}}{f_{SHEAR}} = \frac{FL^2/3EI}{FL/Gbt} = \frac{GL^2bt}{3E(1/12)tb^3} = \frac{4EL^2}{2E(1+\nu)b^2} = \frac{2L^2}{(1+\nu)b^2}. \tag{11.10}$$

$F_{BENDING}$ is dominant when

$$L > b\sqrt{\frac{1+\nu}{2}}. \tag{11.11}$$

Note: For most materials, $\nu = 0.3$; hence, $L < 0.8b$ to minimize the deflection.

The shear and bending stresses are given by

$$\tau = \frac{F}{bt}, \tag{11.12}$$

$$\sigma = \frac{FL}{bt^2/6}. \tag{11.13}$$

11.3.3 Constricted Leaf Spring (Two DOFs)

A constricted leaf spring is an elastic hinge that can be obtained simply by machining two holes in a strip of sheet material with a small distance between them as shown in Figure 11.46. Between the two holes, a small section of material remains. This strip can bend in two directions (about the X- and Z-axes) and twist about the X-axis. The holes can be of equal or unequal diameters, so two configurations apply.

Equations pertaining to each motion represented in these figures are as follows:

$$\frac{2}{D} = \frac{1}{D_1} + \frac{1}{D_2}, \tag{11.14}$$

$$C_X = (0.48)\left(\sqrt{\frac{h}{D}}\right)(E)(t), \tag{11.15}$$

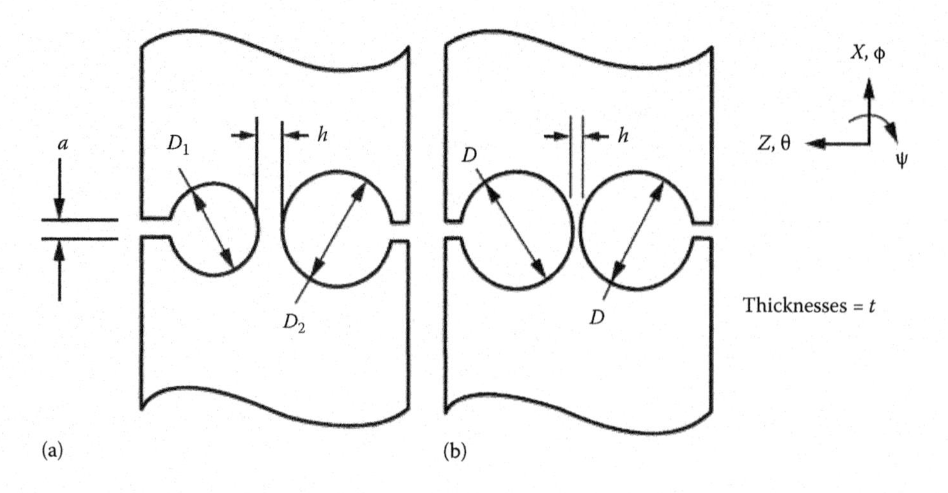

FIGURE 11.46 Dimension designations for constricted leaf springs: (a) with unequal diameter holes and (b) with equal diameter holes.

$$C_Z = (0.56)\left(\sqrt{\frac{h}{D}}\right)(E)(t)\left(1.2 + \frac{D}{h}\right)^{-1}, \tag{11.16}$$

$$k_\psi = (0.093)\left(\sqrt{\frac{h}{D}}\right)(E)(t)(h^2), \tag{11.17}$$

$$\Psi \approx 10.7\sqrt{\left(\frac{D}{h}\right)\left(\frac{T_\psi}{Eh^2 t}\right)}, \tag{11.18}$$

$$\sigma_\Psi \approx 0.58\Psi E\sqrt{\frac{h}{D}}, \tag{11.19}$$

where C_X is the tension/compression stiffness in the X direction (in newtons per meter); C_Z is the lateral stiffness in the Z direction; k_ψ is the rotation stiffness about the Y-axis (in newton-meters per radian); ψ is the rotation about the Y-axis (in radians); σ_ψ is the maximum stress level in the elastic hinge (in megapascals); h is the width of the elastic hinge; D, D_1, and D_2 are the circle diameters (in meters); E is the Young's modulus of the material in gigapascals; t is the thickness of the sheet material (in meters); and $T\psi$ is the torque applied to obtain rotation ψ (in newton-meters).

11.3.4 Elastic Hinge (One DOF)

If the thickness t of a constricted leaf spring is increased by a considerable amount, the resulting elastic hinge (shown in Figure 11.47) has only one DOF left because the out-of-plane DOFs are constrained. Only the rotation ψ on the plane of the figure is then possible. Previously given Equations 11.14 through 11.19 allow the user to calculate the various stiffness properties and stress level for

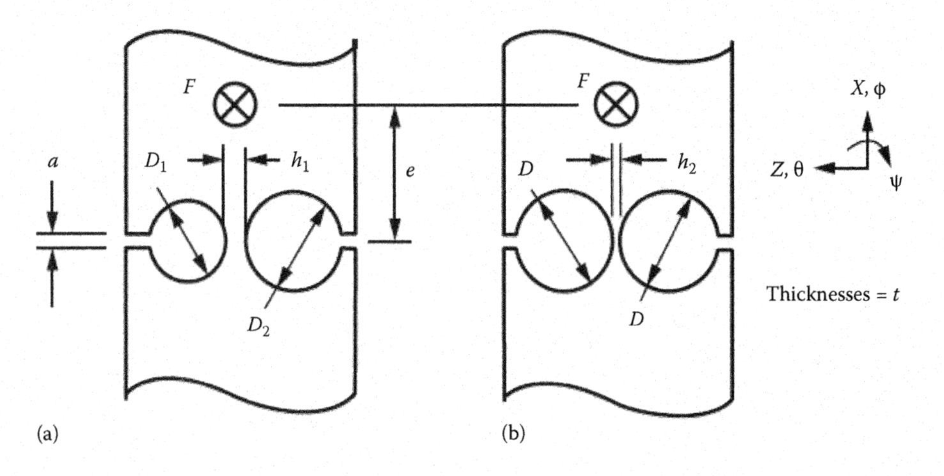

FIGURE 11.47 Constricted leaf spring (a) with unequal hole diameters and (b) with equal hole diameters. A perpendicular force applied at X bends the spring into the page and flexes the bridge between the circles subjected to a shear force F parallel to the dihedral edge.

a given rotation ω. In addition, Equations 11.20 and 11.21 allow the bending stiffnesses about the X- and Y-axes to be calculated because they are no longer negligible:

$$k_\theta = C_X t^2/12, \tag{11.20}$$

$$k_\varphi = C_Z t^2/12. \tag{11.21}$$

Calculating the stiffness in the Y direction is of little use for practical reasons. That stiffness is determined by the shear load and the associated bending load F. The bending deformation will dominate the shear deformation. The stiffness k_θ in newton-meters per radian can be transferred to a linear stiffness in newtons per meter at the location of F by this equation:

$$C_Y = C_X t^2/(12e^2). \tag{11.22}$$

11.3.5 Approximated Elastic Hinges (Two DOFs or Five DOFs)

Instead of using a constricted leaf spring with circular holes (or the elastic hinge of the previous section), one with a simple rectangular beam of length L and cross-sectional area h by t could be used. (See Figure 11.48.) Two DOFs or five DOFs are constrained depending on the thickness t. If the length L equals $2.1\sqrt{(D)(h_2)}$, its tension stiffness will equal that of the corresponding constricted leaf spring. Furthermore, k_ψ will be reduced by a factor of 2.3. This may be very beneficial, although one should be aware that the rotation center of a leaf spring is not fixed. When undeflected, it is located at the midpoint of the length of the leaf spring. With increasing deflection, it shifts in the lateral direction (Z). Depending on the application, this may or may not be acceptable. The bending stiffness of a rectangular section leaf spring hinge can be made to equal that of a constricted leaf spring hinge by making its length L equal to $0.9\sqrt{(Dh)}$. The tension stiffness will then be greater by a factor of 2.3. A rectangular leaf spring with length $L = 1.2\sqrt{Dh}$ will have the same shear stiffness as the constricted leaf spring.

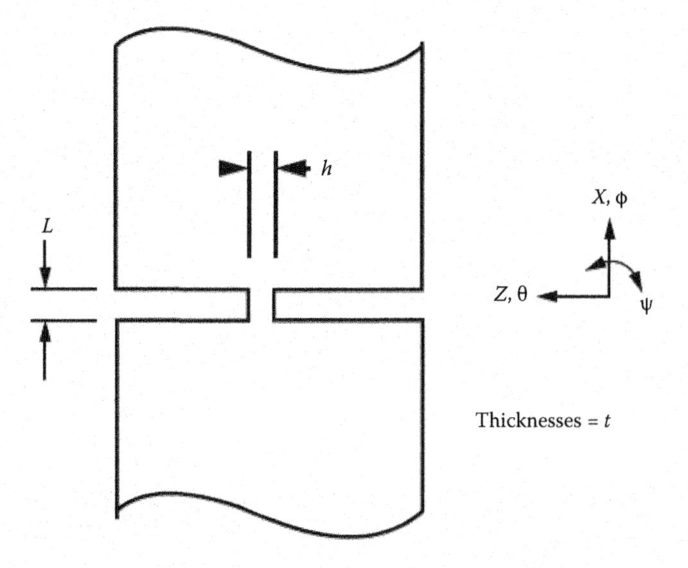

FIGURE 11.48 An elastic hinge with a rectangular cross-sectional flexible bridge member of dimensions h by t and length L can flex in three directions.

Although an elastic hinge is easy to manufacture by drilling two holes, three issues arise. These are the following:

1. To keep the rotation stiffness k_ψ small, it is necessary to make the width h also small. Manufacturing tolerances may then become a problem if h reduces below 1 mm (0.039 in.).
2. The stress level in the thin bridge between the two main parts is high.
3. The large holes occupy a relatively large space.

Application of wire erosion as the manufacturing technique can reduce these potential problems because it allows a partial circle hole to be cut. See Figure 11.49. Note that the stop that limits the allowed rotation ψ (closure of space a at its extremity) may be inadequate and an alternate way to limit that motion may be needed.

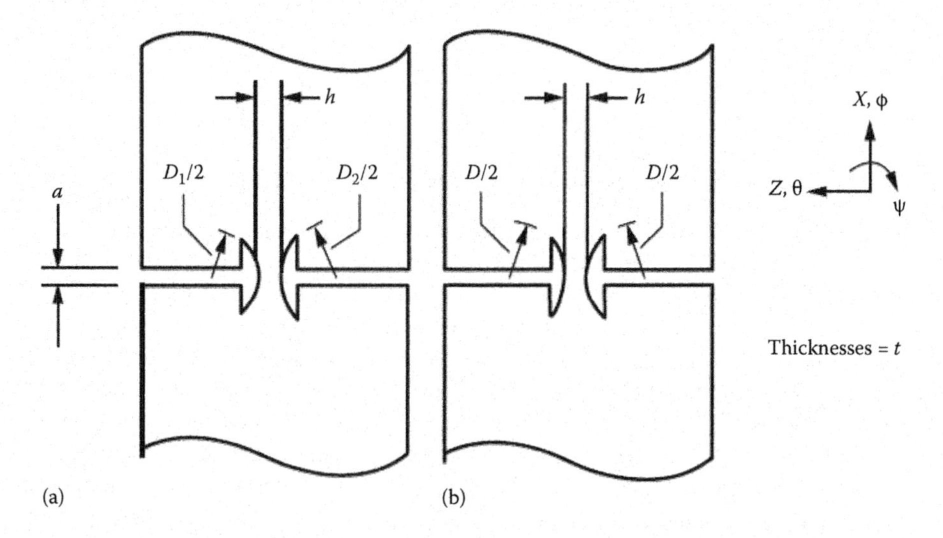

FIGURE 11.49 Modified constricted leaf springs with (a) unequal partial hole radii and (b) equal partial hole radii. This design is easily machined by the wire erosion technique.

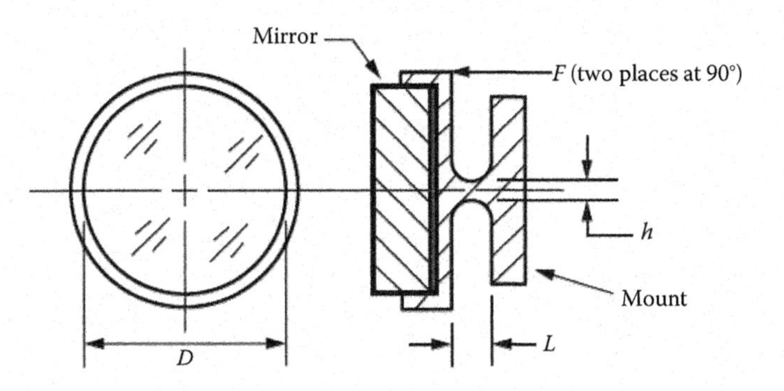

FIGURE 11.50 Mirror mounted on a stalk of diameter h and length L achieves two-axis tilts from actuation forces F_X and F_Y applied at orthogonal points around the mirror rim and normal to that surface. This type of mounting is sometimes called a *mushroom mount.*

Example 11.5 illustrates the design of a tilting mount for a small mirror by using the elastic hinge principle.* Figure 11.50 shows the configuration.

Example 11.5

A circular mirror is mounted in a cylindrical housing which is connected to ground by a constricted leaf spring. See Figure 11.50. An external force F is applied transversely by a screw (not shown). Assume that the screw is self-locking so any adjusted position of the mirror is stable. The mirror is mounted in the housing to guarantee that it is not deformed by force F. The mirror is to be tilted by 1°. Calculate F and the stress in the elastic hinge. The material is aluminum (E = 70 GPa, σ_{yield} = 276 MPa), with h = 0.4 mm and L = 10 mm.

Rearranging Equation 11.18,

$$T_\psi = \left(\frac{\psi}{10.7}\right) Eh^2 t \sqrt{\frac{h}{L}} = \left(\frac{\pi}{(10.7)(180)}\right)(70\times 10^9)(0.4\times 10^{-3})^2(50\times 10^{-3})\sqrt{\frac{0.4\times 10^{-3}}{10\times 10^{-3}}}$$
$$= 0.183\text{Nm}.$$

The associated adjustment force is

$$F = \frac{(2T_\psi)}{t} = \frac{(2)(0.183)}{50\times 10^{-3}} = 7.3\,\text{N}.$$

The bending stress in the hinge is given by

$$\sigma_{\text{BENDING}} = (0.58)(\psi)(E)\sqrt{\frac{h}{D}} = (0.58)\left(\frac{\pi}{180}\right)(70\times 10^{-9})\sqrt{\frac{(0.4)(10^{-3})}{(10)(10^{-3})}} = 142\,\text{MPa}.$$

This stress is smaller than σ_{yield} for aluminum and, therefore, acceptable.

* In solving such a problem, one might apply a small piece of freeware called FlexHinge, which is available from http://en.vinksda.nl/flexure-hinges-theory-and-practice.

The tensile stress of the hinge (normal to the mirror) is given by

$$C_{TENSILE} = (0.48)(E)(t)\sqrt{\frac{h}{D}} = (0.48)(70)(10^{-9})(50)(10^{-3})\sqrt{\frac{(0.4)(10^{-3})}{(10)(10^{-3})}} = 3.4\times10^{8}\ \text{N/m}.$$

This value is quite large. It indicates that the mirror will have a high resonant frequency, as would be desired for practically any application.

11.3.6 Folded Leaf Spring (One DOF)

A folded leaf spring is a combination of two leaf springs that are attached to each other and that usually are perpendicular to each other (see Figure 11.51). Although its function is identical to that of a strut, the way it is mounted is totally different and offers certain advantages.

The fixed section of the folded leaf spring cannot be loaded with out-of-plane forces or moments. This means that the external loading conditions must be equivalent to a load located at the folding line of the leaf springs. Hence, the free section will experience a bending moment M equal to $(F)(L)$ together with a load F (oriented parallel to the folding line) at its attachment point. The same applies at the fixed side of the leaf spring.

From a functional point of view, the folded leaf spring could be replaced by a strut that is located at the intersection of the two planes of the leaf springs. Its attachments to the rigid body and to the "ground" would be at A and B. That is why using the folding leaf spring provides new options for design.

In Section 11.3, it is mentioned that the two sections of the leaf spring do not have to be oriented perpendicular to each other. However, one should be aware of the parasitic sag effect that occurs when this option is implemented. This is illustrated in Figure 11.52.

As a result of the shear deformations of the folded leaf, force F will cause sag that is twice that of a single leaf spring (see Equation 11.11). The stiffness of a folded leaf spring is halved compared to that of a single one. Please note that a situation as shown in Figure 11.52 does not affect the stiffness.

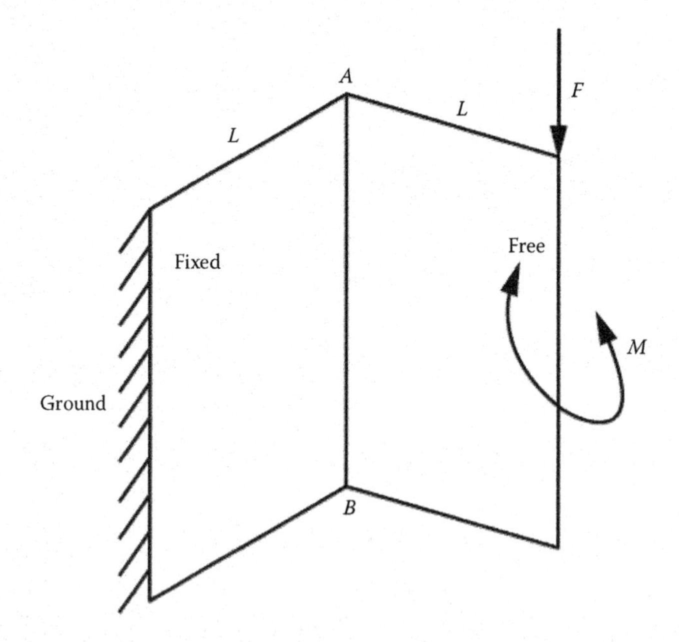

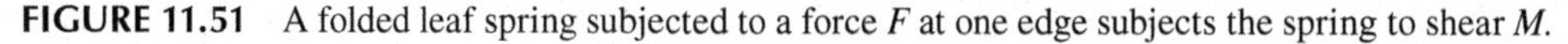

FIGURE 11.51 A folded leaf spring subjected to a force F at one edge subjects the spring to shear M.

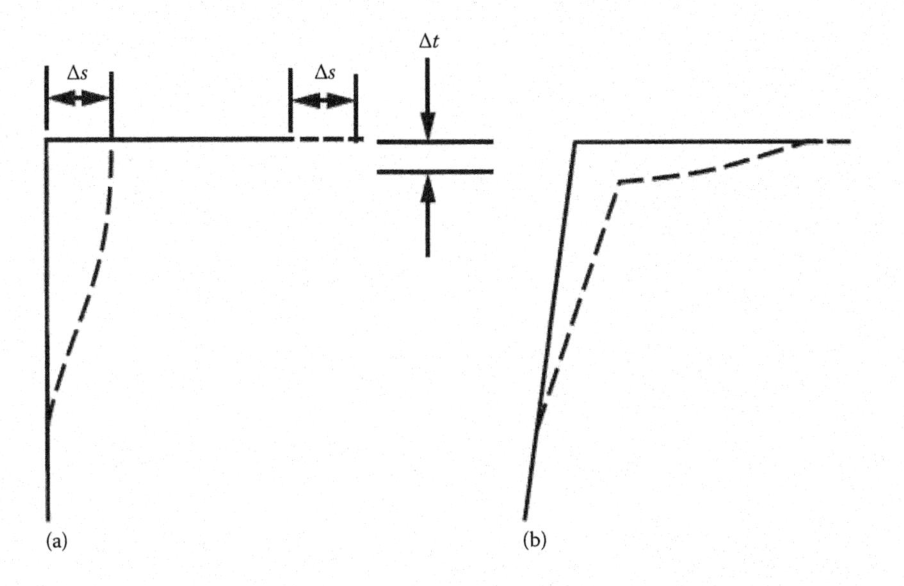

FIGURE 11.52 Parasitic deformation of magnitude Δt resulting from deformation Δs in a folded leaf spring with a non-90° dihedral angle. Drawing scale is graphic. (a) Right angle and (b) obtuse angle.

It is interesting to compare the tension stiffness of the folded leaf spring with that of an equivalent rectangular section strut with length L. The stiffness of the folded leaf spring is given by Equation 11.23, while that of a strut is given by Equation 11.1. The latter equation is repeated here for convenience.

$$C_{\mathrm{F}} = \frac{1}{2\left(\dfrac{L^3}{3EI} + \dfrac{L}{Gbt}\right)}, \tag{11.23}$$

$$C_{\mathrm{AXIAL}} = \frac{EA}{L}. \tag{11.1}$$

Applying the following equations for the moment of inertia I and the shear modulus G of the strut, we can define the cross-sectional area of a strut that has the same tensile stiffness as the folded leaf spring as follows:

$$I = tb^3/12, \tag{11.24}$$

$$G = E\Big/2(1+\nu), \tag{11.25}$$

$$A = \frac{tb^4}{(2)\left[4L^4 + (2)(1+\nu)(Lb^2)\right]}. \tag{11.26}$$

If the bending deflection is negligible compared to the shear deflection, this equation becomes

$$A = \frac{tb^2}{4(1+\nu)(L)}. \tag{11.27}$$

The section ratios of the folded leaf spring and of the conventional strut are given by

$$\frac{A_{\text{FLS}}}{A_{\text{strut}}} = 8\left(\frac{L}{b}\right)^2\left[2\left(\frac{L}{b}\right)^2 + (1+\nu)\right], \tag{11.28}$$

$$\frac{A_{\text{FLS}}}{A_{\text{strut}}} = 8\left(\frac{L}{b}\right)^2(1+\nu). \tag{11.29}$$

We now will apply these equations in Example 11.6.

Example 11.6

Determine the cross-sectional area of a folded leaf spring that has the same tensile stiffness as a conventional strut assuming that the materials are the same. The struts are both made of aluminum with Poisson's ratio $\nu = 0.300$ and length L. Assume that $b = 2L$.

From Equation 11.28,

$$\frac{A_{\text{FLS}}}{A_{\text{strut}}} = 8\left(\frac{L}{b}\right)^2\left[2\left(\frac{L}{b}\right)^2 + (1+\nu)\right] = (8)(1/2)^2\left[(2)(1/2)^2 + 1.3\right] = 3.6.$$

Because the leaf spring has the greater cross-sectional area, it is concluded that more material is required to make it than to make the strut. This is not efficient, so the strut would be the better choice.

11.3.7 Contact Stresses with Various Interface Geometries

11.3.7.1 Sphere-on-a-Flat Plate

Hertz[2] showed long ago what stress levels are to be expected at the contact interface between a sphere and a flat plate. See Figure 11.38a in Section 11.2.9. Other people, such as Johnson (1985), have extended this theory by showing the importance of friction and surface roughness to the obtained results.

The major equations are listed in the following to enable the reader to calculate contact stresses and contact stiffnesses:

$$\text{Contact modulus: } \frac{1}{E_C} = \frac{1-\nu_1^2}{E_1} + \frac{1-\nu_2}{E_2}, \tag{11.30}$$

$$\text{Relative radius: } \frac{1}{R_C} = \frac{1}{R_1} + \frac{1}{R_2}, \tag{11.31}$$

$$\text{Radius of contact circle: } a = \left(\frac{3FR_C}{4E_C}\right)^{1/3}, \tag{11.32}$$

$$\text{Maximum pressure: } p = \frac{3F}{2\pi a^2} = \frac{1}{\pi}\left(\frac{6FR_C^2}{R_C^2}\right)^{1/3}, \tag{11.33}$$

$$\text{Maximum shear stress: } \tau = (0.31p_Z), \text{ where } z = 0.48a, \tag{11.34}$$

$$\text{Maximum tensile stress: } \sigma = \left(\frac{P}{3}\right)(1-2\nu), \text{ at } r = a \tag{11.35}$$

$$\text{Normal displacement: } \delta = \frac{a^2}{R_C} = \left(\frac{3F}{4E_C}\right)^{2/3}\left(\frac{1}{R_C}\right)^{1/3}, \tag{11.36}$$

$$\text{Normal stiffness: } c = 2E_C a = \left(6FR_C E_C^2\right)^{1/3}. \tag{11.37}$$

Please note that Equations 11.30 through 11.37 apply to the contact of spherical surfaces (one of which may be flat, i.e., has an infinite radius). That particular situation can be handled by assuming a very long radius for that surface).

The contact area in the interface is a circle. These are the most commonly used equations. Other equations exist for line contacts and contacts between two surfaces each with two orthogonal radii. A good summary is given in Appendix C of Hale.[3]

A useful analytical tool that can be used to calculate Hertz stress values and stiffnesses in the contact region may be found at http://en.vinksda.nl/toolkit-mechanical-calculations/hertz-contact-stress-calculations. This freeware program can take into account the effects of surface roughness, if desired. The results are generally conservative. An example of a printout from this program (without considering roughness) is shown in Figure 11.53. Further calculations (not shown) indicate that increasing the surface roughness of the fused silica sphere to 1.0 μm (3.9×10^{-5} in.) would reduce the stress levels and contact stiffness by a factor of 3. This indicates the importance of roughness to the calculation results.

Vukobratovich (2014) summarized the use of an approximate method for solving Hertzian contact stress calculations as described by Brewe and Hamrock.[4] The results from using this method are reported to be accurate within about 4%.

11.3.7.2 Sphere in a V-Groove

The sphere in a V-groove contact constrains two DOFs. It is illustrated in Figure 11.38c for the case where the groove consists of two perpendicular surfaces. This is generally done for manufacturing reasons. There is no reason why this angle has to be 90°. Figure 11.38d illustrates the more general design with angle α variable (as shown, it is ~240°). Changing the angle affects the stiffness in the interface. Combining three connections of different types results in a so-called *kinematic mount.*

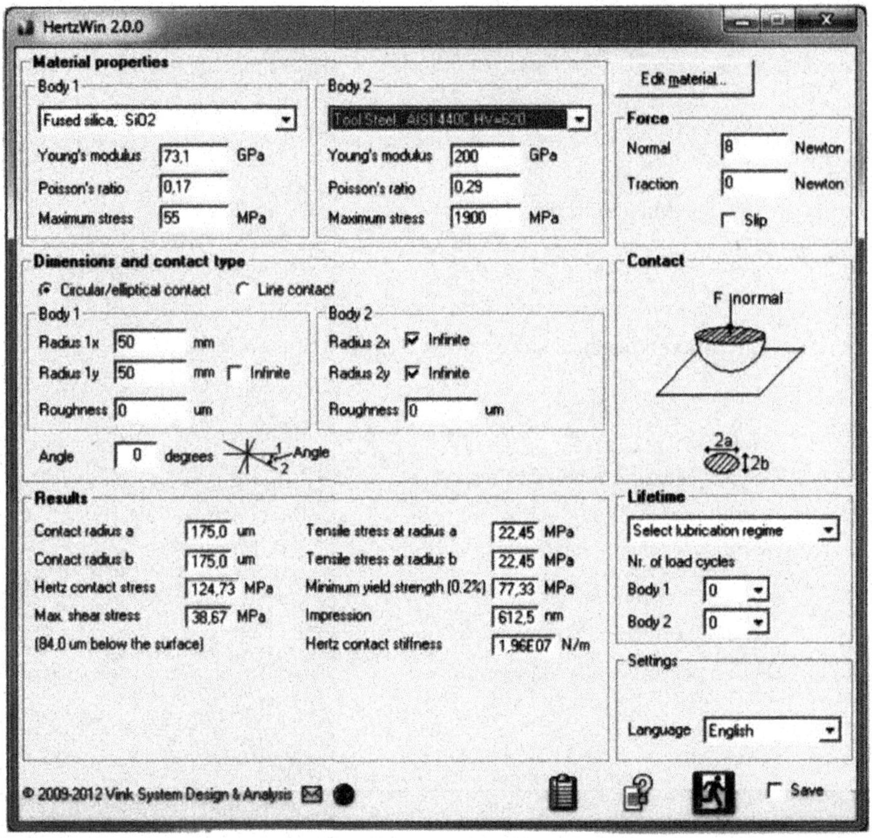

FIGURE 11.53 Computer printout of a typical calculation of Hertz contact stress for a sphere pressed against a flat surface using a freeware product of Vink System Design and Analysis.

A kinematic mount is often used to mount optical components so that they can be removed and replaced with constant positional accuracy.

The stiffnesses C_V and C_H in the vertical and horizontal directions for a sphere in a V-groove can be calculated using the following equations:

$$C_V = (C)(\sin\alpha)^2, \tag{11.38}$$

$$C_H = (C)(\cos\alpha)^2. \tag{11.39}$$

For $\alpha = 45°$, these stiffnesses are equal. However, if needed, one can select a different value and obtain higher stiffness in a preferred direction. Please note that friction can have an important impact on the performance of the sphere in a V-groove connection. This is discussed in Section 11.4.2.

11.3.7.3 Sphere in a Cone

The stiffnesses in the vertical and horizontal directions of a sphere in a cone connection are given by Equations 11.41 and 11.42. These are greater than in the case of a sphere in a V-groove because there is a line contact instead of two point contacts. This stiffness can be calculated when one

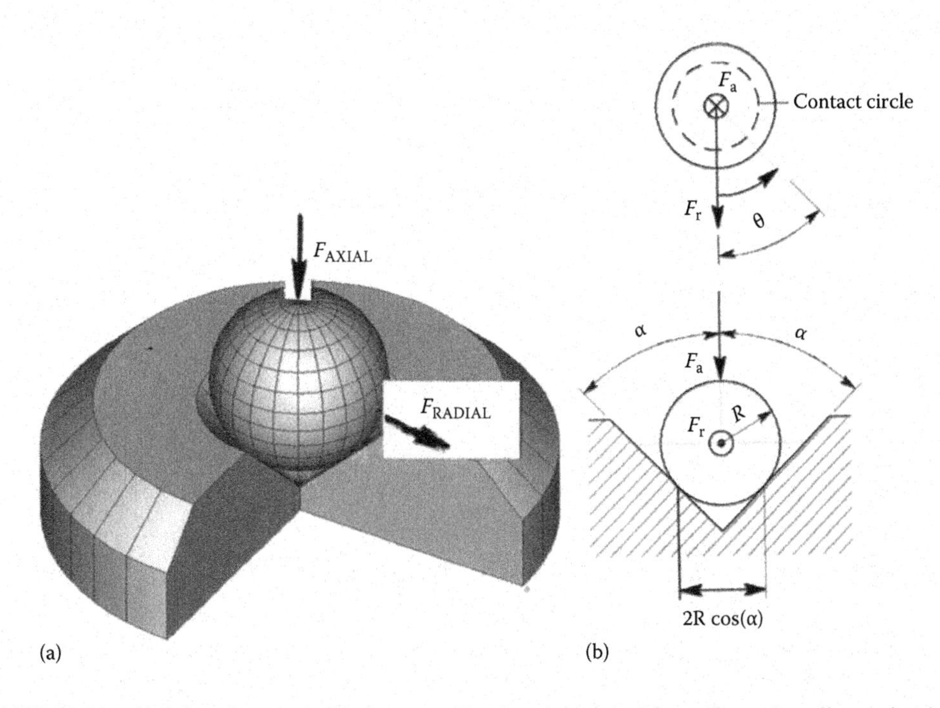

FIGURE 11.54 (a) Sphere in a conical hole pressed in place by vertical force F_{AXIAL} (usually gravity) is constrained radially by force F_{RADIAL} around the circular line contact zone. (b) Geometry of the sphere in a cone showing circular ring contact.

realizes that the physical loading condition between a sphere and a cone is basically similar to that of a cylinder and a flat plate. The cylinder has radius R and contact length $l = 2\pi R \sin \alpha$. Loads on the sphere can be defined as (F_{AXIAL}) and radial (F_{RADIAL}), as shown in Figure 11.54a. Equations for these parameters, as well as several others of importance in the design of this type of interface, are provided in the following. The geometry is shown in Figure 11.54b:

$$\text{Line load: } P = \left(\frac{1}{2\pi R \cos\alpha}\right)\left(\frac{F_a}{\sin\alpha} + \frac{2F_r \cos\phi}{\cos\alpha}\right). \tag{11.40}$$

Note that ($\cos \phi$) varies from −1 to +1. By substituting those values into Equation 11.40, one can establish the maximum and minimum line loads P_{max} and P_{min}:

$$\text{Axial stiffness: } C_a = \left(\frac{2\pi^2 E_C R(\cos\alpha)(\sin^2\alpha)}{\ln\left(\frac{4d_1}{b_{max}}\right) + \ln\left(\frac{4d_2}{b_{min}}\right) - 2}\right). \tag{11.41}$$

To close approximation, $d_1 = R$ and $d_2 = 10R$:

$$\text{Radial stiffness: } C_r = \left(\frac{\pi^2 E_C R \cos^3\alpha}{\ln\left(\frac{4d_1}{b_{max}}\right) + \ln\left(\frac{4d_2}{b_{min}}\right) - 2}\right) \tag{11.42}$$

Half width of line contact:

$$b_{\text{min}} = \left(\frac{4P_{\text{min}} R_{\text{C}}}{\pi E_{\text{C}}} \right)^{1/2} \tag{11.43}$$

$$b_{\text{max}} = \left(\frac{4P_{\text{max}} R_{\text{C}}}{\pi E_{\text{C}}} \right)^{1/2} \tag{11.44}$$

$$\text{Maximum pressure: } p = \frac{2P_{\text{max}}}{\pi b} = \left(\frac{1}{\pi} \right) \left(\frac{PE_{\text{C}}}{\pi R_{\text{C}}} \right)^{1/2} \tag{11.45}$$

$$\text{Maximum shear stress: } \tau = 0.30p \quad \text{at } z = 0.78b \tag{11.46}$$

An important consideration regarding the line contact of a sphere in a cone interface is that the contact quality depends upon the surface geometry and qualities of both the cone and the sphere. For example, if the cone is slightly oval in shape, line contact cannot exist. Rather, there are two point contacts. The potential consequences are discussed in the next section. Please note that the sphere in a cone solution is used in both the kinematic mount and the Kelvin clamp, which are used widely in optomechanical designs. These devices are described in Sections 11.4.2.2 and 11.4.2.3.

11.4 MOUNTING AND CONSTRAINING DOFs OF OPTOMECHANICAL COMPONENTS

Optical components must be mounted inside the instrument in such a way that environmental deformations (mechanical or thermal) do not deform or misalign those components because that would reduce the optical performance and could even cause mechanical failure. In the two previous sections, we have explained the principles for constraining the DOFs and the required specific qualities of the structural elements. Now, this knowledge needs to be applied to the mounting of optical components. We also show how adjusting some of the DOFs in very specific ways allows the component to be aligned relative to other components in the instrument.

11.4.1 Mounting Optomechanical Components

Controlling the six DOFs of a component is often very much determined by the environment, as explained in Chapter 2 of this book. Many optical systems have to be mounted in a very compact way. This can restrain optical mounting to only a few options. For instance, lenses mounted inside the objective of a high-performance photographic camera are usually stacked in a barrel. Mounting them in a different way, e.g., individually attached to a flat baseplate, might be possible functionally, but would be cumbersome and more complex mechanically because adjustment of the interrelationships between elements would need to be accomplished individually. Providing multiple mechanisms for adjustment could become very expensive.

In many other optomechanical instruments, the structural requirements may accommodate entirely different mounting arrangements compared to the camera lens. Then, components or their mountings may interfere with each other or with the structure in the early design concepts.

For instance, if we try to use struts to support a prism, it may be that other components are located at the places where one would have liked to attach the struts to the structure. Hence, compromises are needed and an inconvenient arrangement may result.

The environment also strongly influences the optomechanical design. Systems to be used in space must survive launch and conditions such as depressurization, extreme temperatures, radiation, and gravity release. Instruments to be used in a military environment often are exposed to extreme vibration and shock, temperature extremes, humidity, rainfall, or even submersion in water. These demand different internal mounting provisions and material choices than would the camera lens mentioned previously.

11.4.1.1 Component Shape

Mounting round optical components is usually easier and cheaper than mounting irregularly shaped ones. A clear example of this is once again the photographic camera lens. Here, the mounting cells and housing (barrel) can be made on a lathe, which keeps the optical components concentric because their interfaces are machined in the same setup. In the case of rectangular or irregularly shaped components (e.g., prisms), this is more difficult to realize. Often the surfaces that can be used for mounting are parallel, which easily leads one to suggest that leaf springs might be a good solution for mounting. These parts tend to be more expensive to manufacture and the alignment of the components may be more difficult than the cylindrical barrel.

11.4.1.2 Mounting Transmitting or Reflecting Optics

Transmissive components such as lenses can generally be mounted only on their outer cylindrical surface or by contacting the very edges of the optical surfaces. With small mirrors, usually the rim is available for mounting purposes. Using the mirror back side could be tricky because of the possible thermal expansion differences (especially when bonding is used). Mechanical clamping or bonding on ground surfaces are preferred means for mounting prisms. Large mirrors, such as those considered in several chapters of this book, require complex mounting arrangements due to their extreme sizes and their great sensitivities to gravity, thermal effects, and vibration. In all cases, excessive mechanical stressing of the optics must be avoided. The key goals of optomechanical design are that the various parts of the instrument and of its mount are interconnected in the most direct way and that the interfaces all the way to the ground are designed in a straightforward manner.

11.4.1.3 Thermal Stresses due to Temperature Changes

Mounting brittle components (glass, crystals, or ceramics) require special care. Temperature changes will cause thermal stress in the components depending on the mismatch in the coefficient of thermal expansion (CTE) with their mounts and with any bonding materials used. These stress levels can easily result in failures of the optical components. Examples of such failures of a lens and of a prism from this cause are shown in Figure 11.55. All these failures occurred at low temperatures because parts with widely differing CTEs were improperly bonded together or improperly mounted.

Metal mirrors are generally more ductile than glass, so they are less sensitive to brittleness problems. In smaller sizes, it may be possible to bolt or bond metal optics directly to their mechanical mounts. Stress can be introduced by differences in thermal expansion, but no catastrophic failure should result. The optical surface may, of course, be distorted if the mirror is improperly interfaced to its mount. If possible, each optic should be mounted in a static determined way.

11.4.1.4 Thermal Isolation

Using bonding material (or metal parts with low thermal conductance) to connect an optical component to its mechanical surround may result in thermal isolation. When the components are then subjected to sudden temperature changes, this can result in temperature-induced stress. This may

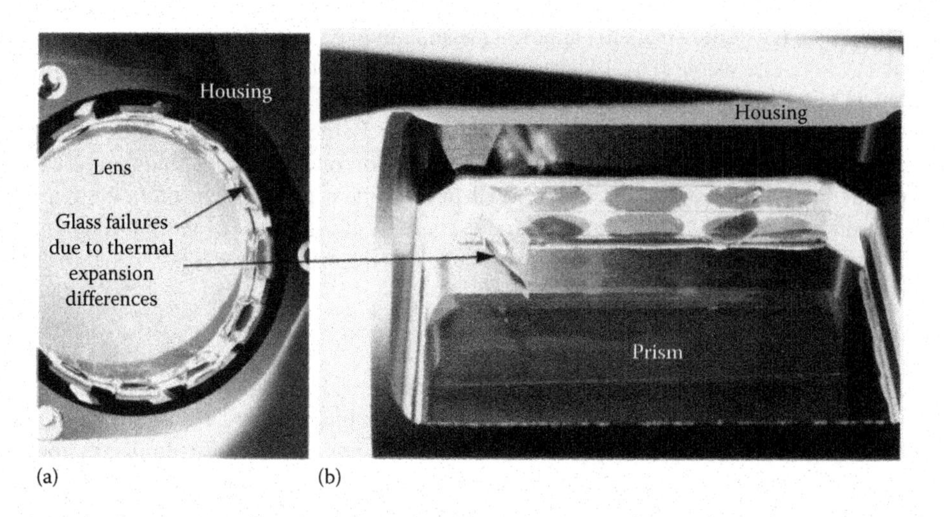

FIGURE 11.55 Stress fractures in glass components (a) lens and (b) prism under temperature change caused by a mismatch between the CTEs of the glasses and metal mounts.

happen, e.g., during the launch of a satellite, when an instrument is taken from a warm enclosure into the cold outdoors, or during thermal vacuum testing.

11.4.1.5 Quantity or One-Off Production

In some ways, the number of instruments to be produced from a given design influences the design itself. This thought needs explanation in at least two ways. First, designing and building a single instrument, such as an astronomical telescope, to be used for as long as possible without on-site maintenance requires careful selection of materials, extremely stable component alignment, remote testing capability (if possible), and intricate temperature controls. The one and only James Webb Space Telescope, which will not be accessible for servicing by astronauts because of its high orbit, requires all alignment, testing, and maintenance features to be anticipated during design and operated remotely. Reliable function and achievement of long-term high-performance goals are mandatory. This adds up to specification complexity, attention to details, thorough planning, and execution of equipment designs and procedures for each step in the development, fabrication, qualification, assembly, installation, launch, alignment, performance assessment, and operation from Earth. It is no wonder that such a system is expensive and takes a long time to bring to operational mode.

At the other end of the design spectrum, we find a multitude of mass-produced but very simple optical instruments such as cameras built into ubiquitous telephones, computers, etc. These cameras perform adequately for their sizes and do not require (or at least do not receive) maintenance. The lifetimes of the devices into which these optics are installed are relatively short because they tend to be replaced within a year or so by new models offering apparently greater capabilities. Extremely long lifetimes are not required of the optics in such cases. If they fail, the whole device may be discarded and replaced.

Intermediate to these extremes are astronomical telescopes, binoculars, digital and video cameras, rifle telescopes, surveyor's transits, etc., that are designed for use by professional and amateur astronomers, bird-watchers, yachtsmen, nature study groups, huntsmen, etc. These instruments usually are built to either low or high performance standards as appropriate to meet the prospective customer's willingness or ability to pay. The degrees of design perfection in all these cases are related to the purchase prices. Lower-cost instruments may be marginally designed, assembled, and aligned, while the most expensive versions reflect genuine technical capability and the desire on the part of the manufacturer to deliver the highest-quality products.

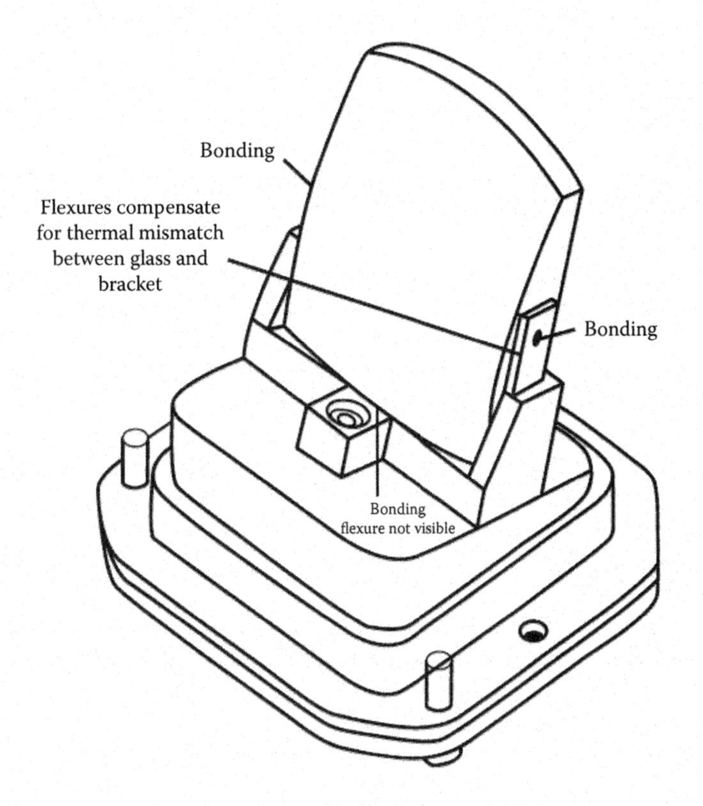

FIGURE 11.56 Mirror mounted on three flexure blades (one not visible) to prevent excessive stress under temperature change.

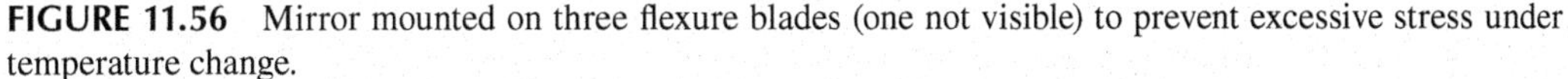

11.4.2 Alternate Mounting Methods

11.4.2.1 Three Leaf Springs

A common way to mount some optical components is to use three leaf springs. These are attached to the optic or to a mechanical cell containing the optic so all six DOFs are constrained once. An example is shown in Figure 11.56. One leaf spring is not visible here. Two parallel leaf springs are used. The third is mounted symmetrically* between the other two and perpendicular. For practical reasons, one may deviate from this arrangement by choosing another angle. This allows the stiffnesses of the mounting to be optimized. When mounting circular lenses, or small mirrors, three leaf springs at 120° interval might be best. The stiffnesses in any lateral direction would then be equal. The stiffness normal to the lens would then be twice as high as the lateral stiffness.

Examining Figure 11.56, one might argue that each leaf spring constrains three DOFs, causing the mount to be three times overconstrained. To overcome that problem, one might substitute three constricted leaf springs. However, using three normal leaf springs is not so bad. This is because the bonded joints are slightly flexible and ease the overconstraint. Usually this is true when the bonding spot is rather small (typically <8 mm diameter). Nominally, the leaf spring should have the same width as the bonding spot because it then is easier to verify that the correct amount of bonding material is injected through the small tapered access holes. The main reason for keeping the bonding spot diameter small is the necessity to keep the thermal stresses low when the temperature changes.

The three leaf springs also provide a TC. This is located at the intersection of a horizontal line that connects the centers of the bonding spots on the sides of the mirror and the vertical passing

* The position of the third leaf spring is important with respect to the TC (Section 11.4.2.2.2) that is created and that should coincide with the optic axis. Usually, this means that the third leaf spring is mounted symmetrically.

through the center of the third bonding spot. Typically, this TC should be located on the optical axis. It remains in place even if the optic is subjected to thermal gradients because its position is defined by the stiffnesses of the leaf springs.

The leaf springs in this design usually do not need to be very thin. This is because the thermal mismatch causes no more than a few tens of micrometer deflection. Such deflections are easily absorbed by flexures of a few millimeters in thicknesses. They can be machined using conventional machining methods. It is advantageous if the center of gravity of the optical component is centered between the leaf springs because this symmetry helps to balance the forces acting on the bonding spots. As indicated in the discussion of bonding spot sizes in Section 9.4.1, only in cases of relatively heavy components or extreme shock/vibration conditions would large spots be needed. Caution should be exercised in choosing large bonding spots because of the potential for aggravating thermal mismatch problems between the component and the mount.

Sometimes optical components are very small. In such cases, it could be acceptable to omit the third (lower) leaf spring. This is because the side bonding spots provide adequate stability. The optic is still mounted in a static determined way (2 × 3 DOFs are constrained). Note that the torsion stiffnesses of the bonding spots are no longer negligible and that there is no longer a TC.

11.4.2.2 Kinematic Mount

A kinematic mount consists of one body to which three portions of spheres or protuberances machined as spheres are attached so that they fit into three V-grooves on a separate body. It is a combination of structural elements introduced in Section 11.3.7.2. The spheres are usually attached to a removable component, while the component containing the V-grooves is attached to the ground. However, there is nothing against reversing this arrangement. From Figure 11.57, one can understand that the ring with the three spheres could actually be a cell holding a glass lens, a mirror, a mechanical actuator, etc.

An important feature of a precisely made kinematic mount (shown in the views in Figure 11.58a and b) is that it can be designed symmetrically so as to fit together in any 120° rotational orientation. It also can be designed in a slightly asymmetric manner (as shown in the view in Figure 11.58c) so as to fit in only one way; i.e., it is static determined. This means that the parts bearing the optical components can be dismounted and remounted to the instrument without losing their relative

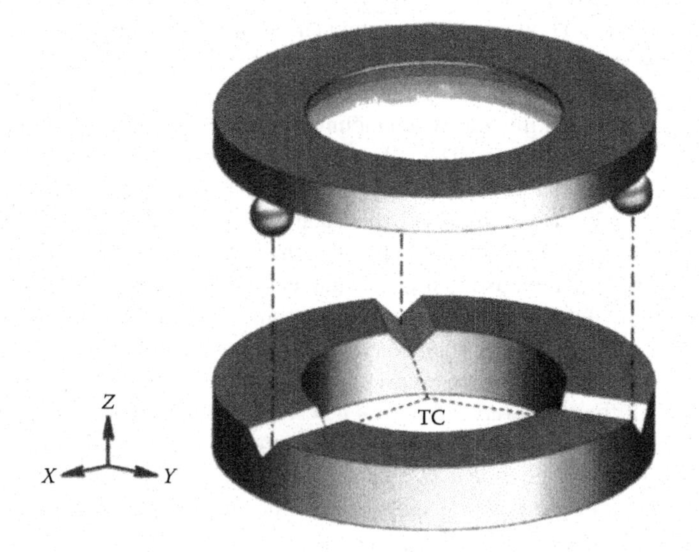

FIGURE 11.57 Kinematic mount that can be used to support an optical component in a manner that allows removal and replacement without misalignment. Means for preload not shown.

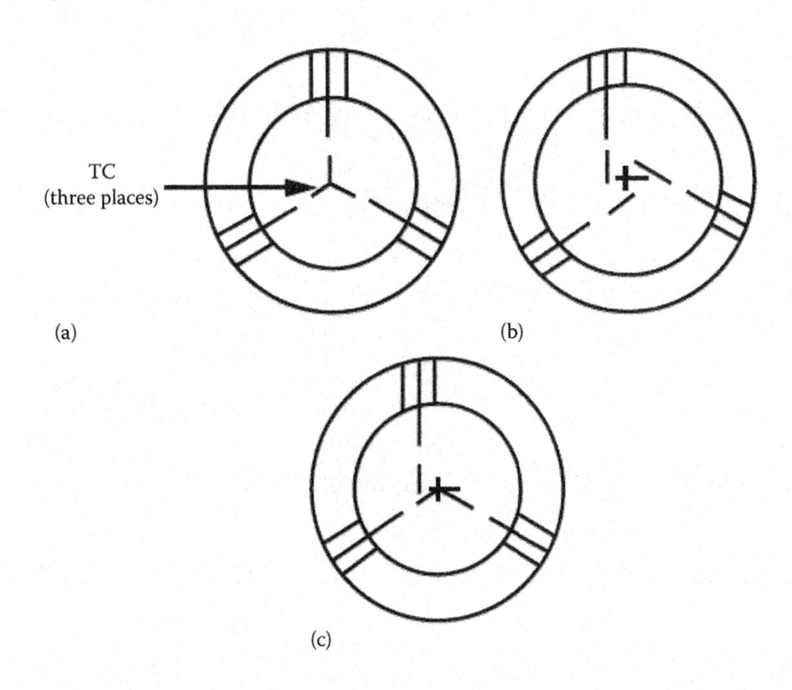

FIGURE 11.58 Configurations for three V-grooves in a kinematic mount base: (a) symmetrical case with dihedral edges at 120° intervals and intersecting at the geometric center; (b) grooves symmetrically arranged, but extended dihedral edges not intersecting at the center; (c) unsymmetrical case that fits together in only one rotational orientation.

rotational alignment. This attribute could be important if delicate components or subassemblies of an optical instrument need to be removed for separate packaging and shipping. If the interfaces and the components attached thereto are closely dimensioned and tight manufacturing tolerances are met, the use of the kinematic mount also allows interchanging those components or subassemblies with new ones during maintenance.

It should be noted that only the configuration in Figure 11.58a has a TC. The configurations in Figure 11.58b and c have geometric centers. The mounted component does not move laterally when the temperature changes. It does, however, make a small rotation. For rotationally symmetric components, this is acceptable. It would not be acceptable for an off-axis parabola and similar optics.

The preload of the kinematic mount usually is provided by gravity and/or by springs (see Figure 11.59). In the latter case, springs ensure that the repeatability of the preload is very high because the contacting surfaces will be slightly deformed to a predefined deflection. The use of screws to hold the fixed and removable parts of a kinematic mount together is discouraged because a specific applied preload is then hard to achieve.

Most often, a 90° angle is chosen for the V-groove. Provided that the friction coefficient is not equal to or higher than unity, this is perfectly acceptable. However, parts are usually cleaned very thoroughly before they are built into an optical system, especially if meant for vacuum operation. This increases the friction coefficient considerably, and friction coefficients greater than unity can easily be obtained. Selecting proper material combinations will help. Figure 11.60 shows a three-V-groove base for a kinematic mount in which thin ceramic plates have been attached to the three V-grooves. This enables different surface hardness and friction characteristics. The angle between the plates shown is >90°, which provides another variable for design optimization. Another option would be to reduce the V-groove angle (i.e., $\alpha < 90°$). This would increase the lateral stiffness and decrease the axial stiffness within the interface. Choosing the appropriate angle would allow to make the stiffness components equal. Selecting 70° would do so. Then, a maximum friction coefficient

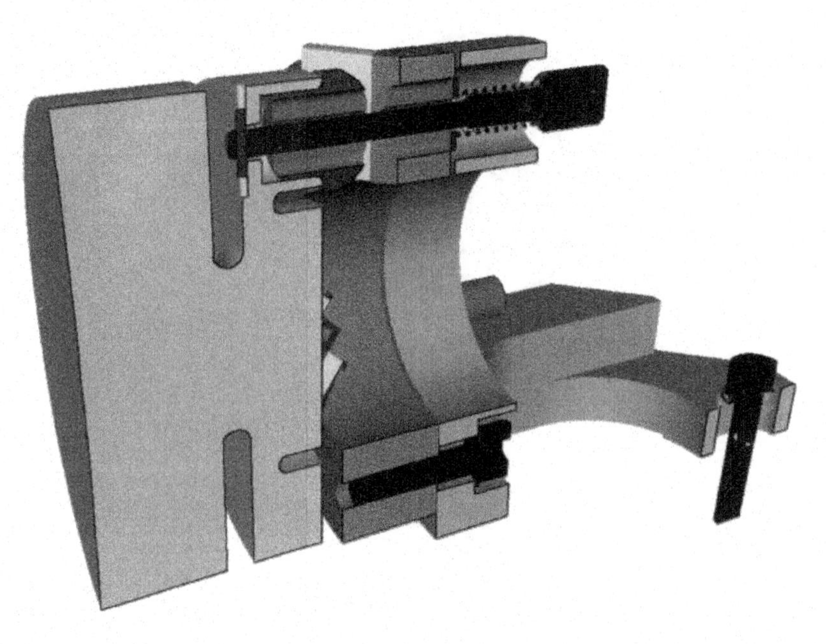

FIGURE 11.59 Application of preload to a kinematic mount by three spring-loaded bayonet pins, one of which is visible.

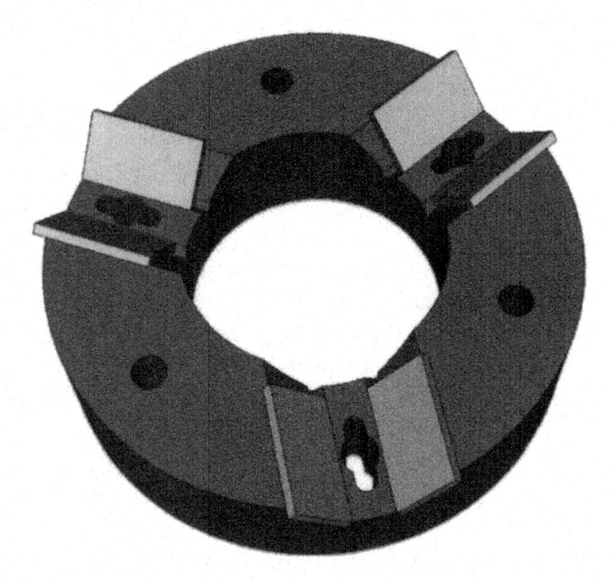

FIGURE 11.60 Base for a kinematic mount with dissimilar material pads attached to the groove faces to modify hardness or friction effects.

of 1.4 would be allowed. Usually that is sufficient. The stiffness in any direction of the component/unit will be equal to the stiffness normal to one of the V-groove surfaces. The stiffness ratio is then 100%. If the V-groove angle is 90°, this ratio would be 0.75 in the lateral direction and 1.5 in the axial direction. These relationships result from using Equations 11.47 and 11.42 through 11.44.

Applying a kinematic mount requires one to be aware of its critical characteristics. Each of these is summarized next.

11.4.2.2.1 Contact Stress

Between the sphere and the V-groove, two point contacts exist. Because the contact areas are very small, high Hertzian stress levels will occur. The methods from Section 11.3.6 can be used to estimate stress. Obviously, no plastic deformation of the contact points is wanted, so suitable materials must be selected. Usually, materials with high strength limits are used. A way to reduce the contact stress is to provide V-grooves with curved contact profiles as shown in Figure 11.61. This concept, known by optomechanical engineers as the "Gothic arch interface," can also be applied to creating a conical seat for a sphere.

A kinematic mount has certain limitations with respect to its applicability due to the possible high surface contact stress levels. The following can be concluded:

- The acceleration to which a kinematic mount can be subjected is limited. Usually, that means that it should be used in relatively static applications.
- Space applications will be difficult or even impossible due to the high loads during launch.
- Impact loads are to be avoided. They will easily damage the contact areas. In a case where brittle materials are used, impacts could cause failure of the sphere or the V-groove.

However, a case where one or more of the mating surfaces is merely plastically deformed from one of these exposures does not mean that the kinematic mount cannot be used any more. Its performance will be slightly degraded, but its mounting reproducibility should still be excellent. It may be possible for any resultant misalignment to be compensated for by readjustment. If subjected to large temperature changes, one may find that the kinematic mount has been somewhat degraded. The material of the V-groove should be harder than that of the sphere to avoid an impact on the sliding capabilities of the kinematic mount. For instance, hardened steel and ZiO_2 have been found to make a good combination.

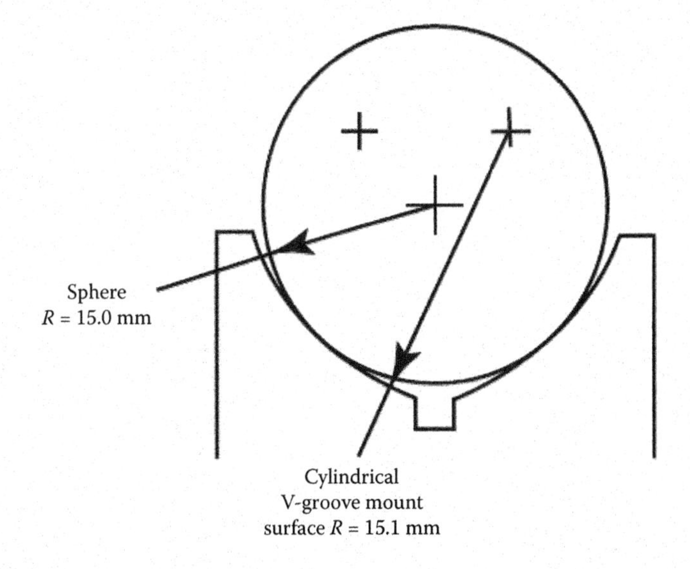

FIGURE 11.61 Suggested means for reducing the Hertzian stress in the contact between a sphere (or cylinder) resting in a V-groove by substituting a long-radius, nonconcentric curve rather than a flat surface, commonly called the Gothic arch interface.

11.4.2.2.2 Thermal Center

Each kinematic mount has a TC. This is a very useful feature in optical applications. Its position is identified in Figure 11.58a and is at the center of the groove pattern. Note that the TC is a two-dimensional property, not a three-dimensional one.

A change in temperature will change the sizes of the component and its mount when materials with different CTEs are used. A relative movement between the sphere and the V-groove will occur, but the transverse location of the TC will not change. In general, a displacement in the axial direction also will occur. This displacement depends on the materials that have been used. Although important, axial changes of optical systems are of less importance than lateral changes. The most important requirement is to keep the optical components centered on the optical axis.

11.4.2.2.3 Thermal Gradients

Thermal gradients can be minimized, but cannot be completely eliminated. The consequences are that the shapes of the optic and of its mount change very slightly. These changes may not be symmetrical, and they may not scale uniformly. The center of the component may shift slightly if the temperature gradient across the mount is large. Gradients frequently are temporary because the temperature tends to equalize.

11.4.2.2.4 Friction

Friction exists in a kinematic mount and influences its reproducibility. When inserting the component, there is always one sphere that contacts one of the V-groove surfaces first. The sphere will then slide down until the other contact is also made. Friction has maximum effect during this sliding action. The next time that the kinematic mount is used, another point will first make contact, thereby creating a different stress distribution. Hence, it can be concluded that friction may affect the reproducibility of the mount. Generally, this effect is small.

Friction effects can be minimized by providing elastic flexibility in the mount. An example of how this can be implemented is shown in Figure 11.62. Under preload, the contact points tend to remain together. This configuration has been tested and shows that the reproducibility of the optical mount improves to less than 1 μm. Because the dimensions of the lens (or mirror) and its mount change when the local temperature changes and because their CTEs will usually be different, the spheres may shift in the V-grooves. The resulting friction force may disturb the optical surface form of the lens/mirror.

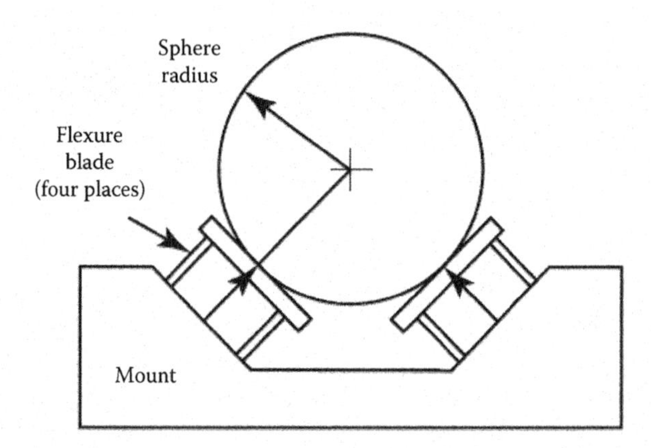

FIGURE 11.62 Suggested means for reducing surface contact friction by adding compliance to V-groove surfaces with flexures.

11.4.2.2.5 Position Uncertainty

In case that no specific measures are taken to reduce the friction effects, an uncertainty remains concerning the position repeatability of the kinematic mount. The uncertainty of location in a spherical contact assuming that the same material is used in both sphere and contact is given approximately by

$$\delta \approx \mu\left[2/(3R)\right]^{1/3}(F/E)^{2/3}, \tag{11.47}$$

where δ is the positional uncertainty, μ is the friction coefficient, R is the sphere radius, F is the force acting at the contact, and E is the elastic modulus of the sphere and of the mating material. To quantify this uncertainty approximately, let us assume the contacting materials to be steel ($E = 200$ GPa), the sphere radius to be 10 mm, the friction coefficient to be 1.0, and the force to be 75 N. The resulting uncertainty is 2.1 μm. To reduce this value, one would need to increase the radius of the sphere or reduce the force.

11.4.2.2.6 Some Variations on the Kinematic Mount

The kinematic mount of Figure 11.57 has lateral stiffnesses that are equal in all lateral directions because the V-grooves are located at 120° intervals. The vertical stiffness is twice the lateral stiffness. Changing the angular intervals will influence the stiffness. Physical limitations may prevent the 120° orientation of the grooves. This is not considered to be a severe consequence, but the actual instrument requirements will prove whether that is true. A rather common configuration is when two V-grooves are oriented in line with each other. The best orientation of the third will then be perpendicular to the other two. Each groove interface is conventional in this configuration.

The cone for a conventional kinematic mount is shown in Figure 11.63a. A variation results when the V-grooves are rotated by 45° about the axis defined by the dihedral edge of the V-groove. Three V-groove surfaces are now located on a common plane. The other three surfaces are oriented perpendicular to this plane. (See Figure 11.63b.) Some form of force F (gravity or a spring in the views in Figure 11.63a and c) is needed to guarantee the physical contact between the mount and all three V-grooves. The TC still exists, and the stiffness properties of the kinematic mount are not changed. There are no strong reasons why this version of the kinematic mount should be used instead of the normal one. It depends mainly on the specific environmental conditions. From the mechanical point of view, the kinematic mount in Figure 11.63b is easier to manufacture, which tends to compensate for the additional spring.

Instead of using three V-grooves in a kinematic mount, six spheres can be used. Two spheres make contact with a third one—as shown in Figure 11.63c. The nice thing about this solution is

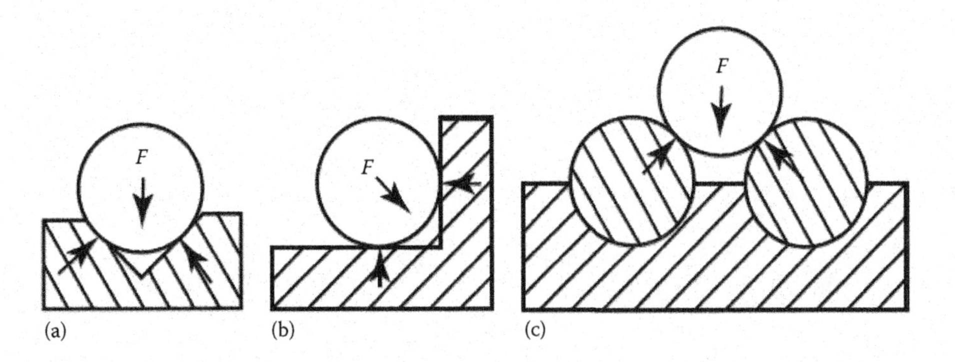

FIGURE 11.63 (a) Normal configuration for a sphere in a V-groove. (b) Alternate configuration with the V-groove rotated 90°. (c) Alternate arrangement with two spheres replacing the groove. F indicates applied preload in all cases.

that it is very inexpensive. Steel and ceramic balls are readily available. For an ordinary kinematic mount, it is recommended that different materials be used for the spheres and for the V-grooves to keep friction low in the contacts. In vacuum applications, this will also prevent cold welding.

If a spring is used to preload the optical mount on its kinematic mount, as in the view in Figure 11.63b, the preload is chosen so accelerations or external loads cannot cause loss of contact between each sphere and its V-groove. Provided that the external load direction on the optical components does not change, it is possible to remove and reinstall it well within micrometer and microradian ranges. Two important conditions have to be fulfilled: (1) External loading before and after reinstallation must be identical in magnitude and direction. Gravity and spring loads are adequate in meeting this requirement. Bolts are clearly not suitable because the preload that they cause is highly variable and depends on friction. (2) Friction between the spheres and the V-groove surfaces must be small. If not, friction may prevent both surfaces from making contact with the sphere.

11.4.2.3 Kelvin Clamp

Instead of a kinematic mount, a Kelvin clamp might be used (see Figure 11.64a). It provides six points of contact between the removable part and the fixed base. One contact (a sphere in a cone) constrains three DOFs, one contact (a sphere in a V-groove) constrains two DOFs, and the third contact (a sphere touching a flat surface) constrains one DOF. This Kelvin clamp is not strictly kinematic because the contacts are made over areas, rather than true point contacts. In a true kinematic design, the contact surface deformation and stress are determined by geometry and the elastic properties of the materials being independent of area.

The conventional Kelvin clamp is potentially subject to instability because the loads are distributed over areas, and the qualities of the contact areas determine the repeatability. For example, when a sphere is placed in a cone, a circular line of contact develops. Repeatability is then partially a function of the roundness of both the cone and sphere. In an alternative design, one movable sphere rests against three adjacent fixed spheres (the close-packed spheres are an alternative to a conical socket). The V-groove interface is provided by a pair of spheres. The second movable sphere contacts these fixed spheres and constrains two DOFs. Point contacts develop and the only factors determining repeatability are friction, the sizes of the spheres, and the elastic properties of the materials. Figure 11.64b shows a Kelvin clamp using three spheres as the cone and two spheres as the V-groove along with a flat. It is truly kinematic, constraining all six DOFs, and is less expensive than the traditional design.

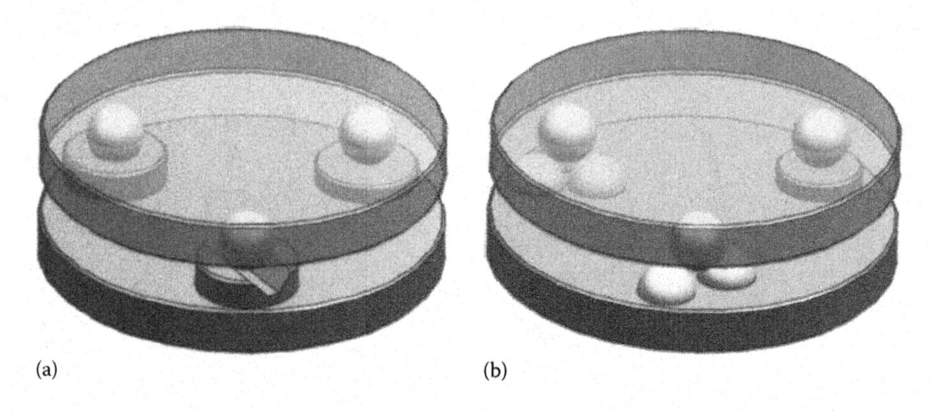

FIGURE 11.64 Semitransparent renderings of (a) a classical kinematic mount with spheres resting in or on a cone, a V-groove, and a flat. (b) Alternate configuration with three spheres acting as the cone and two spheres acting as the V-groove.

11.4.2.4 Six Struts

Mounting a unit using six connecting struts is a technique often used to provide a static determined interface between two units—presumably one fixed and one removable. It is not generally used to mount a single component. This is very convenient in a client–supplier relation because the interface load will be well defined. Hence, the development of both units will have minimal impact on each other. It is not generally used to mount a single component.

The best way to connect the struts to the two structures is by the use of threaded joints coaxial with the strut centerlines. This means that the contact areas will be normal to any applied prestress and the actual loads during use. This way, minor slippages at the contact areas will have negligible effects.

The dimensions of the struts depend on the weight of the supported component. The selected diameter could be as small as 0.5 mm (depending on what can be manufactured) or as large as 5 mm or possibly more depending on the applied loads. Achieving the proper length-to-diameter ratio also must be taken into account.

A good example of a unit mounted using six struts is described next. Figure 11.65 depicts the subassemblies that fit together to create this unit. The "K" values associated with each subassembly indicate the temperature in kelvins during use. This is a cryogenic stage with an inner section (upper right) that was cooled to 2 K. The intermediate stage was cooled to 4 K, while the outer structure was cooled to 10 K. Each stage was connected to the other by using six struts made of medical injection needles. These provided excellent thermal isolation because they were hollow as well as the poor thermal conductivity of stainless steel. The cylinders also provided considerable resistance to buckling load.

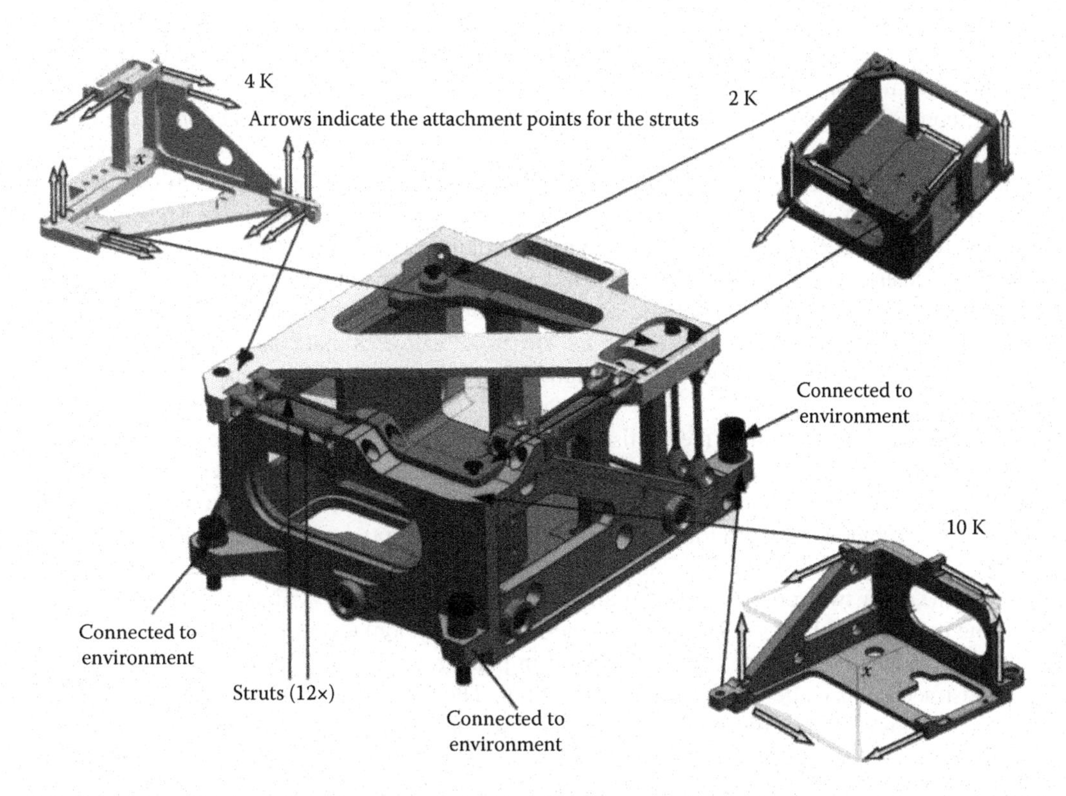

FIGURE 11.65 Configuration of a cryogenic structural unit in which subassemblies are interconnected with struts. (Courtesy of SRON Netherlands Institute for Space Research, Utrecht.)

11.5 ALIGNMENT OF OPTICS BY CONTROLLING THE DOFs

Most, if not all of the components that are mounted in a static determined way using the methods described in this and other sections can be adjusted by changing one or more of their DOFs. Adjusting a strut can cause a connected component to translate or rotate. Adjusting a leaf spring can accomplish the same results. This would be called *alignment* when it is done (presumably once) prior to the delivery of the unit to the customer. In case that the adjustment results from some mechanism that is available and that the user can operate and given the availability of a suitable error sensing means, it can be checked on-site. Theoretically, alignment is permanent and should result in a stable condition for the hardware. However, alignment is more or less subject to the influences of the environment. This is especially true for vibration, shock, and temperature effects. Usually, means for measuring errors and performance-specific tooling are needed to perform the alignment. Generally, these are adjuncts to the hardware—available when needed, but not part of the hardware itself. On the contrary, apparatus such as spaceborne optical instruments that are not accessible for human use on-site would need to be built into the flight apparatus or brought to that location from Earth. It should be obvious that providing actuators and power for making adjustments would significantly increase the complexity and cost of the instrumentation needing adjustment. Defining and providing these "tools" is beyond the scope of this chapter.

11.5.1 ALIGNMENT

Prior to performing an alignment, it should be known exactly which DOFs are to be adjusted and to what accuracy. In some types of optical devices, this can be done by one of the optical analysis software packages that are available in the market. The achievable test accuracy is very high.

It will always be best to mount components based on mechanical tolerances because that is the easiest and therefore the cheapest method. However, in general, some of the components are designed to be used as optical compensators. These components have to be adjusted in a very specific way to compensate for the manufacturing errors of the components and its support structure(s). Usually, components such as lenses or mirrors need to be adjusted in the direction of the optical axis or in a lateral direction. Rotation around one of the lateral axes can often be exchanged with a translation in the lateral direction depending on the optical aberrations that this might cause. In some cases, rotation and translation are fully exchangeable (e.g., spherical mirrors). The mechanism design with the smallest but adequate sensitivity for adjustment is usually chosen as the alignment parameter because it should produce the best and quickest alignment result.

One of the prime qualities of a static determined support structure is that adjusting one of the DOFs will hardly affect the stress levels in the structural elements that constrain the other DOFs. Optimizing the DOF decoupling during design will minimize the consequences for these stress levels.

An example of an alignment mechanism is shown in Figure 11.66. One plate is to be adjusted relative to another in piston and tip/tilt. Both plates have specific mechanical provisions to facilitate this alignment. Their relative positions for both these directions are controlled by three screws arranged in a triangular pattern (only two are shown). These motions are spring loaded using disk springs.* Adjusting one screw will cause a rotation around a line that is defined by the contact points of the other two screws. Adjusting all three in an identical way will change the piston position. The three lateral DOFs (*XY* translation and in-plane rotation) are restrained by friction.

The transverse motions (two translations and in-plane rotation DOF) can be adjusted by using two adjustment tools (called handles in Figure 11.66). These tools are relatively long and have two spherical sections at a specific axial separation. These spheres fit into two through holes in the top plate and two blind holes in the lower plate. Tilting the tools will cause in-plane motions of the

* Otherwise known as Belleville washers.

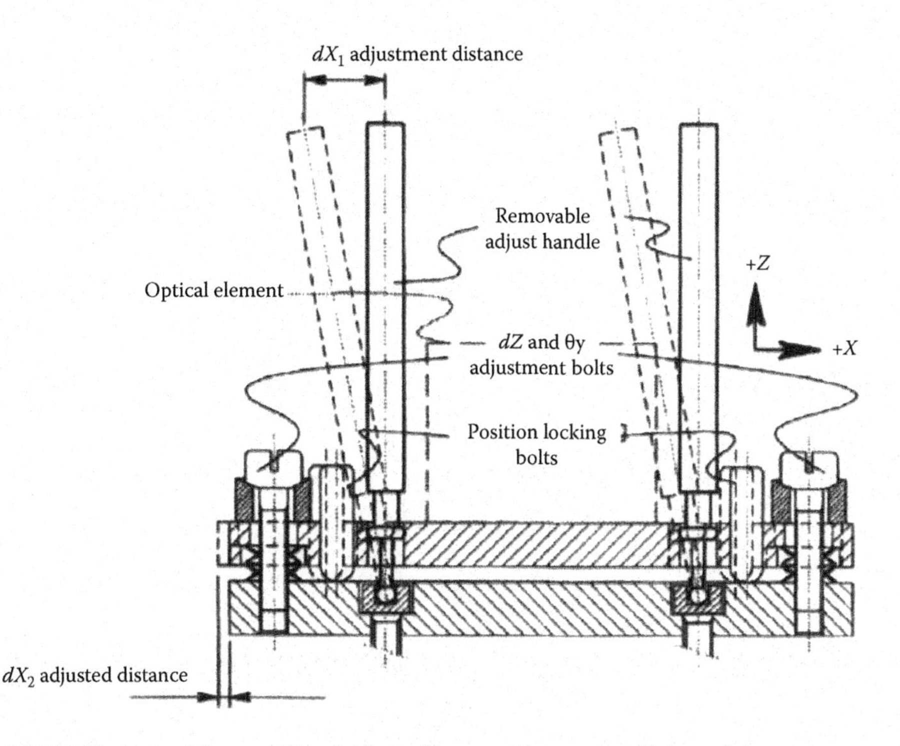

FIGURE 11.66 Example of the use of two tools to align an optic mount on its baseplate.

upper plate relative to the bottom plate. Having visible feedback (such as may come from a separate camera and monitor) displaying the consequences of the tool motions allows adjusting the upper plate to its best in-plane position. Obviously, one has to think ahead in what order the six DOFs are to be adjusted to avoid making too many iterations.

After having found the best position for the upper plate, it is now possible to lock its position by tightening the three locking screws that are located immediately next to the adjustment screws. It would be best if these locking screws could be made concentric with the adjustment screws because that would minimize localized bending stress within the plates.

Another type of adjustment between two plates is shown in Figure 11.67. This is a simple elastic hinge. Adjusting the screw will result in a small rotation. A constricted leaf spring attached to one plate is used to lock the obtained position. This is done by clamping the leaf spring against the second plate by using ordinary screws. The result of this clamping action on the location of the upper plate relative to that of the lower plate will be negligible (although not zero) because the clamping action is oriented perpendicular to the direction of motion of the leaf spring when the upper plate rotates. The change in the relative location of the plates can be measured using test instrumentation such as the capacitive position sensor shown.

Both these examples illustrate that alignment and subsequent locking might be subject to iterations in case that the locking has noticeable impact on the alignment. Bonding a constricted leaf spring to the block might improve the locking but may introduce thermal sensitivity because of the much different reaction of the bonding material to temperature changes compared to that of metals or glass. One could also fill the gap on either side of the elastic hinge with the bonding material. This is much worse because the bonding material shrinks during curing and has a relatively high CTE and so will change dimensions as the temperature changes. These effects could affect the spatial relationship of the fixed and adjustable components.

It is useful to point out the difference between accuracy and precision in any adjustment. These factors determine repeatability. See Figure 11.68, which should be self-explanatory.

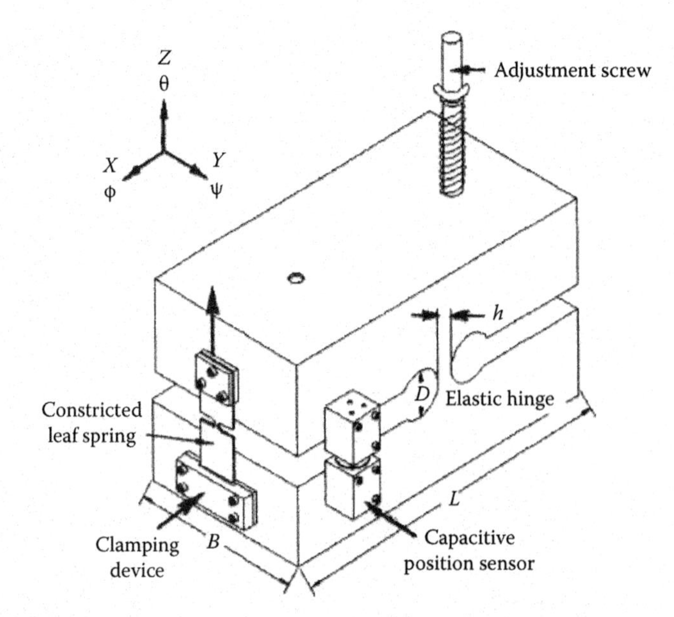

FIGURE 11.67 Means for adjusting and locking angular alignment between an upper assembly and a baseplate.

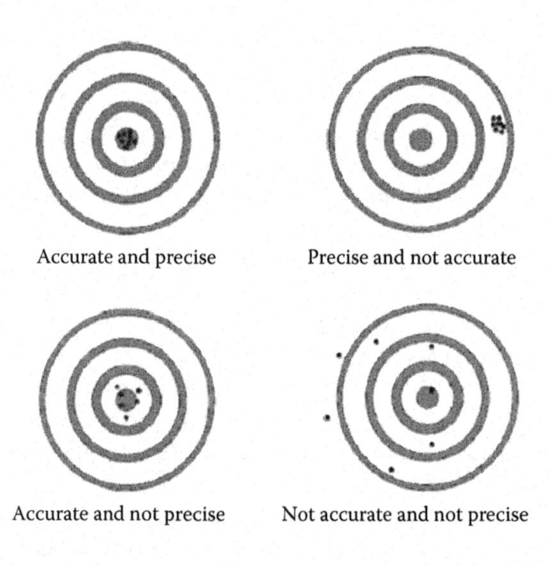

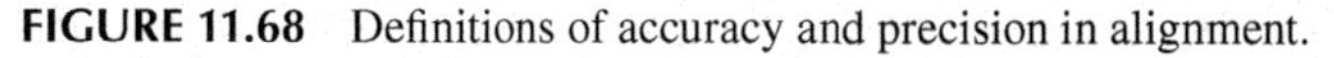

FIGURE 11.68 Definitions of accuracy and precision in alignment.

11.5.2 Alignment Stability

After having performed an alignment, it is important that the obtained result does not change with time. We distinguish between short-term drift and long-term stability. Their causes are different. The main causes of short-term drift are (1) changes in torque setting of bolts and screws that are used during alignment. The internal stress distribution will be affected and therefore also the structural deformation. Releasing these settings also can affect the alignment. (2) Releasing bolts and screws will do the same. (3) Bolts show settling effects; i.e., the internal stress of the bolts tends to decrease within a short period. This is a well-known phenomenon that necessitates verification of the torque values. Torquing and releasing the bolts several times within a short time interval will significantly reduce the settling effect.

The main causes of long-term instability are (1) temperature changes, especially when various materials are combined and no proper thermal compensation has been performed. The best result is obtained in the few cases where the same material is used for the optical component and for the structure. An example is the use of aluminum for both a mirror and its structure. (2) Not all materials have good creep properties; e.g., Invar is known for its poor stability properties. It necessitates use of the proper heat treatment and careful handling. Bolts used to clamp Invar parts may introduce creep. By proper mechanical design, it is possible to minimize these consequences. (3) Stress introduced into optical components (especially metal ones) may relax, thereby causing changes in the optical surface(s). Keeping the stress levels well below the microyield stress level will ensure minimal stress relaxation.

PROBLEMS

1. The stability of a camera tripod is determined by its settling time, which is the time required for vibration to decay to 1% of the initial displacement after impulsive excitation, as discussed in Chapter 8; see Equation 8.46. Rank the following extruded aluminum alloys in terms of settling time (hint: see Equations 11.4 and 11.5 to determine natural frequency dependence on materials properties): 7075 aluminum, with elastic modulus $E_{7075} = 71.7$ GPa and yield strength $\sigma_{7075} = 503$ MPa; 6061-T62, with elastic modulus $E_{6061} = 68.3$ GPa and yield strength $\sigma_{6061} = 241$ MPa; and 5083, with elastic modulus $E_{5083} = 70.3$ GPa and yield strength $\sigma_{5083} = 165$ MPa. *Answer:* Settling time is inversely proportional to natural frequency, and natural frequency is proportional to the elastic modulus of the material; yield strength does not influence frequency. So the best material is 7075, followed by 5083, and then 6061. Note that the settling time for 6061, the worst of the three materials, is only about 2.5% longer than that of the best material, 7075; $(71.7 \text{ GPa}/68.3 \text{ GPa})^{½} = 1.025$.
2. Find the ratio of tension to bending stiffness, $C_{TENSION}/C_{BENDING}$, for a hollow cylindrical cantilever strut with a thin wall. For a thin-walled cylindrical section, $A = 2\pi Rt$ and $I = \pi R_3 t$, where $R = d/2$ and t is the section wall thickness. *Answer:* $C_{TENSION}/C_{BENDING} = (E2\pi Rt/L)/(3E\pi R_3 t/L^3) = (2/3)(L^2/R^2) = (16/3)(L^2/d^2)$. Note that this ratio is independent of the wall thickness t.
3. A cylindrical titanium strut with geometry similar to that of Figure 11.44b has the following dimensions: $d = 3$ mm, $L = 100$ mm, and $b = 12$ mm. For this strut, the lengths of the thin sections are $L/10$, not $L/7$. The elastic modulus of titanium is $E = 114$ GPa. Find the stiffness in tension and bending. What is the decoupling ratio of this strut? *Answer:* $C_{TENSILE} = 80.6 \times 10^6$ N/m, $C_{BENDING} = 1.36 \times 10^3$ N/m, and $C_{TENSILE}/C_{BENDING} = 59.3 \times 10^3$.
4. A constricted leaf spring similar to the one shown in Figure 11.46b is made of AISI 302B stainless steel, with a thickness $t = 10$ mm. A rotation of $\psi = 15 \times 10^{-3}$ rad about the Y-axis is specified. The properties of 302B are an elastic modulus of $E = 193$ GPa and a yield strength $\sigma_\psi = 703$ MPa. Assume that the width of the elastic hinge $h = 1$ mm. What is the diameter D of the circular cutout in the hinge? What torque T_ψ is required for the maximum rotation of 15 mrad? What is the tension/compression stiffness C_X and lateral stiffness C_Z in the Z direction? *Answer:* $D = 5.705$ mm, the torque $T_\psi = 665 \times 10^{-6}$ N/m, the stiffness $C_X = 388 \times 10^6$ N/m, and the stiffness $C_Z = 65.5 \times 10^6$ N/m.
5. A constricted leaf spring similar to the one shown in Figure 11.46b is made of 7075 aluminum, with a thickness of $t = 12$ mm. The specified rotation is $\psi = 1°$ ($\approx 17.45 \times 10^{-3}$ rad), and the specified torque $T_\psi = 50 \times 10^{-3}$ N/m. The properties of 7075 aluminum are an elastic modulus of $E = 71.7$ GPa and a yield strength of $\sigma_\psi = 503.3$ MPa. What are the dimensions h and D of the spring? *Answer:* The width $h = 6.74 \times 10^{-3}$ m, and $D = 14.02 \times 10^{-3}$ m.

6. If the constricted leaf spring of problem 4 is replaced with a rectangular elastic hinge, what must be the length L of the hinge for the same bending stiffness? What is the ratio of tension C_X to lateral stiffness C_Z of the equivalent rectangular elastic hinge? *Answer:* The length $L = 2.15 \times 10^{-3}$ m, the ratio $C_X/C_Z = 13.6$.
7. A folded leaf spring similar to the one shown in Figure 11.51 is made of 302 stainless steel with dimensions $L = 20$ mm and $b = 10$ mm. The properties of 302 stainless steel are an elastic modulus $E = 193$ GPa and a Poisson's ratio of $\nu = 0.31$. If the desired stiffness of the leaf spring is $C_F = 1.5 \times 10^6$ N/m, what is the thickness t of the spring? *Answer:* The thickness $t = 579 \times 10^{-6}$ m.
8. A prism made of Schott N-BAK4 is held in place with type 302B stainless steel clips (see Chapter 7); each clip exerts a force of $F = 50$ N on the prism. What should be the radius R_C of the part of each clip pressing on the glass to limit the tensile stress σ to 10 MPa? For Schott N-BAK4 the elastic modulus is 77 GPa and the Poisson's ratio is 0.240. For 302B stainless steel, the elastic modulus is 193 GPa and Poisson's ratio is 0.25. *Answer:* The minimum radius is $R_C = 415 \times 10^{-3}$ m; the spring clips might as well be left flat.
9. A small mirror with a mass of 1 kg is supported by three sphere/flat interfaces, with each sphere and flat taking one-third of the mirror weight. The mirror is in a laboratory environment, with $a_G = 1g$. The sphere and flat interfaces are made of stainless steel, with an elastic modulus of $E = 193$ GPa, Poisson's ratio of $\nu = 0.27$, and yield strength of $\sigma = 600$ MPa. What is the minimum radius of the spherical part of the contact to keep the tensile stress below yield? *Answer:* The minimum radius is $R = 5.76 \times 10^{-3}$ m.
10. The interfaces of a kinematic mount are made of steel, with an elastic modulus of about 200 GPa. Uncertainty of position of the mount is about ±3 μm. Sapphire is suggested as a substitute for the steel; the elastic modulus of sapphire is 400 GPa. Assuming that the radius of contact, force, and friction in the contact are unchanged, what is the uncertainty in position if sapphire is used instead of steel? *Answer:* The uncertainty is $(3 \times 10^{-6}\ \text{m})(200\ \text{GPa}/400\ \text{GPa})^{1/2} = 2.12 \times 10^{-6}$ m.
11. A small mirror is mounted kinematically with three spheres resting in V-grooves, as shown in Figure 11.57. The radius of the spheres is R = 10 mm, and the angle of the V-groove is $\alpha = 45°$. The spheres and V-grooves are made of stainless steel, with an elastic modulus of 193 GPa and a Poisson's ratio of 0.27. The mirror mass is 500×10^{-3} kg. What are the vertical and natural frequencies of the mirror in its mount? Hint: Use Equation 8.34 in Chapter 8, and assume that the radial stiffness of the V-groove contact is zero; for F, use one-third of the mirror mass. Assume that gravity is acting perpendicular to the mirror optical surface. *Answer:* The radial natural frequency is $f_R = 666$ Hz, and the vertical natural frequency is $f_Z = 942$ Hz.
12. For maximum stability of a kinematic mount consisting of a sphere in a V-groove or cone, the sphere must be able to slide down to the bottom of the interface. What is the relationship between the friction coefficient μ for the sphere and V-groove or cone and the half angle α to ensure sliding? What angle is required for sliding if the friction coefficient $\mu = 1.0$? *Answer:* $\mu < \tan(90° - \alpha)$; when $\mu = 1.0$, $\alpha \leq 45°$.

ACKNOWLEDGMENT

For the past 25 years, I have worked for The Netherlands Organization for Applied Scientific Research (TNO) as systems engineer and optomechanical specialist. TNO is a large contract research organization that develops and builds precision equipment for space and astronomical applications as well as for companies active in the semiconductor sector. The experience gained from these activities forms a large part of the technical basis for writing this chapter about kinematic design and flexure applications. I am very grateful to TNO for granting me the time and access to the technology applied in the preparation of this chapter. I hope that it will prove useful to the readers of this book.

REFERENCES

1. Cardano, G., specific source unknown, circa 1545.
2. Hertz, H.R., *Über die berührung fester elastischer Körper (On the Contact of Rigid Elastic Solids)*, Macmillan, London, 1896.
3. Hale, L.C., *Principles and Techniques for Designing Precision Mechanisms*, Lawrence Livermore National Laboratory, University of California, Livermore, CA, 1999.
4. Brewe, D.E., and Hamrock, B.J., Simplified solution for elliptical-contact deformation between two elastic solids, *J. Lubr. Technol.*, 99, 485, 1977.

FURTHER READING

Braddick, H.J.J., *The Physics of Experimental Method*, Chapman & Hall, London, 1963.

Cacace, L.A., *An Optical Distance Sensor*, University Press TU-Eindhoven, Eindhoven, Netherlands, 2009.

Gloess, R., Challenges of extreme load hexapod design and modularization for large ground-based telescopes, *Proc. SPIE*, 7739, 2010.

Hamelinck, R.F., *Adaptive Deformable Mirror based on Electromagnetic Actuators*, ORO Grafisch Project Management, Koekange, Netherlands, 2010.

Henselmans, R., *Non-contact Measurement Machine*, Ponsen & Looije, Wageningen, Netherlands, 2009.

Johnson, K.L., *Contact Mechanics*, Cambridge University Press, Cambridge, MA, 1985.

Moore, J.H., Davis, C.C., and Coplan, M.A., *Building Scientific Apparatus*, Addison Wesley Press, Boston, MA, 1983.

Nijenhuis, J.R., and Hamelinck, R.F., Meeting highest performance for lowest price and mass for the M1 segment support unit for E-ELT, *Proc. SPIE*, 7735, 2010.

Nijenhuis, J.R., and Hamelinck, R.F., The optimization of the opto-mechanical performance of the mirror segments for the E-ELT, *Proc. SPIE*, 8336, 2011.

Seggelen, J., *A 3D coordinate measuring machine with low moving mass for measuring small products in array with nanometer uncertainty*, Doctoral dissertation, Technical University, Eindhoven, Netherlands, 2007.

Slocum, A.H., The design of three groove kinematic couplings, *Precis. Eng.*, 14, 67, 1992.

Soemers, H.M., *Design Principles for Precision Mechanisms*, T-Point Print, Enschede, Netherlands, 2011.

Vermeulen, J., *Ceramic optical diamond turning machine: Design and development*, Doctoral dissertation, Technical University, Eindhoven, Netherlands, 1999.

Vermeulen, M., *High-precision 3D-coordinate measuring machine: Design and prototype-development*, Doctoral dissertation, Technical University, Eindhoven, Netherlands, 1999.

Vukobratovich, D., Introduction to opto-mechanical design, SPIE short course SC014, 2014.

Werner, C., *A 3D Translation Stage for Metrological AFM*, Ipskamp Drukkers, Enschede, Netherlands, 2010.

Appendix 1: Optical Glasses

TABLE A1.1
Key Mechanical Properties of Current Schott Optical Glasses

Rank and Reference	Glass	Glass Code	Young's Modulus E (GPa)	Poisson's Ratio ν	$K_G = (1 - \nu_G^2)/E_G$ ($\times 10^{-11}$ Pa^{-1})	Thermal Expansion Coefficient α_G ($\times 10^{-6}$ °C^{-1})	Density ρ (g/cm^3)	Stress Optical Coefficient K_s ($\times 10^{-6}$ MPa^{-1})
1[a]	N-FK5	487704.245	62	0.232	1.53	9.20	2.45	2.91
2[a]	K10	501564.252	65	0.190	1.48	6.50	2.52	3.12
3[a]	N-ZK7	508612.249	70	0.214	1.36	4.50	2.49	3.63
4[a]	K7	511604.253	69	0.214	1.38	8.40	2.53	2.95
5[a]	N-BK7	517642.251	82	0.206	1.17	7.10	2.51	2.77
6[a]	N-K5	522595.259	71	0.224	1.34	8.20	2.59	3.03
7[a,b]	N-LLF6				No longer available			
8[a]	N-BAK2	540597.286	71	0.233	1.33	8.00	2.86	2.60
9[a]	LLF1	548458.294	60	0.208	1.59	8.10	2.94	3.05
10[a]	N-PSK3	552635.291	84	0.226	1.13	6.20	2.91	2.48
11[a]	N-SK11	564608.308	79	0.239	1.19	6.50	3.08	2.45
12[a]	N-BAK1	573576.319	73	0.252	1.28	7.60	3.19	2.62
13[a]	N-BALF4	580539.311	77	0.245	1.22	6.52	3.11	3.01
14[a]	LF5	581409.322	59	0.223	1.61	9.10	3.22	2.83
15[a]	N-BAF3				No longer available			
16[a]	F5	603380.347	58	0.220	1.64	8.00	3.47	2.92
17[a]	N-BAF4	606437.289	85	0.231	1.11	7.24	2.89	2.58
18[a]	F4				No longer available			
19[a]	N-SSK8	618498.327	84	0.251	1.12	7.21	3.27	2.36
20[a]	F2	620364.360	57	0.220	1.67	8.20	3.60	2.81
21[a]	N-F2	620364.265	82	0.228	1.16	7.84	2.65	3.03
22[b]	N-SK16	620603.358	89	0.264	1.05	6.30	3.58	1.90
23[a]	SF2	648339.386	55	0.227	1.72	8.40	3.86	2.62
24[a]	N-LAK22	651559.377	90	0.266	1.03	6.60	3.77	1.82
25[b]	N-BAF51	652450.333	91	0.262	1.02	8.37	3.33	2.22

(Continued)

TABLE A1.1 (CONTINUED)
Key Mechanical Properties of Current Schott Optical Glasses

Rank and Reference	Glass	Glass Code	Young's Modulus E (GPa)	Poisson's Ratio ν	$K_G = (1 - \nu_G^2)/E_G$ ($\times 10^{-11}$ Pa^{-1})	Thermal Expansion Coefficient α_G ($\times 10^{-6}$ °C^{-1})	Density ρ (g/cm^3)	Stress Optical Coefficient K_s ($\times 10^{-6}$ MPa^{-1})
26[b]	N-SSK5	658509.371	88	0.278	1.05	6.80	3.71	1.90
27[a]	N-BASF2	664360.315	84	0.247	1.12	7.12	3.15	3.04
28[a]	SF5	673322.407	56	0.233	1.69	8.20	4.07	2.28
29[a]	N-SF5	673323.286	87	0.237	1.08	7.94	2.86	2.99
30[a]	N-SF8	689313.290	88	0.245	1.07	8.56	2.90	2.95
31[a]	SF15				No longer available			
32[a]	N-SF15	699302.292	90	0.243	1.05	8.04	2.92	3.04
33[a]	SF1	717295.446	56	0.232	1.69	8.10	4.46	1.80
34[a]	N-SF1	717296.303	90	0.250	1.04	9.13	3.03	2.72
35[b]	N-LAF3				No longer available			
36[a]	SF10	728284.428	64	0.232	1.48	7.50	4.28	1.95
37[a]	N-SF10	728285.305	87	0.252	1.08	9.40	3.05	2.92
38[b]	N-LAF2	744449.430	94	0.288	0.98	8.06	4.30	1.42
39[b]	LAFN7	750350.438	80	0.280	1.15	5.30	4.38	1.77
40[b]	N-LAF7	749348.373	96	0.271	0.97	7.30	3.73	2.57
41[b]	SF4	755276.479	56	0.241	1.68	8.00	4.79	1.36
42[b]	N-SF4	755274.315	90	0.256	1.04	9.45	3.15	2.76
43[a]	SF14				No longer available			
44[a]	SF11	785258.474	66	0.235	1.43	6.10	4.74	1.33
45[a]	SF56A	785261.492	57	0.239	1.65	7.90	4.92	1.10
46[a]	N-SF56				No longer available			
47[a]	SF6	805254.518	55	0.244	1.71	8.10	5.18	0.65
48[a]	N-SF6	805254.337	93	0.262	1.00	9.03	3.37	2.82
49[a]	LASFN9				No longer available			
	P-SF68	005210.619	79	0.275	1.17	8.43	6.19	1.61

(*Continued*)

TABLE A1.1 (CONTINUED)
Key Mechanical Properties of Current Schott Optical Glasses

Rank and Reference	Glass	Glass Code	Young's Modulus E (GPa)	Poisson's Ratio ν	$K_G = (1 - \nu_G^2)/E_G$ ($\times 10^{-11}$ Pa^{-1})	Thermal Expansion Coefficient α_G ($\times 10^{-6}$ °C^{-1})	Density ρ (g/cm^3)	Stress Optical Coefficient K_s ($\times 10^{-6}$ MPa^{-1})
	LASF35	022291.541	132	0.303	0.69	7.40	5.41	0.73
	FK5HTi	487705.245	62	0.232	1.53	9.20	2.45	2.91
	N-FK51A	487845.368	73	0.302	1.24	12.74	3.68	0.70
	N-PK52A	497816.370	71	0.298	1.28	13.01	3.70	0.67
	N-BK10	498670.239	71	0.203	1.35	5.80	2.39	3.21
	P-BK7	516641.243	85	0.202	1.13	5.99	2.43	2.77
	N-BK7HT	517642.251	82	0.206	1.17	7.10	2.51	2.77
	N-BK7HTi	517642.251	82	0.206	1.17	7.10	2.51	2.77
	N-KF9	523515.250	66	0.225	1.44	9.61	2.50	2.74
	N-PK51	529770.386	74	0.295	1.23	12.35	3.86	0.54
	N-BALF5	547536.261	81	0.214	1.18	7.34	2.61	2.76
	LLF1HTi	548459.294	60	0.208	1.59	8.10	2.94	3.05
	N-KZFS2	558540.255	66	0.266	1.41	4.44	2.54	4.02
	N-BAK4	569560.305	77	0.240	1.22	6.99	3.05	2.90
	N-BAK4HT	569560.305	77	0.240	1.22	6.99	3.05	2.90
	N-BAK4HT	569560.305	77	0.240	1.22	6.99	3.05	2.90
	LF5HTi	581409.322	59	0.223	1.61	9.10	3.22	2.83
	P-SK57Q1	586595.301	93	0.249	1.01	7.23	3.01	2.17
	P-SK57	587596.301	93	0.249	1.01	7.23	3.01	2.17
	P-SK58A	589612.297	97	0.245	0.97	6.82	2.97	2.12
	N-SK5	589613.330	84	0.256	1.11	5.50	3.30	2.16
	N-SK14	603606.344	86	0.261	1.08	6.00	3.44	2.00
	N-SK2	607567.355	78	0.263	1.19	6.00	3.55	2.31

(Continued)

TABLE A1.1 (CONTINUED)
Key Mechanical Properties of Current Schott Optical Glasses

Rank and Reference	Glass	Glass Code	Young's Modulus E (GPa)	Poisson's Ratio ν	$K_G = (1 - \nu_G^2)/E_G$ ($\times 10^{-11}$ Pa^{-1})	Thermal Expansion Coefficient α_G ($\times 10^{-6}$ °C^{-1})	Density ρ (g/cm^3)	Stress Optical Coefficient K_s ($\times 10^{-6}$ MPa^{-1})
	N-SK2HT	607567.355	78	0.263	1.19	6.00	3.55	2.31
	N-BAF52	609466.305	86	0.237	1.10	6.86	3.05	2.42
	P-SK60	610579.308	99	0.253	0.95	7.06	3.08	2.04
	N-KZFS4	613445.300	78	0.241	1.21	7.30	3.00	3.90
	N-KZFS4HT	613445.300	78	0.241	1.21	7.30	3.00	3.90
	N-SK4	613586.354	84	0.261	1.11	6.46	3.54	1.92
	N-PSK53A	618634.357	76	0.288	1.20	9.56	3.57	1.16
	F2HT	620364.360	57	0.220	1.67	8.20	3.60	2.81
	N-SSK2	622533.353	82	0.261	1.14	5.81	3.53	2.51
	N-KZFS11	638424.320	79	0.251	1.19	6.56	3.20	4.21
	N-LAK21	640601.374	91	0.272	1.02	6.08	3.74	1.74
	N-SF2	648338.272	86	0.231	1.10	6.68	2.72	3.06
	N-LAK7	652585.384	90	0.277	1.03	7.10	3.84	1.65
	N-KZFS5	654397.304	89	0.243	1.05	6.38	3.04	3.57
	N-BAF10	670471.375	89	0.271	1.04	6.18	3.75	2.37
	N-LAK12	678552.410	87	0.288	1.05	7.60	4.10	1.44
	P-SF8	689313.290	86	0.253	1.09	9.41	2.90	2.73
	N-LAK9	691547.351	110	0.285	0.84	6.30	3.15	1.83
	P-LAK35	693532.385	101	0.289	0.91	8.13	3.85	1.76
	N-LAK14	697554.363	111	0.283	0.83	5.50	3.63	1.73
	N-BASF64	704394.320	105	0.264	0.89	7.30	3.20	2.38
	N-LAK8	713538.375	115	0.289	0.80	5.60	3.75	1.81
	N-KZFS8	720347.320	103	0.248	0.91	7.77	3.20	2.94

(Continued)

TABLE A1.1 (CONTINUED)
Key Mechanical Properties of Current Schott Optical Glasses

Rank and Reference	Glass	Glass Code	Young's Modulus E (GPa)	Poisson's Ratio ν	$K_G = (1 - \nu_G^2)/E_G$ ($\times 10^{-11}$ Pa^{-1})	Thermal Expansion Coefficient α_G ($\times 10^{-6}$ °C^{-1})	Density ρ (g/cm^3)	Stress Optical Coefficient K_s ($\times 10^{-6}$ MPa^{-1})
	N-LAK10	720506.369	116	0.286	0.79	5.68	3.69	1.97
	P-SF69	723292.293	96	0.251	0.98	8.99	2.93	2.66
	N-LAK34	729545.402	117	0.290	0.78	5.81	4.02	1.52
	N-LAF35	743494.412	109	0.301	0.83	5.27	4.12	2.29
	P-LAF37	755457.399	115	0.296	0.79	6.26	3.99	2.26
	N-LAK33B	755523.422	122	0.295	0.75	5.83	4.22	1.43
	N-SF14	762265.312	88	0.259	1.06	9.41	3.12	2.89
	N-LAF34	773496.424	123	0.292	0.74	5.80	4.24	1.44
	N-SF11	785251.322	92	0.257	1.02	8.52	3.22	2.94
	N-LAF33	786441.436	111	0.301	0.82	5.60	4.36	2.21
	N-LAF21	788475.428	124	0.245	0.74	5.99	4.28	1.46
	N-LASF45	801350.363	116	0.281	0.79	7.36	3.63	2.01
	N-LASF45HT	801350.363	116	0.281	0.79	7.36	3.63	2.01
	N-LASF44	804465.444	124	0.293	0.74	6.21	4.44	1.41
	N-SF6HT	805254.337	93	0.262	1.00	9.03	3.37	2.82
	N-SF6HTultra	805254.337	93	0.262	1.00	9.03	3.37	2.82
	SF6HT	805254.518	55	0.244	1.71	8.10	5.18	0.65
	N-LASF43	806406.426	114	0.290	0.80	5.49	4.26	1.92
	P-LASF47	806409.454	120	0.298	0.76	6.04	4.54	2.39
	P-LASF50	809405.454	119	0.298	0.76	5.90	4.54	2.41
	P-LASF51	810409.458	119	0.299	0.76	6.01	4.58	2.32
	N-LASF40	834373.443	111	0.304	0.82	5.84	4.43	2.19
	N-LASF41	835431.485	124	0.294	0.74	6.19	4.85	1.57

(*Continued*)

TABLE A1.1 (CONTINUED)
Key Mechanical Properties of Current Schott Optical Glasses

Rank and Reference	Glass	Glass Code	Young's Modulus E (GPa)	Poisson's Ratio ν	$K_G = (1 - \nu_G^2)/E_G$ ($\times 10^{-11}$ Pa^{-1})	Thermal Expansion Coefficient α_G ($\times 10^{-6}$ °C^{-1})	Density ρ (g/cm^3)	Stress Optical Coefficient K_s ($\times 10^{-6}$ MPa^{-1})
	N-SF57	847238.353	96	0.260	0.97	8.46	3.53	2.78
	N-SF57HT	847238.353	96	0.260	0.97	8.46	3.53	2.78
	N-SF57HTultra	847238.353	96	0.260	0.97	8.46	3.53	2.78
	SF57	847238.551	54	0.248	1.74	8.30	5.51	0.02
	SF57HTultra	847238.551	54	0.248	1.74	8.30	5.51	0.02
	N-LASF9	850322.441	109	0.288	0.84	7.37	4.41	1.72
	N-LASF9HT	850322.441	109	0.288	0.84	7.37	4.41	1.72
	N-LASF31A	883408.551	126	0.301	0.72	6.74	5.51	1.18
	N-LASF46A	904313.445	124	0.298	0.73	6.00	4.45	1.64
	N-LASF46B	904313.451	121	0.303	0.75	5.97	4.51	1.87
	P-SF67	907214.424	90	0.248	1.04	6.23	4.24	2.96
	N-SF66	923209.400	95	0.259	0.98	5.90	4.00	2.86
Inquiry as of January 2015	N-LAK33A1	754523.422	121	0.292	0.76	5.80	4.22	1.49
Inquiry glass[c]	BAFN6	589485.317	77	0.234	1.23	7.80	3.17	2.50
Inquiry glass[c]	FK3	464658.227	46	0.243	2.05	8.20	2.27	3.71
Inquiry glass[c]	KZFS12	696363.384	66	0.279	1.40	5.20	3.84	2.35
Inquiry glass[c]	N-BAF3	583466.279	82	0.226	1.16	7.20	2.79	2.73
Inquiry glass[c]	N-LAF3	717480.414	95	0.286	0.97	7.60	4.14	1.53
Inquiry glass[c]	N-LAF36	800424.443	110	0.305	0.82	5.70	4.43	2.25
Inquiry glass[c]	N-PSK53	620635.360	78	0.288	1.18	9.40	3.60	1.16
Inquiry glass[c]	N-SF19	667331.290	88	0.231	1.08	7.20	2.90	2.93
Inquiry glass[c]	N-SF56	785261.323	91	0.255	1.03	8.70	3.28	2.87
Inquiry glass[c]	N-SF64	706302.299	88	0.245	1.07	8.50	2.99	2.95

(Continued)

TABLE A1.1 (CONTINUED)
Key Mechanical Properties of Current Schott Optical Glasses

Rank and Reference	Glass	Glass Code	Young's Modulus E (GPa)	Poisson's Ratio ν	$K_G = (1 - \nu_G^2)/E_G$ ($\times 10^{-11}$ Pa^{-1})	Thermal Expansion Coefficient α_G ($\times 10^{-6}$ °C^{-1})	Density ρ (g/cm^3)	Stress Optical Coefficient K_s ($\times 10^{-6}$ MPa^{-1})
Inquiry glass[c]	N-SK10	623570.364	81	0.266	1.15	6.80	3.64	2.25
Inquiry glass[c]	N-SK15	623580.362	84	0.265	1.11	6.70	3.62	1.93
Inquiry glass[c]	P-PK53	527662.283	59	0.271	1.57	13.30	2.83	2.06
Inquiry glass[c]	SF57HT	847238.551	54	0.248	1.74	8.30	5.51	0.02
Inquiry glass[c]	SFL6	805254.337	93	0.260	1.00	9.00	3.37	2.79
Inquiry glass[c]	SFL57	847236.355	97	0.261	0.96	8.70	3.55	2.73

Source: Data (except for K_G) are from Schott, *Schott Optical Glass Catalog*, Schott North America, Inc., Duryea, PA, 2014.

[a] Glass ranking is from Walker, B.H., *The Photonics Design and Applications Handbook*, Lauren, Pittsfield, MA, H-356, 1993.

[b] Glass ranking is from Zhang, S., and Shannon, R.R., Lens design using a minimum number of glasses. *Opt. Eng.*, 34, 3536, 1995.

[c] Inquiry glasses are available by special order only.

Appendix 2: Selected Alkali Infrared Materials

TABLE A2.1
Optomechanical Properties of Selected Alkali Halides and Alkaline Earth Halides

Material Name (Symbol)	Refractive Index n at λ (μm)	dn/dT at λ (μm) (× 10^{-6} °C^{-1})	CTE α (× 10^{-6} °C)	Young's Modulus E (× 10^{10} Pa)	Poisson's Ratio ν	Density ρ (g/cm^3)	Knoop Hardness (kg/mm^2)	$K_G = (1-\nu_G^2)/E_G$ (× 10^{-11} Pa^{-1})
Barium fluoride	1.463 at 0.63	−16.0 at 0.6	6.7 at 75 K	5.32	0.343	4.89	82	1.659
(BaF_2)	1.458 at 3.8	−15.9 at 3.4	18.4 at 300 K				(500 g load)	
	1.449 at 5.3	−14.5 at 10.6						
	1.396 at 10.6							
Calcium fluoride	1.431 at 0.7	−10.4 at 0.66	18.9 at 300 K	9.6	0.29	3.18	160–178	0.954
(CaF_2)	1.420 at 2.7	−8.1 at 3.4						
	1.411 at 3.8							
	1.395 at 5.3							
Potassium bromide	1.555 at 0.6	−41.9 at 1.15	25.0 at 75 K	2.69	0.203	2.75	7	3.564
(KBr)	1.537 at 2.7	−41.1 at 10.6					(200 g load)	
	1.529 at 8.7							
	1.515 at 14							
Potassium chloride	1.474 at 2.7	−36.2 at 1.15	36.5	2.97	0.216	1.98	7.2	3.210
(KCl)	1.472 at 3.8	−34.8 at 10.6					(200 g load)	
	1.469 at 5.3							
	1.454 at 10.6							
Lithium fluoride	1.394 at 0.5	−16.0 at 0.46	55 at 77 K	6.48	0.225	2.63	102–113	1.465
(LiF)	1.367 at 3.0	−16.9 at 1.15	−37 at 20°C				(600 g load)	
	1.327 at 5.0	−14.5 at 3.39						

(*Continued*)

TABLE A2.1 (CONTINUED)
Optomechanical Properties of Selected Alkali Halides and Alkaline Earth Halides

Material Name (Symbol)	Refractive Index n at λ (μm)	dn/dT at λ (μm) (× 10^{-6} °C^{-1})	CTE α (× 10^{-6} °C)	Young's Modulus E (× 10^{10} Pa)	Poisson's Ratio ν	Density ρ (g/cm^3)	Knoop Hardness (kg/mm^2)	$K_G = (1 - v_G^2)/E_G$ (× 10^{-11} Pa^{-1})
Magnesium fluoride	1.384 at 0.46o[a]	+0.88 at 1.15	14.0 (‖) 8.9	16.9	0.269	3.18	415	0.549
(MgF_2)	1.356 at 3.8o	+1.19 at 3.39	(⊥)					
	1.333 at 5.3o							
Sodium chloride	1.525 at 2.7	− 36.3 at 0.39	39.6	4.01	0.28	2.16	15.2	2.298
(NaCl)	1.522 at 3.8						(200 g load)	
	1.517 at 5.3							
Thallium	2.602 at 0.6	−254 at 0.6	58	1.58	0.369	7.37	40.2	5.467
bromoiodide	2.446 at 1.0	−240 at 1.1					(200 g load)	
(KRS5)	2.369 at 10.6	−233 at 10.6						
	2.289 at 30	−152 at 40						

Source: Yoder, P.R. Jr., *Mounting Optics in Optical Instruments*, 2nd ed., SPIE Press, Bellingham, WA, 2008, p. 717.

Note: o = ordinary axis.

[a] Birefringent material.

Appendix 3: Selected Infrared-Transmitting Glasses

TABLE A3.1
Optomechanical Properties of Selected Infrared-Transmitting Glass and Other Oxides

Material Name (Symbol)	Refractive Index n at λ (μm)	dn/dT at λ (μm) (× 10^{-6} °C^{-1})	CTE α (× 10^{-6} °C^{-1})	Young's Modulus E (× 10^{10} Pa)	Poisson's Ratio ν	Density ρ (g/cm^3)	Knoop Hardness (kg/mm^2)	$K_G = (1 - \nu_G^2)/E_G$ (× 10^{-11} Pa^{-1})
Aluminum oxynitride (ALON)	1.793 at 0.6 1.66 at 4.0		5.8	32.2	0.24	3.71	1970	0.293
Calcium aluminosilicate (Schott IRG11)	1.684 at 0.55 1.635 at 3.3 1.608 at 4.6		8.2 at 293–573 K	10.8	0.284	3.12	608	0.851
Calcium aluminosilicate (Corning 9753)	1.61 at 0.5 1.57 at 2.5		6.0 at 293–573 K	9.86	0.28	2.798	600 (500 g load)	0.935
Calcium aluminosilicate (Schott IRGN6)	1.592 at 0.55 1.562 at 2.3 1.521 at 4.3		6.3 at 293–573 K	10.8	0.284	3.12	608	0.851
Fluoride glass (Ohara HTF1)	1.51 at 1.0 1.49 at 3.0	−8.19	16.1	6.42	0.28	3.88	311	1.436
Fluorophosphate glass (Schott IRG9)	1.488 at 0.55 1.469 at 2.3 1.458 at 3.3		16.1 at 293–573 K	7.7	0.288	3.63	346 (200 g load)	1.191
Germanate (Corning 9754)	1.67 at 0.5 1.63 at 2.5 1.61 at 4.0		6.2 at 293–573 K	8.41	0.290	3.581	560 (100 g load)	1.089

(Continued)

TABLE A3.1 (CONTINUED)
Optomechanical Properties of Selected Infrared-Transmitting Glass and Other Oxides

Material Name (Symbol)	Refractive Index n at λ (μm)	dn/dT at λ (μm) ($\times 10^{-6}$ °C^{-1})	CTE α ($\times 10^{-6}$ °C^{-1})	Young's Modulus E ($\times 10^{10}$ Pa)	Poisson's Ratio ν	Density ρ (g/cm^3)	Knoop Hardness (kg/mm^2)	$K_G = (1-\nu_G^2)/E_G$ ($\times 10^{-11}$ Pa^{-1})
Germanate (Schott IRG2)	1.899 at 0.55 1.841 at 2.3		8.8 at 293–573 K	9.59	0.282	5.00	481 (200 g load)	0.960
Lanthanum dense flint (Schott IRG3)	1.851 at 0.55 1.796 at 2.3 1.776 at 3.3		8.1 at 293–573 K	9.99	0.287	4.47	541 (200 g load)	0.918
Lead silicate (Schott IRG7)	1.573 at 0.55 1.534 at 2.3		9.6 at 293–573 K	5.97	0.216	3.06	379	1.597
Sapphire[a] (Al_2O_3)	1.684 at 3.8 1.586 at 5.8	13.7	5.6 (‖) 5.0 (⊥)	40.0	0.27	3.97	1370 (1000 g load)	0.232
Fused silica (Corning 7940)	1.561 at 0.19 1.460 at 0.55 1.433 at 2.3 1.412 at 3.3	10–11.2 at 0.5–2.5	0.6 at 73 K 0.58 at 273–473 K	7.3	0.17	2.202	500 (200 g load)	1.333

Source: Yoder, P.R. Jr., *Mounting Optics in Optical Instruments*, 2nd ed., SPIE Press, Bellingham, WA, 2008, p. 719.

[a] Birefringent material.

Appendix 4: Selected Semiconductor Infrared Materials

TABLE A4.1

Optomechanical Properties of Diamond and Selected Infrared-Transmitting Semiconductor Materials

Material Name (Symbol)	Refractive Index n at λ (μm)	dn/dT at λ (μm) ($\times 10^{-6}$ °C^{-1})	CTE α ($\times 10^{-6}$ °C^{-1})	Young's Modulus E ($\times 10^{10}$ Pa)	Poisson's Ratio ν	Density ρ (g/cm^3)	Knoop Hardness (kg/mm^2)	$K_G = (1-\nu_G^2)/E_G$ ($\times 10^{-11}$ Pa^{-1})
Diamond (C)	2.382 at 2.5 2.391 at 5.0 2.381 at 10.6		−0.1 at 25 K 0.8 at 2913 K 5.8 at 1600 K	114.3	0.069 (for CVD)	3.51	9000	0.094
Indium antimonide (InSb)	3.99 at 8.0	4.7	4.9	4.3	5.78	225		
Gallium arsenide (GaAs)	3.1 at 10.6	1.5	5.7	8.29	0.31	5.32	721	1.090
Germanium (Ge)	4.055 at 2.7 4.026 at 3.8 4.015 at 5.3 4.00 at 10.6	424 at 250–350 K	2.3 at 100 K 5.0 at 200 K 6.0 at 300 K	10.37	0.278	5.323	800	0.890
Silicon (Si)	3.436 at 2.7 3.427 at 3.8 3.422 at 5.3 3.148 at 10.6	1.3	2.7–3.1	13.1	0.279	2.329	1150	0.704

Source: Yoder, P.R. Jr., *Mounting Optics in Optical Instruments*, 2nd ed., SPIE Press, Bellingham, WA, 200, p. 720.

Appendix 5: Selected Chalcogenide Infrared Materials

TABLE A5.1
Optomechanical Properties of Selected Infrared-Transmitting Chalcogenide Materials

Material Name (Symbol)	Refractive Index n at λ (μm)	dn/dT at λ (μm) ($\times 10^{-6}$ °C^{-1})	CTE α ($\times 10^{-6}$ °C^{-1})	Young's Modulus E ($\times 10^{10}$ Pa)	Poisson's Ratio ν	Density ρ (g/cm^3)	Knoop Hardness (kg/mm^2)	$K_G = (1-\nu_G^2)/E_G$ ($\times 10^{-11}$ Pa^{-1})
Arsenic trisulfide (AsS_3)	2.521 at 0.8 2.412 at 3.8 2.407 at 5.0	85 at 0.69 17 at 1.0	26.1	1.58	0.295	3.43	180	5.778
$Ge_{33}As_{12}SE_{56}$ (AMTIR-1)	2.605 at 1.0 2.503 at 8.0	101 at 1.0 72 at 10.0	12.0	2.2	0.266	4.4	170	4.224
Zinc sulfide (ZnS)	2.36 at 0.6 2.257 at 3.0 2.246 at 5.0 2.192 at 10.6	63.5 at 0.63 49.8 at 1.15 46.3 at 10.6	4.6	7.45	0.29	4.08	230	1.229
Zinc selenide (ZnSe)	2.16 at 0.6 2.438 at 3.0 2.429 at 5.0 2.403 at 10.6	91.1 at 0.63 59.7 at 1.15 52.0 at 10.6	5.6 at 163 K 7.1 at 273 K 8.3 at 473 K	7.03	0.28	5.27	105	1.311

Source: Yoder, P.R., Jr., *Mounting Optics in Optical Instruments*, 2nd ed., SPIE Press, Bellingham, WA, 2008, p. 721.

Appendix 6: Nonmetallic Mirror Substrate Materials

TABLE A6.1
Mechanical Properties of Selected Nonmetallic Mirror Substrate Materials

Material Name and Symbol	Source	CTE $\alpha \times 10^{-6}$ °C^{-1} ($\times 10^{-6}$ °F^{-1})	Young's Modulus E $\times 10^{10}$ Pa ($\times 10^{6}$ lb/in.^{2})	Poisson's Ratio ν	Density ρ g/cm^3 (lb/in.^{3})	Specific Heat c_p J/kg K (Btu/lb °F)	Thermal Conductivity k W/m K (Btu/hour ft °F)	Knoop Hardness (kg/mm^2)	Best Surface Smoothness (Å RMS)
Duran 50	Schott	3.2 (1.8)	6.17 (8.9)	0.20	2.23 (0.081)	835 (0.20)	1.02 (0.59)		~5
Pyrex 7740	Corning	3.3 (1.86)	6.30 (9.1)	0.2	2.23 (0.081)	1050 (0.25)	1.13 (0.65)		
Borosilicate crown E6	Ohara	2.8 (1.5)	5.86 (8.5)	0.195	2.18 (0.079)				
Fused silica 7940	Corning	0.58 (0.32)	7.3 (10.6)	0.17	2.205 (0.080)	741 (0.177)	1.37 (0.8)	500	~5
ULE 7971	Corning	0.015 (0.008)	6.76 (9.8)	0.17	2.205 (0.080)	766 (0.183)	1.31 (0.76)	460	~5
Zerodur	Schott	0 ± 0.05 (0 ± 0.03)	9.06 (13.6)	0.24	2.53 (0.091)	821 (0.196)	1.64 (0.95)	60	~5
Zerodur M	Schott	0 ± 0.05 (0 ± 0.03)	28.9 (12.9)	0.25	2.57 (0.093)	810 (0.194)	1.6 (0.92)	540	~5

Source: Yoder, P.R. Jr., *Mounting Optics in Optical Instruments*, 2nd ed., SPIE Press, Bellingham, WA, 2008, p. 722.

Appendix 7: Metallic Mirror Substrate Materials

TABLE A7.1
Mechanical Properties of Selected Metallic and Composite Mirror Substrate Materials

Material Name and Symbol	CTE $\alpha \times 10^{-6}$ °C^{-1} ($\times 10^{-6}$ °F^{-1})	Young's Modulus $E \times 10^{10}$ Pa ($\times 10^{6}$ lb/in.^{2})	Poisson's Ratio ν	Density ρ g/cm^3 (lb^2/in.^{3})	Specific Heat c_p J/kg K (Btu/lb °F)	Thermal Conductivity k W/m K (Btu/hour ft °F)	Hardness	Best Surface Smoothness (Å RMS)
Aluminum (6061-T6)	23.6 (13.1)	6.82 (9.9)	0.332	2.68 (0.100)	960 (0.23)	167 (96)	30–95 Brinell	~200
Beryllium (I-70H)	11.3 (6.3)	28.9 (42)	0.08	1.85 (0.067)	1820 (0.436)	216 (125)		60–80 (sputtered)
Beryllium (S-200-FH)	11.3 (6.3)	30.3 (44)	0.08	1.85 (0.067)	1820 (0.436)	216 (125)		
Beryllium (O-30H)	11.46 (6.37)	30.3 (44)	0.08	1.85 (0.067)	1820 (0.436)	215/365[a] (125/211)	80 Rockwell B	15–25
Copper (OFHC[b])	16.7 (9.3)	11.7 (17)	0.35	8.94 (0.323)	385 (0.092)	392 (226)	40 Rockwell F	40
Glidcop	18.4 (10.3)	13.0 (18.9)	0.33	8.84 (0.321)	380 (211)	216 (125)		
Molybdenum (TZM)	5.0 (2.8)	31.8 (2.8)	0.32 (46)	10.2 (0.371)	272 (0.368)	146 (0.065)	200 (84.5)	10 Vickers
Silicon	2.6 (1.4)	13.1 (19.0)	0.42	2.33 (0.085)	710 (0.170)	137 (79)		4 to 1
Silicon carbide (RB–30% Si)	2.64 (1.47)	31.0 (45)		2.92 (0.106)	660 (0.16)	158 (91)		
Silicon carbide (RB–12% Si)	2.68 (1.49)	37.3 (54.1)		3.11 (0.113)	680 (0.16)	147 (85)		
Silicon carbide CVD	2.4 (1.3)	46.6 (67.6)	0.21	3.21 (0.117)	700 (0.17)	146 (84)	2540 Knoop (500 g)	
CESIC®	2.6 at 300 K (1.4 at 68°F) <0.5 at 90–20 K	23.5 (34.1)		2.65 (0.096)	660 (0.16)	~135 (~78)		

(*Continued*)

TABLE A7.1 (CONTINUED)
Mechanical Properties of Selected Metallic and Composite Mirror Substrate Materials

Material Name and Symbol	CTE $\alpha \times 10^{-6}$ °C^{-1} ($\times 10^{-6}$ °F^{-1})	Young's Modulus $E \times 10^{10}$ Pa ($\times 10^{6}$ lb/in.^{2})	Poisson's Ratio ν	Density ρ g/cm^3 (lb^2/in.^{3})	Specific Heat c_p J/kg K (Btu/lb °F)	Thermal Conductivity k W/m K (Btu/hour ft °F)	Hardness	Best Surface Smoothness (Å RMS)
SXA metal matrix of 30% SiC in 1214 Al[c]	12.4 (6.9)	11.7 (17)		2.90 (0.105)	770 (0.18)	130 (75)		
AlBeMet 162	13.9 (7.7)	19.3 (28)		2.10 (0.076)	1560 (0.373)	(210) (121)		
Berylcast 191	13.3 (7.4)	20.1 (29.3)		2.15 (0.078)	1454 (0.34)	178 (103)		
Al-BeCast 910	13.8 (7.7)	19.2 (28.0)		2.09 (0.076)	1539 (0.36)	104 (60)		
AI-Si alloy 393-T6 (22% Si)	16.1 (9.0)	10.3 (15)		2.64 (0.096)	898 (0.21)	15.6 (9.0)		
Graphite epoxy GY-70/X30	0.02 (0.01)	9.3 (13.5)		1.78 (0.064)		35 (20)		

Source: Paquin, R.A., Advanced materials: an overview, *Proc. SPIE*, CR67, 3, 1997; Ahmad, A. et al., Design of a lightweight telescope with highly stable line of sight, *Proc. SPIE*, 306, 66, 1981; Yoder, P.R., Jr., *Mounting Optics in Instruments*, 2nd ed., SPIE Press, Bellingham, WA, 2008, p. 723; Müller, C. et al., C/SiC high precision lightweight components for opto-mechanical applications, *Proc. SPIE*, 4198, 249, 2001; Parsonage, T., JWST beryllium telescope: Material and substrate fabrication, *Proc. SPIE*, 5494, 39, 2004; Materion. Available at https://materion.com/-/media/files/pdfs/beryllium/beryllium-materials/mb-001designingandfabricatingberyllium.pdf.

[a] Measured at 25°C/–166°C.

[b] Oxygen-free high thermal conductivity.

[c] With SiC particles of mean size 3.5 μm (0.014 in.) per Advanced Composite Materials Corp., Greer, SC.

Index

L

M

N

O

P

R

S

T

U

V

W

Z